Military Aspects of Geology: Fortification, Excavation and Terrain Evaluation

Geological Society books refereeing procedures

The Society makes every effort to ensure that the scientific and production quality of its books matches that of its journals. Since 1997, all book proposals have been refereed by specialist reviewers as well as by the Society's Books Editorial Committee. If the referees identify weaknesses in the proposal, these must be addressed before the proposal is accepted.

Once the book is accepted, the Society Book Editors ensure that the volume editors follow strict guidelines on refereeing and quality control. We insist that individual papers can only be accepted after satisfactory review by two independent referees. The questions on the review forms are similar to those for *Journal of the Geological Society*. The referees' forms and comments must be available to the Society's Book Editors on request.

Although many of the books result from meetings, the editors are expected to commission papers that were not presented at the meeting to ensure that the book provides a balanced coverage of the subject. Being accepted for presentation at the meeting does not guarantee inclusion in the book.

More information about submitting a proposal and producing a book for the Society can be found on its website: www.geolsoc.org.uk.

It is recommended that reference to all or part of this book should be made in one of the following ways:

ROSE, E. P. F., EHLEN, J. & LAWRENCE, U. L. (eds) 2019. *Military Aspects of Geology: Fortification, Excavation and Terrain Evaluation*. Geological Society, London, Special Publications, **473**.

ROSE, E. P. F., EHLEN, J. & LAWRENCE, U. L. 2018. Military use of geologists and geology: a historical overview and introduction. *In*: ROSE, E. P. F., EHLEN, J. & LAWRENCE, U. L. (eds) 2019. *Military Aspects of Geology: Fortification, Excavation and Terrain Evaluation*. Geological Society, London, Special Publications, **473**, 1–29, https://doi.org/10.1144/SP473.15

GEOLOGICAL SOCIETY SPECIAL PUBLICATION NO. 473

Military Aspects of Geology: Fortification, Excavation and Terrain Evaluation

EDITED BY

E. P. F. ROSE
Royal Holloway, University of London, UK

J. EHLEN
formerly US Army Engineer Research and Development Center, USA

and

U. L. LAWRENCE
Capita Property & Infrastructure Ltd, UK

2019
Published by
The Geological Society
London

THE GEOLOGICAL SOCIETY

The Geological Society of London (GSL) was founded in 1807. It is the oldest national geological society in the world and the largest in Europe. It was incorporated under Royal Charter in 1825 and is Registered Charity 210161.

The Society is the UK national learned and professional society for geology with a worldwide Fellowship (FGS) of over 10 000. The Society has the power to confer Chartered status on suitably qualified Fellows, and about 2000 of the Fellowship carry the title (CGeol). Chartered Geologists may also obtain the equivalent European title, European Geologist (EurGeol). One fifth of the Society's fellowship resides outside the UK. To find out more about the Society, log on to www.geolsoc.org.uk.

The Geological Society Publishing House (Bath, UK) produces the Society's international journals and books, and acts as European distributor for selected publications of the American Association of Petroleum Geologists (AAPG), the Indonesian Petroleum Association (IPA), the Geological Society of America (GSA), the Society for Sedimentary Geology (SEPM) and the Geologists' Association (GA). Joint marketing agreements ensure that GSL Fellows may purchase these societies' publications at a discount. The Society's online bookshop (accessible from www.geolsoc.org.uk) offers secure book purchasing with your credit or debit card.

To find out about joining the Society and benefiting from substantial discounts on publications of GSL and other societies worldwide, consult www.geolsoc.org.uk, or contact the Fellowship Department at: The Geological Society, Burlington House, Piccadilly, London W1J 0BG: Tel. +44 (0)20 7434 9944; Fax +44 (0)20 7439 8975; E-mail: enquiries@geolsoc.org.uk.

For information about the Society's meetings, consult *Events* on www.geolsoc.org.uk. To find out more about the Society's Corporate Affiliates Scheme, write to enquiries@geolsoc.org.uk.

Published by The Geological Society from:
The Geological Society Publishing House, Unit 7, Brassmill Enterprise Centre, Brassmill Lane, Bath BA1 3JN, UK

The Lyell Collection: www.lyellcollection.org
Online bookshop: www.geolsoc.org.uk/bookshop
Orders: Tel. +44 (0)1225 445046, Fax +44 (0)1225 442836

British Library Cataloguing in Publication Data

A catalogue record for this book is available from the British Library.
ISBN 978-1-78620-394-6
ISSN 0305-8719

Distributors
For details of international agents and distributors see:
www.geolsoc.org.uk/agentsdistributors

Typeset by Nova Techset Private Limited, Bengaluru & Chennai, India
Printed and bound by CPI Group (UK) Ltd, Croydon CR0 4YY

Contents

Preface

The core chapters of this book were derived from 12 papers offered for presentation at a joint meeting of the History of Geology Group (HOGG) and the Engineering Group of the Geological Society (EGGS). Held at the Society's rooms in Burlington House, London, on 16 November 2016, this meeting featured 'Military Aspects of Engineering Geology, Past and Present'. It formed the sequel to a joint meeting of HOGG with the Society's Hydrogeological Group held on 18 November 2009 that generated a companion volume (Rose, E. P. F. & Mather, J. D. (eds) 2012. *Military Aspects of Hydrogeology.* Geological Society, London, Special Publications, **362**).

The year 2016 was chosen to help mark the centenary of first deployment by the British Army of an engineering geologist to serve as such on a battlefield: Major (later Lieutenant-Colonel) T. W. Edgeworth David, who arrived in France with the Australian Mining 'Corps' in May 1916 to support the British Expeditionary Force on the Western Front. David was invited to give an account of his service during the 'Great War' of 1914–18 at a meeting of the Geological Society on 26 February 1919, after the close of hostilities – whilst he was detained in London for nearly two weeks by the delayed sailing of the troopship in which he was scheduled to return home to Australia. That account was not published, but David's handwritten notes were discovered in the Society's library archives many years later. They formed the basis of a re-enactment given at Burlington House on 10 July 2014 to help mark the centenary of the start of the war (for the UK, on 4 August 1914), and the core of a consequent explanatory article (Rose, E. P. F. 2015. 'Abstract from Geology at the Western Front' by T. W. Edgeworth David. *Earth Sciences History*, **34**, 1–22).

Most of the papers presented at the 2016 meeting have been developed, and supplemented by additional papers, to generate the present book. They were published 'Online First' in 2018, to help mark the centenary of the official end of conflict between Germany and the Allies on 11 November 1918: the final armistice of the First World War. Hardback publication in 2019 marks the centenary of three of the following peace treaties; of David's lecture to the Society; and of the first account by a British geologist of geological work in support of combat operations (King, W. B. R. 1919. Geological work on the Western Front. *Geographical Journal*, **54**, 201–215). It was work for which 'Bill' King was awarded the Society's Wollaston Fund in 1920.

After a historical introduction to the military use of geologists and of geology, the following chapters have been grouped into theme sets that focus on the particular aspects indicated in the book title: fortification, excavation and terrain evaluation. In accordance with standard Geological Society practice, each chapter has been peer-reviewed by at least two referees, anonymous to the authors (initially and, where requested, continuously). The following friends and colleagues assisted in this way (some for more than one paper) and, having allowed their names to be included, are thanked for their help and support: P. H. Banham, David Branagan, Martin Brown, Trevor Davenport, Peter Doyle, T. E. Eastler, F. A. Galgano, J. S. Griffiths, Hermann Häusler, S. W. Henderson, Nigel Hoad, Mike King, P. R. La Pointe, U. L. Lawrence, J. D. Mather, R. T. J. Moody, V. D. Moore, John Peaty, J. D. Priddy, M. S. Rosenbaum, Paul Stewart, Kevin Styles, Geoff Walton, Nick Walton and R. C. Whisonant. We also thank P. L. Guth of the US Naval Academy, Annapolis, and an anonymous reviewer, for their helpful comments on our initial book proposal, and the Society's Book Editor for the volume, J. S. Griffiths, for reviewing our introductory chapter.

Edward P. F. Rose
Honorary Research Fellow,
Department of Earth Sciences, Royal Holloway,
University of London, Egham,
Surrey TW20 0EX, UK

Judy Ehlen
retired from the US Army Engineer Research and
Development Center, Alexandria, Virginia, USA;
now at 3 Haytor Vale, Haytor, Newton Abbot,
Devon TQ13 9XP, UK

Ursula L. Lawrence
Associate Engineering Geologist,
Capita Property & Infrastructure Ltd, Capita House,
Wood Street, East Grinstead,
West Sussex RH19 1UU, UK

Military use of geologists and geology: a historical overview and introduction

EDWARD P. F. ROSE[1]*, JUDY EHLEN[2,3] & URSULA L. LAWRENCE[4]

[1]*Department of Earth Sciences, Royal Holloway, University of London, Egham, Surrey TW20 0EX, UK*

[2]*US Army Engineer Research and Development Center, Alexandria, VA, USA (retired)*

[3]*Present address: 3 Haytor Vale, Haytor, Newton Abbot, Devon TQ13 9XP, UK*

[4]*Capita Property and Infrastructure, Capita House, Wood Street, East Grinstead, West Sussex RH19 1UU, UK*

E.P.F.R., 0000-0003-4182-6426; J.E., 0000-0002-1595-7820; U.L.L., 0000-0001-8820-1699

**Correspondence: ted.rose@earth.oxon.org*

Abstract: Napoleon Bonaparte was, in 1798, the first general to include geologists as such on a military operation. Within the UK, the following century saw geology taught, and national geological mapping initiated, as a military science. Nevertheless, military geologists were not deployed on a battlefield until World War I, first by the German and Austro-Hungarian armies and later and less intensively those of the UK and USA. Geologists were used primarily to guide abstraction of groundwater, construction of 'mine' tunnels and dug-outs, development of fortifications and quarrying of natural resources to enhance or repair supply routes. Only the USSR and Germany entered World War II with organized military geological expertise, but the UK and later the USA made significant use of military geologists, albeit far fewer than the *c.* 400 in total used by German forces. Military geologist roles in World War II included most of those of World War I, but were extended to other aspects of terrain evaluation, notably the rapid construction of temporary airfields and factors affecting cross-country vehicular movement ('going'). After 1945, more military geologists were used in the USA than Germany or the UK, in these and wider roles, but mostly as civilians or reservists.

Although there had been an association between military men, mineralogy and mining in parts of Europe from at least the mid-eighteenth century, military applications of geology as such became apparent only from Napoleonic times onwards (Rose 2009*a*). During the nineteenth century, at least within the UK, geology itself was initially perceived to be a military science. This perception had lapsed by the end of the century, but the Great War of 1914–18 saw 'military geology' developed as a new discipline within geoscience. Stimulated by the two World Wars and succeeding Cold War of the twentieth century, similar applications of military geology were rapidly developed in parallel, but mostly independently, by the armed forces of the opposing nations. This article provides a historical account from the Napoleonic origins of the discipline to the present day as the foundation for an introduction to the papers that comprise the rest of the volume.

A benchmark book on military geology (Underwood & Guth 1998), the first to be published following the end of the Cold War of 1947–91, was introduced by George Kiersch, a distinguished American engineering geologist and former World War II military engineer. His historical overview (Kiersch & Underwood 1998) reflected understanding of the subject in the USA at that time and was also published, in extended form, as a journal article (Kiersch 1998). The book was generated by a symposium that became the first in a series of meetings now entitled the International Conferences on Military Geosciences. These meetings take place near-biennially and alternated between the USA and Europe until the twelfth, which was held in South Africa in 2017 (Rose 2018*a*). Data from other books generated by the early conferences (Rose & Nathanail 2000; Ehlen & Harmon 2001; Doyle & Bennett 2002; Caldwell *et al.* 2004) helped an encyclopedia summary by Rose (2004*a*) to be extended

From: Rose, E. P. F., Ehlen, J. & Lawrence, U. L. (eds) 2019. *Military Aspects of Geology: Fortification, Excavation and Terrain Evaluation*. Geological Society, London, Special Publications, **473**, 1–29.
First published online November 8, 2018, https://doi.org/10.1144/SP473.15

into a historical overview to complement those by Kiersch by adding a European perspective (Rose 2008*a*). The present article enhances previous overviews with data from books generated by subsequent conferences (Nathanail *et al.* 2008; Häusler & Mang 2011; McDonald & Bullard 2016) plus other recent books (Rose & Mather 2012; Harmon *et al.* 2014) and ten more years of ancillary published research.

Nineteenth-century foundations

For the nineteenth century, significant perceptions of military applications of geology can be demonstrated from France, the German-speaking states of mainland Europe and, most importantly, the UK. Relevant literature from other countries is relatively sparse, but an article in Spanish that was translated into French (Quijano y Arroquia 1876) is evidence of wider European interest.

Napoleonic France

Napoleon Bonaparte was the first general to take geologists as such on a military expedition, when he led a French invasion of Egypt in July 1798 (Rose 2004*b*, 2005*a*, 2008*b*). His army was accompanied by a Commission of Sciences and Arts, *c.* 150 engineers and other technical experts whose task was to examine almost every aspect of ancient and contemporary Egypt so as to enable the French to occupy and govern the country effectively. Four members of the Commission were listed as '*minéralogistes*' (geologists): Déodat de Dolomieu together with three of his former pupils at the School of Mines then being redeveloped in Paris, Louis Cordier, François-Michel de Rozière and Victor Dupuy (alias Dupuis).

Dolomieu (Fig. 1), from whom the mineral dolomite is named and consequently the Dolomite mountain range of NE Italy, was one of the four most senior members of the Commission. An aristocrat with distinguished geological achievements to his credit, his initial career within the Sovereign Military Order of Malta had ended as a consequence of the French revolution of 1789. Following seizure of his family estates and suppression of the Order in France, he had gained employment as an inspector of mines and a visiting professor of geology at the nascent School of Mines in Paris. He had volunteered to join the expeditionary force to Egypt through love of adventure and the chance to explore a relatively unknown country, gaining a place through his status at the School of Mines and friendship with other senior members of the Commission. However, illness forced him to leave Egypt in March 1799, before the French army had gained access to

Fig. 1. Déodat de Dolomieu (1750–1801), the senior geologist who accompanied the French invasion of Egypt in 1798: one of a series of portraits of influential members of Napoleon Bonaparte's expeditionary force by its artist André Dutertre. From Rose (2004*b*); © The British Library Board and reproduced by permission from the portrait supplement to Saintine *et al.* (1830–36), shelfmark 1311.h.2.

the country as a whole and much geological work had been done.

Cordier left Egypt to accompany Dolomieu on his homeward journey. He subsequently developed a distinguished career in the French government's *Corps des Mines* and as a Professor of Geology at the Natural History Museum in Paris. Rozière and Dupuy remained in Egypt until the French army was defeated and expelled by a British expeditionary force in 1801. Both returned to France and to careers in the *Corps des Mines*, Rozière, from 1820 to 1824, concurrently as Professor of Geology and Mineralogy at the Saint-Étienne School of Mines, near Lyon. He was almost entirely responsible for publication of French geological achievements in Egypt, but these featured pure rather than applied aspects of geology and made no contribution to military geology as such.

Napoleon returned to France in late 1799 to seize power as First Consul and in 1804 to change the Republic into an Empire, with himself as Emperor. The French Revolutionary Wars of 1792–1802 continued from 1803 as the Napoleonic Wars. From

1792 until 1815 conflict blazed at times across almost the whole of Europe, from Portugal in the west across Spain, France and central Europe into Russia in the east, and extended into Africa, North America and the Far East (most notably India). Although Napoleon himself made no use of geology or geologists in his campaigns after Egypt, even for his final battle, at Waterloo in Belgium, in 1815 (Rose 2015*a*), a precedent had been set. Scientific investigations made in connection with military expeditions were adopted by many nations in the following century. Much of the early knowledge of the western USA was gained in this manner (Betz 1984).

German-speaking European states

Karl Georg von Raumer (1783–1865) was one of the geologists in mainland Europe caught up in the Napoleonic Wars (Rose 2005*b*). A student of Abraham Gottlob Werner (1749–1817) at the prestigious Freiberg Mining Academy in the German state of Saxony from 1805 to 1806, his fieldwork scheduled for 1806 had to be deferred to 1807–08 to avoid Napoleon's campaigns in Germany. He became Professor of Mineralogy at the University of Breslau in the then Prussian province of Silesia (now part of Poland) in 1811, and volunteered for service with the Prussian army during the 1813–14 war of German liberation from French Napoleonic domination. His knowledge of terrain led to his appointment to the staff of the Prussian General (later Field Marshal) Gebhard Leberecht von Blücher. Some scholars (e.g. Betz 1984; Kiersch 1998; Kiersch & Underwood 1998; Rose *et al.* 2000) have followed earlier German authors in reporting that Blücher made use of Raumer's terrain expertise before triumphing over French forces at the second Battle of the Katzbach, on 26 August 1813, and that Raumer should therefore be regarded as a military geologist. However, this seems to have been a misconception (Rose 2005*b*). Raumer was indeed used as a staff officer during his operational military service, but there is no mention in his autobiography that he made use of geology at this time.

Johann Samuel Gruner (alias von Grouner: Fig. 2) (1766–1824) was, in contrast, arguably the first true military geologist (Häusler & Kohler 2003). A distinguished Swiss mining engineer, he had also studied at Freiberg, taking courses in mining, mineralogy and geology from 1787 to 1790. He later became Director-General of all Swiss mines in the Helvetic Republic, short-lived as a French satellite state. He participated in military operations towards the end of the French Revolutionary Wars, when French troops were campaigning in Switzerland against an Austro-Russian army. He later emigrated to the German state of Bavaria and when, in 1814, Bavaria joined Prussia in the German War of Liberation from

Fig. 2. Johann Samuel Gruner (alias von Grouner), arguably the world's first military geologist. From Häusler & Kohler (2003); pen-and-ink portrait, negative 4101 reproduced courtesy of the *Burgerbibliothek* Bern, Switzerland, by kind permission of Stephanie Gropp, per Hermann Häusler.

Napoleonic domination, he commanded a volunteer rifle battalion in the campaign to drive Napoleon's army back to Paris. Drawing on his experience both as a geologist and a soldier, he wrote a memorandum in 1820 on the relationship between geology and the science of war, the first such work of its kind, published posthumously (von Grouner 1826). However, the German states were not to be united into a single nation until 1871 and the memorandum made no discernable impact either on geology or military training, in Germany or elsewhere.

It was an Austrian career soldier, Rudolf Freiherr von Schmidburg (1810–1902) who was to draw more significant attention to the relevance of geology to terrain studies and their implications for war, but not until the second half of the nineteenth century (Rose *et al.* 2000). By the end of the century, military geographers in both France and Italy were also using geology to explain terrain features and land surface conditions (Betz 1984).

The UK

The Geological Society of London, the world's oldest national geological society, was founded in the midst of the Napoleonic Wars by 11 men dining together on 13 November 1807. Of these, three had military associations (Rose 2009*a*, 2014). Jacques Louis, Comte de Bournon, a royalist refugee from republican France, was a former Captain of artillery. James Frank had served as a physician with the British Army since 1794, in the higher rank of Inspector of Hospitals since 1800, and with the British expeditionary force in Egypt in 1801. George Bellas Greenough, who became the Society's founding president, had served part-time since 1804 in the Light Horse Volunteers of London and Westminster, one of the reserve units raised to defend the UK from potential French invasion. However, their interests were in pure rather than applied science, and none is known to have noted potential military applications for geology.

About this time, French wartime domination of Europe prevented importation of limestone pure enough to make the millstones used for grinding gunpowder and its constituents. An explosion in a powder mill was ascribed to use of a substitute limestone containing silica that generated sparks. It was military funding and imperative that consequently initiated geological reconnaissance to find a source of more suitable limestone within the UK. Surveys by John MacCulloch, an employee of the Board of Ordnance which had responsibility for the Royal Artillery and the Royal Engineers, began in 1809 but peace was declared in 1814 before the project came to an end (Rose 1996).

National geological mapping in the UK soon began, under military auspices (Rose 1996). From 1814, MacCulloch received military (Board of Ordnance) funding for geological surveys in Scotland. Funding was transferred to civilian government responsibility in 1826, but the surveys eventually generated a geological map of Scotland, published in 1836. The (Ordnance) Geological Survey of Ireland was founded in 1826 and led until 1828 by Captain J.W. Pringle of the Royal Engineers, a veteran of the Battle of Waterloo (Rose 1999). He was assisted by other Royal Engineers officers: Lieutenants J.E. Portlock, G.F. Bordes, R.S. Fenwick, W. Lancey and A.W. Robe. The Survey was revived in the 1830s under the leadership of Portlock, by then promoted to the rank of Captain. Funding was transferred to civilian auspices in 1845, but the Survey was led by another Royal Engineer, Captain Henry James, until 1846 (Rose 1996). The future British Geological Survey was founded in 1835, again with Ordnance funding, to map initially in England and Wales. It too became a civilian government agency in 1845. By coincidence, the Geological Survey of Great Britain, as it became, was to be led until 1871 by two successive Directors, Henry Thomas De la Beche and Roderick Impey Murchison, who had received military rather than university educations. Murchison, a former infantry officer, was a veteran of the Napoleonic Wars.

During the nineteenth century, geology was among the subjects taught intermittently at all army officer training establishments in the UK (Rose 1997, 2009*a*, 2014). A course was available from the Professor of Chemistry at the Royal Military College, Sandhurst – the geologist Richard Phillips – for a few years after 1818. Geology was taught at Addiscombe in Surrey to officer cadets of the East India Company's army, by John MacCulloch from 1819 to 1835 and by D.T. Ansted from 1845 to 1861. For the British Army: James Tennant lectured on geology to Royal Engineer and Royal Artillery cadets at the Royal Military Academy, Woolwich, from 1848 to 1868; T. Rupert Jones to cadets of the infantry and cavalry at the Royal Military College, Sandhurst, from 1858 to 1870; Jones also to young officers of all arms at the Staff College, Camberley, until 1882; and A.H. Green at the School of Military Engineering, Chatham, from *c.* 1888 to 1896. However, by the end of the century, attempts to teach elements of geology to officers of the British Army in general had largely been abandoned.

By the mid-nineteenth century, military aspects of geology were featuring in publications. A graduate of Addiscombe, Richard Baird Smith, published an 'Essay on Geology, as a branch of study especially meriting the attention of the Corps of Engineers' (Smith 1849). Portlock (1849, p. 14), now in the rank of Colonel, published a *Rudimentary Treatise on Geology*, noting that 'the Soldier … may find in Geology a valuable guide in tracing his lines both of attack and defence'. He also contributed (Portlock 1850) a 116-page article on geology to the authoritative *Aide-Mémoire to the Military Sciences* (Lewis *et al.* 1846–52), a three-volume textbook published for the 'scientific corps' of the British and East India Company's armies. A journal article on the 'Importance of a knowledge of geology to military men' was published by Captain F.W. Hutton (1862) on his graduation from the Staff College, Camberley.

Increasing awareness of the military relevance of geology can largely be credited, directly or indirectly, to the influence of Joseph Ellison Portlock (1794–1864: Fig. 3). Portlock was a respected veteran, as a combat Royal Engineer officer, of active service in 1812 to defend Canada from invasion by the US Army (Rose 2009*a*). Later, as already noted, he had helped to found the Geological Survey of Ireland in 1826 and then re-found and lead it in the 1830s. His major publications (e.g. Portlock 1849, 1850) had substantial print runs and so presumably high impact on their readers. Retiring with the rank of Major General, he became President of the Geological Society of London for 1856–58. Since Portlock was at that time influential in planning developments in military education, it was presumably no coincidence that a full-time resident lecturer in geology was appointed to the staff of the Royal Military College, Sandhurst, in 1858 and a Professor of Geology at the Staff College after its re-location

Fig. 3. Joseph Ellison Portlock, pioneer of British military geology, 1850s photograph. From Rose (1996); reproduced by kind permission of the Geological Society of London.

Fig. 4. (Left to right) Charles Barrois, Professor of Geology in the University of Lille, Captain W.B.R. King and Lieutenant-Colonel T.W. Edgeworth David on 26 November 1918 following the Allied liberation of Lille in late October. Extract from a photograph shown in full by Branagan (2005); reproduced by kind permission of King's granddaughter Jane Ritchie.

to Camberley nearby soon afterwards, in 1862 (Rose 1997). Both appointments were filled by Thomas Rupert Jones (1819–1911), Assistant Secretary to the Geological Society of London at the time of Portlock's presidency. Portlock's own geological expertise assisted military engineering work that included a major fortification programme implemented mid-century in the UK (Mather 2018).

World War I (1914–18)

Military geology as a discipline and military geologists as such were an innovation of World War I, the Great War whose hostilities lasted from August 1914 to November 1918. Most military geologists supported operations on the Western Front, across Belgium and northern France. However, neither the Belgian nor French armies were to include geologists to serve in specialist geotechnical roles. Fighting on their own terrain, their forces were able to draw on civilian expertise in geology as and when required (Hanot *et al.* 2017; Hubé 2018). The same was true for Russian forces on the Eastern Front (Sychev 1979). Italian forces faced Austro-Hungarian troops in the Dolomite Mountains of NE Italy without the support of a military geological service but were urged to create one post-war (Fossa-Mancini 1926; Fabbi & Romano 2015). Strictly, military geologists were an innovation of the British, American, Austro-Hungarian and especially German armies, and were principally a consequence of the problems caused by large numbers of troops being concentrated into relatively static positions because of extensive trench warfare.

British military geologists

The British Expeditionary Force on the Western Front at its peak consisted of five armies and so *c.* 1 500 000 troops equipped with *c.* 500 000 horses and mules, mostly to provide transport of supplies forward from the nearest railheads. Authoritative near-contemporary accounts (King 1919; Anon 1922; Edgeworth David in Rose 2015*b*) describe how the Force was eventually supported by two military geologists (Fig. 4) based at its General Headquarters, Lieutenant (later Captain) W.B.R. King and Major (later Lieutenant-Colonel) T.W. Edgeworth David. A third geologist supervised an exploratory boring unit, Lieutenant Loftus Hills (see Rose 2015*b*, fig. 7).

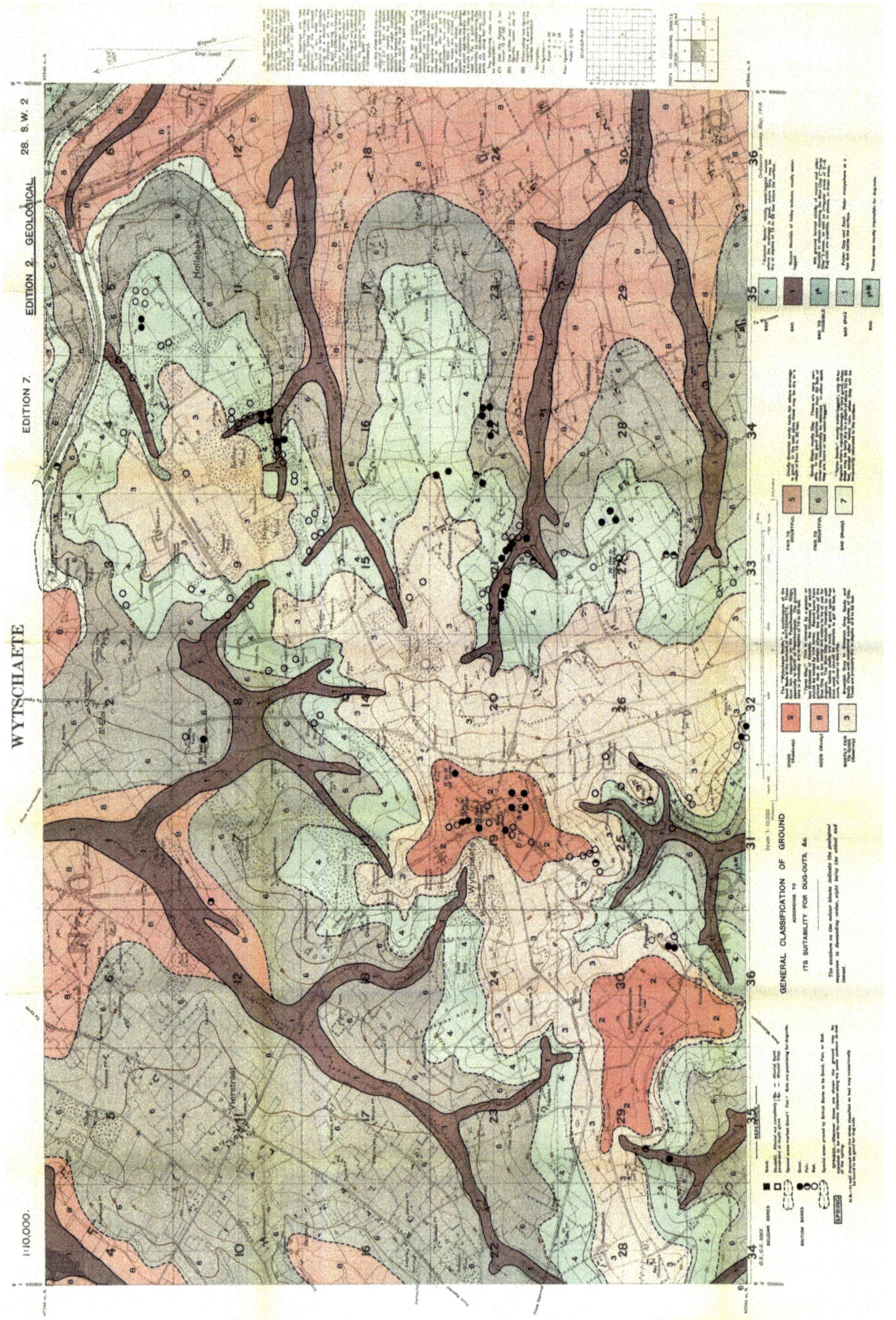
WYTSCHAETE
EDITION 7.
EDITION 2. GEOLOGICAL
28. S.W. 2
1:10,000.
GENERAL CLASSIFICATION OF GROUND
ACCORDING TO
ITS SUITABILITY FOR DUG-OUTS, &c.

William Bernard Robinson King (1888–1963) had graduated from the University of Cambridge in 1912 and joined the Geological Survey of Great Britain (Rose & Rosenbaum 1993*a*; Rose 2015*b*). He was mapping in the Welsh borderland as war loomed, quickly volunteered for military service and was, on 21 September 1914, commissioned as an infantry officer, a Second Lieutenant in the 7th (Merioneth and Montgomery) Battalion of the Royal Welsh Fusiliers. When the Director of the Geological Survey was asked to nominate a geologist to serve as a 'water supply officer' to guide well drilling for the British Expeditionary Force, he nominated King and provided some six weeks of preparatory training in hydrogeology. King moved to France to serve as a Staff Lieutenant to the Force's Chief Engineer (later re-titled Engineer-in-Chief) on 1 June 1915. He was to remain in post until hostilities ended. Traditionally, the British Army on campaign had obtained water from surface sources or shallow wells. However, the British Expeditionary Force soon needed to exploit groundwater from deep aquifers in French Cretaceous chalk and Belgian Paleogene sands, and eventually raised five Water Boring Sections Royal Engineers to do this, one for each army (Rose 2012*a*). King was innovative in generating a variety of water prospect maps, principally at scales of 1:100 000 and 1:250 000, for Belgium and adjacent parts of northern France, and in guiding the emplacement of >470 boreholes (Rose 2009*b*, 2012*a*).

Tannatt William Edgeworth David (1858–1934) had graduated from the University of Oxford in 1880, studied briefly at the Royal School of Mines in London, and emigrated to Australia in 1882 to take up appointment as an Assistant Geological Surveyor to the Government (Geological Survey) of New South Wales (Branagan 2005; Rose 2015*b*). His geological mapping over the next nine years contributed to the developing mining industry within the state. In 1891 he was appointed Professor of Geology in the University of Sydney, initiating a distinguished academic career. Deemed to be too old to volunteer for a commission on the outbreak of war, David was later to help found and raise an Australian Mining Corps (Battalion), in which he was granted a commission with the rank of Major, at the age of 57, with seniority from 25 October 1915. He arrived in France in May 1916 and served initially as Geological Adviser to the Controllers of Mines of the British First, Second and Third Armies. Later that year he was posted to the Inspector of Mines Office at the British Expeditionary Force General Headquarters. David was to guide work of the Tunnelling Companies Royal Engineers as these expanded from nine to a total of 25 between 1916 and 1918, together with three Australian companies, a further three from Canada, and one from New Zealand. By the middle of 1916 the British Expeditionary Force had *c.* 25 000 men actively engaged in tunnelling. Initially, their role was to tunnel mines beneath the fortified front line of the enemy and charge these with explosives that would be detonated to create a breach in the fortifications immediately prior to infantry assault. Later, as the front line came to be held more by artillery fire power than infantry manpower, mining was replaced by the construction of dug-outs to protect troops from artillery or aerial bombardment. David prepared geological cross-sections and innovative thematic maps, notably a series at scale of 1:10 000 (e.g. Fig. 5), that guided the work underground (Rose & Rosenbaum 2011).

Although King as a hydrogeologist and David as an engineering geologist initially worked in separate Engineer branches at General Headquarters, they co-operated on their own initiative and were eventually allowed to work together. Additional to their priority tasks, they contributed some advice used in preparation of 'Tank maps' to guide cross-county movement of the newly invented armoured vehicles (Anon 1922). They also provided advice on the winning of raw materials (Rose 2018*b*).

American military geologists

The US Army Corps of Engineers had developed from officers whose primary military duties centred on the construction and maintenance of fortifications, such as those described by Henderson (2018). These officers were combined with those of a, sometimes, separate branch that focused on topographical surveying into an organization with considerable experience in construction works. 'The British and French governments made the arrival of American engineers in France their top priority after the United States declared war [on Germany] on April 6, 1917' (US Army Corps of Engineers 2008, pp. 123–124). A brochure was quickly prepared (by 24 April 1917) on behalf of the Geology Committee of the National Research Council of the USA with 'the purpose of stating succinctly and clearly the competency of the geologist in war service' and duly 'commended to the attention of commanding officers'

Fig. 5. Geological map of Wytschaete, Belgium, one of a set of 12 British maps at scale of 1:10 000, showing classification of ground according to its suitability for the construction of dug-outs. Shades of red indicate relatively 'good' (dry) strata, shades of blue-green 'bad' (wet) units, purple indicates alluvium in river valleys and black spots the sites of exploratory boreholes. From Anon (1922); see also Rose & Rosenbaum (2011) and Willig *et al.* (2015).

(Penrose 1917, p. 5), but this was a short and simplistic document of unknown influence.

Later that year, in June 1917, Lieutenant General (later General of the Armies) John J. Pershing and his headquarters staff of an American Expeditionary Force (AEF) arrived in France. As part of the staff's preparations for deploying the US 1st Infantry Division and subsequent combat and support units, Brigadier General Sherwood A. Chester and Colonel Ernest Graves of the US Army Corps of Engineers evaluated Allied military mining and water-supply operations on the Western Front and then asked for geologists to be attached to the AEF (Brooks 1920; Nelson & Rose 2012). Accordingly, Alfred H. Brooks took leave from the US Geological Survey to lead the requested effort in military geology and was commissioned as a Major (later Lieutenant Colonel) in the Engineer Officers Reserve Corps, followed by Captain (later Major) Edwin C. Eckel. In September 1917, they established a Section of Geology in General Pershing's headquarters. As the AEF expanded, in July 1918 it approved Brooks's plan to have six geologist officers with him at Pershing's headquarters, place five more with each of two US armies, and assign two others for work along the lines of communication. However, hostilities ended on 11 November 1918 before Brooks could fill most of these appointments. By that time, four geologists served with Brooks at AEF Headquarters, two with the US 1st Army, one with the 2nd Army, and one within a Water-Supply Section: nine in overall total.

These geologists supplied information and advice about the soils and rocks to be encountered by the AEF in mining operations and in building and maintaining trench systems and other fortifications, constructing roads and railways, and seeking supplies of potable water. Active within only the final 14 months of the war, they followed French and British precedents, especially for water supply, but pioneered their own style in preparing geotechnical maps and reports. Deployed to the SE of the British operational area, they faced similar challenges in trench warfare, but these were commonly associated with Jurassic sedimentary rocks (Fig. 6) rather than the Cretaceous and Paleogene bedrock of the British region described by Doyle (2018).

The geologists were demobilized like other troops when the war came to an end. Brooks himself remained in France as geologist to the American Commission to Negotiate Peace and did not return to the US Geological Survey until August 1919.

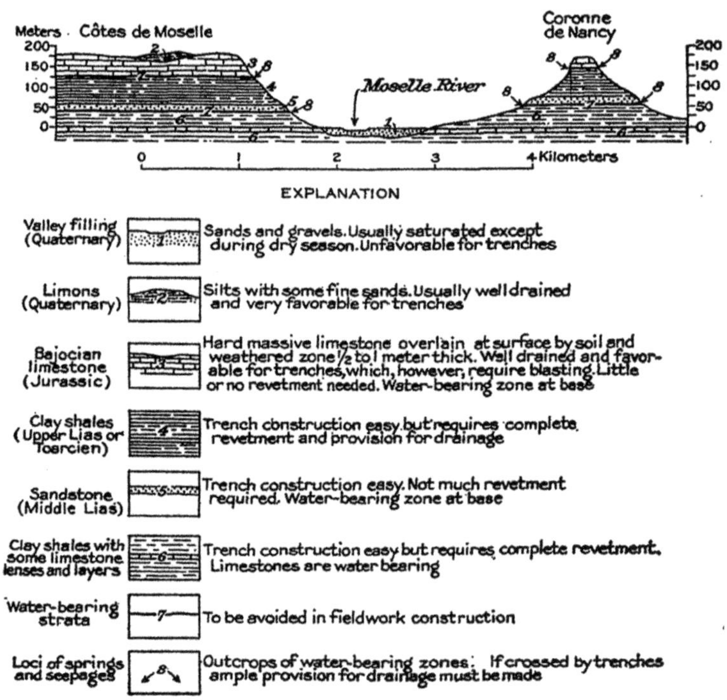

Fig. 6. American diagrammatic cross-section across part of the Moselle valley near Pont-à-Mousson in NE France, showing Jurassic bedrock and Quaternary superficial cover, with indications of suitability for trench construction. From Brooks (1920).

He promptly published a benchmark analysis of how the AEF had used geology on the Western Front (Brooks 1920) and recommended that geology be included in future military training. He suggested peacetime collection of geological information for potential theatres of war and proposed that a 'staff of geologic engineer reserve officers' be organized within the US Army (Brooks 1920, p. 124), but that proposal was not adopted. However, the term 'military geology' to distinguish a new discipline had now entered American scientific literature (Pogue 1917) and thus the English-speaking world. It was quickly incorporated into the title of a book (Gregory 1918).

German military geologists

German forces were to make much greater use of geologists during the war, and from an earlier date (Häusler 2000*a*). Their use was pioneered by Captain (later Major) Walter Kranz (1873–1953; Fig. 7), who effectively founded twentieth-century military geology within the German armed forces (Rose *et al.* 2000; Häusler 2003; Willig 2018). Kranz had retired from service within the Engineer corps of the German Army on 22 March 1913, as a Captain, and been appointed to the Fortification Service of the Strasbourg Fortress. A border city long fortified by the French against potential German attack, Strasbourg had been regarded at the time of the Franco-Prussian War in 1871 as one of the strongest fortresses in France. However, German troops besieged and captured it. In a province subsequently annexed by Germany, it was re-fortified against re-capture by the French. Kranz published articles specifically on *Militärgeologie* (military geology) that initiated recognition of the subject (Kranz 1913, 1914) and in 1917 he completed studies for a doctoral degree in military hydrogeology.

Fig. 7. Walter Kranz, pioneer of twentieth-century German military geology. From Rose *et al.* (2000); reproduced courtesy of the Archive for the History of Geology in Freiburg (*Geologen-Archiv*) at the *Universitätsbibliothek* Freiburg-im-Breisgau, by kind permission of Andreas Hoppe, per Hermann Häusler.

Mobilized from the army reserve on the outbreak of war in August 1914, Kranz was assigned to government military geological work at Strasbourg in November 1914. There he suggested to his superior headquarters on 16 December 1914 that he should be entrusted with the investigation of problems related to the drainage and water supply of field positions, and to mine warfare. Soon he was promoting his subject as *Kriegsgeologie* (war geology) rather than military geology, to make its immediate relevance more apparent. He published several articles on this topic (e.g. Kranz 1915). His persistent persuasiveness achieved the desired result. In a circular dated 2 April 1915 the General commanding the Engineer Corps in the German Army's supreme headquarters invited all Engineer Generals to make use of Kranz for advice. Moreover, on 21 May 1915 the Chief of Staff of the Field Army ordered that 'the generals of the Engineers at the [individual] army headquarters should be informed that they could consult geologists, and that these should be made available to them on request' (translated from Kranz 1927, p. 9).

In 1916, a geological organization as such was formed within the German Army, under the auspices of its (essentially geographical) Mapping and Survey branch (Häusler 2000*a*). Training for military geologists was provided from December 1917. By late 1918 a military geologist was attached to each army high command and a military geology unit to each of 28 Survey units (*Vermessungsabteilungen*). By the end of the War the German Army had made use of *c.* 200 men for their geological expertise and they had generated over 5500 geotechnical reports. A leaflet issued to guide the use of military geologists had a print run of 4000 copies.

According to Willig (1999) and Häusler (2000*a*), military geologist tasks in general related to the construction of field positions (especially trenches, dug-outs and mine tunnels, to minimize the use of time, labour and materials, and the risk of excavation collapse or water inflow); water supply (especially the siting of new boreholes and dams); quarrying or mining of raw materials (both aggregates for construction projects and reserves of economic significance, e.g. ores and coal); hygienic and technical problems (notably advice on problems of drainage, earthing of cables and route alignment for roads and railways); and other military problems (e.g. site selection for standing camps, ammunition dumps, heavy gun positions and airfields).

Specialist maps were innovatively generated from conventional geological maps for many key areas, primarily at the scale of 1:80 000 but later 1:25 000 for local use and 1:250 000 for more general planning. Typically these were *Minierkarten* (mining maps: see Willig 2018, fig. 6) to guide underground excavation, *Wasserversorgungskarten* to guide groundwater abstraction (water supply maps: see Rose 2009*b*), and *Bau-* or *Rohstoffkarten* to guide quarrying for construction materials (construction or raw materials maps: see Fig. 8). At least one of the 28 geological units, that attached to Survey Department 13 (*Geologengruppe de Vermessungsabteilung 13*, Württemberg) is known to have compiled more detailed maps (13 sheets at a scale of 1:10 000) as well as 11 sheets at 1:25 000 plus two more general maps (Häusler 2000*a*). The German Army had intended to transfer the geological organization from Survey to Engineer command during 1918, so as to relate geologists more closely to their primary military clients, but hostilities ended in November 1918 before the change in command could be effected.

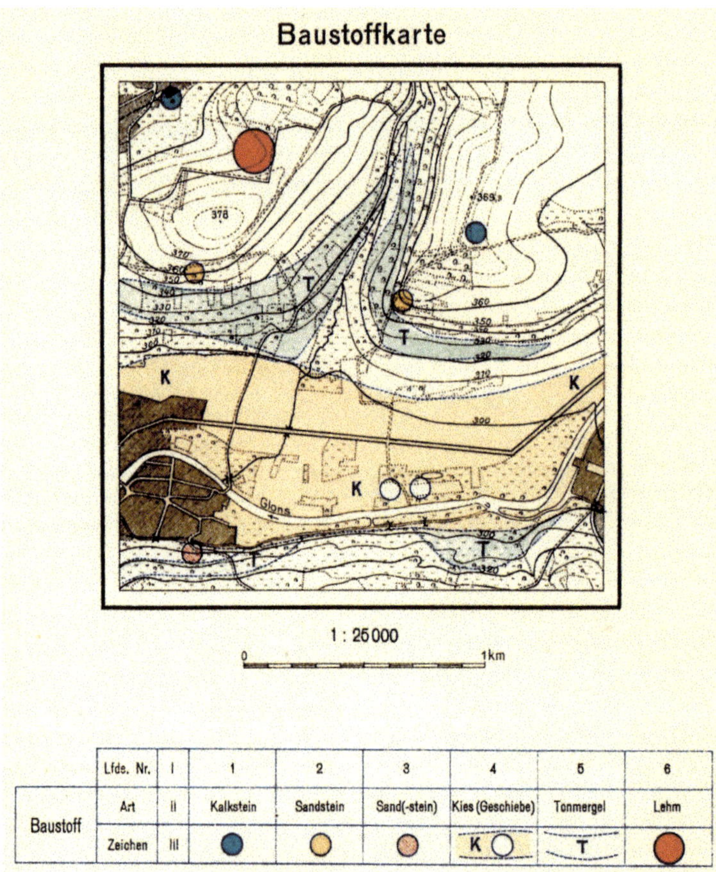

Fig. 8. Construction materials map, one of a set of maps illustrating how the same piece of ground can be mapped at a scale of 1:25 000 to depict also rock types (*Gesteinsartenkarte*), stratigraphic units (*Geologische Karte*), potential water supply (*Wasserversorgungskarte*) or water table depth and tunnelling conditions (*Grundwasser- und Minier-Karte*). Key: rock type indicated by colour (legend translation: (1) limestone; (2) sandstone; (3) sand; (4) gravel; (5) clay or marl; (6) loam); potential extraction site by circle; relative potential yield by circle size. The numbers 1–6 correspond with those in an accompanying table, where rock properties and potential uses are summarized for each rock type. From von Bülow *et al.* (1938).

German military geological work was to stimulate an extensive literature post-war. Kranz himself was to generate 170 published articles in his lifetime, about a third of them relating to military geology and most of these based on his personal experience during the war (Häusler 2003). Wilser (1921) quickly generated a textbook based on his own experience as the senior geologist within the wartime organization and edited (Wilser 1923–29) a series of 14 volumes that described the geology of the battlefields contested by German troops, on all fronts. A textbook by Kranz (1927) focused on engineering rather than military geology but was derived from his wartime military work. Further textbooks were to follow, as World War II loomed: by Wasmund (1937), based on theory rather than experience, and Kranz (1938), Mordziol (1938) and von Bülow *et al.* (1938), all based on skills learnt during World War I. More recently, Willig *et al.* (2015) have compiled an innovative geological World War I battlefield guide specifically for the Ypres region of Belgium, sourced from records both of the German forces and their opponents.

Austro-Hungarian military geologists

A presidential address in February 1916 to the Hungarian Geological Society (Schafarzig 1916), published in both Hungarian (for local readership) and German (for Austrian readership), contained a significant section on *Kriegsgeologie*. It is evident from this that the relevance of geology to military operations was well understood in Austria-Hungary as well as Germany by this time – when the British Army possessed but a single military geologist, and the US Army none at all. By February 1918, the Austro-Hungarian Army had developed a military geological organization similar to that of the German Army and by the end of the war had made use of around 60 men as military geologists (Häusler 2000*a*). Ultimately there were 17 military geology units supporting Austro-Hungarian Military Survey units, each geological unit typically consisting of three to five geologists. A guidebook on *Kriegsgeologie* was printed on 3 March 1918 and the first of five courses to train military geologists was held in Vienna later that month. Other courses followed in May, June and August and a fifth was planned for the final months of the war.

Austro-Hungarian geologists, like their German counterparts, generated reports and thematic maps relating primarily to field fortifications, mine warfare, water supply and the quarrying of construction materials. They served on both the Western and Eastern fronts, in the Balkans, and on the 600 km-long and largely mountainous ‘Southern Front’ that extended from the Swiss–Austrian border eastwards across the Tyrol, the Dolomites and the Julian Alps to the Adriatic Sea near the port of Trieste. Its crucial section lay near the River Isonzo, and 12 Battles of the Isonzo were fought between Austro-Hungarian and Italian forces between June 1915 and November 1917, with over 500 000 casualties. The roles of Austro-Hungarian geologists operating in this region have been described extensively by Häusler (2000*a*, 2013) and more briefly by Angetter & Hubmann (2015).

World War II (1939–45)

World War II dawned with a greater perception of military applications of geology in the Union of Soviet Socialist Republics (USSR) and in Germany than elsewhere. From 1934 forces of the USSR had the use of a 123-page military geological textbook: *Primenenie geologii na voine* (*Application of geology in war*) that consisted of translations of the benchmark publications of King (1919) and Brooks (1920) published together (Sabine 1992). Sychev (1979) in *The Great Soviet Encyclopedia*, based on Ovchinnikov *et al.* (1945) and Popov (1958), noted that work was done in the USSR to study and generalize the military geological experience gained in World War I with regard to the location of defensive works and the carrying out of various types of military engineer work. Also, that military geological services were formed in virtually all the warring countries, implying their existence in the USSR although not recording specific details. Ovchinnikov *et al.* (1945) described how geology could be applied to fortification, mine warfare, water supply, roads, airfields and camouflage. They also provided an introduction to geological surveying, mapping and aerial photographic interpretation, in a book of nearly 400 pages that provided evidence of considerable experience in military geology. Smith (1964, p. 319) cited an American account claiming that ‘German intelligence sources reported that over 15 000 persons were employed in military geology in the USSR during the war years’. However, little record of this work is currently in the public domain and Betz (1984, p. 357) has inferred from the large number ‘that many [of the supposed military geologists] were actually working on problems assigned to branches of economic geology’.

Other countries were less well prepared. It is known that a Royal Geological Unit (*Regio Reparto Geologico*) was formed within the Engineer branch (*Genio Militare*) of the Italian Army and that it compiled a water supply map for use in North Africa showing the positions of wells and the depth to water, but its roles have yet to be described (Aldino Bondesan, pers. comm. 2018). Little is known of Japanese military use of geology (Yajima 2006). The work of military geologists is known, as in

World War I, mostly from British, American and German sources. However, whereas in World War I it was the relatively static trench warfare of the Western Front that required greatest geotechnical input, World War II came to necessitate support for more mobile forms of conflict and in a wider range of countries.

British military geologists

Initially, the British again made use of very few military geologists. Effectively 'field force' geologists as in World War I, single geologists served as staff officers within the Engineer branch at general headquarters – but in more than one theatre of operation (Rose & Rosenbaum 1993*b*; Rose 2011). Major W.B.R King (Fig. 9) served from September 1939 as the water supply officer with the new British Expeditionary Force sent to northern France, the same man in the same place and role as in World War I, until the Force was defeated and evacuated at the end of May 1940. His university and military protégé Captain F.W. Shotton also served as a water supply officer, from 1941 to 1943 with Middle East Command, based in Egypt but with responsibilities ranging widely across the Middle East and North Africa (Rose & Clatworthy 2008*a*). Captain (later Major) J.V. Stephens, recruited from the Geological Survey of Great Britain, served from 1943 to 1945 in Sicily and Italy (Rose & Clatworthy 2007*a*); see Rose (2018*c*, fig. 8). Also from 1943 to 1945, Major (later Lieutenant-Colonel) King succeeded by Captain (later Major) Shotton served with 21st Army Group preparing for or participating in the June 1944 liberation of Normandy and the subsequent NW Europe campaign that ended with Allied victory in Europe (Rose *et al.* 2006, 2010).

Fig. 9. Major W.B.R. King, Royal Engineers, the senior British military geologist of World War II; from Rose & Rosenbaum (1993*b*), reproduced by kind permission of his daughter Professor C.A.M. King and granddaughter Jane Ritchie.

Their work focused initially on water supply (Rose 2012*b*) but broadened as conflict developed to include many aspects of engineering geology (such as site selection for the rapid construction of temporary airfields) and terrain evaluation (for cross-beach or cross-country trafficability). Thematic geotechnical maps were compiled to communicate technical information to non-geologist staff officers involved with the planning of military operations. For the amphibious and airborne landings in Normandy, these included beach trafficability maps at the scale of 1:5000 (Fig. 10), groundwater prospect maps at 1:50 000 and 1:250 000, and soils maps (to guide site selection for airfields and to indicate potential cross-country trafficability) at 1:250 000 (Rose *et al.* 2006, 2010).

Work by the small number of geologists was facilitated by resources made available at the Geological Survey of Great Britain and by reprinting, by the Geographical Section General Staff, of geological maps selected from the many published locally before the war. From October 1940 it was complemented by geophysical surveys made by detachments of 42nd Geological Section of the South African Engineer Corps: a 38-man unit that helped geologists to guide military well drilling for potable water in East Africa, the Mediterranean region and the Middle East (Rose 2018*d*).

In 1943, as British forces moved increasingly from defence to counter-attack, geologists were used to found two units to compile terrain intelligence for military use, the only occasions (other than as the South African section) when the British Army was to use geologists as teams rather than as individual staff officers. The Strategic Branch of the Geological Survey of India, based in the Survey's headquarters at Calcutta (present-day Kolkata: capital of the Indian state of West Bengal), grew from its foundation in June 1943 to comprise two officers at the rank of Major (or equivalent) and six Captains, plus a Staff Captain as administrator and appropriate clerical, drawing office and ancillary staff (Rose 2005*c*). At Oxford in England, a Geological Section within the Inter-Service Topographical Department (Fig. 11) grew from four men appointed as Royal

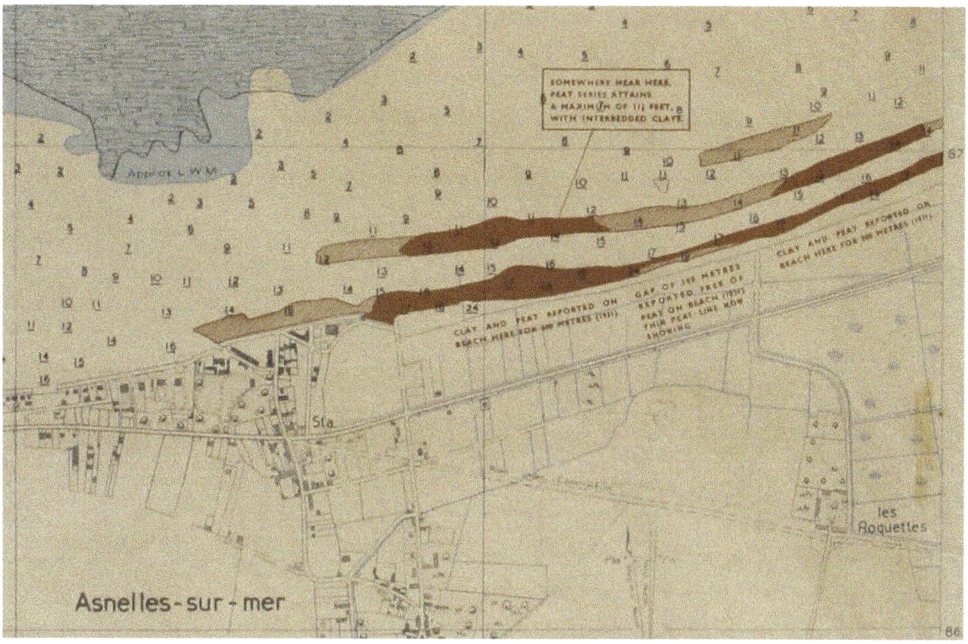

Fig. 10. Part of Sheet 74 Asnelles-sur-Mer of France 1:5000 map series, 1st edition January 1944; one of a set of 'secret' maps, compiled and printed for Allied forces, depicting beaches in Normandy selected for invasion, on D-Day in June 1944. Whole map illustrated and described by Rose *et al.* (2006); reproduced from the Shotton Archive by kind permission of the Director, Lapworth Museum of Geology, University of Birmingham.

Fig. 11. Members of the Geological Section of the Inter-Service Topographical Department, photographed in March 1945. From left to right, standing: Captain M.R.F. Ashworth (soil chemist in civilian life), Lieutenant (later Captain) J.R. Foster-Smith (mining engineer/geologist), Lieutenant (later Captain) W.R. Williams (economic geologist), Lieutenant (later Captain) A. Ludford (geologist), Captain W.E. Fraser (petroleum geologist), an unidentified RAF draughtsman, another draughtsman (probably Sapper Hunt) and the Section head: Major J.L. Farrington (economic geologist). Sitting: Captain A.D.M. Bell (petroleum geologist), Miss Jane Labey (Section typist) and Mr L.C. Huff (one of two geologists of the US Geological Survey's Military Geology Unit who served with the Section from January to March 1945). Photograph and identifications courtesy of the late Dr Albert Ludford; from Rose & Clatworthy (2008*b*).

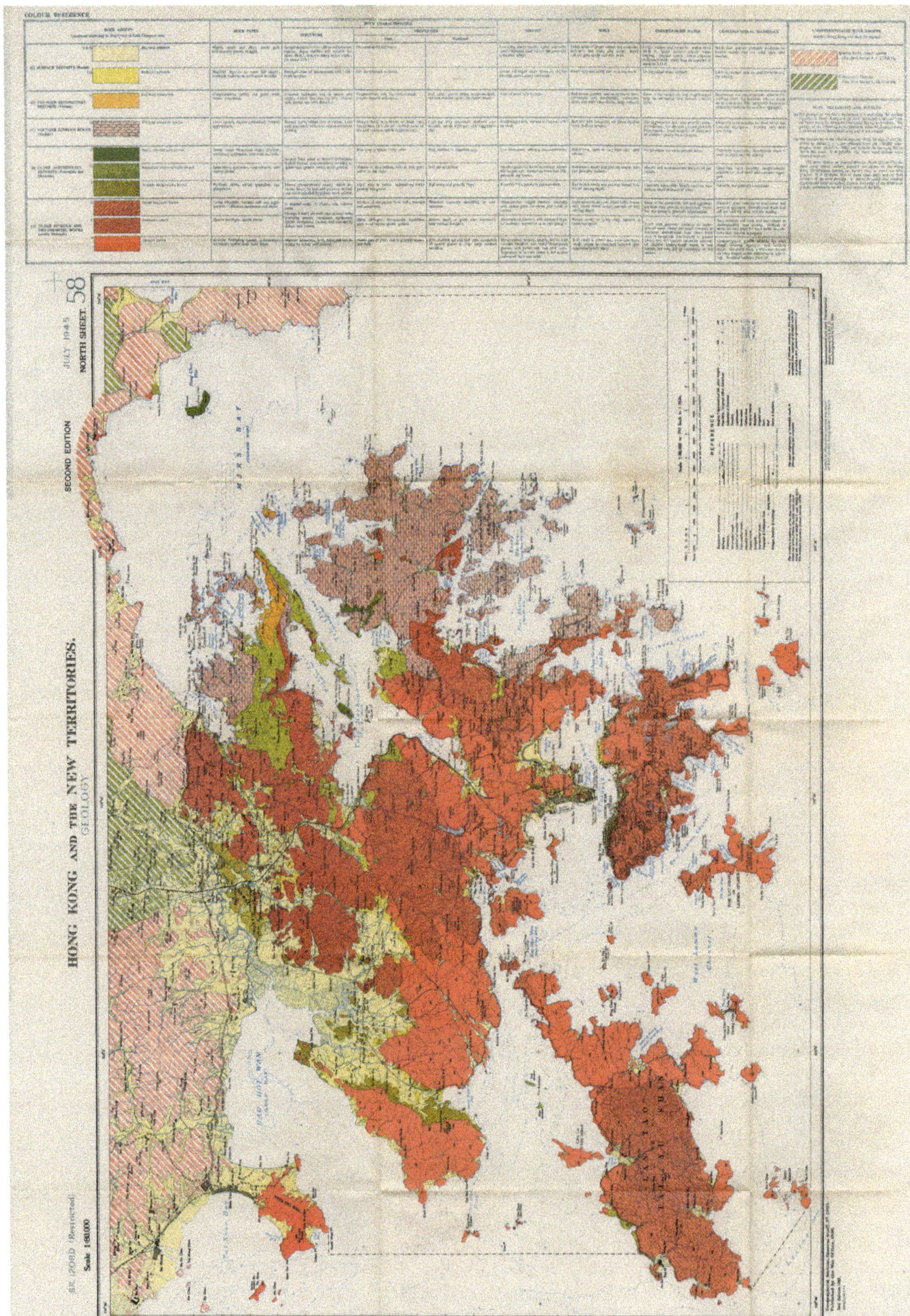
COLOUR REFERENCE
HONG KONG AND THE NEW TERRITORIES.
GEOLOGY
SECOND EDITION
JULY 1945
NORTH SHEET 58
Scale 1:80,000

Engineers' Captains in November (later one Major plus three Captains) to a maximum effective strength of 12 geoscientists by the end of hostilities (Rose & Clatworthy 2007*b*, 2008*b*). These two units were to generate a large number of specialist maps (e.g. Fig. 12) and reports, building on earlier such work by organizations staffed primarily by geographers.

Thus, in the first year of the war, the British Army made use of only one uniformed military geologist, Major 'Bill' King (Fig. 9). In the final year it made use of at least 24: 12 at the Inter-Service Topographical Department in England at Oxford, eight at the Strategic Branch of the Geological Survey of India at Calcutta, two in Italy (Majors 'Steve' Stephens and 'Bill' Macfadyen: Rose 2018*c*), and two in NW Europe (Majors 'Fred' Shotton and 'Dai' Ponsford: Rose *et al.* 2006). Bill King was to draw on their experience to compile a textbook for the Royal Engineers published soon after the war ended (Anon 1949).

American military geologists

The Japanese attack on Pearl Harbor on 7 December 1941 brought the USA into World War II. The Geological Society of America, meeting at Boston later that month, provided a forum for geologists determined to contribute to the war effort (Nelson & Rose 2012). Ultimately this determination led to foundation of the Military Geology Unit of the US Geological Survey in July 1942. Based in Washington DC, this comprised in-house and other earth scientists and engineers tasked to gather terrain and related strategic intelligence. Price & Woodward (1942) urged in detail that geologists should be used for the interpretation of terrain for military operations, conversion of ground for military use, maintenance of supplies and communications, and miscellaneous other applications. Erdmann (1943, p. 1190) later recommended that a military geologist should be 'a professional geologist of the widest range of sound training and field experience in areal mapping, engineering geology, geomorphology and the geology of unconsolidated materials' and subsequently advocated applications of geology to terrain intelligence (Erdmann 1944).

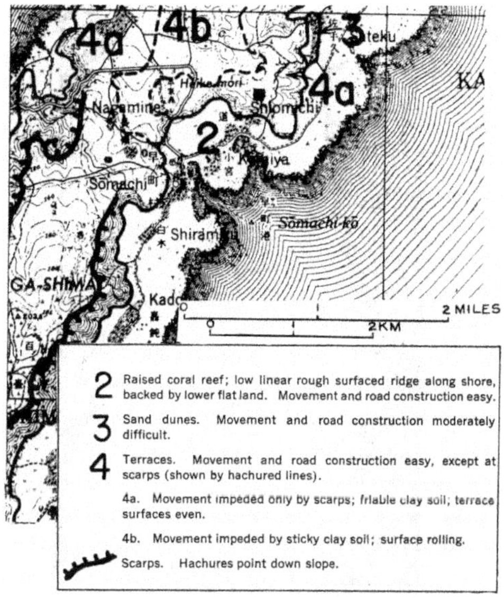

Fig. 13. Terrain map (original at scale *c.* 1:55 000), prepared in World War II by the Military Geological Unit of the US Geological Survey, of Kikaiga-shima, in the Ryukyu Islands NE of Okinawa, Japan, showing terraces and inferred inland trafficability. Cropped from US Geological Survey & US Army Corps of Engineers (1945); see also Nelson & Rose (2012).

As authoritatively described by Hunt (1950) and Terman (1998*a*), the Military Geology Unit compiled reports containing data about regions outside the USA as tables, text and maps (e.g. Fig. 13) for use by Allied forces, especially American and British. About 50 'Strategic Engineering Studies' were completed in 1942, and at least 50 more in 1943. The Unit's products contributed significantly to the Allied campaigns in Sicily and Italy from 1943, and complemented British geotechnical work associated with Normandy and the NW Europe Campaign in 1944–45. In 1944–45, the Unit deployed teams from Washington to US Army theatre headquarters in Europe and, more substantially, to the SW Pacific and Pacific Ocean theatres, principally to provide intelligence for strategic and tactical planning and

Fig. 12. Simplified geological map of Hong Kong, at scale of 1:80 000, compiled at Oxford in 1944 by the Geological Section of the Inter-Service Topographical Department to accompany a report completed in January 1945. The key groups rocks by use of three principal colours, with colour intensity varied so as to increase to five the main units depicted: (1) pale yellow (surface deposits: Recent); (2) deep yellow (younger sedimentary deposits: Tertiary); (3) pale red (younger igneous rocks: Tertiary); (4) green/blue (older sedimentary deposits: Palaeozoic and Mesozoic); and (5) deep red (older igneous and metamorphic rocks: mostly Mesozoic). The coloured 'rock groups' are subdivided by shading or overprint to differentiate 10 categories and the key tabulates data for each of these categories in seven columns: (1) rock types; (2) rock characteristics (three sub-columns: structure, properties fresh, properties weathered); (3) terrain; (4) soils; (5) underground water; (6) construction materials; and (7) notes on undifferentiated rock groups, reliability, and sources. From Rose (2013), cf. Mackay (2018); reproduced by permission of the Royal Geographical Society's Library.

advice to and (in the Pacific) participation in combat operations.

Numbers serving in the Unit increased from 10 professional personnel and one technical/clerical staff member in July 1942 to 80 professionals and 20 support staff in January 1945, as the war neared its end. Overall the total involved at least 88 geologists, 11 soil scientists, 15 other professionals and 43 support staff. Also, in total, the Unit produced 140 major terrain folios, 42 other major reports and 131 minor studies, in aggregate containing about 5000 maps, 4000 photographs and figures, 2500 large tables of data, and 140 terrain diagrams. Key components comprised terrain analysis for military purposes and data concerning rivers, potential road and airfield construction or repair, construction materials and water supply.

Like the British Army (Anon 1949), the US Army post-war was soon to publish a manual to show from experience how geology could be applied to military operations (Department of the Army 1952).

German military geologists

From 1937, as World War II loomed, a new military geological organization was developed in Germany (Häusler & Willig 2000), and in 1938 Germany occupied and annexed Austria. During the war, as comprehensively described by Häusler (1995*a*, *b*, 2000*b*) and summarized by Häusler (2018), this enlarged Germany was to make use of about 400 geologists to support its armed forces, most in the German Army but many also in the air force (the *Luftwaffe*, which operated all sites related to aerial defence, e.g. anti-aircraft gun batteries, as well as airfields), the paramilitary construction agency *Organisation Todt* (which built the fortification systems known as the West Wall and the Atlantic Wall, involving a peak workforce of about 1.5 million men), the *Waffen-SS* (which had its own military geological organization, with distinctive insignia), and even some in the German Navy (which operated batteries of coastal artillery as well as harbour and other shore installations, and planned for and against amphibious operations). Smaller numbers of geologists served within *Milgeo* (the military geographic branch of the Army, which prepared handbooks and maps for countries of actual or potential operational interest) and within an inter-service *Forschungsstaffel zur besonderen Verwendung [z.b.V.]* (research team for special use), a multidisciplinary unit developed for terrain analysis (Häusler 2007).

The establishment for the German Army's military geology organization expanded to comprise 40 centres or teams (*Wehrgeologenstellen*) by November 1943, each team typically comprising two officers or officials qualified to post-doctoral status (Fig. 14), supported by three non-commissioned officers as technicians or assistant geologists, and four other ranks for less-skilled tasks, such as driving or typing. Team deployments by the end of the war had in total ranged from North Africa to northern Norway, and from France eastward far into the Soviet Union. The library and archive of the organization (including copies of over 5400 typewritten reports) was held near Berlin until the closing months of the war, when these documents were stored at depth in a salt mine at Heringen in central Germany to preserve them from Allied bombing. The mine was captured by American forces and its contents transferred to the United States, first to the US Geological Survey, some later to the Federal Records Center and finally to the National Archives and Records Administration at College Park near Washington DC. The typewritten reports

Fig. 14. Ferdinand Trusheim in the uniform of a *Technischer Kriegsverwaltungsrat*, (Technical War Administration Officer) the status usually assigned to military geologists given commissioned rank to serve as such with the German Army in World War II: the national emblem is worn above the right breast pocket, as by other members of the contemporary German Army, rather than on the sleeve (as by auxiliary forces). During the Battle of France, from May to June 1940, Trusheim served as the 'Corps geologist' with the German 7th Corps, part of the 16th Army, and later assisted Otto Burre in geological preparation for 16th Army's role in Operation Sea Lion, the invasion of England scheduled for September 1940. Photograph taken at Berlin-Wannsee in 1942. From Rose & Willig (2002), reproduced courtesy of Drs Ursula and Hans Trusheim, per Dierk Willig.

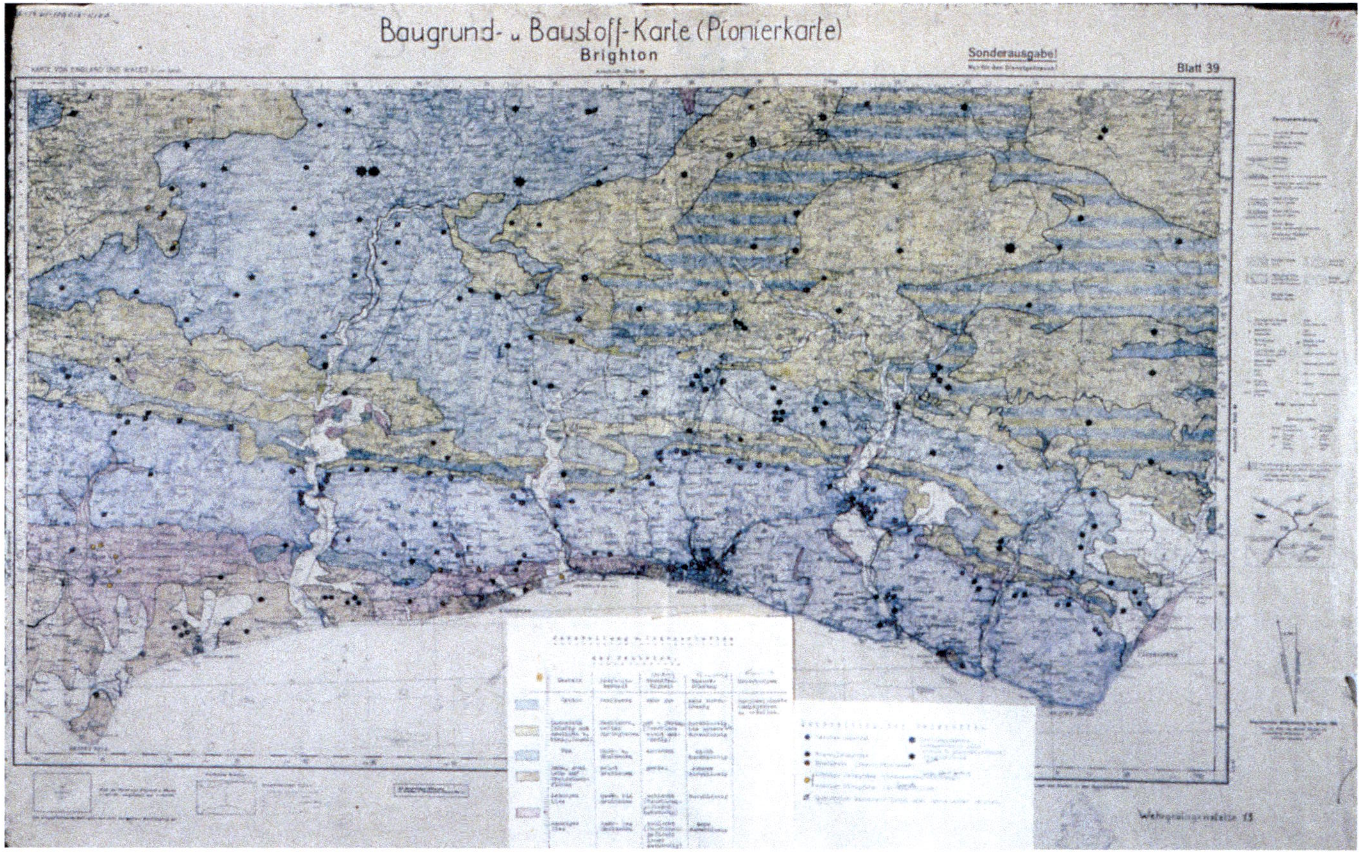

Fig. 15. Map of the Brighton area of southern England, classifying ground in terms of its suitability for various construction purposes and as a source of construction materials (*Pionierkarte* = engineering map); prepared at a scale of 1:100 000 by German military geologists to assist planning for Operation Sea Lion, the German invasion of England scheduled for September 1940. Quarry sites are plotted by spots colour-coded to indicate rock type (key to bottom right), on a simplified geological map. Colours used on the map as a whole are defined in a five-column key (bottom left of map) which indicates the corresponding rock type, its workability, stability and permeability, plus any remarks (such as suitability for trenches and dug-outs). From Rose & Willig (2004); reproduced courtesy of the US National Archives and Records Administration, College Park, MD, USA.

were mostly returned to Germany in the 1960s and are now preserved either in the *Bundesarchiv-Militärarchiv* at Freiburg-im-Breisgau or within the Bundeswehr Geoinformation Center of the modern German Army (Willig & Häusler 2012).

The range of case histories thus accessible to scholars includes examples of geological work done in haste to aid operational planning (e.g. for an invasion of England, scheduled for September 1940; Rose & Willig 2002, 2004, see Fig. 15) as well as longer-term projects to aid occupation and fortification (e.g. on the Channel Islands off the French coast; Robins *et al.* 2012). Häusler (2018) uses these and other sources to describe how over 60 geologists in total from the German Army, the *Luftwaffe* and the *Organisation Todt* contributed to the German occupation and fortification of Norway: a massive effort that helped to pioneer aspects both of rock mechanics and soil mechanics.

The Cold War (1947–91)

During the following Cold War, it is again the armed forces of the UK, USA and Germany for which military geological information is most accessible. Even so, relatively little has yet been placed in the public domain.

British military geologists

All the British military geologists who had served as such in World War II were demobilized at the end of hostilities. However, from 1949 a few geologists were recruited to serve as Royal Engineer officers in the British reserve army (the Territorial Army, Army Emergency Reserve or successor organizations) as systematically described by Rose & Hughes (1993*a*, *b*) and Rose & Rosenbaum (2017). Never exceeding eight in total in any year, they were committed to unlimited call-out liability in war and to a minimum of 15 days' annual service in peacetime. They served on attachment to British regular forces worldwide, providing technical guidance for a wide range of military projects, exercises and operations, particularly with regard to groundwater abstraction, site investigation for construction projects, quarrying, stability of slopes and tunnels, and cross-terrain 'going'. Several of them were to contribute to a new engineering geological textbook, sponsored jointly by the UK Ministry of Defence and the Institution of Civil Engineers (Anon 1976).

American military geologists

Post-war, the Military Geology Unit of the US Geological Survey did not disband, but transformed into a Section (and subsequently, in 1949, a Branch) of the Survey (Terman 1998*b*). Former Unit members were replaced as they returned to pre-war duties by younger scientists, to maintain continuity of expertise. The Branch continued to compile terrain intelligence on a global scale for much of the Cold War and during major periods of armed conflict. It was eventually discontinued in 1972, after making use of *c.* 150 geoscientists in total, but most of its functions passed to other US Geological Survey units or to the US Department of Defense.

Bonham (1997) compiled a list of *c.* 1500 US military geology reports and maps prepared between 1942 and 1975. In addition to these, it was noted that the Survey's military geologists generated more than 200 reports of more general geological interest that were published as Survey bulletins and professional papers or in non-Survey journals. The reports listed generally include basic subjects such as rock types, soils, water resources, landforms and vegetation, as well as terrain interpretation for cross-country movement and for the construction of roads and airfields throughout the world. Reports on specific areas range from generalized texts with small-scale maps derived from published sources to detailed texts with large-scale maps commonly based on aerial photographic interpretation and, especially for Alaska and western Pacific islands, involving field mapping. Other reports deal with topics of interest in military geology without reference to specific areas. Most of the reports were prepared by teams, primarily of geologists but commonly also including soils scientists, botanists, climatologists and geographers, and were published anonymously.

Terman's (1998*b*) analysis distinguished the evolution of seven principal administrative units and research programmes that evolved within the Military Geology Section and its successor organizations after 1945:

(1) The Strategic Studies Section, operational 1945–72 and funded by the US Army Corps of Engineers and later by the US Defense Intelligence Agency, was the major production unit. Notably, it contributed to a comprehensive small-scale National Intelligence Surveys Program and to a Pacific Engineer Intelligence Program. The latter involved the compilation in Washington DC of seven series of thematic maps for most of SE Asia at a scale of 1:250 000.

(2) The Pacific Field Program for the US Army Corps of Engineers 1945–62 was a research and mapping programme in areas formerly occupied by Japanese troops.

(3) The Alaska Terrain and Permafrost Section of the US Army Corps of Engineers and the US Geological Survey carried out field studies 1947–65 on the surface geology and

permafrost features in Alaska and other Arctic areas. It compiled reports with maps and engineering interpretations, mostly of Alaska and at a scale of 1:250 000.

(4) The European Field Program, principally undertaken by a US Geological Survey team in association with the US Army Europe 1953–64, generated 24 military engineering geology maps at 1:250 000 for western Germany and 131 cross-country movement maps at 1:100 000 for the whole of Germany.

(5) The Austere Landing Site Program 1956–70 for the US Air Force compiled large-scale studies on arid lands, both inside and outside the USA, and on Arctic ice-free land.

(6) The Special Intelligence Element, established in 1959 by the US Army Corps of Engineers and the Defense Intelligence Agency, provided geoscience consultants to the special intelligence community.

(7) The Nuclear-Test Detection Program 1962–72 compiled studies assisting in the interpretation of global seismic signals, notably a five-volume Atlas of Asia and Eastern Europe at scale of 1:5 000 000.

The US Army Corps of Engineers also used geological expertise throughout the Cold War. Its geologists, civilians like those of the US Geological Survey, also often worked in teams comprised of geologists, hydrologists, soil scientists, engineers and other professionals, and often in collaboration with US Geological Survey geologists or those from other Department of Defense agencies. However, during the late 1960s and early 1970s some officers served in uniform with the job title 'geologist'. Most military geology was, and still is, accomplished at the various Corps research laboratories, now combined as the Engineer Research and Development Center and with headquarters in Vicksburg, Mississippi.

Characterization of terrain has been an ongoing task for Corps research since at least the 1960s, when a series of terrain analogues was developed for regions outside the USA where it was thought the army could become involved, e.g. deserts or tropical areas (e.g. Kolb & Dornbusch 1966). This led to detailed 'country studies', characterizations of terrain, including geology, hydrology and geomorphology, compiled primarily by multi-disciplinary teams from existing data. Terrain information was provided as guidance for troops in the field as well as for training on US military bases. Analysis and interpretation of aerial photography were important parts of these efforts, but after the ERTS-1 satellite (Landsat-1) was launched in 1972, emphasis rapidly moved to the use of satellite imagery and all that it offered.

Other tasks undertaken by Corps geologists included tunnel detection, notably in Vietnam and Korea (Anon 1979; Eastler 2004), identification of potential drilling sites for water, and the location of engineering materials. Environmental assessment of army facilities within the USA also began during the later stages of the Cold War.

Articles in Underwood & Guth (1998), Ehlen & Harmon (2001), Harmon *et al.* (2014) and McDonald & Bullard (2016) provide many specific examples of the wide range of military geoscience applications that developed in the USA within the Cold War period and subsequently.

German military geologists

During the Cold War, following creation of an army (the *Bundeswehr*) within the Federal Republic of Germany in 1956, a new 'geophysical' service evolved, combining meteorological and geological expertise so as to provide advice on 'going', i.e. cross-country mobility, as discussed by Malm (2018), as well as a range of other geoscience-related tasks (Rose *et al.* 2000; Willig 2012). At its peak the service included about 20 full-time geologists, all civilians but with reserve army ranks for use on military exercises or operational deployment – including a geologist team specifically assigned to each corps of the German Army. Geological tasks varied with the region and, therefore, the geology concerned, but typically involved environmental protection, assessment of natural obstacles to cross-country movement, construction projects and water supply and drainage. In some areas, tasks involved with environmental protection became so numerous that the function of the geologists was more to help protect the environment than it was to help protect the nation state.

The Geophysical Service merged with the Army's Geographical Service in 2003 to form the current Bundeswehr Geoinformation Center. Until 1993, German military geologists were active only within the borders of Germany, but thereafter they were deployed overseas, to support United Nations or Allied forces, e.g. in Somalia, Kosovo and (most recently) Afghanistan, typically to help develop secure water supplies by the appropriate siting of boreholes (Willig 2012).

Principal applications of military geology

'Military geology may be defined simply enough as the application of geology to the art of war' (Erdmann 1943, p. 1169). From the range of applications, noted above, that have emerged over *c.* 200 years, four stand out as of particular significance: those relating to (1) water supply, (2) construction of fortifications, (3) ground excavation by trenching, tunnelling and quarrying, and (4) terrain evaluation for military operations.

Applications relating to water supply in general lie outside the scope of this book. They have already been extensively documented in a companion volume (Rose & Mather 2012). However, the sections above show how potable water has long been recognized as a vital military resource and problems of water supply were amongst the earliest and most frequent tasks assigned to military geologists serving on both sides in World War I. They were still important in World War II and subsequent conflicts. Additional to the challenge of providing water to large numbers of troops concentrated on campaign, without a secure water supply no defended locality, however well-fortified and garrisoned, can withstand siege for long. Examples of problems encountered in providing water supply to fortified localities are the only aspect of military hydrogeology discussed in this volume. **Mather (2018)** provides some examples of how water supply was achieved in southern England during the fortification works of the eighteenth and nineteenth centuries, revealing innovative means of exploiting groundwater from sites underlain by Cretaceous chalk or Paleogene clastic sediments potentially subject to seawater intrusion. Papers on coastal fortification, ground excavation and terrain evaluation constitute the rest of the book.

Coastal fortification

Fortification is an ancient skill. Some fortifications in the Middle East range back to about 5000 years in age (Toy 2006). However, the earliest known military defences to be erected in Great Britain are late-Bronze Age and Iron Age hillforts: earthworks, originally reinforced or topped by timber palisades, which were constructed in large numbers during the thousand years before the beginning of the Christian era (Johnson 1978; Williams 1999). The Romans invaded Britain in AD 43, progressively captured the hillforts, and in the next *c.* three centuries initiated a programme of substantial stone-built fortifications both inland and at key coastal sites. **Bromhead (2018)** describes how a coastal fort from this period, constructed on an abandoned marine cliff eroded into a near-horizontal sequence of Early Cretaceous sandstones overlying mudrocks, has suffered damage in the 1700 years since its completion. This damage is inferred to be the consequence of a few major landslips rather than continuous creep downslope, and so to provide unusually long-term evidence concerning the nature and frequency of natural slope failure.

In the eleventh century, Norman conquest of England initiated a new phase of inland fortification that included castle building and extended through the medieval period. However, the next national programme of coastal fortification in England was initiated only in the sixteenth century, by King Henry VIII (Fig. 16). His scheme was enhanced in subsequent centuries, culminating in the mid-nineteenth century with fortification of the major naval bases in southern England at Portsmouth (**Mather 2018**), Plymouth, Portland, Dover, Chatham and the mouth of the River Medway, and Pembroke in south Wales: the largest scheme of fortification ever seen in Britain (Porter 1977; Crick 2012).

Fig. 16. Lindisfarne Castle, constructed in the sixteenth century on the coast of NE England on a site fortified by order of King Henry VIII. It occupies an eroded remnant of the Great Whin Sill, a dolerite intrusion of late Carboniferous age, in places >70 m thick, whose outcrop scarp has been partly fortified by Hadrian's Wall of AD 122 and by several medieval castles (cf. Rose 2005*d*). Photo courtesy of Claire Rose.

Henderson (2018) provides an account of coeval fortification in the USA, to provide a defence against potential invasion by the British. Following national independence from colonial rule in 1784, the government of the USA soon approved construction of a 'first system' of forts at key locations along the country's eastern seaboard, followed by a 'second system' early in the nineteenth century and, from 1816, an ambitious 'third system'. Only 42 'third system' forts were actually built, primarily on weak Quaternary sediments: some on Pleistocene drumlins or moraine, many on Holocene barrier islands, their geological setting being therefore very different to the sites of nineteenth-century forts in England described by **Mather (2018)**.

The final article in this section of the book (**Häusler 2018**) moves the theme of coastal fortification into the twentieth century. It looks at Norway, whose rugged coastline of over 25 000 km has been eroded principally into a strong Precambrian–Paleozoic terrain, albeit with a locally thick Quaternary clay cover. During World War II, at least 63 German and Austrian geoscientists were incorporated into the German armed forces occupying Norway: an exceptional concentration of expertise. Most assisted construction of the northern part of the 'Atlantic Wall': in total a massive system of intermittent coastal fortifications that extended *c.* 2600 km in a near direct line from northern Norway south to the border of France with Spain, to defend the western boundary of German-occupied Europe against an expected Allied invasion. The professional experience they gained was to give significant impetus postwar to international developments in both rock mechanics and soil mechanics.

Excavation

For World War I, **Doyle (2018)** describes herein how geology influenced trench construction by British troops and the Germans facing them in the northern part of the Western Front: terrain across Belgium and part of northern France underlain by Cenozoic sands and clays or by Cretaceous chalk with a variable Quaternary cover. This was but one part of a complex of trenches (e.g. Fig. 17) that extended from the North Sea southwestwards to the border of France with Switzerland, and so crossing older and stronger rocks outside the operational area of the British Expeditionary Force. **Willig (2018)** goes deeper, to provide an overview of tunnelling by German and Austro-Hungarian military engineers

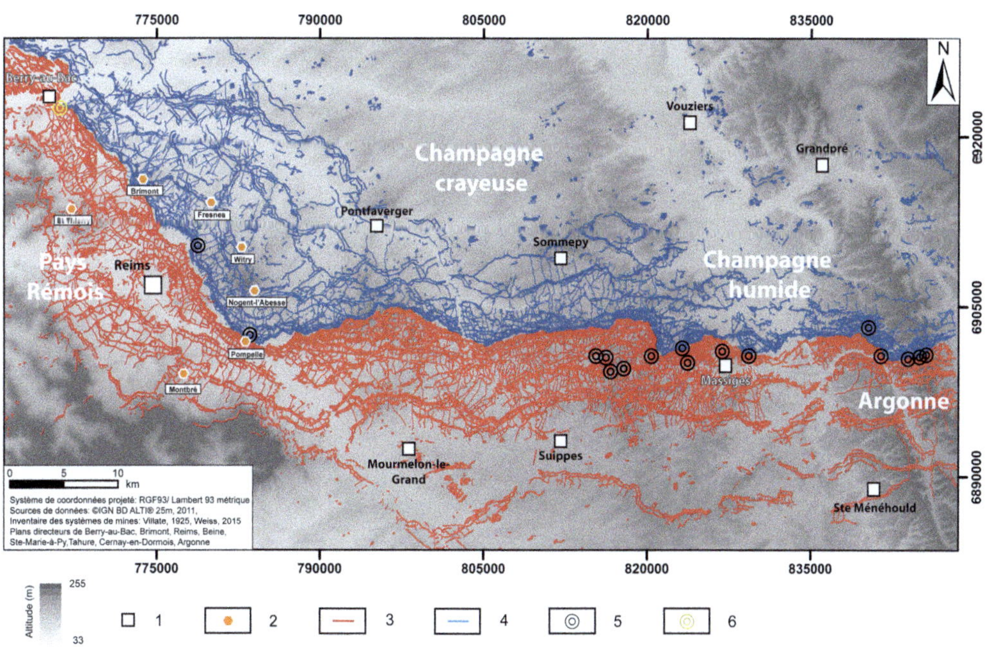

Fig. 17. Map of Allied defence excavations (trenches, dug-outs and mine systems: in red) and the opposing German works (in blue) for the Champagne-Argonne region of the Western Front in 1918. Numbers in key at base indicate: (1) locality, (2) fort of Séré de Rivières system, (3) Allied trenches and communications, (4) German trenches and communications, (5) mine system, (6) mine system of Berry-au-Bac. Reproduced by kind permission of Pierre Taborelli from Taborelli *et al.* (2017).

not on the Western Front (for which tunnelling by both sides has already been extensively documented, e.g. by Jones 2010), but on the relatively less well-known Eastern Front, facing Russia and its satellite countries. Twelve teams of military geologists assisted this work, comprising over 60 men in total. **Rose (2018*b*)** returns the theme to the Western Front, describing the formation and role of Quarrying Companies raised by the Royal Engineers. A wartime innovation, they produced the large quantities of 'stone' required for the enhancement or repair of the road and rail infrastructure essential to the British Expeditionary Force: at its peak a force of *c.* 1.5 million troops.

For World War II, **Rose (2018*c*)** extends his account of Quarrying Companies by describing their re-creation in 1940 and their wide deployment to support more mobile operations of the British Army, in circum-Mediterranean countries as well as western Europe. This is followed by a complementary account (**Rose 2018*e***) for Tunnelling Companies of the Royal Engineers, focused on tasks in northern France, the UK, Gibraltar and Malta and so including rocks stronger than those encountered by such companies when these were first raised for service, for the Western Front in World War I. **Mackay (2018)** continues the tunnelling theme. He describes a railway tunnel through granite in Hong Kong, deliberately sealed by British troops during World War II as part of a programme of 'route denial' intended to inhibit access by invading Japanese forces. The tunnel has been subject to recent remedial work: a case history in a region noted for the intensive studies in engineering geology that now guide its major construction works. Route denial and route maintenance have both proved to be historically significant aspects of wartime military geology.

Finally, **Bulmer (2018)** brings the military tunnelling theme up to date by an extensive account of the tunnels recently created by insurgents in weak Cenozoic sedimentary rocks in the Mosul region of northern Iraq, to hide combatants and their materiel from aerial observation and attack, and of attempts by Iraqi government forces and their allies to locate them.

Terrain evaluation

The final section of this book contains articles that might loosely be grouped under the heading of terrain evaluation (terms such as terrain analysis or terrain appreciation being sometimes understood to imply a more specific scope when used in a British military context; Americans use only the term terrain analysis).

> Military interest in terrain goes back to the earliest times … [but] It was not until after World War II that serious and sustained attention was given to the problem of evaluating and predicting terrain conditions [for military purposes]
>
> (Parry 1984, p. 570).

In former centuries, use of military geographers as staff officers to guide commanders in their use of terrain during campaigns was pioneered in the armies of France and Prussia, from the eighteenth century onwards. Topographical maps that illustrated terrain features in detail became an increasingly important military resource from this time. In the UK, the Ordnance Survey was founded in 1791 to undertake a national survey of Great Britain in the face of a perceived growing threat of invasion from revolutionary France (a survey completed in the nineteenth century and revised subsequently), but its origins arguably lie in military surveys of Scotland begun in 1747 following the Jacobite rebellion of 1745–46 (Seymour 1980).

In the American Civil War (1861–65), several generals, both Union and Confederate (e.g. William S. Rosecrans and Thomas 'Stonewall' Jackson) had topographers on their staffs who provided maps of the terrain to aid troop movements as well as to subsequently document the battlefields (Pittman 2000). **Harrelson *et al.* (2018)** provide a case history in which Union military commanders made tactical errors in planning their Red River Campaign in the southern USA by overlooking unique geological aspects of the terrain, including the river's flashy nature, high sediment load, anastomosing channels and rapids.

From World War I onwards, topographical maps came to be supplemented by thematic maps generated by military geographers and military geologists. Examples from World War I include water supply and 'mining' maps generated by both sides on the Western Front (Brooks 1920; Anon 1922; Häusler 2000*a*; Rose 2009*b*; Rose & Rosenbaum 2011). The German and Austro-Hungarian armies, supported by larger military geological staffs than the Allies, created thematic maps of great variety and number (Häusler 2000*a*). In World War II, German military geologists again compiled thematic maps of many kinds for many regions (Häusler 1995*a*, *b*), and as Allied forces moved increasingly from defence to counter-attack, both the US Army (Nelson & Rose 2012) and the British armed forces (Rose 2005*c*; Rose & Clatworthy 2007*b*, 2008*b*) created military geologist units specifically to compile thematic maps and reports. Moreover, **Rose (2018*f*)** describes how Allied terrain evaluation in World War II was significantly enhanced by the interpretation of aerial photographs, a technique pioneered for the Royal Air Force and the British Army (and so for Allied forces in general) by a unit that included some distinguished geoscientists amongst its most

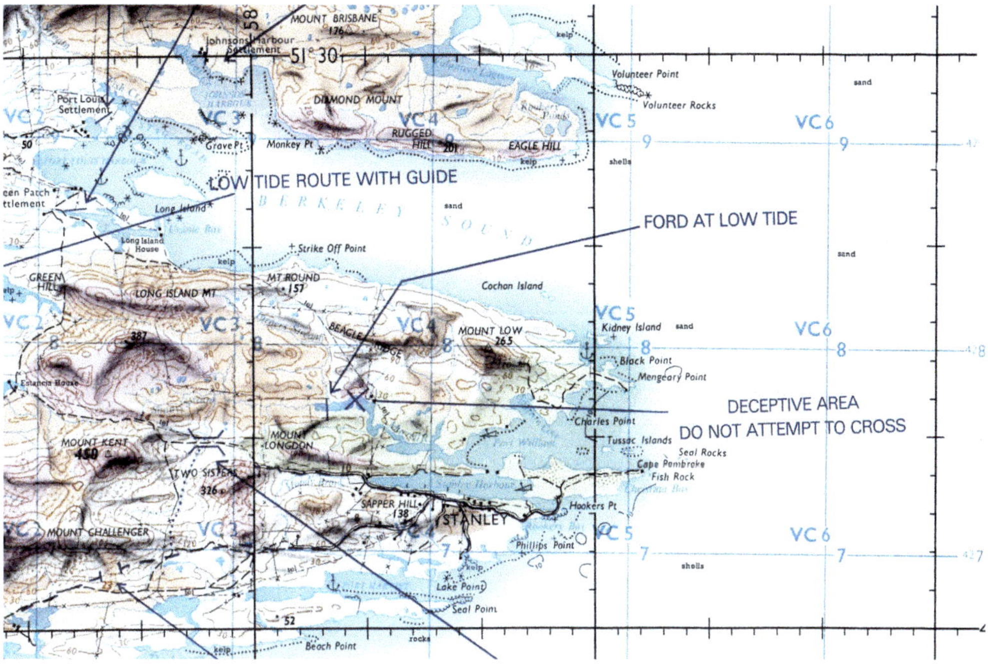

Fig. 18. Part of the Falklands Islands Royal Engineers Briefing Map series GSGS 5453 edition 3, compiled at scale of 1:250 000 under the direction of the Engineer-in-Chief (Army), primarily to indicate 'going' in 1982. Grid intervals: blue at 10 km, black at 15′. Green = firm ground usually passable, white = light peat passable with care, orange = bad going passable with local advice, magenta = bog or rock not passable to vehicles. The 'derivation and reliability' shown in the map margin indicates a contribution 'by RE [i.e. Royal Engineers] geologists'. In fact, only one geologist was involved, from the British reserve army, working for two days with an agronomist who had recently returned from the islands. They drafted a first edition which was printed in full colour 36 hours later. A refined second edition was printed within a week (L.R.M. Cocks, pers. comm. 2018.) © Crown copyright, reproduced by kind permission of the Copyright and Release Branch, Defence Geographic Centre, Feltham, UK.

influential officers. **Malm (2018)** brings the book to a close with a review, from a German perspective, of the techniques developed over the last 100 years in order to predict cross-country mobility for wheeled and tracked vehicles, now a major factor in military operational planning and so a key component of modern terrain evaluation. Such terrain studies formed part of the planning for British operations to liberate the Falkland Islands in 1982 (Fig. 18), although then (as often in World War II) the thematic map was compiled urgently and in haste rather than as the result of a long-term study.

Concluding remarks

These few examples serve to introduce potentially wide-ranging military aspects of geology. During the nineteenth century the British Army included successive Inspectors-General of Fortifications (or senior officers of similar title: Porter 1977) responsible for major construction works both in the UK and overseas whose staff occasionally had access to geological information and military geologists such as J.E. Portlock (**Mather 2018**). Smith (1964, p. 313) accepted that during the Russo-Japanese War of 1904–05 (cf. **Willig 2018**) the Russian Army 'used a number of geologists primarily as advisers in constructing fortifications'. Forts and fortifications were being constructed across much of Europe between 1815 and 1945 (Williams 1999; Kaufmann & Kaufmann 2014*a*, *b*) and Germany, at least from World War I onwards, was to make major use of military geologists in their development (**Willig 2018**; **Häusler 2018**). However, as throughout history, developments in weapon systems continued to make old fortifications obsolete, and to serve as catalysts for the development of new types of defensive structures.

Ditches have been dug to define regional boundaries or impede military movement for several millennia, and stone has sometimes been quarried in impressively large quantities for fortress construction. For example, the 135 km-long Hadrian's Wall built by some 30 000 soldiers and labourers in the ten years following AD 122 to mark the northern boundary of the Roman province of Britannia, required more than 24 million stones and was flanked by a ditch (the Vallum) 5.4–5.9 m wide at the top and 2.6–2.9 m deep, part of a zone of entrenchment 37 m across and larger than any prehistoric earthwork (Moffat 2009). The Romans frequently created small-scale earthworks during siege operations, but large-scale excavation actually during a time of conflict was a major feature of the two World Wars rather than earlier history.

Sapping (i.e. trenching) and mining (i.e. tunnelling) are ancient military engineering skills long used in siege operations, particularly after the development of artillery able to breach fortress walls and consequent changes in fortress design intended to withstand bombardment. Of these, tunnelling to develop facilities underground has many implications for military forces of the present day. Even before recent conflicts in Iraq that involved tunnelling (**Bulmer 2018**), Eastler (2004, p. 35) predicted ‘underground terrain may very well be the pivotal battlefield of the not so distant future’. Indeed, *The Times* newspaper (Anon 2018, p. 30) recently alleged, albeit without substantive evidence, that:

> Of the 10 000 large-scale [underground] military facilities estimated to exist, North Korea has almost half … [and] the capability to move up to 30 000 troops per hour through a network of deep tunnels into South Korea.

US military training is currently being revised in the light of experience in Iraq and Afghanistan and such perceptions: ‘The Pentagon is spending $500 million preparing American soldiers to fight underground as military chiefs reassess the likely battlefields of the future’.

Quarrying is a more recent significant military engineering skill. It became important in World War I because of the scale of near-static troop confrontation on the Western Front. German military geologists were then innovative in developing thematic resource maps, showing sites or regions for the quarrying of construction materials (building stone, sand and aggregates) and material of economic significance (e.g. coal and metalliferous ores). These maps helped to guide the construction of fortification systems by the Engineer corps. A significant proportion of German military geologist tasks was related to the winning of raw materials in occupied land (Rose *et al.* 2000). The Allies, fighting to defend their own terrain, made use more of local civilian than military geologist expertise for such tasks.

Raw materials also formed a significant proportion of German military geologist tasks in World War II. Fortification systems, such as the West Wall (bordering Germany and France) and the Atlantic Wall (that marked the western limit of German-occupied Europe), consumed vast quantities of aggregate for concrete in 1938–40 and 1940–44 respectively. The Atlantic Wall alone included *c.* 15 000 bunker and casemate complexes with outer walls of reinforced concrete 2 m thick, eventually absorbing *c.* 17 600 000 tonnes of concrete (Rolf 1988). More recent operations have been less demanding on construction materials, but still placed demands on military geologists, e.g. following conflict in the Falkland Islands of the South Atlantic in 1982, when the British Army operated a quarry near Port Stanley (Fig. 18) (Rosenbaum 1985).

Much military geology is but hydrogeology and engineering geology undertaken for military clients, although in wartime speed of completion is typically more important than cost-effectiveness. Terrain evaluation, however, remains an aspect of geology that has particular military as well as civilian significance.

> Since the early 1950s the procedures for terrain evaluation have developed far beyond the ‘no-go’ outline overlays of World Wars I and II and this portion of the military art has been finally transformed into a military science (quoted by Parry 1984, p. 579).

Acknowledgements Permission to re-use illustrations is indicated in figure captions where appropriate. The authors are grateful to Jim Griffiths for reviewing this paper.

Funding This research received no specific grant from any funding agency in the public, commercial, or not-for-profit sectors.

References

Angetter, D. & Hubmann, B. 2015. Important Austrian war geologists and their tasks at the southern front of World War I. *Rendiconti online della Società Geologica Italiana*, **36**, 7–9.

Anon 1922. *The Work of the Royal Engineers in the European War, 1914–19: Geological Work on the Western Front.* Institution of Royal Engineers, Chatham.

Anon 1949. *Military Engineering Vol. XV. Application of Geology. War Office Code 8287.* HMSO, London.

Anon 1976. *Military Engineering Vol. XV. Applied Geology for Engineers. Army Code No. 71044.* HMSO, London.

Anon 1979. *A History of the Waterways Experiment Station 1929–1979.* http://www.dtic.mil/dtic/tr/fulltext/u2/a635951.pdf [last accessed April 2018].

Anon 2018. Troops sent underground as US prepares for battlefields of the future. *The Times*, 27 June 2018, 30.

Betz, F., Jr 1984. Military geoscience. *In*: Finkl, C.W. (ed.) *The Encyclopedia of Applied Geology.* Van Nostrand Reinhold, New York, 355–358.

Bonham, S. 1997. *Reports and Maps of the Military Geology Unit 1942–1975.* US Geological Survey Open File Report, 97–175.

BRANAGAN, D. 2005. *T. W. Edgeworth David: A Life: Geologist, Adventurer, and Knight in the Old Brown Hat.* National Library of Australia, Canberra.

BROMHEAD, E.N. 2018. The landslip-damaged Roman fort at Lympne in SE England. *In*: ROSE, E.P.F., EHLEN, J. & LAWRENCE, U.L. (eds) *Military Aspects of Geology: Fortification, Excavation and Terrain Evaluation.* Geological Society, London, Special Publications, **473**, https://doi.org/10.1144/SP473.9

BROOKS, A.H. 1920. The use of geology on the Western Front. *US Geological Survey Professional Paper*, **128-D**, 85–124.

BULMER, M.H. 2018. Geological considerations of contemporary military tunnelling near Mosul, northern Iraq. *In*: ROSE, E.P.F., EHLEN, J. & LAWRENCE, U.L. (eds) *Military Aspects of Geology: Fortification, Excavation and Terrain Evaluation.* Geological Society, London, Special Publications, **473**, https://doi.org/10.1144/SP473.11

CALDWELL, D.R., EHLEN, J. & HARMON, R.S. (eds) 2004. *Studies in Military Geography and Geology.* Kluwer Academic, Dordrecht.

CRICK, T.J. 2012. *Ramparts of Empire: The Fortifications of Sir William Jervois, Royal Engineer, 1821–1897.* The Exeter Press, Exeter.

DEPARTMENT OF THE ARMY 1952. *Geology and Its Military Applications. Department of the Army Technical Manual TM 5-545.* Department of the Army, Washington, DC.

DOYLE, P. 2018. Trench construction and engineering geology on the Western Front, 1914–18. *In*: ROSE, E.P.F., EHLEN, J. & LAWRENCE, U.L. (eds) *Military Aspects of Geology: Fortification, Excavation and Terrain Evaluation.* Geological Society, London, Special Publications, **473**, https://doi.org/10.1144/SP473.6

DOYLE, P. & BENNETT, M.R. (eds) 2002. *Fields of Battle: Terrain in Military History.* Kluwer Academic, Dordrecht.

EASTLER, T.E. 2004. Military use of underground terrain – a brief historical perspective. *In*: CALDWELL, D.R., EHLEN, J. & HARMON, R.S. (eds) *Studies in Military Geography and Geology.* Kluwer Academic, Dordrecht, 21–37.

EHLEN, J. & HARMON, R.S. (eds) 2001. *The Environmental Legacy of Military Operations.* Reviews in Engineering Geology **XIV**. Geological Society of America, Boulder, CO.

ERDMANN, C.E. 1943. Applications of geology to the principles of war. *Bulletin of the Geological Society of America*, **54**, 1169–1194.

ERDMANN, C.E. 1944. Applications of geology to terrain intelligence. *Bulletin of the Geological Society of America*, **55**, 783–788.

FABBI, S. & ROMANO, M. 2015. The First World War of Italian geologists: between patriotic interventionism and objective pragmatism. *Rendiconti online della Società Geologica Italiana*, **36**, 67–71.

FOSSA-MANCINI, E. 1926. Storia della geologia militare. *Giornale di geologia pratica*, **20**, 37–110.

GREGORY, H.E. (ed.) 1918. *Military Geology and Topography.* Yale University Press, New Haven.

HANOT, F., SIMIEN, F., CHARLES, N. & VICELLI, J. 2017. *Sous les lignes de front: Regards géologiques sur la grande guerre.* BRGM éditions, Orléans.

HARMON, R.S., BAKER, S.E. & MCDONALD, E.V. (eds) 2014. *Military Geosciences in the Twenty-first Century.* Reviews in Engineering Geology **XXII**. Geological Society of America, Boulder, CO.

HARRELSON, D.W., TORRES, N., TILLOTSON, A. & ZAKIKHANI, M. 2018. Geological influence of the Great Red River Raft on the Red River Campaign of the American Civil War. *In*: ROSE, E.P.F., EHLEN, J. & LAWRENCE, U.L. (eds) *Military Aspects of Geology: Fortification, Excavation and Terrain Evaluation.* Geological Society, London, Special Publications, **473**, https://doi.org/10.1144/SP473.5

HÄUSLER, H. 1995*a*. Die Wehrgeologie im Rahmen der Deutschen Wehrmacht und Kriegswirtschaft. Teil 1: Entwicklung und Organisation. *Informationen des Militärischen Geo-Dienstes*, Vienna, **47**, 1–155.

HÄUSLER, H. 1995*b*. Die Wehrgeologie im Rahmen der Deutschen Wehrmacht und Kriegswirtschaft. Teil 2: Verzeichnis der Wehrgeologen. *Informationen des Militärischen Geo-Dienstes*, Vienna, **48**, 1–119.

HÄUSLER, H. 2000*a*. Die Österreichische und Deutsche Kriegsgeologie 1914–1918. *Informationen des Militärischen Geo-Dienstes*, Vienna, **75**, 1–161.

HÄUSLER, H. 2000*b*. Deployment and role of military geology teams in the German army 1941–45. *In*: ROSE, E.P.F. & NATHANAIL, C.P. (eds) *Geology and Warfare: Examples of the Influence of Terrain and Geologists on Military Operations.* Geological Society, London, 159–175.

HÄUSLER, H. 2003. Dr. Walter Kranz (1873–1953) – Der erste Militärgeologe des 20. Jahrhunderts. *MILGEO: Organ des Militärischen Geowesens des Österreichischen Bundesheeres*, Bundesministerium für Landesverteidigung, Vienna, **12**, 1–80.

HÄUSLER, H. 2007. Forschungsstaffel z.b.V. Eine Sondereinheit zur militärgeografischen Beurteilung des Geländes im 2. Weltkrieg. *MILGEO: Schriftenreihe des Militärischen Geowesens*, Bundesministerium für Landesverteidigung, Vienna, **21**, 1–209.

HÄUSLER, H. 2013. Oberleutnant in der Reserve Dr. Artur Winkler von Hermaden – Leiter der Geologengruppe 'Isonzo' im Jahr 1918. *Berichte der Geologischen Bundesanstalt*, Vienna, **103**, 24–51.

HÄUSLER, H. 2018. The northern Atlantic Wall: German engineering geology work in Norway during World War II. *In*: ROSE, E.P.F., EHLEN, J. & LAWRENCE, U.L. (eds) *Military Aspects of Geology: Fortification, Excavation and Terrain Evaluation.* Geological Society, London, Special Publications, **473**, https://doi.org/10.1144/SP473.4

HÄUSLER, H. & KOHLER, E. 2003. Der Schweitzer Geologe, Oberberghauptmann und Major Johann Samuel Gruner (1766–1824): Begründer der Militärgeologie. *Minaria Helvetica*, **23a**, 47–102.

HÄUSLER, H. & MANG, R. (eds) 2011. *International Handbook Military Geography Vol. 2. Proceedings of the 8th International Conference on Military Geosciences, 15–19 June 2009, Vienna, Austria.* Truppendienst, Ministry of Defence, Vienna.

HÄUSLER, H. & WILLIG, D. 2000. Development of military geology in the German Wehrmacht 1939–45. *In*: ROSE, E.P.F. & NATHANAIL, C.P. (eds) *Geology and Warfare: Examples of the Influence of Terrain and Geologists on Military Operations.* Geological Society, London, 141–158.

Henderson, S.W. 2018. American coastal defence Third System forts: how geomorphology and geology dictated placement and influenced history. *In*: Rose, E.P.F., Ehlen, J. & Lawrence, U.L. (eds) *Military Aspects of Geology: Fortification, Excavation and Terrain Evaluation.* Geological Society, London, Special Publications, **473**, https://doi.org/10.1144/SP473.10

Hubé, D. 2018. *14–18 Tremblements de Guerre: Les géologues au coeur de l'histoire.* BRGM éditions, Orléans.

Hunt, C.B. 1950. Military geology. *In*: Paige, S. (ed.) *Application of Geology to Engineering Practice. Berkey Volume.* Geological Society of America, New York, 295–327.

Hutton, F.W. 1862. Importance of a knowledge of geology to military men. *Journal of the Royal United Service Institution*, **6**, 342–360.

Johnson, P. 1978. *The National Trust Book of British Castles.* Book Club Associates, London.

Jones, S. 2010. *Underground Warfare 1914–1918.* Pen & Sword Military, Barnsley.

Kaufmann, J.E. & Kaufmann, H.W. 2014*a*. *The Forts and Fortifications of Europe, 1815–1945: The Central States: Germany, Austria-Hungary and Czechoslovakia.* Pen & Sword Military, Barnsley.

Kaufmann, J.E. & Kaufmann, H.W. 2014*b*. *The Forts and Fortifications of Europe, 1815–1945: The Neutral States: The Netherlands, Belgium and Switzerland.* Pen & Sword Military, Barnsley.

Kiersch, G.A. 1998. Engineering geosciences and military operations. *Engineering Geology*, **49**, 123–176.

Kiersch, G.A. & Underwood, J.R. 1998. Geology and military operations, 1800–1960: An overview. *In*: Underwood, J.R., Jr & Guth, P.L. (eds) *Military Geology in War and Peace.* Reviews in Engineering Geology **XIII**. Geological Society of America, Boulder, CO, 5–27.

King, W.B.R. 1919. Geological work on the Western Front. *Geographical Journal*, **54**, 201–215; discussion 215–221.

Kolb, C.R. & Dornbusch, W.K., Jr 1966. *Analogs of Yuma Terrain in the Middle East Desert*, **I**. Technical Report, 3-360. Waterways Experiment Station, Vicksburg, MS.

Kranz, W. 1913. Militärgeologie. *Kriegstechnische Zeitschrift*, **16**, 464–471.

Kranz, W. 1914. Militärgeologie. *Strassburger Post*, **163,** 10 February.

Kranz, W. 1915. Kriegsgeologie. *Strassburger Post*, **173,** 12 March.

Kranz, W. 1927. *Die Geologie im Ingenieur-Baufach.* Enke-Verlag, Stuttgart.

Kranz, W. 1938. *Technische Wehrgeologie - Wegweiser für Soldaten, Geologen, Techniker, Ärzte, Chemiker und andere Fachleute.* Jänecke-Verlag, Leipzig.

Lewis, G.G., Jones, H.D., Nelson, R.J., Larcom, T.A., De Moleyns, E.C. & Williams, J. (eds) 1846–52. *Aide-Mémoire to the Military Sciences.* Weale, London.

Mackay, A.D. 2018. Engineering geological considerations for the 'Old' Beacon Hill Railway Tunnel, Hong Kong Special Administrative Region. *In*: Rose, E.P.F., Ehlen, J. & Lawrence, U.L. (eds) *Military Aspects of Geology: Fortification, Excavation and Terrain Evaluation.* Geological Society, London, Special Publications, **473**, https://doi.org/10.1144/SP473.12

Malm, F. 2018. One hundred years of cross-country mobility prediction in Germany. *In*: Rose, E.P.F., Ehlen, J. & Lawrence, U.L. (eds) *Military Aspects of Geology: Fortification, Excavation and Terrain Evaluation.* Geological Society, London, Special Publications, **473**, https://doi.org/10.1144/SP473.7

Mather, J.D. 2018. Groundwater supplies to maritime and coastal defences in southern England: a story of risk and innovation. *In*: Rose, E.P.F., Ehlen, J. & Lawrence, U.L. (eds) *Military Aspects of Geology: Fortification, Excavation and Terrain Evaluation.* Geological Society, London, Special Publications, **473**, https://doi.org/10.1144/SP473.8

McDonald, E.V. & Bullard, T. (eds) 2016. *Military Geosciences and Desert Warfare: Past Lessons and Modern Challenges.* Springer Science + Business Media, New York.

Moffat, A. 2009. *The Wall: Rome's Greatest Frontier.* Birlinn Limited, Edinburgh.

Mordziol, C. 1938. *Einführung in die Wehrgeologie.* Verlag Otto Salle, Frankfurt.

Nathanail, C.P., Abrahart, R.J. & Bradshaw, R.P. (eds) 2008. *Military Geography and Geology: History and Technology.* Land Quality Press, Nottingham.

Nelson, C.M. & Rose, E.P.F. 2012. The US Geological Survey's Military Geology Unit in World War II: 'the Army's pet prophets'. *Quarterly Journal of Engineering Geology and Hydrogeology*, **45**, 349–367, https://doi.org/10.1144/1470-9236/11-054

Ovchinnikov, A.M., Popov, V.V. & Grigoryev, I.F. (eds) 1945. *Voennaia geologiia (Military geology).* Gosgeolitizdat, Moscow-Leningrad.

Parry, J.T. 1984. Terrain evaluation, military purposes. *In*: Finkl, C.W. (ed.) *The Encyclopedia of Applied Geology.* Van Nostrand Reinhold, New York, 570–580.

Penrose, R.A.F. 1917. *What a Geologist can do in War.* Lippincott, Philadelphia, PA.

Pittman, W.E. 2000. Geologists and the American Civil War. *In*: Rose, E.P.F. & Nathanail, C.P. (eds) *Geology and Warfare: Examples of the Influence of Terrain and Geologists on Military Operations.* Geological Society, London, 84–103.

Pogue, J.E. 1917. Military geology. *Science*, **46**, 8–10.

Popov, V.V. 1958. *Geologiia v voenno-inzhemnom dele (Geology in Military Engineering).* V. V. Kuibyshava Military Engineering Academy, Moscow.

Porter, W. 1977. *History of the Corps of Royal Engineers*, **II**. Institution of Royal Engineers, Chatham.

Portlock, J.E. 1849. *A Rudimentary Treatise on Geology: For the Use of Beginners.* Weale, London.

Portlock, J.E. 1850. Geognosy and geology. *In*: Lewis, G.G., Jones, H.D., Nelson, R.J., Larcom, T.A., De Moleyns, E.C. & Williams, J. (eds) *1846–52 Aide-Mémoire to the Military Sciences.* Weale, London, **2**, 77–182.

Price, P.H. & Woodward, H.P. 1942. Geology and war. *American Association of Petroleum Geologists Bulletin*, **26**, 1832–1838.

Quijano y Arroquia, A.R. (translated by A. Joly) 1876. *La guerre et la géologie.* Librairie Militaire de J. Dumaine, Paris.

Robins, N.S., Rose, E.P.F. & Cheney, C.S. 2012. Basement hydrogeology and fortification of the Channel Islands: legacies of British and German military engineering.

In: Rose, E.P.F. & Mather, J.D. (eds) *Military Aspects of Hydrogeology*. Geological Society, London, Special Publications, **362**, 203–222, https://doi.org/10.1144/SP362.11

Rolf, R. 1988. *Atlantic Wall Typology*. AMA, Beetserwaag, The Netherlands.

Rose, E.P.F. 1996. Geology and the army in nineteenth century Britain: a scientific and educational symbiosis? *Proceedings of the Geologists' Association*, **107**, 129–141.

Rose, E.P.F. 1997. Geological training for British army officers: a long-lost cause? *Royal Engineers Journal*, **111**, 23–29.

Rose, E.P.F. 1999. The military background of John W. Pringle, in 1826 founding superintendent of the Geological Survey of Ireland. *Irish Journal of Earth Sciences*, **17**, 61–70.

Rose, E.P.F. 2004*a*. Military geology. *In*: Selley, R.C., Cocks, L.R.M. & Plimer, I.R. (eds) *Encyclopedia of Geology*. Elsevier, Oxford, **3**, 475–487.

Rose, E.P.F. 2004*b*. Napoleon Bonaparte's Egyptian campaign of 1798: the first military operation to be assisted by geologists? *Geology Today*, **20**, 24–29.

Rose, E.P.F. 2005*a*. Napoleon Bonaparte's invasion of Egypt 1798–1801 the first military operation assisted by both geographers and geologists – and its defeat by the British. *Royal Engineers Journal*, **119**, 109–116.

Rose, E.P.F. 2005*b*. Karl von Raumer: a pioneer of German military geology. *Geology Today*, **21**, 182–186.

Rose, E.P.F. 2005*c*. British military geology in India: its beginning and ending. *Royal Engineers Journal*, **119**, 46–53.

Rose, E.P.F. 2005*d*. Impact of military activities on local and regional geologic conditions. *In*: Ehlen, J., Haneberg, W.C. & Larson, R.A. (eds) *Humans as Geologic Agents*. Reviews in Engineering Geology **XVI**. Geological Society of America, Boulder, CO, 51–66.

Rose, E.P.F. 2008*a*. Military applications of geology: A historical perspective. *In*: Nathanail, C.P., Abrahart, R.J. & Bradshaw, R.P. (eds) *Military Geography and Geology: History and Technology*. Land Quality Press, Nottingham, 17–34.

Rose, E.P.F. 2008*b*. The French invasion of Egypt led by Napoleon Bonaparte in 1798: Its pioneers of military and Egyptian geology. *In*: Nathanail, C.P., Abrahart, R.J. & Bradshaw, R.P. (eds) *Military Geography and Geology: History and Technology*. Land Quality Press, Nottingham, 45–60.

Rose, E.P.F. 2009*a*. Military men: Napoleonic warfare and early members of the Geological Society. *In*: Lewis, C.L.E. & Knell, S.J. (eds) *The Making of the Geological Society of London*. Geological Society, London, Special Publications, **317**, 219–241, https://doi.org/10.1144/SP317.13

Rose, E.P.F. 2009*b*. Water supply maps for the Western Front (Belgium and northern France) developed by British, German and American military geologists during World War I: pioneering studies in hydrogeology from trench warfare. *The Cartographic Journal*, **46**, 76–103.

Rose, E.P.F. 2011. Credit due to the few: British field force geologists of World War II. *In*: Häusler, H. & Mang, R. (eds) *International Handbook Military Geography Vol. 2. Proceedings of the 8th International Conference on Military Geosciences, 15–19 June 2009, Vienna, Austria*. Truppendienst, Ministry of Defence, Vienna, 429–442.

Rose, E.P.F. 2012*a*. Groundwater as a military resource: pioneering British military well boring and hydrogeology in World War I. *In*: Rose, E.P.F. & Mather, J.D. (eds) *Military Aspects of Hydrogeology*. Geological Society, London, Special Publications, **362**, 49–72, https://doi.org/10.1144/SP362.4

Rose, E.P.F. 2012*b*. Groundwater as a military resource: development of Royal Engineers Boring Sections and British military hydrogeology in World War II. *In*: Rose, E.P.F. & Mather, J.D. (eds) *Military Aspects of Hydrogeology*. Geological Society, London, Special Publications, **362**, 105–138, https://doi.org/10.1144/SP362.7

Rose, E.P.F. 2013. Wartime geotechnical maps: applied geology and airfield construction suitability maps compiled remotely for Hong Kong in WW2 illustrate techniques pioneered by military geologists. *Geoscientist*, **23**(1), 10–15.

Rose, E.P.F. 2014. Military geosciences before the twenty-first century. *In*: Harmon, R.S., Baker, S.E. & McDonald, E.V. (eds) *Military Geosciences in the Twenty-first Century*. Reviews in Engineering Geology **XXII**. Geological Society of America, Boulder, CO, 19–26.

Rose, E.P.F. 2015*a*. The Battle of Waterloo, 18 June 1815: some geological reflections to mark the bicentenary. *Geology Today*, **31**, 103–108.

Rose, E.P.F. 2015*b*. 'Abstract from Geology at the Western Front' by T. W. Edgeworth David. *Earth Sciences History*, **34**, 1–22.

Rose, E.P.F. 2018*a*. The International Association for Military Geosciences: a history to 2017. *Scientia Militaria: South African Journal of Military Studies*, **46**, in press.

Rose, E.P.F. 2018*b*. Quarrying Companies Royal Engineers in World War I: a geologically constrained innovation to support British armies on the Western Front. *In*: Rose, E.P.F., Ehlen, J. & Lawrence, U.L. (eds) *Military Aspects of Geology: Fortification, Excavation and Terrain Evaluation*. Geological Society, London, Special Publications, **473**, https://doi.org/10.1144/SP473.1

Rose, E.P.F. 2018*c*. Quarrying Companies Royal Engineers in World War II: contributions to military infrastructure within the UK and to Allied forces during the North African, Italian and NW Europe campaigns. *In*: Rose, E.P.F., Ehlen, J. & Lawrence, U.L. (eds) *Military Aspects of Geology: Fortification, Excavation and Terrain Evaluation*. Geological Society, London, Special Publications, **473**, https://doi.org/10.1144/SP473.2

Rose, E.P.F. 2018*d*. Officers of 42nd Geological Section, South African Engineer Corps: geologists and geophysicists who created a unique unit that supported the British army during the Second World War. *Scientia Militaria: South African Journal of Military Studies*, **46**, in press.

Rose, E.P.F. 2018*e*. Tunnelling Companies Royal Engineers in World War II: excavation of bomb-proof facilities in France, Gibraltar, Malta and the UK. *In*: Rose, E.P.F., Ehlen, J. & Lawrence, U.L. (eds) *Military Aspects of Geology: Fortification, Excavation and*

Terrain Evaluation. Geological Society, London, Special Publications, **473**, https://doi.org/10.1144/SP473.3
Rose, E.P.F. 2018*f*. Aerial photographic intelligence during World War II: contributions by some distinguished British geologists. *In*: Rose, E.P.F., Ehlen, J. & Lawrence, U.L. (eds) *Military Aspects of Geology: Fortification, Excavation and Terrain Evaluation*. Geological Society, London, Special Publications, **473**, https://doi.org/10.1144/SP473.13
Rose, E.P.F. & Clatworthy, J.C. 2007*a*. The Sicilian and Italian campaigns of World War II: roles of British military geologists in Allied engineer 'Intelligence' and 'Works'. *Royal Engineers Journal*, **121**, 94–103.
Rose, E.P.F. & Clatworthy, J.C. 2007*b*. Specialist maps of the Geological Section, Inter-Service Topographical Department: aids to British military planning during World War II. *The Cartographic Journal*, **44**, 13–43.
Rose, E.P.F. & Clatworthy, J.C. 2008*a*. Fred Shotton: a 'hero' of military applications of geology during World War II. *Quarterly Journal of Engineering Geology and Hydrogeology*, **41**, 171–188, https://doi.org/10.1144/1470-9236/07-034
Rose, E.P.F. & Clatworthy, J.C. 2008*b*. Terrain evaluation for Allied military operations in Europe and the Far East during World War II: 'Secret' British reports and specialist maps generated by the Geological Section, Inter-Service Topographical Department. *Quarterly Journal of Engineering Geology and Hydrogeology*, **41**, 237–256, https://doi.org/10.1144/1470-9236/07-016
Rose, E.P.F. & Hughes, N.F. 1993*a*. Sapper Geology: part 2. Geologist pools in the reserve army. *Royal Engineers Journal*, **107**, 173–181.
Rose, E.P.F. & Hughes, N.F. 1993*b*. Sapper Geology: part 3. Engineer Specialist Pool geologists. *Royal Engineers Journal*, **107**, 306–316.
Rose, E.P.F. & Mather, J.D. (eds) 2012. *Military Aspects of Hydrogeology*. Geological Society, London, Special Publications, **362**, https://doi.org/10.1144/SP362.1
Rose, E.P.F. & Nathanail, C.P. (eds) 2000. *Geology and Warfare: Examples of the Influence of Terrain and Geologists on Military Operations*. Geological Society, London.
Rose, E.P.F. & Rosenbaum, M.S. 1993*a*. British military geologists: the formative years to the end of the First World War. *Proceedings of the Geologists' Association*, **104**, 41–49.
Rose, E.P.F. & Rosenbaum, M.S. 1993*b*. British military geologists: through the Second World War to the end of the Cold War. *Proceedings of the Geologists' Association*, **104**, 95–108.
Rose, E.P.F. & Rosenbaum, M.S. 2011. British geological maps that guided excavation of military dug-outs in Belgium during World War I. *Quarterly Journal of Engineering Geology and Hydrogeology*, **44**, 293–306, https://doi.org/10.1144/1470-9236/08-111
Rose, E.P.F. & Rosenbaum, M.S. 2017. Sapper Geology: part 4. Royal Engineer Specialist Advisory Team geologists. *Royal Engineers Journal*, **131**, 95–102.
Rose, E.P.F. & Willig, D. 2002. German military geologists and terrain analysis for Operation Sealion: the invasion of England scheduled for September 1940. *Royal Engineers Journal*, **116**, 265–273.
Rose, E.P.F. & Willig, D. 2004. Specialist maps prepared by German military geologists for Operation 'Sealion': the invasion of England scheduled for September 1940. *The Cartographic Journal*, **41**, 13–35.
Rose, E.P.F., Häusler, H. & Willig, D. 2000. Comparison of British and German military applications of geology in world war. *In*: Rose, E.P.F. & Nathanail, C.P. (eds) *Geology and Warfare: Examples of the Influence of Terrain and Geologists on Military Operations*. Geological Society, London, 107–140.
Rose, E.P.F., Clatworthy, J.C. & Nathanail, C.P. 2006. Specialist maps prepared by British military geologists for the D-Day landings and operations in Normandy, 1944. *The Cartographic Journal*, **43**, 117–143.
Rose, E.P.F., Clatworthy, J.C. & Robins, N.S. 2010. Water supply maps for Northwest Europe developed by British military geologists during World War II: innovative mapping for mobile warfare. *The Cartographic Journal*, **47**, 55–91.
Rosenbaum, M.S. 1985. Engineering geology related to quarrying at Port Stanley, Falkland Islands. *Quarterly Journal of Engineering Geology and Hydrogeology*, **18**, 253–260, https://doi.org/10.1144/GSL.QJEG.1985.018.03.06
Sabine, P. 1992. Geology in the First World War. *Geologists' Association Circular*, **893** (August 1992), 10.
Saintine, X.-B., Marcel, J.J. & Reybaud, L. 1830–36. *Histoire Scientifique et Militaire de l'Éxpédition Française en Égypte*. Dénain & Delamare, Paris.
Schafarzig, F. 1916. Eröffnungs-Vortrag anlässlich der am 9 Februar 1916 abgehaltenen Hauptversammlung der Ungarischen Geologischen Gesellschaft. *Földtani Közlöny*, **46**(Suppl.), 43–56 [text in Hungarian **46**, 1–22].
Seymour, W.A. (ed.) 1980. *A History of the Ordnance Survey*. Dawson & Sons, Folkestone.
Smith, J.S.C. 1964. Military applications of geology. *Transactions of the Kansas Academy of Science*, **67**, 311–336.
Smith, R.B. 1849. Essay on geology, as a branch of study especially meriting the attention of the Corps of Engineers. *Corps Papers, and Memoirs on Military Subjects; of the Royal Engineers and the East India Company's Engineers*, **1**, 27–34.
Sychev, A.K. 1979. Geology, Military. *In*: Prokhorov, A.M. (ed.) *The Great Soviet Encyclopedia*. 3rd edn. English translation. Collier Macmillan, New York & Macmillan, London, **6**, 305.
Taborelli, P., Devos, A., Laratte, S., Brenot, J., Bollot, N., Cancès, B. & Desfossés, Y. 2017. Apport des plans directeurs et de l'outil LiDAR aéroporté pour la caractérisation des impacts morphologiques de la Grande Guerre: example de la cote 108 (Berry-au-Bac, France). *Géomorphologie: Relief, Process, Environment*, **23**, https://geomorphologie.revues.org/11711
Terman, M.J. 1998*a*. Military Geology Unit of the U.S. Geological Survey during World War II. *In*: Underwood, J.R., Jr & Guth, P.L. (eds) *Military Geology in War and Peace*. Reviews in Engineering Geology **XIII**. Geological Society of America, Boulder, CO, 67–74.
Terman, M.J. 1998*b*. Military Geology Branch of the U.S. Geological Survey from 1945 to 1972. *In*: Underwood, J.R., Jr & Guth, P.L. (eds) *Military Geology in War*

and Peace. Reviews in Engineering Geology **XIII**. Geological Society of America, Boulder, CO, 75–81.

Toy, S. 2006. *A History of Fortification from 3000 BC to AD 1700*. Pen & Sword Military Classics, Barnsley.

Underwood, J.R., Jr & Guth, P.L. (eds) 1998. *Military Geology in War and Peace*. Reviews in Engineering Geology **XIII**. Geological Society of America, Boulder, CO.

US Army Corps of Engineers 2008. *The U.S. Army Corps of Engineers: A History*. Headquarters US Army Corps of Engineers Office of History, Alexandria, VA.

US Geological Survey & US Army Corps of Engineers 1945. *The Military Geology Unit* (pamphlet). Geological Society of America, New York.

von Bülow, K., Kranz, W. & Sonne, E. 1938. *Wehrgeologie*. Quelle & Meyer, Leipzig.

von Grouner, J.S. 1826. Verhältnis der Geognosie zur Kriegs-Wissenschaft. *Moll's neue Jahrbücher der Berg- und Hüttenkunde*, Nürnberg, **6**, 187–233.

Wasmund, E. 1937. *Wehrgeologie in ihrer Bedeutung fur die Landesverteidigung*. Mittler & Sohn, Berlin.

Williams, G. 1999. *Stronghold Britain: Four Thousand Years of British Fortifications*. Sutton, Stroud.

Willig, D. 1999. Entwicklung der Wehrgeologie: Aufgabenspektrum und Beispiele. 1. Von den Anfängen bis 1918. *Fachliche Mitteilungen,* Amt für Wehrgeophysik, Traben-Trarbach, **225**, 1–116.

Willig, D. 2012. Hydrogeology and the *Bundeswehr*: water supply to German armed forces in Somalia, Kosovo and Afghanistan between 1993 and 2010. *In*: Rose, E.P.F. & Mather, J.D. (eds) *Military Aspects of Hydrogeology*. Geological Society, London, Special Publications, **362**, 253–265, https://doi.org/10.1144/SP362.14

Willig, D. 2018. German military geology and military mining on the Eastern Front in World War I. *In*: Rose, E.P.F., Ehlen, J. & Lawrence, U.L. (eds) *Military Aspects of Geology: Fortification, Excavation and Terrain Evaluation*. Geological Society, London, Special Publications, **473**, https://doi.org/10.1144/SP473.14

Willig, D. & Häusler, H. 2012. Aspects of German military geology and groundwater development in World War II. *In*: Rose, E.P.F. & Mather, J.D. (eds) *Military Aspects of Hydrogeology*. Geological Society, London, Special Publications, **362**, 187–202, https://doi.org/10.1144/SP362.10

Willig, D., Rose, E.P.F., Heyse, I., Allersmeier, C. & Doyle, P. 2015. *Militärhistorisch-kriegsgeologischer Reiseführer zum Wytschaete-Bogen (Messines Ridge) bei Ypern (Belgien)*. Schriftenreihe Geoinformationsdienst der Bundeswehr Heft 4, 2015. Geoinformationsdienst der Bundeswehr, Euskirchen.

Wilser, J. 1921. *Grundriss der angewandten Geologie unter Berücksichtigung der Kriegserfahrungen für Geologen und Techniker*. Borntraeger, Berlin.

Wilser,. J. (ed.) 1923–29. *Die Kriegsschauplätze 1914–1918 geologisch dargestellt in 14 Heften*. Borntraeger, Berlin.

Yajima, M. 2006. Japanese wartime geology: a case study in northeast China. *Historia Scientiarum*, **15**, 222–232.

The landslip-damaged Roman fort at Lympne in SE England

EDWARD NICHOLAS BROMHEAD

9 Kevins Drive, Yateley, Hampshire GU46 7TL, UK
edward.bromhead@btinternet.com

Abstract: The late third century Roman fort at Lympne in Kent, one of a series known as the Saxon Shore forts, is one of the earliest substantial stone-built fortifications in SE England. It was sited on the lower slopes of an escarpment, the remnant of a former coastal slope, eroded principally into Early Cretaceous clays capped with sandy limestones. At this location, the escarpment rises to *c.* 100 m above sea-level. Several episodes of landslip damage during the 17 centuries since its construction can be distinguished, based on evidence of various kinds: notably, the reconstruction of its original layout; the results of both geotechnical and archaeological site investigations; historical evidence; and analogy with sites and experience in adjacent lengths of the escarpment. The landslip damage manifests itself through the toppling and displacement of the curtain wall and projecting towers or bastions of the fort. The specific engineering geological interest arising from discoveries on site is the attribution of the damage to a small number of episodes of major movement rather than to creep; the magnitude of the movements associated with each episode; and the evidence as to their nature and date.

The stability of natural slopes is a matter of considerable interest to engineering geologists and geotechnical engineers, especially in terms of the processes and mechanics of landsliding, the rates and magnitudes of movement, the controls on the process of the underlying geology, the influence of climate and weather, and, in particular, what human influences can affect the stability of those slopes. It is often impractical to put dates to slope movements, but at Lympne (pronounced 'Lim') in southern Kent there are the remains of a masonry and concrete fort dating back to the Roman occupation of Britain that has been affected by landsliding. By combining archaeological and engineering investigations, insights into the evolution of natural slopes have been gained. The fort is thought on numismatical evidence (Smith 1852; Cunliffe *et al.* 1980) to have been built in the late third century AD. In the 17 centuries since then, it has been damaged on more than one occasion by landsliding – a consequence of its construction on the lower part of a former coastal cliff abandoned by the sea, the retreat of which formed Romney Marsh at the toe of the slope.

The following account describes the history of the fort at Lympne, the underlying geology and geography of the site, the damage sustained to the fort and therefore the nature and extent of the landslides that have affected it. It also describes geotechnical investigations made at the site in conjunction with archaeological investigations. Their purpose was to interpret the evolution of the slope and to explore the magnitude of the landslide movements by establishing the shape of the fort in its initial configuration. It is unknown what understanding the ancient Romans in general had of geology or the behaviour of geomaterials, but they were prolific builders, employing masonry, brickwork and unreinforced concrete, and they showed themselves to be capable of constructing extensive earthworks and tunnels (commonly for the supply of water to cities). Their engineering construction techniques may have been empirical, but they were highly effective.

The slopes at Lympne are abandoned cliffs, once subject to marine erosion and now defended from the sea by the growth and development of Romney Marsh, a sparsely populated wetland area bordering part of the counties of Kent and East Sussex in SE England. They are thus at the more stable end of their degradation process and differ from clay slopes subject to contemporary marine erosion, as on the Kent coast further to the NE (e.g. Bromhead 1978; Dixon & Bromhead 2002). The long-term patterns of behaviour in the slopes, including some estimates of the frequency of instabilities, can be deduced from the damage to the fort. There are no recorded accounts of when the landsliding took place and therefore the assignment of dates is a problem and is only possible within a broad timespan based on inference and the use to which the fort was, or was not, put.

Historical setting

It is well known from a prolific literature (e.g. Salway 1981) that Roman forces led by Julius Caesar made two incursions into southern England, in 55 and 54 BC, but various events prevented the preparation of a Roman invasion that would be followed by long-term occupation until AD 40. Even then, an invasion planned by the Emperor Caligula was

From: Rose, E. P. F., Ehlen, J. & Lawrence, U. L. (eds) 2019. *Military Aspects of Geology: Fortification, Excavation and Terrain Evaluation*. Geological Society, London, Special Publications, **473**, 31–45.
First published online March 6, 2018, https://doi.org/10.1144/SP473.9

abandoned and his successor Claudius began a successful invasion only in AD 43, seemingly capitalizing on his predecessor's preparations.

In the wake of the Roman conquest, permanent forts were built to control key points (Falkus & Gillingham 1981). Initially, their positioning was determined by the concentrations of the native population (e.g. at Verulamium, present day St Albans in Hertfortshire, and Gosbecks, near Colchester in Essex), but overall strategy and the protection of road communications were also important (e.g. at Corinium, present day Cirencester in east Gloucestershire). The well-known British revolt led by Queen Boudica of the Iceni in around AD 60–61 is evidence that the Romans were wise early in their occupation to build inland defences against possible insurgency. Ireland was not occupied and, in consequence, coastal forts had ultimately to be built in NW England and parts of Wales to provide some protection from warriors raiding across the Irish Sea. Scotland was the field for numerous military expeditions, but was never fully integrated into the Roman province of Britannia. The Emperor Hadrian began work on his now-famous stone-built wall in AD 122 and, although an Antonine Wall was constructed more simply in a more northerly location in AD 138, it was later abandoned and Hadrian's Wall was firmly established as Britannia's northern frontier. Not until AD 367 was Britannia to be seriously threatened by invasion from Celtic tribes: the Scotti from Ireland and the Picts from Scotland.

The Roman occupation of Britain lasted for nearly four centuries until AD 410. A strong military garrison promoted stability and stability, in turn, promoted prosperity, indicated by a growing density of rural villas and the growth of market towns. However, during the latter three centuries especially, the Roman Empire was subject to frequent internal conflict as its leadership was contested. Some of this involved Britannia. For example, a civil war saw the establishment by Clodius Albinus in AD 193 of an independent state involving Britain, Hispania (Spain) and part of Gaul (France). However, this was ended by the Emperor Septimius Severus, who defeated Albinus in Gaul in the year AD 197. In AD 208 Septimius Severus crossed to Britain to campaign in what is now Scotland, but in AD 211 he died in northern England, in York, and the attempt at Scottish conquest was abandoned before it was complete.

A later secession (AD 260–271) of Britannia, Gaul, Germania (Germany) and Hispania to form a Gallic Empire was led by Postumus, a Roman commander of provincial origin who declared himself Emperor of the West. This ended after Postumus was murdered by his own troops. However, within a short period Britannia and northern Gaul broke away again, this time initially under the leadership of another military leader who usurped power, Carausius, albeit for a similarly short period of independence from Rome from AD 286 to 295. Carausius was assassinated after only seven years in power and the Roman co-Emperor Constantius Chlorus subsequently re-conquered Britain. Constantius himself died in York, just as Septimius Severus had before him, in AD 306.

The Roman occupation of Britain must therefore be seen as a time of repeated, rather than once-only, invasion, with related military conflict. Moreover, apart from periods at the beginning of the occupation and at its end, the conflicts were not Roman legions v. barbarians, but between different factions within a military trained and equipped to comparable standards. Any understanding of military architecture, such as fortification, after the first century in the main part of what is now England has to be understood in that context.

The Carausian Rebellion is of particular interest because, prior to rebellion, Carausius headed the *Classis Britannica*, a fleet of ships based in the English Channel with the responsibility of eliminating Saxon and Frankish pirates, who had been raiding the coasts of northern France (Armorica) and Belgium (Belgica). However, he was accused of letting the pirates plunder without interference and only capturing them on their way back to their bases, then keeping their booty for himself. The Emperor Maximian duly ordered his deposition and execution, a sentence that provoked Carausius to declare himself Emperor in Britannia and northern Gaul. Carausius's forces consisted of not only his fleet, augmented by new ships he had built and three legions stationed in Britannia, but also a legion whose command he had seized in Gaul, several units of foreign auxiliaries, a levy of Gaulish merchant ships and barbarian mercenaries attracted by the prospect of booty. Power over northern Gaul was taken from him in AD 293 by forces loyal to the Emperor in Rome and this defeat provoked his assassination. Two years later, Britannia itself was re-conquered. It was during Carausius's brief period of control that the Roman fort at Lympne was most probably constructed and, from the coin evidence noted above, certainly occupied.

Forts of the Saxon Shore

Lympne is one of 11 defensive sites in SE England commonly known as the forts of the Saxon Shore (Johnson 1976; Wilson 1980; Pearson 2002). The others are Brancaster, Caister-on-Sea and Burgh Castle (all in Norfolk), Walton Castle (Suffolk), Bradwell-on-Sea (Essex), Reculver, Richborough and Dover (Kent), Pevensey (East Sussex) and Portchester (Hampshire) (Fig. 1). They were constructed to several different plans and portray different stages

Fig. 1. Map of southern Britain and northern France showing the three commands named in the *Notitia Dignitatum*, plus the locations of Shore forts and related military sites. After cplakidas (https://en.wikipedia.org/wiki/Saxon_Shore#/media/File:Litus_Saxonicum.png), generously placed in the public domain.

in the development of Roman military architecture during the third and fourth centuries, arguably in response to the appearance of seaborne Saxon raiders from the mid-third century. The forts were built mostly at points along the coast vulnerable to potential penetration into the Roman province, such as inlets or strategic estuaries. As a group, they constitute the most formidable stone-built coastal fortifications in Britain and were to remain so for many centuries. Indeed, they are widely regarded as the most prominent and substantial of all the surviving Roman monuments in Britain.

The name 'forts of the Saxon Shore' derives from a late fourth or early fifth century document, named in Latin the *Notitia Dignitatum* (the Register of Dignitaries). This lists various military dignitaries and the locations and units under their command, including nine forts around the coast of SE England that are recorded as under the command of the Count of the Saxon Shore (*comes litoris Saxonici*). The system of forts has thus conventionally been named after its commander, although whether Saxon Shore should be understood as the shore attacked by Saxons or settled with Saxons (as Germanic elements of the Roman military system), or both, is a matter of dispute. Lympne is almost certainly the site named in the *Notitia* as *Portus Lemanis* and is listed as containing a cavalry unit. Other forts not named in the

Notitia have been ascribed to the system by historians on the basis of similarity in geographical position, construction features and construction date.

Two groups of forts have been differentiated morphologically. An early group (Brancaster, Caister and Reculver) resembles most of the inland forts in Roman Britain, sharing their size and internal layout. They are almost square with rounded corners (the so-called playing card shape common in Roman military architecture) and were constructed in the early to mid-third century. The main group, consisting of the other eight forts in Britain, were different in construction, with thicker (up to 3.5 m thick) walls, greater variability in plan and, most importantly, the presence of semi-circular bastions on the outer faces of the fort walls. These features are common in examples across most of the Roman Empire, bastions providing a defensive capability in forts or town walls by allowing attackers to be subjected to enfilading fire from arrows and/or other missiles, but they are unique to the Saxon Shore forts in Britain. The main forts were later in construction than the early group, dating to the final quarter of the third century.

Lympne belongs to this latter group. It is not unique in having been damaged by coastal erosion. Large parts of the forts at Richborough, Reculver, Burgh Castle and Bradwell-on-Sea have also been lost, and Walton Castle has entirely disappeared; it is known only from historical records.

Related forts in northern France are described in the *Notitia* as being in two separate commands (*dux Belgicae Secundae* and *dux tractus Armoricani et Nervicani*). The Saxon Shore thus represented but one of three separate contemporary command structures recorded along the Channel coast (Fig. 1). Complicating factors to interpretation are that:

- there are more forts of roughly the same age than there are names in the *Notitia*, both in Britain and France;
- some of the attributions of names to forts are occasionally uncertain;
- some of the forts are now damaged or entirely lost;
- nearly two millennia of coastal erosion or accretion have changed the geographical context of many forts, thought to have all been constructed adjacent to sheltered anchorages.

Although it seems probable that the Roman name for the fortification and associated port at Lympne was *Portus Lemanis*, a later name for the ruins is Stutfall Castle. The origins of this are as uncertain as the meaning of the name Saxon Shore.

Geography and geology

Lympne occupies a near-coastal position at the foot of an escarpment at the eastern end of Romney Marsh (Fig. 2), a tract of low-lying reclaimed coastal marsh enclosed on the seaward side by shingle beaches and artificial embankments and on its northern side by the escarpment. The Roman fort (Fig. 3) occupies the lower slopes of part of the escarpment. The escarpment was long supposed to be a former coastal cliff and this supposition was later proved correct by the finding of beach deposits buried

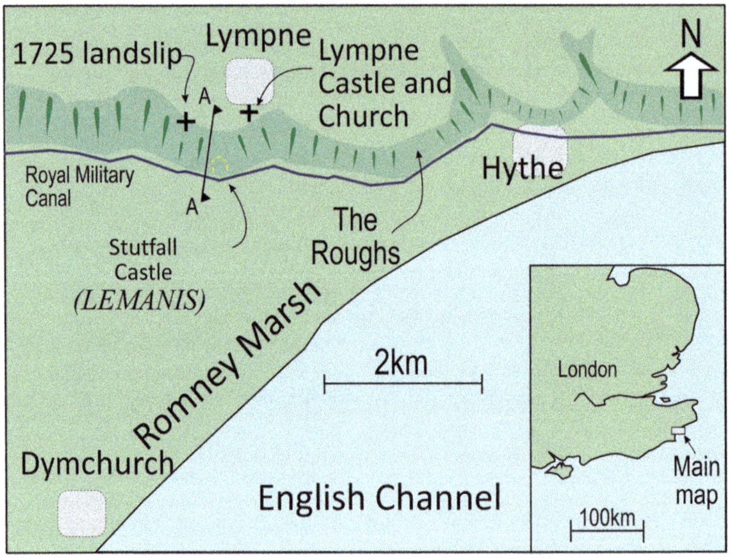

Fig. 2. Location map of Lympne Roman fort (Stutfall Castle, *Lemanis*) and its environs in southern Kent at the east end of Romney Marsh. A–A indicates the line of cross-section illustrated at two scales as Figure 4.

Fig. 3. Drone photograph of the Lympne Roman fort site with the escarpment running eastwards into the distance. Damage to the fort walls is clear, and what appears to be a Roman road leads to the site of the west gate. Published by kind permission of Paul R. Davis Overview Aerial Photography.

under the toe of the escarpment slope. The slope therefore falls into the category of abandoned cliffs, where natural sea defences at the toe allowed them to degrade, a situation described by Hutchinson & Gostelow (1976) for the slopes at Hadleigh Castle in Essex. Hutchinson had identified many slopes of this kind, including the escarpment that runs through Lympne, during a survey of coastal slopes in SE England that formed his PhD thesis (Hutchinson 1966). Slope interpretation was, right up until his death in 2011, a subject of great interest to him.

In principle, the geology is simple (Fig. 4). From top to bottom the exposed sequence of Early Cretaceous sedimentary rocks consists of three near-horizontal formations (named as in Brenchley & Rawson 2006).

- Hythe Formation (Lower Greensand Group), in its easterly facies (in Kent and East Sussex) consisting of alternating layers (0.15–0.6 m thick) of strong, blue-hearted sandy limestones and poorly cemented, grey, glauconitic sandy mudstones and argillaceous sandstones.
- Atherfield Clay Formation (Lower Greensand Group), consisting predominantly of silty clays.
- Weald Clay Formation (Weald Group), a unit consisting mainly of clays and silty clays with subordinate thin bands of sandstone, shelly limestone and clay ironstone.

Both the Hythe Formation and the Atherfield Clay Formation are about 15 m thick at this locality. The Weald Clay Formation forms the remainder of the escarpment above sea-level and continues to *c.* 60 m below it; geotechnical processes are therefore not influenced by deeper deposits. The escarpment rises to a height of *c.* 100 m above sea-level at Lympne and declines gently in height to the east in accordance with the regional dip. It is *c.* 65 m above sea-level at The Roughs, then less accessible, but lower still, through the town of Hythe (Fig. 2).

Dips are mainly to the east and more slightly to the north. Faulting is largely absent. Unlike the inland escarpments of the Hythe Formation, the coastal slopes do not appear to be cambered. Springs from the base of the Hythe Formation are common. Many of them were formerly tapped by the Folkestone and District Water Company, although their works are now abandoned. One stream that passes through the eastern side of the fort probably originally fed the Roman bathhouse. In the early twentieth

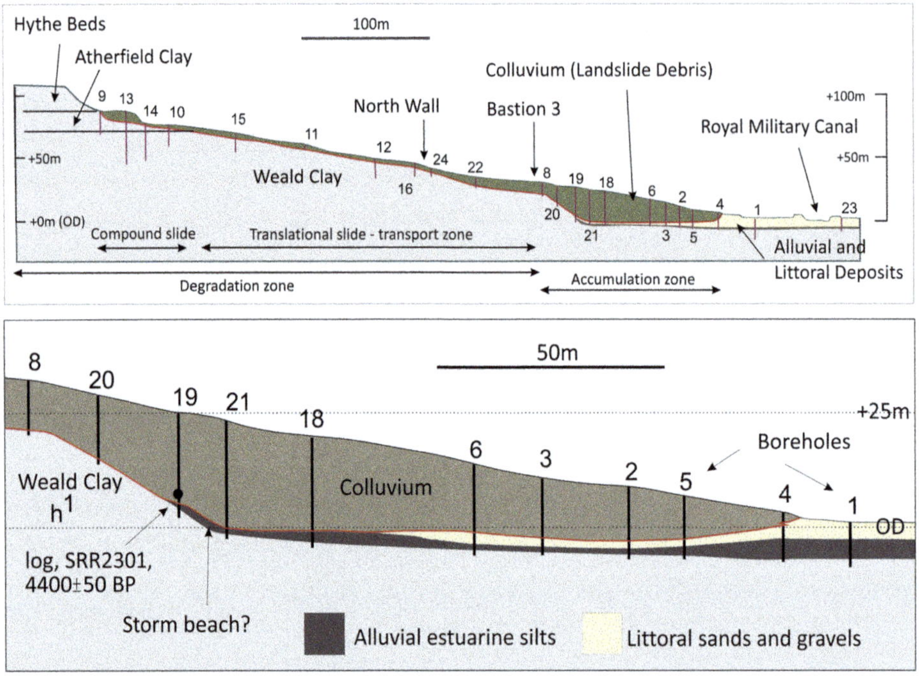

Fig. 4. Cross-section details established by site investigation immediately west of the Roman fort at Lympne.

century it fed the town of Lympne, pumped upslope by a hydraulic ram.

The slopes are analogous to the uppermost 100 m of those at Sevenoaks Weald, *c.* 70 km inland to the WNW, where movements in the pre-existing mantle of landslip debris were triggered by highway construction works (Skempton & Weeks 1976). Compound landslips were associated there with the Atherfield Clay Formation, a sheet of landslide debris and a series of lobate superimposed local slides. However, these occupy the upper slopes at Sevenoaks Weald, whereas in the vicinity of Lympne they reach to the marsh at the toe of the slope.

Coastal environment

At the present day, the slopes are protected from marine erosion by a considerable width of the eastern part of Romney Marsh and gravel beaches. The evolution of this landscape element has been described by Howland (1986) and Cunliffe (1988), but it has long been known that this was a changing environment. For example, the fort, when named as Stutfall Castle, came to the attention of John Leland at the time of King Henry VIII in the sixteenth century, soon followed by William Camden and then to the antiquarians John Aubrey (in the seventeenth century) and William Stukeley (in the eighteenth century), as described by Copley (1977). Leland, quoted by Copley (1977), noted in connection with Hythe, that:

> The havn is a pretty rode (anchorage) and liith (lies) meatly strayt (conveniently opposite) for passage owt of Boleyn (Boulogne). Yt croketh yn (curves inland) so by the shore a long, and is so bakked fro the mayne (open) se with casting of shingil that smaul shippes may cum up a larg myle toward Folkestan as yn a sure gut (safe narrow channel) [Copley's translations of obsolete words and spellings].

This seemingly supports the idea that, running along part of the escarpment then and so possibly also in Roman times, there was a shingle barrier beach analogous to the present day Chesil Beach in Dorset, protecting a landward lagoon. Eddison (1983) discussed the evolution of such a system, based partly on the work of Cunliffe *et al.* (1980). The lagoon would have provided an anchorage for shipping, similar to other Saxon Shore forts. Camden himself, in his rambling account of Romans, Saxons, Normans and the English, with digressions on subjects such as landform evolution and the development of place names, speculates that the River Rother may have been deflected through what is now Romney Marsh and thus provided a port at Lympne (Copley 1977, pp. 79–80). Camden notes also that the settlement of West Hythe was abandoned by the sea 'two centuries ago', i.e. in the fifteenth century. Moreover, Shepway Cross stands at the head of the slopes at which West Hythe lies at the foot. This is the historical meeting place associated with the Cinque Ports and 'Shepway' is probably an altered form of 'Shipway', so a local name unconnected with sheep, but related to naval ports and ships. Both names are given on early editions of Ordnance Survey maps.

Elliott (1852) provides a reconstruction based on the present course of the River Rother, with a postulated canal following the line of the earthwork known as the Rhee Wall.

It therefore appears likely that the Lympne fort is no different from other Saxon Shore forts in having been constructed adjacent to a sheltered harbour, although, at the present day, some of these harbours have silted up and other sites have been exposed to erosion so that a complete circuit of walls does not exist. Although most Saxon Shore forts were square or rectilinear in outline, some were shaped to fit the defences to a particular site; the next nearest fort to the west (at Pevensey) is an irregular oval in plan. Interest in the original shape of the Lympne fort (then named Stutfall Castle) was pioneered by the historian Charles Roach Smith, and shape is indeed central to an understanding of the landslide processes on the site (Smith 1852).

Geotechnical findings

The present state of the slopes on the escarpment has been determined from site investigation to consist of three main zones (Fig. 4), established largely from a series of 24 cable percussion boreholes in which continuous sampling was attempted, supplemented by some trial pits:

- a degradation zone consisting of compound landslides involving the Hythe Formation (Hythe Beds of earlier literature) and the upper part of the Atherfield Clay Formation, with the sub-horizontal basal part of the subsurface following a bedding-controlled layer within the latter;
- a transport zone in which there is translational landsliding on a basal subsurface sub-parallel to ground level on top of *in situ* Weald Clay;
- an accumulation zone where the transported debris has overridden and buried a sea cliff in Weald Clay and also buried beach and marsh deposits.

The degradation zone is poorly developed at Lympne, probably as a result of quarrying in the Hythe Formation to provide building stone for the Roman fort and later for Lympne's medieval castle and church.

In 1988, some years after this work at Lympne, described by Hutchinson *et al.* (1985), landslipping affected a site a few kilometres to the east at a location known as The Roughs. The cause was duly investigated (Bromhead *et al.* 1998).

Unpublished investigations were also carried out at a site identified as a possible highway route up the escarpment between Lympne and The Roughs east of West Hythe, and also when building foundations at Port Lympne Animal Park, west of Lympne. These investigations, involving boreholes with closely spaced undisturbed samples and trial pit work, again showed that the degradation zone was clearly of compound landslide form with a basal shear following along a weak bed or beds in the Atherfield Clay Formation.

Such a bedding-controlled basal shear had previously been identified a few kilometres to the east of Hythe at Sandgate (Foster 1980; Palmer 1991). There the Atherfield Clay Formation is largely below sea-level. Consequently, the compound landslides are significantly larger than where they are perched on top of the escarpment, involving the three formations noted as present at Lympne and the overlying Sandgate Formation (predominantly fine sands, silts and silty clays).

The lower slopes were not investigated at Port Lympne and, if the transport zone is better developed at the Roman fort than at sites further east, it is considered to be a function of the overall height of the slope. The accumulation zone was more comprehensively investigated at the Roman fort than elsewhere to the east and, although in general terms in The Roughs it has also overridden marsh and beach deposits, it is much smaller than at Lympne. It is possible to speculate with some justification that the date of at which marine erosion ceased is older in the west and younger in the east (Howland 1986).

A highly significant finding was of a log, encountered at the junction of the colluvium with the underlying storm beach, which provided a sample for radiocarbon dating (SRR2301; see Fig. 4), yielding a date of 4400 ± 50 BP. It appeared from boreholes 18 and 21 that marsh deposits were absent and that this log was found at the top of a storm beach in borehole 19, suggesting that at that time the accumulation zone was in the process of development and then subsequently grew and advanced across saltmarsh. The radiocarbon date post-dates the recovery of sea-level to about present day levels and so it is likely that the steep slope in Weald Clay was actively eroded at around the radiocarbon date. However, the age is such that it is almost certain that the Roman constructors of the fort encountered much the same general slopes as are seen today.

Nineteenth-century archaeological investigations

In the eighteenth century, Stukeley ([1724] 1776) produced an etching of Stutfall Castle as it appeared in 1722 (Fig. 5). Lines have been drawn in the western

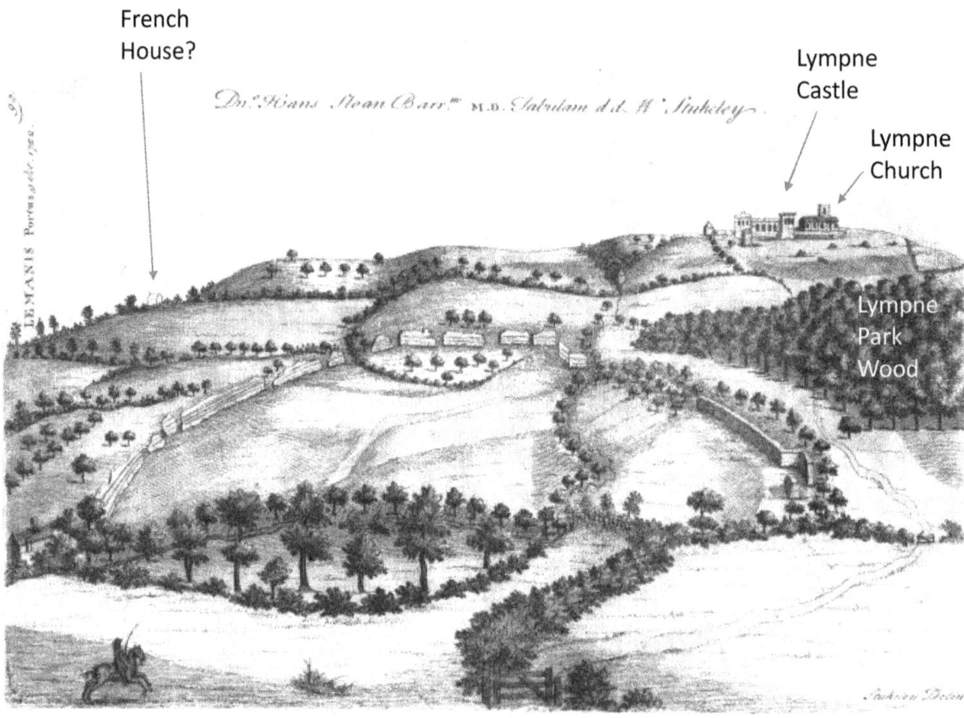

Fig. 5. View of the fort from Stukeley ([1724] 1776).

and northern walls that are clearly intended to represent the tile courses inserted at intervals in the masonry of the fort walls and which show through the finished outer masonry. The outer skin of the walls and bastions is constructed of squared dressed blocks, whereas the inner part of the walls is Roman concrete and, where preserved, the blocks have been pointed with a pinkish mortar containing fragments of crushed tile. It therefore appears that the western, NW and northern walls were broken, but still standing. Stukeley depicts the eastern wall in a different way.

Archaeological investigations at Lympne began in the nineteenth century. Most were carried out in *c.* 1850 by Charles Roach Smith, who published his findings in two volumes, with a great deal of commonality between them (Smith 1850, 1852). The latter publication was published specifically for the subscribers who paid for the work. Relatively minor excavations were conducted in 1890 by Sir Victor Horsley (Horsley 1894).

In his detailed investigations *c.* 1850, Roach Smith uncovered:

- a house (Roach Smith's description), which is now thought to be the *Principia* (the administrative and religious centre of the fort and so its most important building) situated in the north of the fort, the west wing of which was dislocated;
- a second house in the SE corner of the fort, now believed to be a bathhouse, but also showing some dislocation;
- the circuit of the walls (but not their timber piled foundations);
- a postern and gate on the west wall; and
- the East Gate.

In a drawing in Roach Smith's work, it is shown that there is an opening at the bastion immediately south of the East Gate; this also appears to have been depicted by Stukeley.

Serendipitously, a stone dislodged from the paving of the gatehouse was revealed to be an altar that commemorates Aufidius Pantera, Prefect of the *Classis Britannica*. It appears from marine encrustation, including barnacles, that this had spent some time under seawater. Roach Smith sold the altar to the British Museum in London, where it currently resides.

Given Leland's observation about the coast (Copley 1977), the Pantera altar could well have been part of a structure built on the barrier beach and ruined by movements of the beach between the first and third centuries, but exploited as a source of materials when building the fort now known as Stutfall Castle. This would account for the inability of anyone to find the remains of an earlier fort.

Roach Smith's reconstruction of the shape of the fort in 1852 is shown in Figure 6. He concluded that bastions, which he denoted as H and I, were in their original place and he therefore reconstructed the plan form with an apsoidal northern wall. This reconstruction does not show the *Principia*, which is located in James Elliott's reconstruction (from an addendum in the 1852 booklet produced for subscribers), which is more diagrammatic (Fig. 7). Note that the southern, east and west wall bastions are hollow, but those on the north wall are solid. Elliott shows short returns at the southern ends of both the east and west walls, but this may be fanciful. Various drawings in the Roach Smith publications show the farmhouse at the southern end of the west wall, which was in place in the Stukeley etching. This interpretation of the shape of the fort survived the test of time, although occasionally the fort was shown in subsequent illustrations as of playing card form. Had this been the case, it would have caused considerable landslide deformation, with slippage, in places, of *c.* 100 m.

Roach Smith found several coins, most of which were issued by Carausius or later rulers, giving an uncontested late third century date to occupation of the fort and an association with the Carausian Rebellion. Roach Smith also discovered some iron cramps and stamped tiles, but few other artefacts or building foundations.

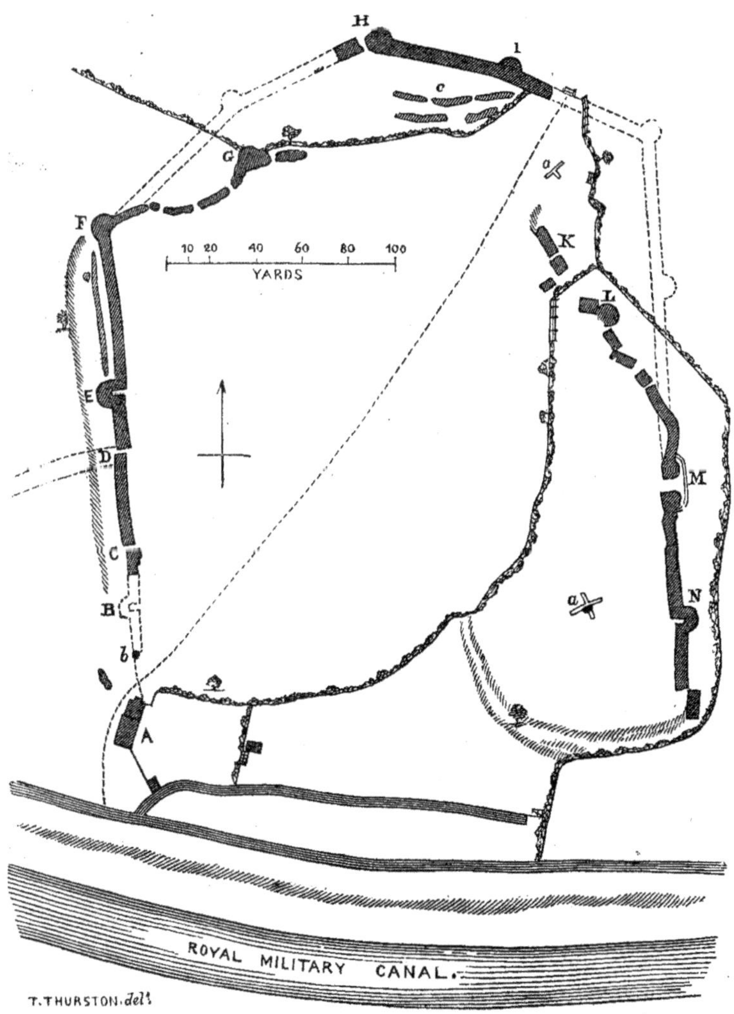

PLAN OF THE CASTRUM AT LYMNE.

Fig. 6. Plan of the fort from Smith (1852).

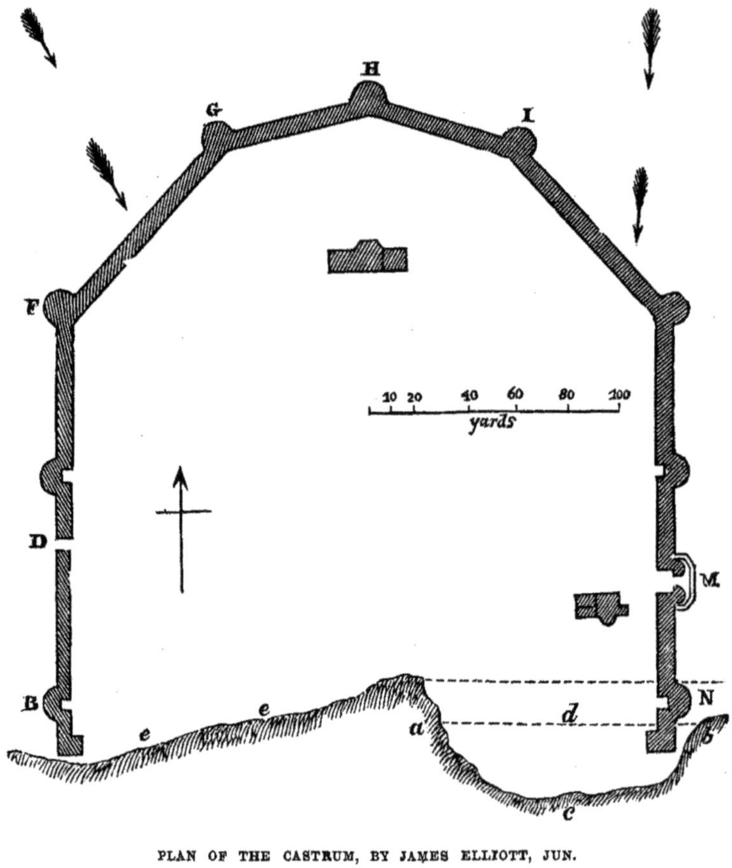

Fig. 7. Plan of the fort from Elliott (1852).

Twentieth-century archaeology

There was no further investigation of substance until 1979, when Professor Barry Cunliffe spent a field season at Lympne (Cunliffe *et al.* 1980). A programme of combined geotechnical and archaeological work was undertaken a few years later (Hutchinson *et al.* 1985).

The East Gate and bathhouse were re-excavated by Cunliffe. His additional work on the wall showed the significant height to which the walls and bastions must originally have risen.

Professor John Hutchinson directed a joint programme in which the archaeological element was undertaken by Dr Cynthia Poole of the Oxford Institute of Archaeology and the geotechnical work was undertaken by Nick Lambert and the present author. The geotechnical work involved exploratory pits dug in 1982 at various locations along the north wall to determine the shape of the original foundations that had proved too deep for Roach Smith to uncover. It is to be regretted that these pits (Fig. 8) show up on aerial photographs even today where the backfill has subsided (e.g. Fig. 3), although Roach Smith's pits were more effectively infilled and are not visible.

On the basis of this work, a simpler apsoidal planform was established (Fig. 9). However, even that required inferred displacements of some bastions of the order of 70 m, an important geotechnical finding for a natural slope of only 10–11° inclination. It appeared from a limited geophysical (electrical resistivity) survey that there had not been any ditches excavated upslope of the fort as part of its original defences.

A geophysical survey of the fort area using a magnetometer was carried out in 2016 by the Kent University Archaeology Unit (Bosworth 2016) and this confirmed the absence of ditches. It also found no trace of a Roman road leading to the East Gate, which is perhaps unsurprising because the East Gate is displaced considerably downslope. Any such road most probably now lies largely within Lympne Park Wood (Fig. 5). A trackway leading to the West Gate appears on aerial photographs as a crop mark.

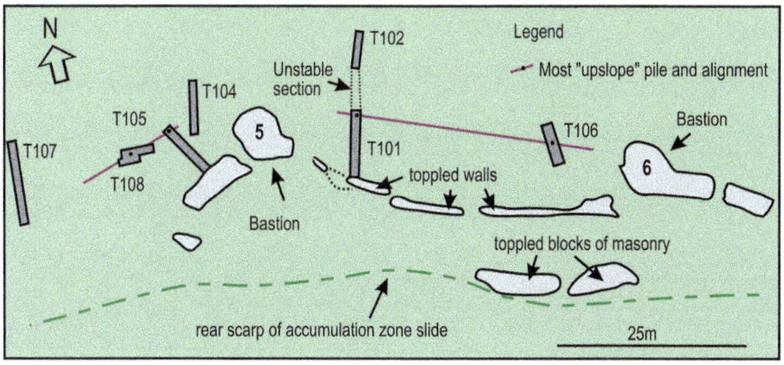

Fig. 8. The furthest upslope north wall foundation pile positions found in trenches numbered T108, T105, T101 and T106. For positions of bastions numbered 5 and 6 (see Fig. 9).

The finding of timber piles, notwithstanding that they are short and formed from split logs with a tapered point, was surprising. They represented a major undertaking if present around the whole periphery of the site. Similar piles have been discovered in association with the Saxon Shore fort at Pevensey and so it seems likely that they were a common feature of Shore fort construction. Those from which the north wall has been displaced are vertical at the upslope extremity of the foundation, but are now flat at the southern extremity. It therefore appears likely that the wall was tilted out of position during landslipping and then slid forward, overturning as it went.

Hutchinson & Bromhead (1996) made an effort to reconstruct the original cross-section of the wall based on field measurements and an estimate of height (Fig. 10). This reconstruction is almost certainly correct as far as the foundations and lower sections are concerned, including the internal offsets. However, the height was determined from the need to infer a substantial walkway. The wall could easily have been higher if it did not feature a walkway. Earth pressure calculations were made, and, even taking into account modified active earth pressures on the upslope side to incorporate the effects of landsliding and the removal of passive restraint on the downslope side from movements of the accumulation zone, the calculation might be best summarized by saying that the collapse of this wall, if solely on the basis of the active earth pressure thrust, would disprove the entirety of modern earth pressure theory. The wall can only be calculated to topple if it has already rotated considerably, and that is a position supported by the orientation of the timber piles where the wall lifted off the upslope piles.

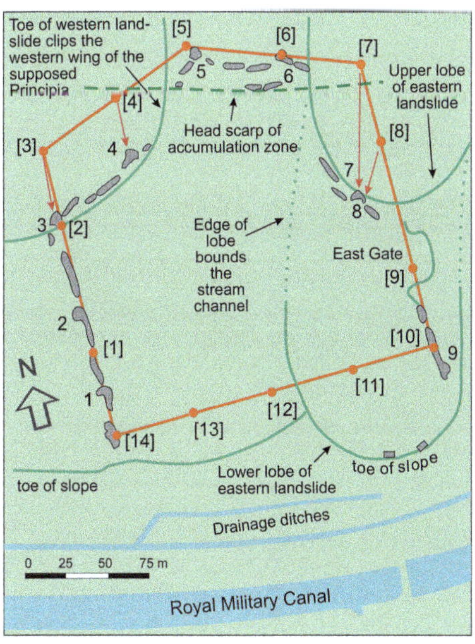

Fig. 9. Plan of the fort after Hutchinson *et al.* (1985).

Landsliding on the site

It appears on geomorphological grounds that a landslide emanating from the foot of the Hythe (Beds) Formation escarpment immediately south of the medieval Lympne Castle (where there is a sag pond in the degradation zone) caused major damage to the NE corner of the fort (Hutchinson *et al.* 1985). A continuation southwards in the form of a secondary lobe appears to have displaced the east wall, East Gate and bathhouse. This secondary lobe has also carried toppled fragments of the south wall with it as encountered by Cunliffe *et al.* (1980), whose investigation showed a double toe, which probably indicates both the pre-Roman shape of the cross-section and also the subsequent failures that obliterated the south wall.

An interesting analogue of a lobate landslide running from the base of the Hythe Formation down the escarpment as far as the marsh was also experienced

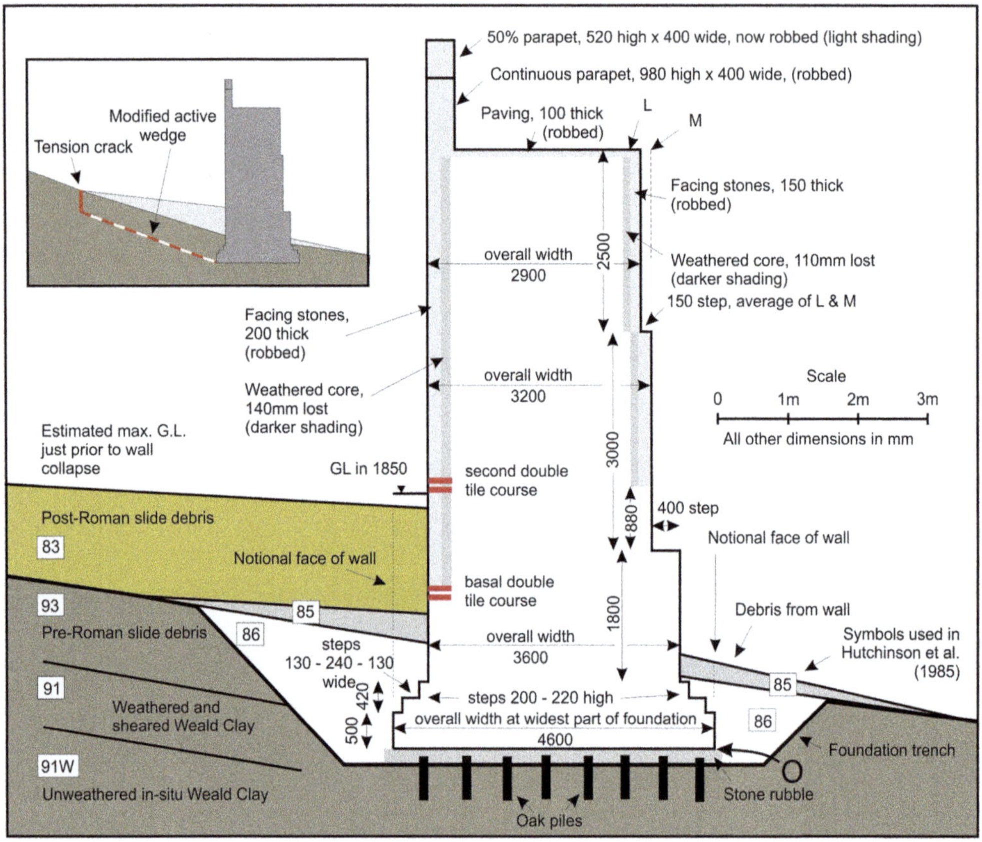

Fig. 10. Wall cross-section, after Hutchinson & Bromhead (1996).

in The Roughs system of landslides (Fig. 2), where the beach deposits underneath the marsh clays were brought to the surface. The western section of The Roughs landslide includes an experimental station that had been constructed in the World War I as the prototype of an acoustic aircraft early warning system to detect engine noise from approaching Zeppelin airships by means of two sound reflectors. The station had a system of septic tank drainage and pipework. This concentrated groundwater was associated with a landslide in 1988, the station having long been abandoned because developments in radar before World War II made it obsolete. One of the reflectors was involved in the 1988 landslide; the other is slowly degrading through a combination of weathering and vandalism.

It is therefore tempting, by analogy, to infer that, at least in part, the similar lobate landslide system affecting the eastern side of the Lympne fort might have been caused by the failure or blockage of some Roman era pipework that fed the bathhouse inside the fort. The date of this eastern landslide is uncertain, but it is probably later than the end of the Roman occupation in Britain because the fort was still in use when the *Notitia* was compiled.

A second major landslide system has affected the NW corner of the Lympne fort without significantly displacing the entirety of the west wall. It is likely that this landslide also has its analogue in the main part of the 1988 Roughs landslide that involved the degradation zone compound landslide and the upper parts of the transport zone. This NW corner landslide must surely post-date the Stukeley etching (Fig. 5). In Stukeley's etching, a small house can be seen on the skyline. Roach Smith's artist drew a similar view in which the house is absent, although the NW fort wall is shown displaced. Although acknowledging that this is weak evidence, it should be noted that in *c.* 1725 a house in this location is described as having suffered an 'uncommon sinking of the ground' (Collinson 1728). If the house is, in fact, the named French House, a landslide of the lateral extent of that experienced in The Roughs would also clip the NW corner of the fort. This postulated landslide system would also account for the deformation seen in the *Principia*.

The combined archaeological and geotechnical investigation uncovered piles from which the north wall has slipped. The most upslope piles were found to be vertical, but other piles had been pulled out of alignment, in the most extreme case aligned with the basal shear. It appears that the Roman constructors had piled into the underlying *in situ* Weald Clay and had built the wall up through the layer of landslide debris. Whatever their geotechnical competence, it was therefore evident that the original constructors had distinguished between the weaker slide debris and the underlying more competent Weald Clay at this location. It was not possible to determine what had been done above the accumulation zone where the landslide debris is thickest because the wall foundations were too deep to be uncovered, or still lay beneath the walls.

From the foundation piles layout (Fig. 8) it can be seen that bastion 5 (Roach Smith's H) is slightly displaced and this is probably enough to have pulled the north wall (5–6 or H–I) out of plumb and off the piles, allowing it to slide and rotate. As the contact between landslide debris and solid Weald Clay is convex upwards at this location, where the transport zone merges into the accumulation zone, it would rotate more until it became unstable and toppled; this has little to do with earth pressures.

The separation of the wall into the segments shown in the Stukeley etching probably reflects the butt joints in the wall, built by different work parties when the fort was constructed, which have eroded differentially.

Again on geomorphological grounds, it appears that the main part of the accumulation zone may well have settled at its head and shunted seawards, probably under the loading of the east side landsliding. This would have had the effect of removing support to the north wall.

Military context

The province of Britannia was involved in major Roman dynastic conflicts at least three times in a little over a century, which makes the existence of defensive structures in SE England unsurprising. The Saxon Shore forts constructed principally at this time have thick walls and, in most cases, closely spaced bastions, features designed to resist siege. Excavation reports have been published that indicate such features for the forts at Richborough (Cunliffe 1968), Portchester (Cunliffe 1975), Burgh Castle (Johnson 1983), Caister-on-Sea (Darling & Gurney 1993) and Reculver (Philp 2005). Two volumes of edited papers illustrate the continuing debate associated with the interpretation of the Saxon Shore fort system (Johnson 1977; Maxfield 1989).

From the historical record it seems that potentially both besiegers and besieged were different factions of Roman armies, thus with equivalent training, command structures, tactics and siege equipment. Construction of the Shore forts is more massive than would be necessary to repel 'Saxon pirates', able to seize shipping and/or raid coastal settlements, but presumably lacking in the numbers, expertise and equipment necessary to undertake siege operations. Such 'pirates' are unlikely to have come up the English Channel from the Bay of Biscay. Moreover, they would have been foolhardy to pass from their homeland through the straits of Dover (Fig. 1) for a significant distance in any strength in the expectation of a safe return, assuming that the eastern *dux* had the same anti-pirate mission as the *comes*.

The thickness of masonry at bastions and walls is strongly indicative of a defensive architecture designed and constructed to resist siege. The close spacing of bastions or towers provides force multipliers to defenders and is indicative that the forts could be garrisoned and held by small forces, an inference supported, in part, by the lack of barracks for large units of men. Indeed, the finding of 'putlog' holes in a bastion at Lympne and central postholes in bastions at Burgh Castle probably indicates timber upperworks on the bastions. These would then rise higher than the adjacent walls and might provide accommodation for defenders as well as defensive positions. It has been speculated that the Burgh Castle postholes were the pivot points for *ballistae*: weapons that launched large missiles at a distant target. However, although such artillery must surely have been used where appropriate, artillery cannot have been the bastions' main purpose. The fort at Lympne provides evidence for this: the most upslope bastion cannot discharge projectiles as far upslope as a similar weapon can project downslope. An assailant's *ballista* might therefore be positioned out of range of that of the fort, but still have the range to attack it. The Lympne fort is the only place where this is apparent, at least for forts on the north side of the English Channel. Further argument against the interpretation of the bastions as artillery positions is that the rate of missile discharge would have been too slow for effective enfilade fire and the depression angle possible would not have allowed such fire to cover the walls between bastions.

The *Notitia* lists the situation as it existed at about the turn of the fourth and fifth centuries. It provides no indication of who constructed the fortifications in the first place, about a century earlier, or what their original purpose was. Indeed, it is likely that their purpose changed over time to meet changes in perceived threat. Whether the construction of the Lympne fort with an apsoidal planform indicates that its designer understood the geotechnics of the slope and this was an attempt to counter instability, or whether this reflects some unknown military purpose, is not apparent. However, the apex does not point directly upslope, but at an angle to the slope dip direction, so the former is unlikely. It is at least as likely that the fort was designed to straddle a road leading down and across the slope, possibly towards a port at the foot of the slope some distance to the west. The forts at Dover and Richborough certainly dominated Roman harbours and, given the location of most Saxon Shore forts in sheltered places, it is probable that most of the forts were associated with harbours (Wilson 1980). Evidence for a port at Lympne might be sought by future archaeological investigation.

Engineering geological significance

The slopes at Lympne had probably evolved quickly when subjected to marine erosion following the well-known recovery of Holocene sea-levels to more or less modern elevations after *c.* 6 ka BP. The radiocarbon-dated log shows the slope to have become abandoned by *c.* 6.5 ka BP. There was therefore plenty of time for the slope to degrade naturally over the next 4.8 ka and to have adopted more or less the present day angle by the time of the Roman occupation of Britain, somewhere between 10° in the lower slopes and 11° in the upper slopes. The abandonment was the result of coastal changes that eventually led to the formation of the headland of Dungeness, largely a shingle beach in the form of a cuspate foreland, *c.* 18 km south of Lympne.

Investigations, most notably at Lympne but also elsewhere in the escarpment, show compound landsliding associated with a weak horizon in the Atherfield Clay Formation, but not elsewhere in the sequence. The strata are sub-horizontal and this therefore gives rise to a 'bedding-controlled landslide' seated on a 'slide-prone horizon' (Bromhead 2013). Lower in the slopes the sliding takes the form of sheets, bounded at the contact with the underlying Weald Clay Formation by well-developed basal shear surfaces. These run into the wedge-shaped accumulation zone that buries marsh and beach deposits. Elongated lobate landslides have taken place that terminate at the foot of the slope; movements of the compound landslides do push the sheet slides, but not as far as the foot of the slope. There are strong similarities with other slopes associated with similar geological features, e.g. at Sevenoaks Weald. Movements of the sheet slides have been of the order of tens of metres, but the lobate slides travel further and are responsible for movements as much as 70 m on these relatively low-angled slopes.

What is perhaps most surprising is not the instability of the slopes in the coastal escarpment, but how long the periods of relative stability have been. The NW walls in Lympne fort appear to have been toppled by a landslide in the Little Ice Age, now well known as a period lasting from *c.* 1300 to 1850 when Europe and North America were subject to much colder winters than subsequently, but not before. Moreover, they appear to have only toppled when pulled out of plumb and not under a landslide-enhanced active pressure. The eastern wall was almost certainly affected by one of the lobate landslides and, in turn, probably due to some influence from the adjacent stream. Dating of this is less certain, but probably post-Roman and earlier than the well-known medieval warming period that began about AD 900.

Statistics for the return period for natural recurrences of instability are difficult to establish; dateable incidents are typically associated with building works. However, instability would be expected to be infrequent in a slope such as this where the degradation zone is nearly exhausted. Evidence is provided here that, since the construction of the fort, the stability of the whole eastern side has been affected once; part of the NW area has been affected once; and the south wall has collapsed, but the main western wall has not moved. No particular part of the fort has been affected more than once in *c.* 1700 years and not all of it has been affected. A return period for landslipping in specific areas of longer than 1700 years therefore seems likely, perhaps of the order of two millennia.

Acknowledgements The author is a consulting geotechnical engineer and was on the staff of Kingston Polytechnic when the fieldwork was carried out. US Army European Research Office funding allowed a series of boreholes to be drilled, some instruments to be installed and some trial pits to be dug, together with detailed archaeological investigations. The geotechnical work was carried out outside the footprint of the fort itself so as not to compromise its archaeology, and the archaeological work was mainly associated with examination of the landslip damage to the north wall. The archaeological work was supervised and largely undertaken by Dr Cynthia Poole, who was then of the Institute of Archaeology in Oxford. The land was part of the Hythe army ranges and drilling formed part of a Royal Engineers' training exercise. Several field investigations were undertaken under the command of Warrant Officer Dave Walker by a crew of junior ranks, and fieldwork of other kinds by Dr Anna Davies (*née* Hopper) for her PhD thesis. Information on other investigations came from Sir Owen Williams & Partners and A.G. Weeks & Partners. Permission to work on the land was given kindly by the late Mr & Mrs H. Margary, then resident in Lympne Castle. Dick Moody, Ursula Lawrence and Ted Rose are thanked for reviews and editorial work that improved the manuscript as first submitted.

Funding Following a period of self-funded work, largely in the form of surveys, a joint geotechnical and archaeological investigation was undertaken at Lympne supported by a generous grant to the late Professor J.N.

Hutchinson from the US Army European Research Office, administered by the late Warren E. Grabau. Investigation of The Roughs was carried out with a variety of funds, not least industrial support in kind from Soils Ltd and also from the Corps of Royal Engineers under a Natural Environment Research Council scheme known as 'Joining Forces For The Environment'.

References

Bosworth, L. 2016. *Report on a Geophysical Survey of Stutfall Castle, the Roman Saxon Shore Fort at Lympne, Kent.* University of Kent, Canterbury.

Brenchley, P.J. & Rawson, P.F. (eds) 2006. *The Geology of England and Wales.* 2nd edn. Geological Society, London.

Bromhead, E.N. 1978. Large landslides in London Clay at Herne Bay, Kent. *Quarterly Journal of Engineering Geology*, **11**, 291–304, https://doi.org/10.1144/GSL.QJEG.1978.011.04.03

Bromhead, E.N. 2013. Reflections on the residual strength of clay soils, with special reference to bedding-controlled landslides. *Quarterly Journal of Engineering Geology and Hydrogeology*, **46**, 132–155, https://doi.org/10.1144/qjegh2012-078

Bromhead, E.N., Hopper, A.C. & Ibsen, M.-L. 1998. Landslides in the Lower Greensand escarpment of South Kent. *Bulletin of Engineering Geology and the Environment*, **57**, 131–144.

Collinson, P. 1728. An uncommon sinking of the ground in Kent. *Philosophical Transactions of the Royal Society of London*, **35**, 551–552.

Copley, G.J. 1977. *Camden's Britannia: Kent.* Hutchinson, London.

Cunliffe, B.W. (ed.) 1968. *Fifth Report on the Excavations of the Roman Fort at Richborough, Kent.* Oxford University Press, Oxford.

Cunliffe, B.W. 1975. *Excavations at Portchester Castle 1: Roman.* Society of Antiquaries of London, London.

Cunliffe, B.W. 1988. Romney Marsh in the Roman period. *In*: Eddison, J. & Green, C. (eds) *Romney Marsh: Evolution, Occupation, Reclamation.* Oxford University Committee for Archaeology, Oxford, 83–87.

Cunliffe, B., Reece, R., Henig, M., Chadwick Hawkes, S., Careand, V. & Young, C.J. 1980. Excavations at the Roman Fort at Lympne, Kent 1976–78. *Britannia*, **11**, 227–288.

Darling, M.J. & Gurney, D. 1993. *Caister-on-Sea: Excavations by Charles Green 1951–55.* East Anglian Archaeology, **60**. Norfolk Museums Service, Norwich.

Dixon, N. & Bromhead, E.N. 2002. Landsliding in London Clay coastal cliffs. *Quarterly Journal of Engineering Geology and Hydrogeology*, **35**, 327–343, https://doi.org/10.1144/1470-9236/2000-53

Eddison, J. 1983. The evolution of the barrier beaches between Fairlight and Hythe. *Geographical Journal*, **149**, 39–53.

Elliott, J. 1852. Notes on the original plan of the Castrum at Lymne, and on the past and present state of Romney Marshes [appendix]. *In*: Smith, C.R. (ed.) *Roman Castrum at Lymne.* Published privately for subscribers to the excavations, London, 35–47. Reprinted by H. Margary, Lympne Castle, undated.

Falkus, M.E. & Gillingham, J.B. 1981. *Historical Atlas of Britain.* Book Club Associates, London.

Foster, C.J. 1980. *A study of coastal landslips at Sandgate, Kent.* MSc thesis, Surrey University.

Horsley, V. 1894. Note in the Athenaeum. *The Athenaeum*, **3491**, 394.

Howland, A.F. 1986. The natural process of degradation of an overconsolidated clay under differing climatic conditions as illustrated by the Weald Clay. *International Geomorphology*, **1986** (Part 1), 139–154.

Hutchinson, J.N. 1966. *The stability of cliffs composed of soft rocks, with particular reference to the coasts of south-east England.* PhD thesis, University of Cambridge.

Hutchinson, J.N. & Bromhead, E.N. 1996. Back analysis of the collapse of a Roman wall built on landslip. *In*: Senneset, K. (ed.) *7th International Symposium on Landslides*, 17–21 June 1996, Trondheim, Norway. Balkema, Rotterdam, 1243–1250.

Hutchinson, J.N. & Gostelow, P. 1976. The development of an abandoned cliff in London Clay at Hadleigh, Essex. *Philosophical Transactions of the Royal Society of London*, **A-283**, 557–604.

Hutchinson, J.N., Poole, C., Lambert, N. & Bromhead, E.N. 1985. Combined archaeological and geotechnical investigations of the Roman Fort at Lympne, Kent. *Britannia*, **16**, 209–236.

Johnson, D. (ed.) 1977. *The Saxon Shore.* Council for British Archaeology Research Report, **18**. Council for British Archaeology, York.

Johnson, S. 1976. *The Roman Forts of the Saxon Shore.* Elek, London.

Johnson, S. 1983. *Burgh Castle Excavations by Charles Green 1958–61.* East Anglian Archaeology, **20**. Norfolk Museums Service, Norwich.

Maxfield, V.A. (ed.) 1989. *The Saxon Shore: a Handbook.* Exeter Studies in History, **25**. University of Exeter, Exeter.

Palmer, M.J. 1991. Ground movements of the Encombe landslide at Sandgate, Kent. *In*: Chandler, R.J. (ed.) *Slope Stability Engineering: Developments and Applications.* Institution of Civil Engineers, London, 291–296.

Pearson, A.F. 2002. *The Roman Shore Forts: Coastal Defences of Southern Britain.* Tempus, Stroud.

Philp, B.J. 2005. *The Excavation of the Roman Fort at Reculver.* Kent Archaeological Rescue Unit, Dover.

Salway, P. 1981. *The Oxford History of England. Roman Britain.* Clarendon Press, Oxford.

Skempton, A.W. & Weeks, A.G. 1976. The Quaternary history of the Lower Greensand escarpment and Weald Clay vale near Sevenoaks, Kent. *Philosophical Transactions of the Royal Society of London*, **A-283**, 493–526.

Smith, C.R. 1850. *The Antiquities of Richborough, Reculver, and Lymne.* John Russell Smith, London.

Smith, C.R. 1852. *Roman Castrum at Lymne.* Published privately for subscribers to the excavations, London. Reprinted by H. Margary, Lympne Castle, undated.

Stukeley, W. [1724] 1776. *Itinerarium Curiosum; or, An Account of the Antiquities, and Remarkable Curiosities in Nature or Art, Observed in Travels through Great Britain.* Baker & Leigh, London. 2nd edn published posthumously.

Wilson, R. 1980. *Roman Forts.* Bergström & Boyle, London.

Groundwater supplies to maritime and coastal defences in southern England: a story of risk and innovation

JOHN D. MATHER

Department of Earth Sciences, Royal Holloway, University of London, Egham, Surrey TW20 0EX, UK

mather@jjgeology.eclipse.co.uk

Abstract: The viability of any fort or garrison depends on the availability of a reliable water supply. The source of choice is an underlying aquifer, reached by a secure on-site well or borehole. Unfortunately, at coastal and maritime sites, seawater intrusion can cause problems. In the late eighteenth century a deep well was sunk to supply the garrison at Sheerness, Kent, which successfully exploited sands beneath the London Clay. At Landguard Fort in Suffolk, a shallow gallery was designed to skim freshwater overlying saline water within loose sand and shingle. In the mid-nineteenth century, a network of forts was built to defend the Royal Dockyard at Portsmouth, which included some offshore forts on the Spithead shoals. Boreholes were drilled beneath these forts to abstract water from the Chalk thought to lie beneath. The Chalk proved to be at too great a depth, but the Bracklesham Group yielded a sustainable supply from <200 m. In carrying out these projects, military engineers sank wells and drilled speculative boreholes, taking financial risks, unacceptable in other parts of the public sector. They developed new technologies and their innovative ideas and discoveries led to an increased understanding of the distribution and use of groundwater.

> In considering the eligibility of a site for a permanent military work or any large public establishment, the means of securing an ample supply of water is one of the first questions that engages attention. The situation may be in all other respects advantageous, but if any difficulty or doubt arises as to a command of water, the idea of occupying it will be generally abandoned
>
> (Jebb 1842, p. 266).

This was written by Major Joshua Jebb (later Sir Joshua and Surveyor General of Prisons) and emphasizes the fundamental importance that military personnel have long attached to water supply and the need for it to be under direct military control. To meet Jebb's requirement, many fortified sites used an on-site well or, from the late eighteenth century, a borehole, as a source of water supply. Apart from the security inherent from their situation within the boundaries of a garrison, water could be extracted on demand and there was little need for extensive storage facilities at the surface. Two other options were available: the collection of rainwater and bringing in water from off-site. Rainwater capture was subject to the vagaries of the weather and required both catchments and storage cisterns within the garrison precincts. Pipelines or leats bringing water from off-site could be cut off by attacking forces and, although resistance might be prolonged by on-site storage, this could only provide a short-term solution.

Although on-site wells and boreholes had the potential to provide the most satisfactory solution to garrison water supplies, they were dependent on the presence of a suitable aquifer beneath the site. In addition, wells were also susceptible to pollution, particularly so at maritime and coastal garrisons, where problems were compounded by proximity to the sea. Here, seawater intrusion often meant that groundwater was brackish and unfit for drinking.

The specific problems of coastal sites are well illustrated by the town and garrison of Gibraltar, situated on a rocky limestone peninsula, *c.* 5 km in length and 1.6 km in maximum width, at the western end of the Mediterranean. An ongoing battle to provide potable water meant that, over the years, recourse was made to many potential sources. Early supplies were obtained from shallow wells in Quaternary sands and rainwater harvested from roofs and hillsides. Groundwater supplies were gradually lost when the aquifer recharge area was developed as a graveyard, an airport runway was built over another potential aquifer and further contamination resulted after seawater was used regularly for washing roads and flushing sewers. Thin freshwater lenses within the limestone core of the peninsula proved difficult to exploit and reliance had to be placed on artificial rainfall catchments, which eventually covered *c.* 243 000 m^2 of the central limestone ridge, with the collected water stored in underground reservoirs. At times, additional water was imported by tanker from places as distant as Southampton and Rotterdam. A separate piped supply of saline sanitary water was developed to reduce the demand for potable water, but shortages persisted as civilian and military populations increased. Problems were

From: Rose, E. P. F., Ehlen, J. & Lawrence, U. L. (eds) 2019. *Military Aspects of Geology: Fortification, Excavation and Terrain Evaluation*. Geological Society, London, Special Publications, **473**, 47–60.
First published online March 6, 2018, https://doi.org/10.1144/SP473.8

solved only in 1993, when a near-total commitment was made to desalination (Rose *et al.* 2004).

A secure supply of potable water within the confines of a garrison was of such importance that considerable expense and financial risk could be justified, particularly where the defence of a dockyard or safe anchorage was involved. This paper discusses two examples: the first involves strategically important garrisons built to protect London and the Thames estuary in the late eighteenth century and the second is an example dating from the mid-nineteenth century and involves the defence of the Royal Naval Dockyard at Portsmouth. In both cases, innovative solutions to the problems of water supply had a significant impact on groundwater development in a wider context.

East coast defences (Thames estuary)

The Thames estuary to the east of London was fortified with a series of blockhouses in the middle part of the sixteenth century by the Tudor monarch Henry VIII. At the same time, Henry built blockhouses to the north on both sides of the joint estuary of the rivers Stour and Orwell (near the later Landguard Fort), which formed a natural harbour where an invading army could land and attack London (Fig. 1). These fortifications were remodelled over the years until, in 1778, fearing a French attack, proposals were made to strengthen them on the condition that work should not begin unless a plentiful and secure supply of freshwater could be assured (Page 1784, 1797; Mather 2012*a*). The task of identifying possible potable water resources was entrusted to the military engineer Thomas Hyde Page (1746–1821), who, following distinguished service in the American War of Independence, had been appointed Commanding Engineer of the Eastern Coastal District (Mather 2012*a*). Page had already supervised alterations to defences at both Sheerness, on the Isle of Sheppey in Kent, and Landguard Fort on the Essex–Suffolk border (Fig. 1) and was ideally placed to undertake water supply investigations at these two sites.

Sheerness

The naval dockyard at Sheerness was established in the seventeenth century to support the larger yard at Chatham, *c.* 24 km to the WSW on the River Medway, for the cleaning and repairing of ships

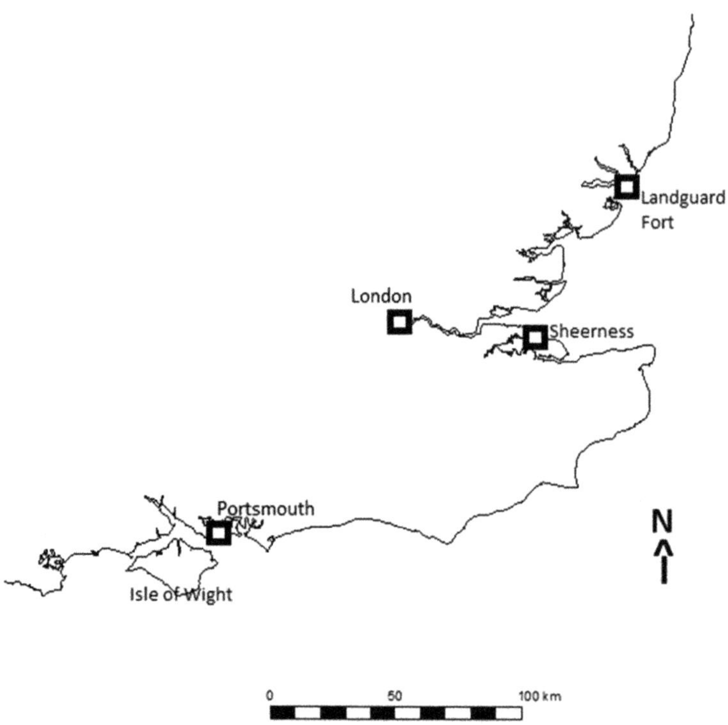

Fig. 1. Map of SE England showing the garrisons and forts discussed in the text and their positions relative to London.

and the building of small warships. It lay in an area of swampland, underlain by the London Clay Formation of Eocene age. There was little freshwater available and much of that was brackish. When Page was appointed, the dockyard was supplied with drinking water by boats from Chatham and London and water was also brought in from a well at Queenborough, *c.* 3 km to the south, probably in casks. The latter well was on the site of the fourteenth century Queenborough Castle, which had been demolished in the seventeenth century. Dry and disused for many years, the well had been deepened by orders of the Commissioners of the Navy in 1723. Using a 6 cm diameter auger, water-bearing strata were penetrated at 85 m below ground level, *c.* 24 m below the original depth of the well, yielding a good and wholesome supply (Collison 1729).

Perhaps it was the successful deepening of this well which suggested to Page that a deep well might be successful at Sheerness, although, according to his account, digging the well was 'at first considered as a mere experiment, the probability of success being much against it' (Page 1784, p. 10). Initial attempts within the garrison ended in disaster, as the well-sinkers were unable to prevent the excavation from continually filling with sand. A second attempt, at Fort Townsend *c.* 180 m to the west, was begun in June 1782, finishing 13 months later. At a depth of 13 m, the unconsolidated alluvial deposits, which had caused the earlier abandonment, were bottomed and a 1.8 m diameter shaft was sunk, by hand, through hard blue clay to 100 m. Just below that, the workmen were almost drowned as water and sand blew into the well, filling the basal 12 m (Fig. 2). Water rose in the well and within a few days stabilized at *c.* 2.4 m below the top. This water was of good quality and was extracted by a series of buckets on an endless chain, lifted either by a wind pump or horses walking around in a circle.

If not the first deep well to be sunk through the London Clay Formation into the underlying sands of the Harwich Formation, it was the first for which a detailed description was published. Page was proud of his achievements at Sheerness and his success was immediately recognized by his peers. In 1783 he was elected a Fellow of the Royal Society and received a knighthood from George III. The well was named the King's Well in honour of the monarch, who had visited the site during a crucial phase in its construction. However, Page considered that his knighthood was in recognition of his military service in America, rather than for his achievement at Sheerness (Page 1797).

Fig. 2. Contemporary(?) drawing of the strata intersected in the deep well at Sheerness. From Geological Society of London archive reference LDGSL – 446.

Landguard Fort

The earliest fortifications at Landguard Fort, near Felixstowe in Suffolk, date from 1534 within the reign of Henry VIII, but the first real fort was completed in 1627–28 (Leslie 1898). In 1626, during construction, a report, containing a detailed survey, was drawn up for the Privy Council. This stated that:

> There is an aqueduct of very good sweet water with a conduit house of bricke, the water is brought in leaden pipes a mile and halfe of the howse unto the ditch side and cannot be finished until the draw-bridge be sett up, but the pipes of leade lye readie to finish it
>
> (Leslie 1898, p. 14).

There are subsequent plans, dated 1732–37, showing the water supply to be drawn from these springs, which were at the village of Walton *c.* 2.4 km to the north (Pitman 1906). They were still in use when Page was appointed to investigate a possible on-site source.

The fort was built on loose sand and shingle and Page thought that there was little chance of finding freshwater. However, he decided to sink a well through the shingle to see if there was a hard bottom, as at Sheerness (Page 1784). At the same time, a ditch was being excavated for one of the batteries and a small quantity of freshwater was discovered leaking into the ditch. Page ordered a well to be sunk at a more convenient place (Fig. 3) and freshwater was found 2.4 m below the ground surface. The quantity was so great that it was difficult to keep the well dry. As the well was deepened, the water entering the hole became saline, so he ordered the well to be backfilled to bring its depth to 3.6 m, a little above what had been the lowest freshwater depth. The proposed deep well was abandoned and he ordered another shallow well to be sunk to the same shallow depth, 12 m away, connecting the two wells by a perforated brick drain (Fig. 4). According to Page, the well was successful, but Pitman (1906) records that the supply was only good enough for cooking purposes; drinking water continued to be drawn from the springs. History shows that Page was fortunate in deciding not to sink a deep well because, when a borehole was drilled to 247 m in 1881–84, it found only saline water (Whitaker 1906).

At Landguard Fort, Page found a thin layer of freshwater overlying saline water, but he was unable to understand how this freshwater got into the sands. He accepted that, on every sandy shore, salt water

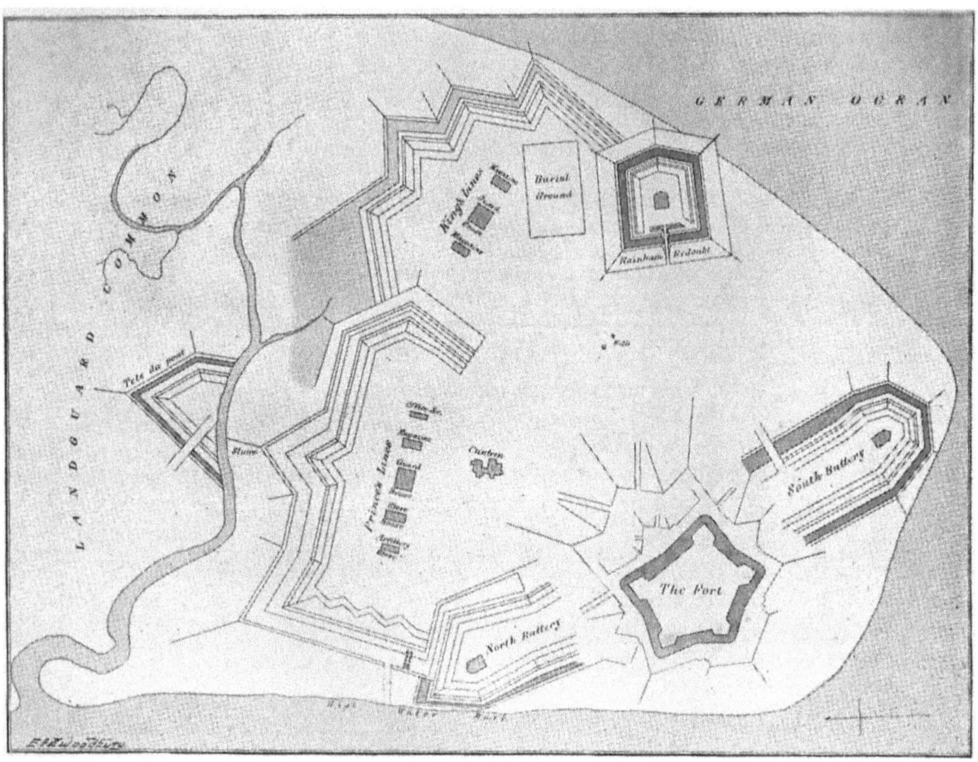

Fig. 3. Plan of Landguard Fort and the outer defence works dated 1785. The two wells sunk by Page are shown between the Fort and Rainham Redoubt. The defence works were maintained until 1815, but all evidence of their existence has now disappeared (from Leslie 1898).

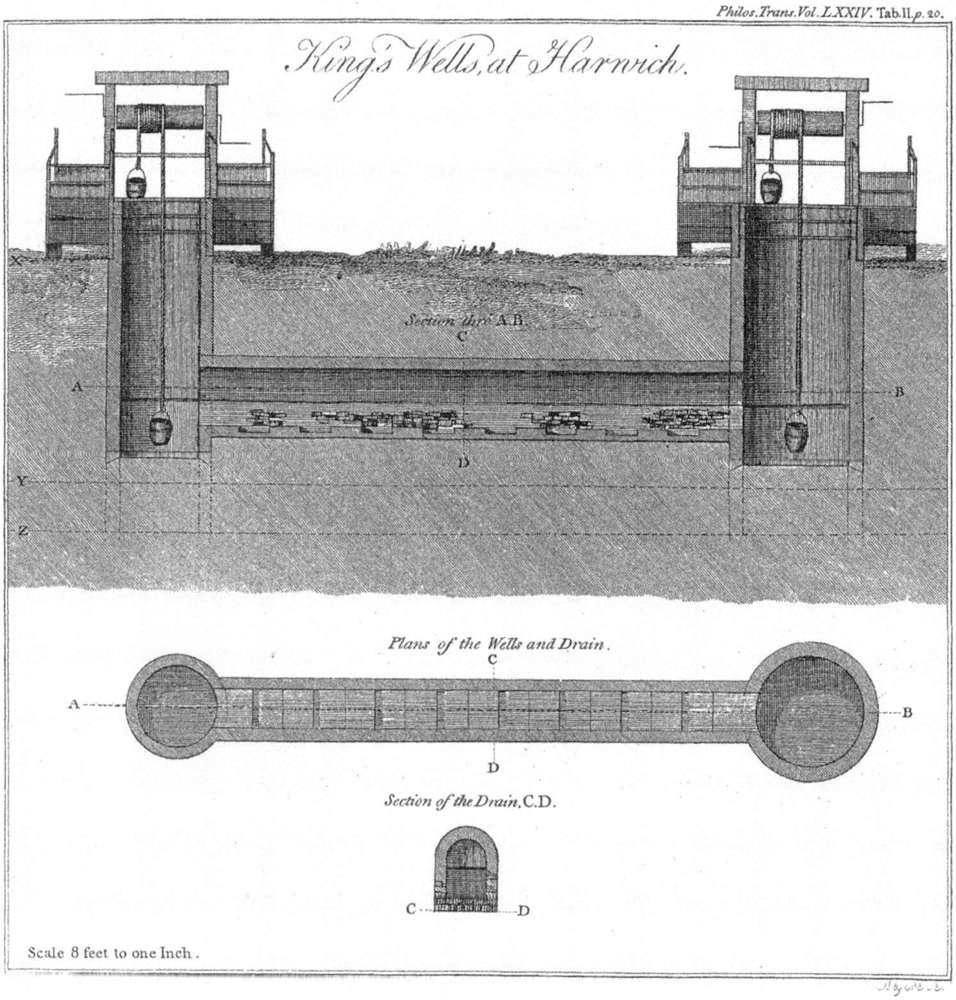

Fig. 4. Section of the completed wells and connecting gallery at Landguard Fort (not at Harwich, as claimed on the figure) reproduced from Page (1784). The internal diameter of the well on the right is 6 ft (1.8 m).

would be forced into the sand at high tide and would move back towards the sea again as the tide fell, and that this would reoccur during every tidal cycle. Whether this movement of salt water within the sand could render it fresh through friction, he was unable to judge. He thought not, and even if such a process did occur, he could not understand how so much freshwater could accumulate in the sand at Landguard Fort. Despite work in France, which had been reviewed in the *Philosophical Transactions of the Royal Society* (Anon 1675), demonstrating that rainfall was more than adequate to account for the flow of rivers and springs, he did not contemplate that the freshwater could have been derived from rainfall. However, he did note that the depth of the plane which separated fresh and salt waters could be 'ascertained to a degree of great accuracy' (Page 1784, p. 18) and speculated that the greater specific gravity of the salt water was sufficient to prevent mixing with the overlying freshwater. It was to be another 100 years before an officer, Captain Willem Badon Ghijben (1845–1907) from the Dutch Army Corps of Engineers, working in coastal dunes near Amsterdam, derived a mathematical relationship for the position of this equilibrium interface (de Vries 1994).

Portsmouth dockyard

The Allied victory in 1815 over the French Emperor Napoleon Bonaparte at Waterloo brought peace to

Europe, but the complacency that followed brought about a decline in Britain's naval power. By 1848 there were fears that the French might take advantage of this and attempt a sudden strike against the coast of southern England using their new steam-powered vessels (Patterson 1967). These fears were increased when Napoleon's nephew seized power, assuming the title of Emperor in 1852. Alarm intensified with the development of iron-clad warships and the Armstrong gun, with a range >7.5 km. This gun put the dockyard at Portsmouth, the most important British naval base, in range of such weapons if situated on Portsdown Hill immediately to the north, and from a hostile fleet of warships if anchored at Spithead to the south (Fig. 5).

In 1857, a committee was set up to investigate the probable implications of the new gun and it proposed the construction of works on the Spithead shoals lying between Portsmouth and the Isle of Wight. In 1859, the work of this committee was subsumed by a Royal Commission set up to report on the 'Defences of the United Kingdom'. A comprehensive report was produced in February 1860 and eventually *c.* 70 forts or batteries were built in England,

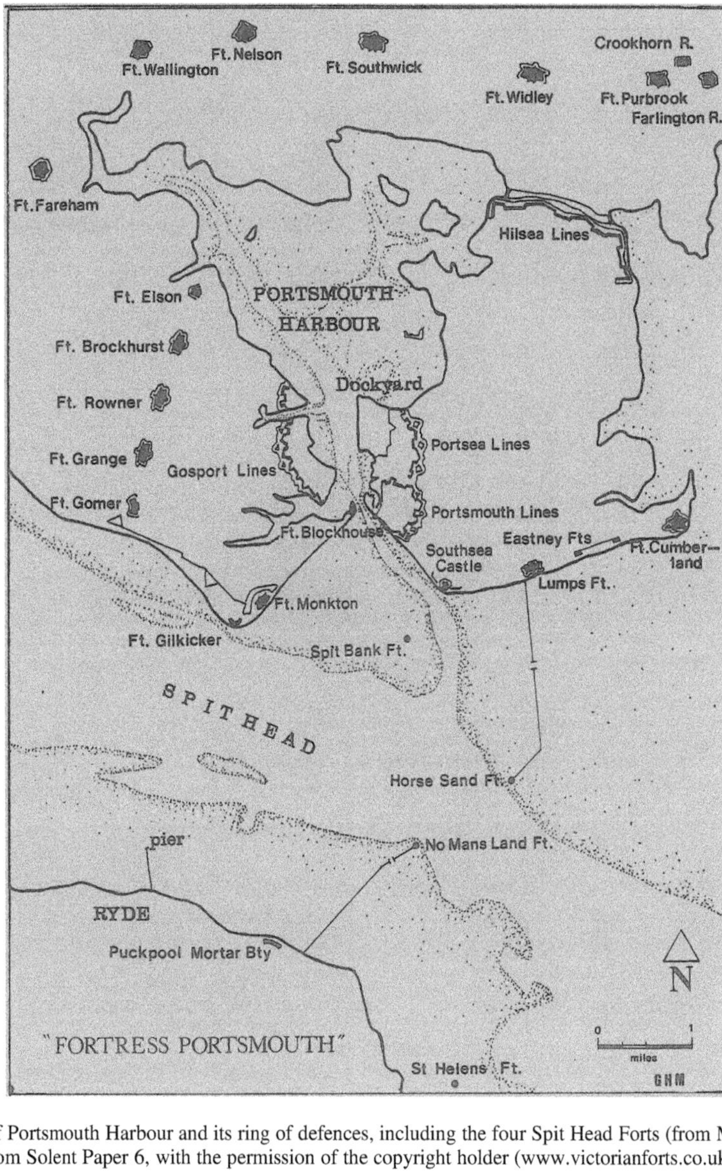

Fig. 5. Map of Portsmouth Harbour and its ring of defences, including the four Spit Head Forts (from Moore 1990). Reproduced from Solent Paper 6, with the permission of the copyright holder (www.victorianforts.co.uk).

wholly or in part in response to the recommendations. Because of an association with the Prime Minister of the time, Viscount Palmerston, these are collectively known as Palmerston Forts. Proposals included the completion of a ring of forts to defend Portsmouth dockyard and harbour against an attack by an enemy force that had landed elsewhere. This involved the building of a line of forts on Portsdown Hill to the north to complement work on forts to the west (the Gomer–Elson Line), which had already begun in the 1850s (Fig. 5). They also supported the previous proposals for forts on the Spithead shoals to protect the main channel into the harbour and the dockyard from offshore bombardment. Considerable debate in Parliament ensued and attempts were made to abandon or dilute the proposals on the grounds of expense, that there was no danger from France, or that the Spithead Forts would be ineffective in preventing iron-clad ships from running past them (Patterson 1967).

Decisions were delayed, but the construction of the land-based Portsdown Forts was well advanced by 1863. The Spithead Forts experienced longer delays and it was not until 1864 that construction of the foundations began. Two sites were abandoned because a firm foundation could not be obtained, but four forts were eventually constructed on shoals at No Man's Land and Horse Sand (Mitchell & Moore 1993), which straddle the main channel into Portsmouth at Spit Bank (Moore 1997), between Horse Sand Fort and the entrance to Portsmouth Harbour, and on shoals close to St Helen's Point on the Isle of Wight (Mitchell 1988) (Fig. 5). Construction difficulties led to further delays and Spit Bank Fort was the first to be completed in 1878, followed closely by St Helen's Fort. Despite an earlier start, No Man's Land and Horse Sand Forts, which were larger, were not completed until 1880. These Victorian forts, on the Spithead shoals and on the crest of Portsdown Hill, were subsequently named Palmerston's Folly because the Prime Minister of the day was thought to have taken needless precautions against a purely imaginary threat. This he had compounded, in the eyes of local people, by building the land-based forts so that they faced inland.

The forts all required a secure water supply, ideally provided by on-site wells and/or boreholes, which would have been relatively straightforward at the Portsdown Forts because they overlie the outcrop of Upper Cretaceous Chalk, the main aquifer of the area. To the south, modern mapping shows that the Chalk is overlain by a thick succession of Paleogene sediments (Bristow *et al.* 1991; Hopson 2000). The Reading Formation, which here rests unconformably on the eroded Chalk surface, consists of *c.* 35 m of colour-mottled clays and silty clays with localized bodies of fine- and medium-grained sands, particularly at the top and base. These sand bodies yield water in small quantities. The overlying London Clay Formation, up to 150 m in thickness, consists of grey, bioturbated silty and sandy clay, with sheet-like and lenticular bodies of fine-grained sand. Where present, moderate yields of groundwater have been obtained from these sands. The overlying Bracklesham Group, 170–230 m thick, contains a varied succession of interbedded clays, silts and sands. The latter yield small supplies, but the interlayered nature of the aquifer means that exploitation is difficult. The Bracklesham Group is overlain by sandy clays of the Barton Clay Formation (40–50 m thick), which have little value as a source of water supply.

Lieutenant-Colonel Portlock

The four sea forts posed different problems of water supply to those on Portsdown Hill. The obvious solutions were rainfall capture and bringing in water by barge, both of which reduced the security and effectiveness of the forts. A decision was therefore made to drill boreholes from the forts, searching for a suitable aquifer beneath. This was a highly speculative and expensive venture and it is pertinent to examine why it was thought that the risk was justified.

Between 1847 and 1849, one of the Royal Engineers officers based in Portsmouth was Lieutenant-Colonel (later Major-General) Joseph Ellison Portlock (1794–1864). Born in Gosport, the only son of a distinguished naval Captain (Vetch 2004), he was an enthusiastic geologist and had been elected a member of the Geological Society of London in 1825 and was to become its President from 1856 to 1858. During his time at Portsmouth, he produced a substantial article on geology for the *Aide Mémoire to the Military Sciences*, a three-volume handbook that advised military engineers on technical aspects of their work (Portlock 1848), and a textbook for more general use (Portlock 1849). By 1850, deep boreholes for water supply had been drilled on both sides of the channel into Portsmouth Harbour, beneath Portsea Island to the east and Gosport to the west (Fig. 5). In 1850 and 1852, following his transfer to Cork, Ireland, Portlock summarized results from these boreholes in two addresses given to the Geological Society of Dublin (Portlock 1851, 1853).

At Gosport, a borehole in the Royal Clarence Victualing Yard had penetrated 36 m of clays and sands, followed by two beds of London Clay, 27 and 31 m thick, respectively, separated by a 1 m thick bed of shelly sand, before a strong flow of water was found in sand at a depth of 95 m. Based on this succession, Portlock and his colleagues proposed to supply Fort Blockhouse, at the southern end of the harbour channel (Fig. 5), by a further deep borehole. In this, the upper clays and sands were found to be 24 m in thickness and the London Clay was divided into three thick beds by sands, which were either dry or yielded brackish water. On penetrating the lower bed of London Clay, clean sands were found that yielded a supply of good water at a depth of 94 m, a similar depth to that in the Royal Clarence well. Buoyed by the success of these boreholes, the Board of Ordnance drilled another deep well, this time at Portsea, to the east of the harbour channel (Fig. 5) and nearly along the strike from the Royal Clarence well, expecting to find water at a comparable depth. However, no water-bearing sand was found within the London Clay and it was not until entering the Chalk at a depth of 186 m that water entered the hole, rising to *c.* 1 m from the surface.

Further boreholes were drilled during the next ten years or so on both sides of the harbour (Pilbrow 1860; Whitaker 1910). The general picture emerged of an upper sequence of clays, grey and green sands and sandy clays, which Portlock thought were post-Tertiary in age, but were subsequently recognized as part of the Early to Mid-Eocene Bracklesham Group. They contained limited groundwater resources and overlaid *c.* 100 m of London Clay, which might or might not contain water-bearing sands. The London Clay was underlain by the Reading Formation, which appeared to contain little water and which, in turn, was underlain by the Chalk at a depth of *c.* 160–190 m below the surface. It seems reasonable to speculate that the military engineers, who had succeeded Portlock, considered that boreholes drilled from the offshore forts would find water – if they were lucky, within sand beds in the London Clay, as at Gosport or, if that source failed, certainly within the Chalk, as at Portsea. This is supported by a description of the work in progress on Horse Sand Fort, which is given in Spon (1875) and which talks about a borehole being sunk into the Chalk.

Spithead forts

Fragmentary records of the water supply boreholes drilled at the forts are accessible from the British Geological Survey website under 'Geology of Britain viewer-borehole scans' (http://mapapps.bgs.ac.uk/geologyofbritain/home.html). Basic information from these records has been abstracted by Whitaker (1910) and Buchan *et al.* (1942). Construction of Horse Sand Fort (National Grid Reference (NGR) SZ 655949, Fig. 5) began in July 1861 when a 6 ft (1.83 m) cast iron cylinder was sunk to 24.5 m on the Horse Sand Shoal to ascertain the nature of the subadjacent strata as a foundation for the proposed fort. This was halted in April 1862, at a depth of 24.5 m, when work was suspended because of political pressure.

It was not until 13 November 1868 that work began on the water supply borehole, which was bored from the base of this cylinder by Mather and Platt of Salford, Lancashire. In parallel with the boring, cast iron tubing, the lower end of which was shod with steel, 18 inches (45.7 cm) in diameter and 1 inch (2.54 cm) thick, was forced, in 9 ft (2.75 m) lengths, through the underlying greenish grey clays and sandy clays (Fig. 6). Traditionally, such tubing was forced down using the weight of the boring machine with stones and pig iron added. A hydraulic method was designed specifically for the work at Spithead, which enabled boring operations to be carried on uninterrupted by the process of driving the tubes. A full description of the methodology in given by Spon (1875). Progress was steady

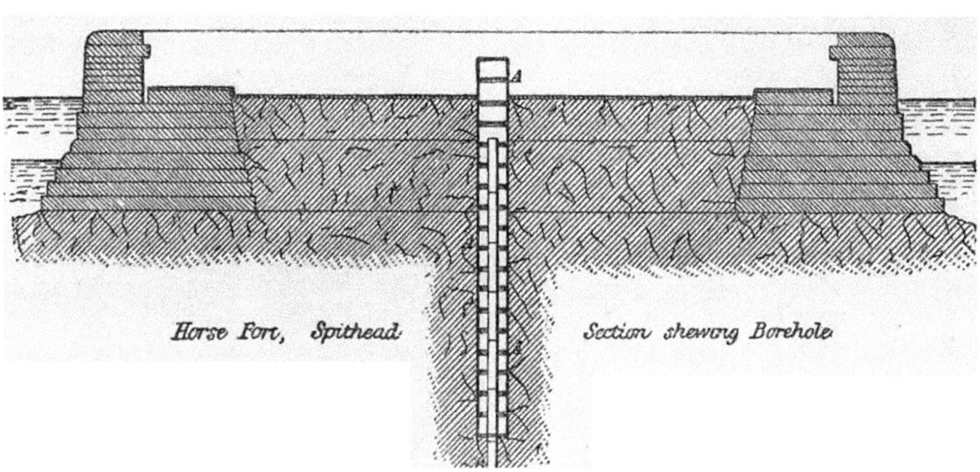

Fig. 6. Horse Sand Fort, partially completed, during the drilling of the centrally placed borehole, showing the cast iron cylinder and the borehole at its base (from Spon 1875).

until the tube was forced through sandstone rock at *c.* 144 m depth, when it fractured and boring had to be continued using similar tubing at a reduced diameter of 15 inches (38.1 cm). Unfortunately, during the breaking up and removal of the fractured 18 inch (45.7 cm) tube, the lower part of the 15 inch (38.1 cm) tube was also broken, resulting in considerable delay in clearing the obstruction and necessitating the deployment of 9 inch (22.9 cm) diameter wrought iron tubes, shod with steel, to complete the hole. Water was eventually found at a depth of 174 m below the basement floor at about −169 m Ordnance Datum (OD) within a clean, sharp, light grey sand, and rose to stand at about mid-tide level, fluctuating with a variation of about one-third that of the tide (Whitaker 1910). Yield information is confusing, but, on completion, the yield seems to have been *c.* 32 m^3 per day with a drawdown of *c.* 30 m.

Borehole records for No Man's Land Fort (NGR SZ 639938) are less complete than those for Horse Sand, although the forts are almost identical and it seems likely that similar techniques were used. Again, a cast iron cylinder was sunk at the surface of the shoal and an 18 inch (45.7 cm) diameter borehole put down to 160 m before a reduction to 15 inch (28.1 cm). The borehole was continued to a depth of *c.* 177 m, with water struck in grey–green sands below 174 m at about −170 m OD. Water rose in the borehole, fluctuating with the tide around the mid-tide level, and the yield was higher than at Horse Sand Fort at *c.* 105 m^3 per day with a drawdown of *c.* 29 m.

The hole at Spit Bank Fort (NGR SZ 636972) was bored between 31 May and 12 October 1877, at the same time driving 18 inch (45.7 cm) diameter wrought iron tubing, shod with steel (Figs 7 & 8). Water was found at a depth of 39 m in a pale green sand, but was unfit for drinking. The tubing was then forced down to 122 m until water was again found, in ample quantity and good quality, within a dark green sand. The rest water level was again close to the surface, coincident with that of the mean tide, and the yield was 152 m^3 per day with a drawdown of *c.* 30 m.

The borehole at St Helen's Fort (NGR SZ 648 898) seems to have been drilled in 1867, before those at the other forts, and was completed at a depth of just 52 m. Freshwater was found in sands, beneath a claystone at a depth of *c.* 46 m, and rose to *c.* 1 m below the level of high water of ordinary spring tides. Yields were restricted to *c.* 13 m^3 per day with a drawdown of 7.5 m.

The boreholes were still in use during the early years of World War II, but were disused by the mid-1950s, when reviewed by the Geological Survey's Water Department during countrywide surveys to collect basic groundwater information (Mather 2012*b*). Pumps had been removed from at least some of the boreholes by 1957.

The three northern boreholes (Spit Bank, Horse Sand and No Man's Land) all proved a sequence of Paleogene strata underlying up to 27 m of recent marine deposits consisting of shingle and sand, with shells, wood and, at No Man's Land, the jaw of a red deer. Beneath this, the boreholes penetrated an interbedded succession of clays, silty and sandy clays, silts and sands, and occasional shelly and pebbly beds with pyrite nodules and pieces of lignite. These fossiliferous sediments have all been referred to the Bracklesham Group (Whitaker 1910), although Osborne White (1915) thought that the upper 27 m, in the Horse Sand well, might belong to the overlying Barton Group.

If, as suggested, the military engineers were aiming to obtain water from either sands in the London Clay or from the Chalk, their failure to reach the surface of the London Clay by the time they had reached a depth of −170 m OD would have been a major setback. They had no idea how much deeper they would need to drill to reach the London Clay surface, but would have recognized that, once there, they would need to deepen the hole by another 130 m or more to reach the Chalk. Their drilling technique had already resulted in problems in the Horse Sand borehole, where the heavy tubing had fractured as they drove it through sandstone, and the likelihood of them being able to force it to depths >300 m would have seemed remote.

Modern geophysical work (Hamblin *et al.* 1992) shows that the forts overlie the western part of the Hampshire–Dieppe Basin. Boreholes would have had to reach depths of *c.* 280 m at Spit Bank Fort, 420 m at Horse Sand Fort and 500 m at No Man's Land Fort to intersect the Chalk surface, which dips to the south at *c.* 3°. This dip means that beds of similar age occur at different depths in each of the boreholes and that the horizon yielding potable water at −169 m OD at Horse Sand Fort is not the same as that yielding water at a similar depth at No Man's Land Fort, where the equivalent bed will occur *c.* 60 m lower. It follows that the horizon yielding water at No Man's Land did not yield potable water when penetrated by the Horse Sand borehole. The evidence suggests that each of the northern boreholes abstracted water from different beds of water-bearing sands within the Bracklesham Group and that individual beds were not laterally extensive. To the south, St Helen's Fort overlies the deepest part of the Hampshire–Dieppe Basin (Hamblin *et al.* 1992) and the on-site borehole probably abstracted water from sands within the mainly non-marine Headon Hill Formation, which overlies the marine sediments of the Barton Group (and so also the underlying Bracklesham) encountered in the northern boreholes.

The groundwater in the Bracklesham Group was confined, with water levels rising to around the level

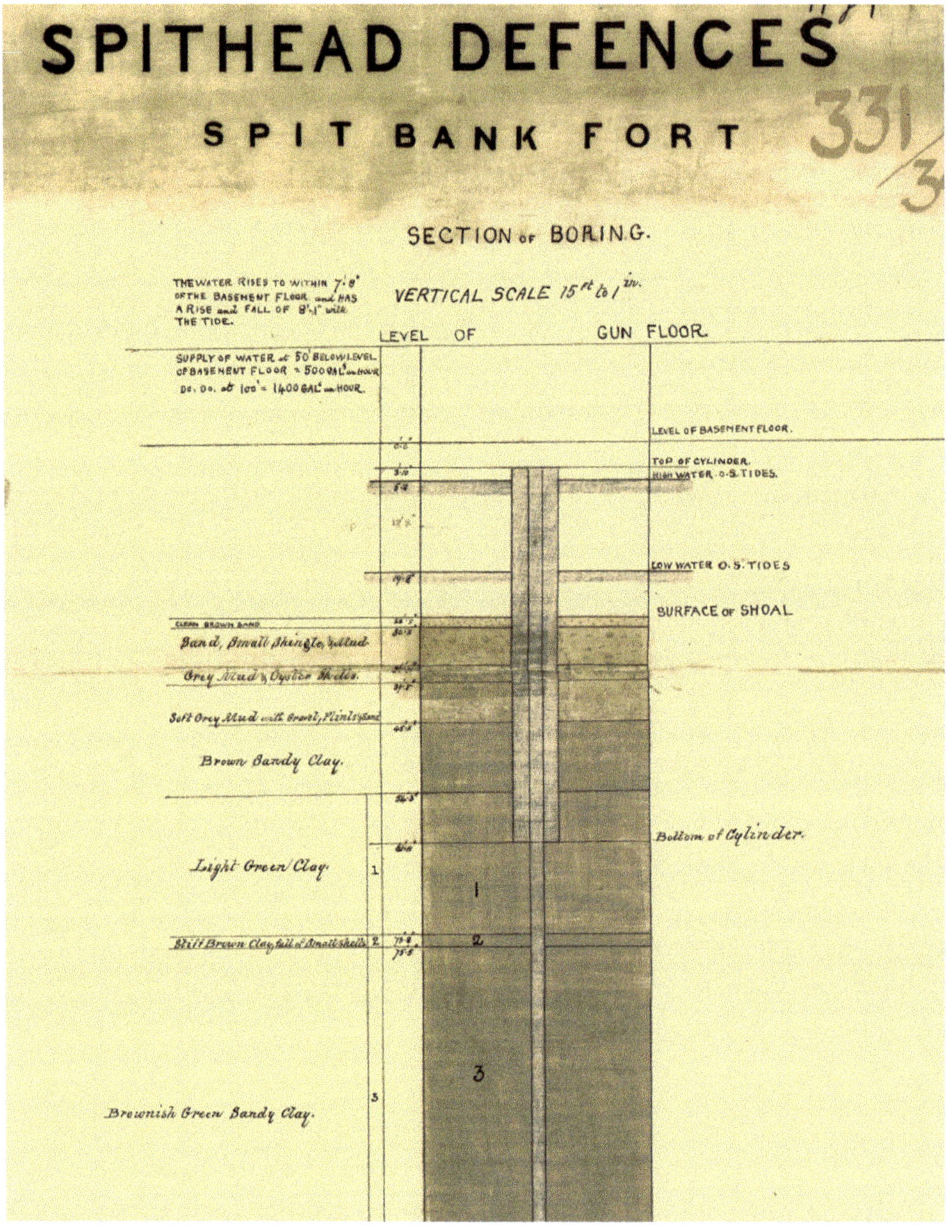

Fig. 7. Log of the upper portion of the Spit Bank borehole showing geological and hydrogeological information. Attached to a letter, dated 14 November 1877, from Clement de B. Carey, Captain RE, to Henry Bristow, then Senior Director, England and Wales at the Geological Survey. Contains British Geological Survey materials © NERC 2017, delivered under the terms of the Open Government Licence.

of mid-tide. With present day sea-levels, recharge is possible only from the limited Bracklesham outcrop on the mainland to the north and it seems probable that the water extracted by the forts was recharged during a period of much lower sea-levels during the Pleistocene. If this is the case, water was effectively mined and was not being replaced by modern recharge. Yields obtained from the boreholes were small, partly because of the nature of the Bracklesham Group itself and partly because of the drilling and completion techniques available at the time. The Group consists of sandy beds

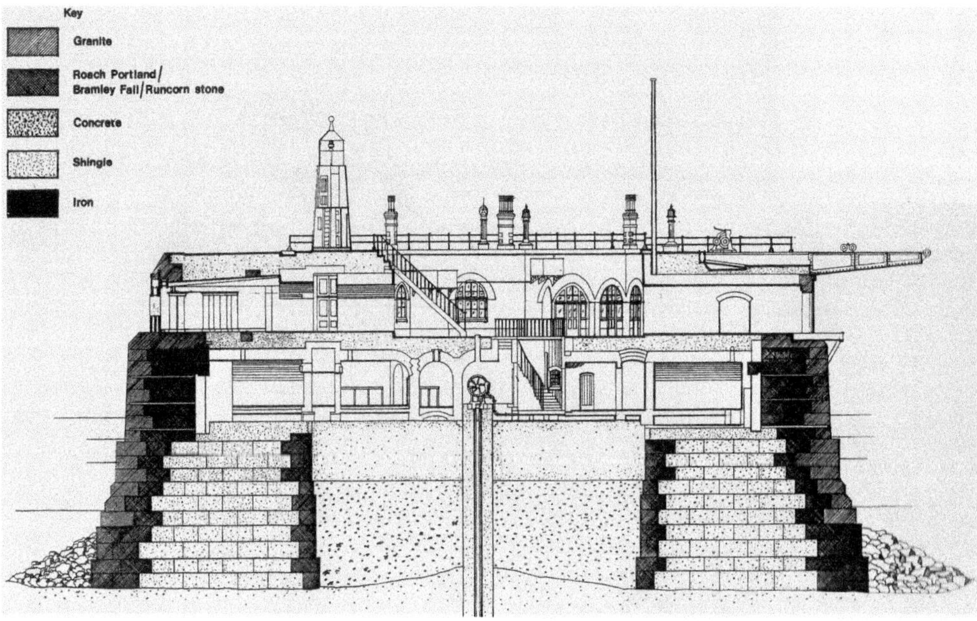

Fig. 8. Cross-section through the Spit Bank Fort showing the completed borehole in its centre (from Mitchell 1988). Reproduced from Solent Paper 1, with the permission of the copyright holder (www.victorianforts.co.uk).

separated by silts and clays and, with modern properly designed filter packs and sand screens, some horizons might be capable of significant yields. However, the well tubing had no slotted sections to allow water to flow into the borehole and the spaces between various diameter tubes were filled with Portland cement or puddled clay and concrete. This meant that the only access point for water was from the bottom of the borehole. The fine-grained nature of the sands meant that sand flowed into the hole, reducing both the rate of water inflow and the borehole depth. However, despite the problems, the discovery of potable groundwater in the Bracklesham Group was a bonus and provided the forts with a potable water supply for the next 70 years.

Discussion and conclusions

Robins & Rose (2009) and Mather & Rose (2012) have identified several areas in which technical developments, driven by military needs, benefited the wider water supply industry. These included the introduction of driven tube wells and mobile drilling rigs, the development of groundwater prospect maps, and the use of horizontal abstraction galleries in coastal and desert environments. In addition, military requirements meant that engineers could invest in speculative schemes that would have been difficult to finance elsewhere in the public sector. The attitude to such schemes was well expressed in comments made on an abortive civilian deep boring at Southampton, drilled to 403 m between 1842 and 1851. When completed, the water was brackish, with a total dissolved solids concentration of *c.* 1300 mg l^{-1}, prompting one writer to comment that,

> Public money should never be expended, unless with a certainty of attaining the end product; it should never be employed in carrying out theoretical views or for the attainment of what may be called hypothetical or eventual advantages
>
> (Swindell & Burnell 1882, p. 85).

The writer thought that some scientific body, or lover of science, should pursue such questions.

Page's initial setback at Sheerness prompted questions in Parliament, describing the Sheerness well as 'not a well for freshwater, but a sink for the money of the public' (Page 1797, p. 6). However, Page persisted with his well, presumably with the support of his superiors, despite the strong possibility of failure. Perhaps the general idea that public money should not be risked on such ventures was circumvented because of the support of the King and the very real fears of an attack along the seaboard of SE England.

The discovery of freshwater in sands beneath the London Clay at Sheerness stimulated further activity and about 25 years later, John Farey (1766–1826), a pupil of the great English geologist William Smith (1769–1839), could claim that 'every considerable brewer in London … has a deep well … for use in brewing his porter' (Farey 1807, p. 212). Page's well was also important in the emerging understanding of the geology of the London Basin and in explaining the source of the deep groundwaters beneath the city (Mather 2016). On the negative side, the discovery of potable water beneath the London Clay led many water engineers to assume that there was freshwater anywhere if one drilled to sufficient depth, a view that was to lead to dry holes, considerable expense and litigation (Farey 1823).

At Landguard Fort, the two wells, connected by a brick drain, acted as a collecting gallery, skimming off the freshwater. It is unclear whether this gallery was Page's own idea or was based on experience elsewhere. Land drainers had been using covered drains for some years (Young 1770) and networks of gathering pipes, laid beneath ground level, had been used to collect water from springs and areas of seepage. However, the Landguard Fort gallery was probably the first time in Britain that such a system had been used to develop a freshwater lens in a coastal environment, an engineering solution that was later widely used by the British army (Addison & Shotton 1946) and elsewhere (De Rijke 1898).

At Portsmouth, holes were bored below the Spithead Forts by the Salford firm of Mather and Platt. Traditionally, the boring tool would have been attached to iron rods and worked into the ground through a vertical jumping motion and a circular motion applied by the workmen. When the accumulated debris needed to be cleared away, the rods had to be removed, which was a time-consuming process, particularly as the hole became deeper. The Spithead boreholes used a method invented by Colin Mather (*c.* 1812–77), which was the precursor to modern percussive drilling techniques and used a flat hemp rope instead of rods, improving the percussive action and the collection of debris (Mather 1855). Drilling continued during the installation of the tubing with a hydraulic method, designed specifically for the work at Spithead (Spon 1875). Innovative technology was being used to drill and line what must have been some of the earliest deep holes to be bored offshore anywhere in the world.

Military engineers were employing drilling contractors with state-of-the-art technology, but, apart from Portlock, the names of most of those providing geological advice remain unknown. One was the civil engineer and geologist Robert William Mylne (1816–90), who was an acknowledged expert on water supply. His obituarist records that 'He obtained a good water supply for one of the sunk forts in the sea off Portsmouth … ' (W.B.D. 1890, p. xx). An acknowledgement to Mylne for supplying an account of the strata towards the base of the borehole at St Helen's Fort, in Whitaker (1910), implies that this was the scheme with which he was involved. In other cases, it appears that no geological advice was sought. Fort Fareham was one of three forts, proposed by the Royal Commission, to the NW of the Elson–Gomer Line (Fig. 5). Cost-cutting meant that Fareham was the only one of these to be built, with work commencing in 1861. The first contract let was for the sinking of five wells to a depth of 135 ft (41 m). These were completed in July 1861, but no water was found. A fresh contract had to be let for sinking them to 200 ft (61 m), a task completed in September 1861 (Moore 2003). The Chalk was not reached, but water was obtained from sands at the top of the Reading Formation, underlying the London Clay (Whitaker 1910). In this situation, the use of geological expertise might have avoided dry wells and expensive remedial work.

The names of engineers and geologists, such as Thomas Hyde Page, Joseph Portlock and Colin Mather, involved in the work described in this paper, are now largely forgotten, but they made a significant contribution to the development of groundwater supplies. Perhaps sheltered from financial constraints because of the strategic significance of their work, they could develop and use innovative technology and take risks that were not possible elsewhere in the public sector.

Acknowledgements Many thanks to Caroline Lam at the Geological Society and Rosemary Fry at the British Geological Survey for help with archive materials and to David Moore, the copyright holder for the Solent Paper Series, for permission to reproduce illustrations from the Series.

References

Addison, H. & Shotton, F.W. 1946. Water supply in the Middle East Campaigns, III. Collecting galleries along the Mediterranean coast of Egypt and Cyrenaica. *Water and Water Engineering*, **49**, 427–436.

Anon 1675. A particular account, given by an anonymous French Author in his book of the Origin of Fountains, printed 1674 at Paris, to shew, that the Rain and Snow-waters are sufficient to make Fountains and Rivers flow perpetually. *Philosophical Transactions*, **10**, 447–450.

Bristow, C.R., Freshney, E.C. & Penn, I.E. 1991. *Geology of the Country Around Bournemouth.* British Geological Survey, Memoirs, Sheet 329 (England and Wales). HMSO, London.

Buchan,. J.A., Butler, A.J., Holmes, S.C.A. & Morris, L.S.O. 1942. *Water Supply from Underground Sources of Reading–Southampton District. Part VIII. Well-Catalogues for Parts of New Series of One-inch Sheets 330 and 331, with 344 and 345 (Isle of Wight).* Wartime

Pamphlet, **15**. Geological Survey of Great Britain, England and Wales, London.

Collison, P. 1729. A letter from the King's Officers at Sheerness and Chatham, to the Honourable the Commissioners of the Navy, giving an account of what they met with in opening an ancient Well near Queenborough in Kent. *Philosophical Transactions*, **36**, 191–194.

de Rijke, J. 1898. On sand dunes and dune sand as a source of water supply. Appendix 1. *In*: Burton, W.K. (ed.) *The Water Supply of Towns and the Construction of Waterworks*. 2nd edn. Crosby, Lockwood, London.

de Vries, J.J. 1994. Willem Badon Ghijben and Johan M.K. Pennick, pioneers of coastal-dune hydrology. *Applied Hydrogeology*, **4**, 55–57.

Farey, J. 1807. On the means of obtaining water. *Monthly Magazine and British Register*, **23**, 211–212.

Farey, J. 1823. On artesian wells and boreholes. *Monthly Magazine and British Register*, **56**, 309.

Hamblin, R.J.O., Crosby, A., Balson, P.S., Jones, S.M., Chadwick, R.A., Penn, I.E. & Arthur, M.J. 1992. *United Kingdom Offshore Regional Report: the Geology of the English Channel*. HMSO, London.

Hopson, P.M. 2000. *Geology of the Fareham and Portsmouth District, a Brief Explanation of the Geological Map*. Sheet Explanation of the British Geological Survey Sheet 316 Fareham and Part of Sheet 331 Portsmouth (England and Wales). British Geological Survey, Keyworth.

Jebb, J. 1842. Notes on the theory and practice of sinking artesian wells. *Papers on Subjects Connected with the Duties of the Corps of Royal Engineers*, **5**, 266–280.

Leslie, J.H. 1898. *The History of Landguard Fort, in Suffolk*. Eyre & Spottiswoode, London.

Mather, C. 1855. On earth-boring machinery. *Journal of the Society of Arts*, **3**, 503–509.

Mather, J.D. 2012*a*. Water supply to Britain's eastern coastal defences in the 18th century and the work of Sir Thomas Hyde Page (1746–1821). *In*: Rose, E.P.F. & Mather, J.D. (eds) *Military Aspects of Hydrogeology*. Geological Society, London, Special Publications, **362**, 35–47, https://doi.org/10.1144/SP362.3

Mather, J.D. 2012*b*. War as a catalyst for change: groundwater studies in the Geological Survey of Great Britain before 1950 and the impact of two world wars. *In*: Rose, E.P.F. & Mather, J.D. (eds) *Military Aspects of Hydrogeology*. Geological Society, London, Special Publications, **362**, 139–160, https://doi.org/10.1144/SP362.8

Mather, J.D. 2016. William Smith: the natural order of strata and the search for underground water supplies. *Earth Science History*, **35**, 124–144.

Mather, J.D. & Rose, E.P.F. 2012. Military aspects of hydrogeology: an introduction and overview. *In*: Rose, E.P.F. & Mather, J.D. (eds) *Military Aspects of Hydrogeology*. Geological Society, London, Special Publications, **362**, 1–17, https://doi.org/10.1144/SP362.1

Mitchell, G.H. 1988. Spit Bank and the Spithead Forts. 3rd edn. Solent Papers, **1**. G. H. Mitchell, West Wickham, Kent.

Mitchell, G. & Moore, D. 1993. Horse Sand and No Man's Land Forts. *The Redan*, **28**, 15–29.

Moore, D. 1990. Fort Brockhurst and the Gomer – Elson Forts. Solent Papers, **6**. David Moore, Gosport, Hampshire.

Moore, D. 1997. Spitbank Fort: a history and description. *The Redan*, **41**, 24–36.

Moore, D. 2003. Fort Fareham. Solent Papers, **7**. David Moore, Gosport, Hampshire.

Osborne White, H.J. 1915. *The Geology of the Country near Lymington and Bournemouth*. Geological Survey, Memoirs, Explanation of Sheets 230 and 331 (Mainland). England and Wales. HMSO, London.

Page, T.H. 1784. Description of the King's Wells at Sheerness, Landguard Fort and Harwich. *Philosophical Transactions of the Royal Society of London*, **74**, 6–19.

Page, T.H. 1797. *An Account of the Commencement and Progress in Sinking Wells at Sheerness, Harwich and Landguard Fort, for Supplying those Dock-Yards and Garrisons with Fresh Water, etc.* John Stockdale, London.

Patterson, A.T. 1967. 'Palmerston's Folly'. The Portsdown and Spithead Forts. The Portsmouth Papers, **3**. Portsmouth City Council, Portsmouth.

Pilbrow, J. 1860. On a well section near Gosport. *Quarterly Journal of the Geological Society of London*, **16**, 447–449, https://doi.org/10.1144/GSL.JGS.1860.016.01-02.57

Pitman, L.C. 1906. *Short History of Landguard Fort Water Supply*. Unpublished typescript, provided by Mr D.A. Wood, Landguard Fort.

Portlock, J.E. 1848. Geognosy and geology. *In*: Lewis, G.G., Jones, H.D., Nelson, R.J., Larcom, T.A., de Moleyns, E.C. & Williams, J. (eds) 1846–52. *Aide-Mémoire to the Military Sciences*. Vol. 2, Part 1. Weale, London, 77–182.

Portlock, J.E. 1849. *Rudimentary Treatise on Geology for the Use of Beginners*. Weale, London.

Portlock, J.E. 1851. On the inequalities of the sea bottom, during the Tertiary Epoch. *Journal of the Geological Society of Dublin*, **4**, 245–248.

Portlock, J.E. 1853. Address delivered at the Annual Meeting, Feb. 11, 1852, by the President, Lieutenant-Colonel Portlock, RE. *Journal of the Geological Society of Dublin*, **5**, 58–59.

Robins, N.S. & Rose, E.P.F. 2009. Military uses of groundwater: a driver of innovation? *Hydrogeology Journal*, **17**, 1275–1287.

Rose, E.P.F., Mather, J.D. & Perez, M. 2004. British attempts to develop groundwater and water supply on Gibraltar 1800–1985. *In*: Mather, J.D. (ed.) *200 Years of British Hydrogeology*. Geological Society, London, Special Publications, **225**, 239–262, https://doi.org/10.1144/GSL.SP.2004.225.01.17

Spon, E. 1875. *Water Supply. The Present Practice of Sinking and Boring Wells, with Geological Considerations and Examples of Wells Executed*. E. & F.N. Spon, London.

Swindell, J.G. & Burnell, G.R. 1882. *Rudimentary Treatise on Wells and Well-Sinking*. New edn. Crosby Lockwood, London.

Vetch, R.H. 2004. Portlock, Joseph Ellison (1794–1864), surveyor, army officer and geologist. (Revised by E. Baigent.) *Oxford Dictionary of National Biography*. Oxford University Press, www.oxforddnb.com/view/article/22585 [last accessed 21 March 2017].

W[illiam] B[oyd] D[awkins] 1890. Obituary notices of fellows deceased; Robert William Mylne. *Proceedings of the Royal Society of London*, **48**, xx–xxi.

WHITAKER, W. 1906. *The Water Supply of Suffolk from Underground Sources with Records of Sinkings and Borings*. Geological Survey of England and Wales, Memoirs. HMSO, London.

WHITAKER, W. 1910. *The Water Supply of Hampshire (including the Isle of Wight), with Records of Sinkings and Borings*. Geological Survey of England and Wales, Memoir. HMSO, London.

YOUNG, A. 1770. *A Six Months' Tour through the North of England Containing, an Account of the Present State of Agriculture, Manufactures and Population, in Several Counties of this Kingdom, 1*. Strahan, London.

American coastal defence Third System forts: how geomorphology and geology dictated placement and influenced history

STEPHEN W. HENDERSON

Department of Geology, Oxford College of Emory University, Oxford, GA 30054, USA

0000-0002-6662-7450

shender@emory.edu

Abstract: In total, 42 Third System forts were built along the east coast of the USA between 1816 and 1867. Because of their purpose, to defend key locations against potential British amphibious assault, their sites were limited to islands, shoals, shorelines, riverbanks or hilltops. Sited along the Atlantic and Gulf of Mexico coastlines, many are constructed upon distinctive geomorphological landforms that are related more to features influenced by Quaternary glaciation or sea-level changes than by pre-Quaternary bedrock. A selection of forts is described to illustrate the principal geological and geomorphological features that influenced their construction: at Boston Harbor in Massachusetts on drumlins, near New York City on moraine and at Fort Pulaski protecting Savannah Harbor in Georgia on saltmarsh. Further south, Fort Clinch in northern Florida, forts guarding Pensacola Bay in western Florida and the entrance to Mobile Bay in the Gulf region, and Fort Massachusetts on Ship Island, Mississippi, were all primarily situated on Holocene barrier islands. Construction, use and 1861–65 Civil War threats to these forts can be directly related to their geomorphological context. Although all are now redundant, most remain as a remarkable and highly visible aspect of the military history of the USA.

In total, 42 Third System coastal defence forts were built in the USA from 1816 to 1867, many of which continued in use to and through World War II. Reflecting a diversity of geography, geology and building styles, this paper discusses a selection of 15 of them on the Atlantic and Gulf of Mexico coasts from Massachusetts south to Mississippi (Fig. 1).

In principle, the geomorphological setting of coastal defence structures might be expected to relate primarily to coastal geology. However, given that most of the world's coastlines, including those of the western Atlantic and Gulf of Mexico, are submergent, other factors are also significant. The American areas exhibit features not only related to coastal processes, but to former glaciation, changing sea levels and sediment supply. Even though the American coastline from New York south to Florida and the Gulf coast is considered to be submergent with well-developed estuaries, it is characterized by the presence of barrier islands due partially to high alluvial sediment supply. A combination of geological events has produced the array of landforms upon which the forts have been constructed. The New England coastline, including Boston, has been glaciated, resulting in the presence of both glacial erosional and depositional features. Boston Harbor is characterized by drowned drumlins lying upon Precambrian bedrock. The New York City area has also been glaciated and has geomorphic features that are largely depositional in nature, consisting primarily of end moraines and outwash, here lying upon bedrock sediments of Cretaceous age. Fort Pulaski in Georgia and Fort Clinch in Florida are sited on the coastal plain, on Holocene sediments overlying Pleistocene deposits bordering a thick Cenozoic sedimentary sequence. The geology of the Gulf Coast is similar: Holocene sediments again overlie Pleistocene deposits. The pattern of Quaternary deposition is indicative of changing sea levels, largely a consequence of waxing and waning global glaciation.

Background to Third System fortification

The earliest coastal defence forts in the USA were built under colonial rule prior to the establishment of the nation in 1776. Several European countries had colonies in North America and many of these early settlements constructed fortifications for coastal defence. Seemingly, the earliest was Charlesfort, built in 1562 by French Huguenots on Parris Island, now in South Carolina. Most of these early forts were earthen with timber, of various shapes, and often bastioned for small cannon (Lewis 1970). In 1626, the Dutch West India Company

From: Rose, E. P. F., Ehlen, J. & Lawrence, U. L. (eds) 2019. *Military Aspects of Geology: Fortification, Excavation and Terrain Evaluation*. Geological Society, London, Special Publications, **473**, 61–82.
First published online May 10, 2018, https://doi.org/10.1144/SP473.10

Fig. 1. Map of the east and Gulf of Mexico coasts of the USA showing the positions of key localities and simplified bedrock geology.

erected Fort Amsterdam at the Battery in lower Manhattan (Burrows & Wallace 1999). In 1638, Willem Kieft, Director of New Netherland, noted that Fort Amsterdam was 'totally and wholly in a ruinous condition so that people could go in and out of said fort on all sides' (Burrows & Wallace 1999, p. 31).

A few of these colonial forts were much more substantially built. The first constructed to defend Boston in the English Massachusetts Bay Colony was begun on Castle Island in 1634 as an earthwork with three cannon. By 1703, the earlier forts on Castle Island had been replaced by a brick, four-bastioned fort armed with 72 cannon (Anon 2016*a*). The Spanish founded St Augustine, Florida, in 1565. To defend their settlement, wooden forts were built starting in 1565 and continuing until 1675. This changed with the foundation of Castillo de San Marcos in 1672 (Bearss & Paige 1983). It was constructed from the Pleistocene Anastasia Formation, a coquina limestone locally quarried on Anastasia Island (Schweitzer 2015*a*). The walls were 3.7 m thick on the landwards side and 5.8 m thick on the seawards front. Their strength was tested in 1702 when a British force from Charleston captured St Augustine and bombarded the fort. Cannon balls bounced off the walls or embedded themselves several inches into the coquina (Anon 1999). The fort continued in use with British occupation from 1763 until 1783, when the British ceded East Florida to Spain. Florida became part of the USA in 1821 and in 1826 the fort was renamed Fort Marion (Bearss & Paige 1983).

During the American Revolutionary War, a conflict from 1775 to 1783 between Great Britain and its 13 colonies otherwise known as the American War of Independence, most of the older colonial forts that were still in existence played a role, but there were no significant advances in fort design. After the Revolution, the older forts fell into disrepair and it was realized that the new country needed to have effective coastal defences to deter the former colonial powers from any attempts to re-impose their rule.

In 1794, the first Federal authorization for new coastal defence was passed by the US Congress and signed into law by the first President of the

United States, George Washington. This led to the First System of fortification from 1794 to 1804. First System forts generally are characterized as being open works with earth parapets. Towards the end of this period, a number of them were strengthened with facing revetments of stone. At first, most were armed with older, smoothbore cannon left over from the Revolution. Soon, newer iron 32- and 24-pounders were employed (Lewis 1970). Examples of this First System include Fort Jay and Fort McHenry. Fort Jay on Governors Island in New York City's Upper Bay was begun in 1794 as a square earthen fort with four bastions, and by 1797 it combined brick and earth (Smith 1923). From 1799 to 1805, Fort McHenry was built to defend Baltimore Harbor on Whetstone Point, site of the earlier Fort Whetstone. It was built as a five-sided earth and masonry fort with five corner bastions, resembling a five-pointed star (Chartrand 2012).

As threats of another war with Britain loomed in 1807, renewed attention was paid to coastal fortifications, resulting in the Second System of fortification built from 1808 to 1816. Some of the features of the Second System were similar to the later modifications of the First System, such as masonry-faced earthen forts, but newer components included all-masonry star-shaped, circular and elliptical-shaped forts with open batteries. These Second System forts represent the first large-scale construction effort planned and carried out by military engineers trained at the US Military Academy established in 1802 at West Point, New York. Several excellent examples of these innovative designs were built in New York City, including Castle Clinton, Fort Columbus, Castle Williams and Fort Wood (Lewis 1970).

Except for the walled counterscarp, the gate, sally port, magazine and two barracks, much of the earlier Fort Jay on Governors Island was demolished in 1806 to make way for a new, larger, five-sided fort with granite and sandstone walls. It was renamed Fort Columbus and is considered to be an early Second System fort (Smith 1923; Chartrand 2012). Built from 1810 to 1811, Fort Wood on Bedloe's Island (now re-named Liberty Island) was a non-bastioned 11-point star-shaped fort made of granite. It later became the base for the Statue of Liberty. The two circular forts defending New York were especially innovative. Colonel Jonathan Williams, US Army Chief of Engineers, and the first superintendent of the US Military Academy, designed them. Built over the period 1807–12, Castle Williams on Governors Island was the first casemated battery (a battery partially enclosed within a thick protective masonry vault) in the USA, setting the stage for the design of Third System forts. Castle Clinton on the Battery in lower Manhattan, begun in 1808 and originally named the West Battery, was similar (Chartrand 2012).

After the War of 1812, fought between the UK and the USA and their respective allies, which ended in 1815, it was apparent that the USA needed an even stronger, more uniform coastal defensive system. The strong fortifications defending New York City had deterred any British plans for invasion there. However, with the notable exception of Fort McHenry in Baltimore Harbor, many of the earlier coastal forts had proved ineffective against British attacks. In 1816, the US Congress therefore authorized the creation of a Fortifications Board with the responsibility for designating fortification sites and overseeing their designs. The French military engineer, Simon Bernard, headed it with other members: Joseph Swift, William McRee and Joseph Totten. Swift, the first graduate of the US Military Academy in 1802, was Chief of Engineers of the US Army, having designed some of New York City's fortifications. McRee and Totten were both 1805 graduates of the US Military Academy. McRee had surveyed sites for coastal fortifications in the southern USA as a member of the Army Corps of Engineers, and Totten had assisted in the construction of Castle Williams and Clinton in New York City. Totten became US Army Chief of Engineers in 1838, holding that position for almost 25 years. The Third System of fortification is largely his legacy (Weaver 2001). Swift and McRee soon resented the fact that a Frenchman headed the board and they both resigned in 1818, leaving Bernard and Totten. Some of the distinguishing characteristics of Third System fortifications include casemated batteries with concentrated armament to produce enormous firepower, and massive vertical-walled brick and stone forts frequently built in a truncated hexagon shape (Weaver 2001).

The Fortifications Board identified nearly 200 sites for potential fortification, but only 42 Third System forts were actually built, at locations extending from the state of Maine on the northern Atlantic coast south to Louisiana on the Gulf of Mexico. Fifteen of these forts are described below, from north to south, to illustrate the principal variations in geography, geology and design. Forts of Boston Harbor in Massachusetts are discussed first, followed by the defences of New York City, and, subsequently, on the SE Atlantic coast, Fort Pulaski in Georgia and Fort Clinch in Florida. Attention then turns to the Gulf coast, to the defences of Pensacola Bay, Mobile Bay and Ship Island.

Defences of Boston, Massachusetts

Three Third System forts were constructed to defend Boston Harbor (Fig. 1) on islands that are drowned glacial drumlins: Fort Independence on Castle Island, Fort Winthrop on Governors Island and

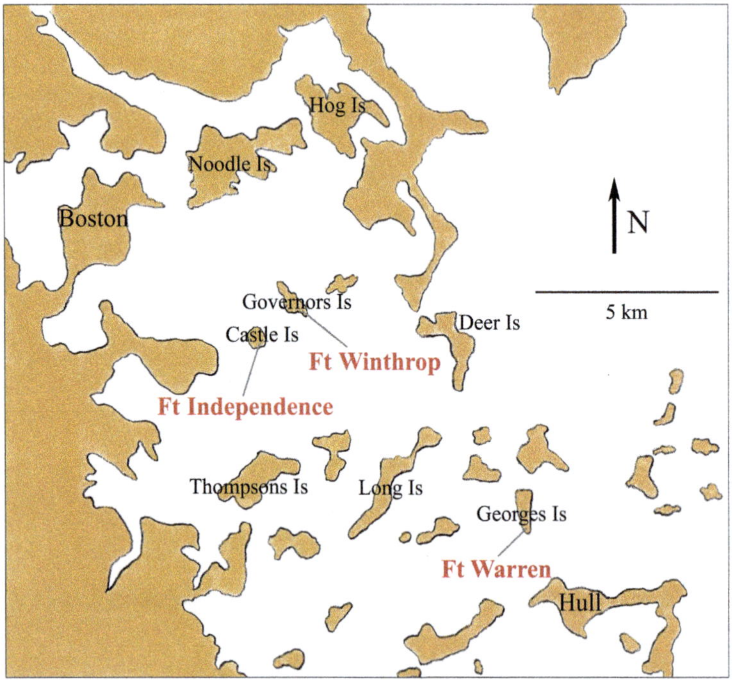

Fig. 2. Map of Boston Harbor showing major islands and positions of Third System forts.

Fort Warren on Georges Island (Fig. 2). The sites lie within a field of approximately 200 drumlins: the only one in the USA that occurs in a coastal setting. It has figured prominently in the nation's military history: the Revolutionary War battlefields of Breed's Hill and Bunker Hill are drumlins in the Charlestown sector of Boston (Newman *et al.* 1993).

Beneath the glacial sedimentary landforms, the bedrock of the Boston area consists of intrusive and volcanic igneous rocks, the Roxbury Conglomerate, and the Cambridge Argillite, all late Precambrian in age. They were part of the early Paleozoic Avalonian microcontinental terrane, and were folded and faulted when it collided with Laurentia in the mid–late Paleozoic Acadian and Alleghenian orogenies, forming the continent of Euramerica. The Cambridge Argillite underlies most of Boston Harbor, a relatively weak rock whose erosion has created the harbour itself and the low hills beside it (Kopera 2011).

Deposits within the Boston Basin represent two major glacial advances during the Pleistocene. The older deposits are dated as Illinoian and the younger as late Wisconsinan, the most recent glacial period of the North American ice-sheet complex (Table 1). The drumlins were created from Illinoian till by the subsequent Wisconsinan glacial advance. Newman *et al.* (1993) have suggested that the drumlins formed by differential erosion of partially or completely frozen sediments and that Wisconsinan till was deposited on the sculpted drumlins during the final glacial retreat. The Boston Basin drumlins generally

Table 1. *Correlation chart of Holocene and late Pleistocene glacial stages*

Glacial/Interglacial	North America	Northern Europe	Great Britain	Approximate age (ka)
Interglacial			Flandrian	<11
Glacial	Wisconsinan	Weichselian	Devensian	115–11
Interglacial	Sangamonian	Eemian	Ipswichian	130–115
Glacial	Illinoian	Saalian	Wolstonian	240–130
Interglacial and glacial	Pre-Illinoian A	Saalian	Wolstonian	380–240
Interglacial	Pre-Illinoian A	Holsteinian	Hoxnian	420–380

Adapted from Gibbard *et al.* (2011).

trend from WNW to ESE (Newman *et al.* 1993). Post-glacial rebound in the late Wisconsinan and early Holocene, concurrent with, and followed by, rapid sea-level rise in the late Wisconsinan and Holocene, created the drumlin islands of Boston Harbor (Newman *et al.* 1993). The harbour area was submerged approximately 5630 years BP and subsequent longshore processes have modified the terrain, creating tombolos and spits and partially eroding some of the drumlins (Luedtke & Rosen 1993).

The Boston Harbor islands have been occupied, for many purposes, over many centuries. Indeed, Native Americans occupied Long Island, the largest of the islands, at least as early as the Early Archaic archaeological period (9000–8000 years BP). With European settlement in the seventeenth century, the islands began to be farmed. It was soon realized that they would serve as important sites to defend the city of Boston and naval installations from attack. As such, the first fortifications were started on Castle Island in 1634. Many of the islands have had forts established on them that were in use from colonial times until the Cold War of 1947–91 (Luedtke & Rosen 1993).

Fort Independence, built on Castle Island, replaced a First System fort of the same name (Fig. 2). The first fort was erected there in 1634 by Massachusetts Bay Colony Governor Thomas Dudley, and Castle Island consequently became the oldest continuously occupied fortification site in British North America. A four-bastion stone fort was built after the earlier fort burned down in 1673. It was expanded from 1701 to 1703 and named Castle William in honour of William III, King of England from 1689 to 1702. When British forces evacuated Boston in March 1776, they blew up the fort. In 1801 the Federal government began construction of a new fort there, to be known as Fort Independence. It stood through the war of 1812 and was not attacked by the British (Roberts 1988).

The Third System Fort Independence was built on the foundations of the previous First System fort (Weaver 2001). It was begun in 1833, but construction was delayed by a lack of funding until 1836; it was essentially complete by 1848 (Parkman 1978). Colonel Sylvanus Thayer, previously Superintendent of the US Military Academy, completely reconstructed the old fort as a five-sided pentagon with bastions at each corner (Fig. 3) (Weaver 2001). Its former height was doubled, casemate interiors were added, its perimeter was expanded and the crumbling brick was replaced with Late Ordovician Cape Ann Granite from the quarries at Rockport, Massachusetts, about 60 km to the NE of Boston (Parkman 1978).

During the American Civil War (1861–65), when the Confederate states of the south attempted to separate from the Federal (Union) states of the north, several regiments of Massachusetts volunteer militia trained and were stationed at Fort Independence (Kales 2007). After the Civil War, the fort was maintained and updated, but in 1880 it was abandoned. In 1898, during the Spanish–American War, it served as a storage site for laying harbour mines and was garrisoned by Massachusetts volunteer militia. Before World War I, Castle Island became the site of Coast Artillery Headquarters Radio Station (Butler 2001). During both world wars, the island served as an important observation post (Humphrey *et al.* 1993); it was connected to the mainland by a concrete causeway in 1928 (Johnson 2005).

The Third System Fort Winthrop was constructed on Governors Island (Fig. 2), a typical drumlin, with a lee slope to the ESE. It replaced the Second System Fort Warren begun in 1808 (and was re-named after Massachusetts Governor John Winthrop in 1834): an earthen and masonry star fort. The Third System fort consisted of four batteries surrounding a defensive tower. The tower comprised three levels of casemates surmounted by a parapet with emplacements for 16 guns (Weaver 2001) built from Cape Ann Granite (Anon 1878). By 1863, during the Civil War, Fort Winthrop was armed with 25 Rodman guns, 11 pieces of varying calibres, and was occupied by Massachusetts state militia and

Fig. 3. Fort Independence from the water (cf. Fig. 2).

volunteers (Anon undated). In the 1870s it was strengthened but soon thereafter placed in caretaker status. Units of Massachusetts volunteer militia were garrisoned there during the Spanish–American War of 1898. A magazine exploded in 1902 (Anon 1902; Schorow 2008) and the fort was abandoned in 1905 (Butler 2001), and subsequently demolished in 1941 (Roberts 1988). From 1943 to 1946, expansion of Boston's Logan International Airport connected Governors Island to the mainland and the site of the fort is now beneath one of Logan's runways (Seasholes 2003).

Georges Island, the site of the Third System Fort Warren (Fig. 2), originally comprised two drumlins: one 14.6 m high and the other 19.5 m. However, before fort construction began in 1837, the surface was levelled. The fort's ramparts currently rise to 15.25 m above sea level (Anon 2015*a*). Fort Warren was the fifth greatest in size of all Third System forts and took its name from Fort Warren on Governors Island when that was re-named Fort Winthrop: Major General Joseph Warren being a physician and President of the Provincial Massachusetts Congress who was killed in the Revolutionary War Battle of Bunker Hill in June 1775. Fort Warren's position on Georges Island defended two major channels of Boston's outer harbour leading into the port of Boston.

Fort Warren was not completed until the beginning of the Civil War in 1861. It had a modified pentagonal shape, with one axis shorter than the other in order to best adapt to the island's shape (Fig. 4) (Weaver 2001). Its construction illustrates some of the most detailed stonework of any of the Third System forts, including spiral staircases, compound arches and door lintels (Weaver 2001). Like Fort Independence and Fort Winthrop, it was built predominately of Cape Ann Granite from Rockport (Anon 1878). The fort contained casemates two storeys tall, topped by a barbette tier (a raised platform) and was designed to mount over 200 guns, many of which were on the barbette. The fort also features a demilune (half-moon battery) protecting the north sally port and a large water battery enclosed within a ravelin (an angled fortification separate from the main fort) pointed at the convergence of the ship channels, effectively covering those approaches (Weaver 2001).

Beginning in October 1861, Fort Warren served as a prison for over 2000 Confederate officers, enlisted men and government officials. James Mason and John Slidell were imprisoned at Fort Warren after they were captured aboard the British mail packet RMS *Trent* in November 1861. Mason and Slidell were Confederate diplomats heading to Britain and France to lobby for recognition of the

Fig. 4. Fort Warren from the air (cf. Fig. 2).

Confederacy by those countries. Their arrest led to tensions between the USA and Britain and they were subsequently released by the Lincoln administration in January 1862. Other Confederate government officials imprisoned at Fort Warren included Vice-President Alexander H. Stephens and Postmaster General John Reagan (Anon 2015*b*).

Fort Warren continued to be garrisoned after the Civil War. In the 1890s, it was upgraded with new, dispersed batteries for more modern breech-loading rifled guns. These batteries were constructed with reinforced concrete and protected by sloped earthworks under the Endicott Program (1890–1905), so called because Secretary of War William C. Endicott was Chair of the Board of Fortifications that recommended the upgrades. In World War II, it was a control centre for Boston Harbor's south mine field. In 1947, after the war, the fort was decommissioned (Anon 2015*b*).

Defences of New York City

The coastline of New York City, at the SE tip of New York State and the most populous city in the USA (Fig. 1), is formed by the Upper and Lower bays, separated at the Narrows (Fig. 5). The Upper Bay, also known as New York Harbor, lies SW of the island of Manhattan between New Jersey on the west and Brooklyn on Long Island on the east. The Lower Bay widens out between Staten Island and Brooklyn, and enters the New York Bight of the Atlantic Ocean between Sandy Hook in New Jersey on the west and Rockaway Point on Long Island on the east.

The early defences of the city were situated on and close to Manhattan Island within the Upper Bay. New York City had been defended since the Dutch West India Company erected Fort Amsterdam at the Battery in lower Manhattan in 1626 (Burrows & Wallace 1999). Later, additional forts were constructed on Governors Island (Fig. 5), Ellis Island and Bedloe's Island (now Liberty Island) in the Upper Bay (Lewis 1970; Weaver 2001). Because of the nature of the approaches to New York City, its defences have been extensive. The wide opening of the Lower Bay into the Atlantic Ocean made this approach difficult to defend (Fig. 5). Sandy Hook was used to defend the deepest channel into the

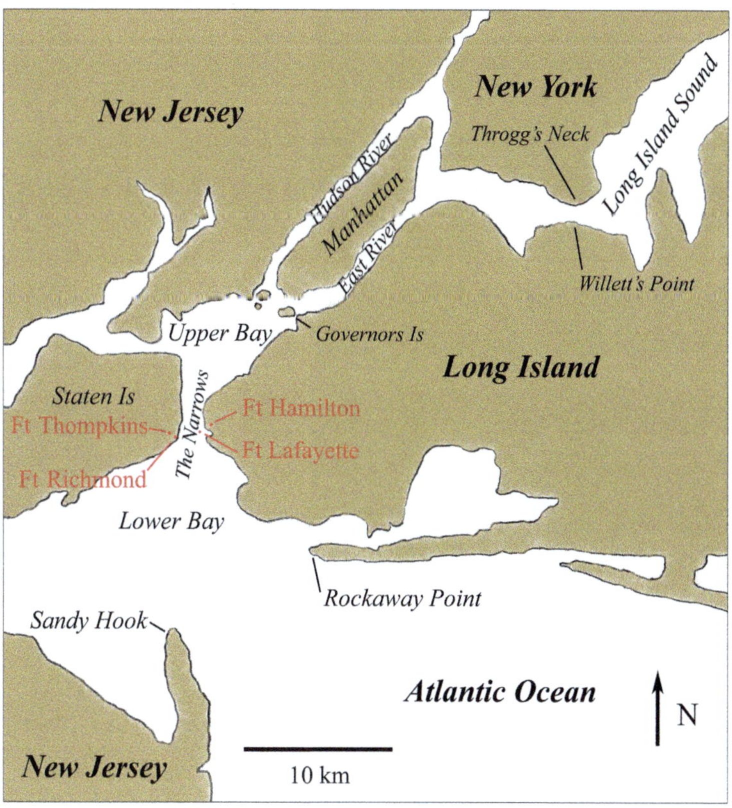

Fig. 5. Map of the New York City area showing key localities and positions of Third System forts.

entrance to the Lower Bay as early as 1813 when Fort Gates was constructed. In 1859, the army began construction of a new fort in this location, but it was abandoned after the Civil War and was never completed. New batteries were constructed in the 1890s and the fort was designated Fort Hancock; it was in use for various purposes until 1974 (Anon 2016*b*). Rockaway Point on Long Island was not fortified until 1917 when Fort Tilden was established, armed with long-range 16-inch guns for coastal defence. After World War II, most of Fort Tilden's guns were removed and it became part of the US air defence system until 1954 (Williford & Polaski 2004).

Another approach to New York City is from Long Island Sound through the East River that runs between Manhattan and Brooklyn. As such, forts were erected to defend this strategic position where the Sound and East River meet (Fig. 5). Fort Schuyler on Throgg's Neck in the Bronx was begun in 1833 and completed in 1856 (Weaver 2001). The fort opposite Fort Schuyler on Willett's Point in Queens on Long Island was started in 1862 but was never completed. Later, the army installation there was known as Fort Totten (Weaver 2001).

Located at the Narrows, the Third System forts Lafayette, Hamilton, Richmond and Tompkins protected the southern approach to the Upper Bay (Fig. 5). These forts are associated with the Late Wisconsinan Pleistocene Harbor Hill Moraine (Fig. 6). Fort Hamilton and Fort Tompkins lie on the moraine, whereas Fort Lafayette and Fort Richmond are water batteries, lying below the moraine (Fig. 7). The geology of the area around the Narrows consists primarily of Cretaceous sediments overlain by Pleistocene glacial deposits. Current evidence and interpretation indicates deposition of the Harbor Hill Moraine approximately 21 ka BP at the end of

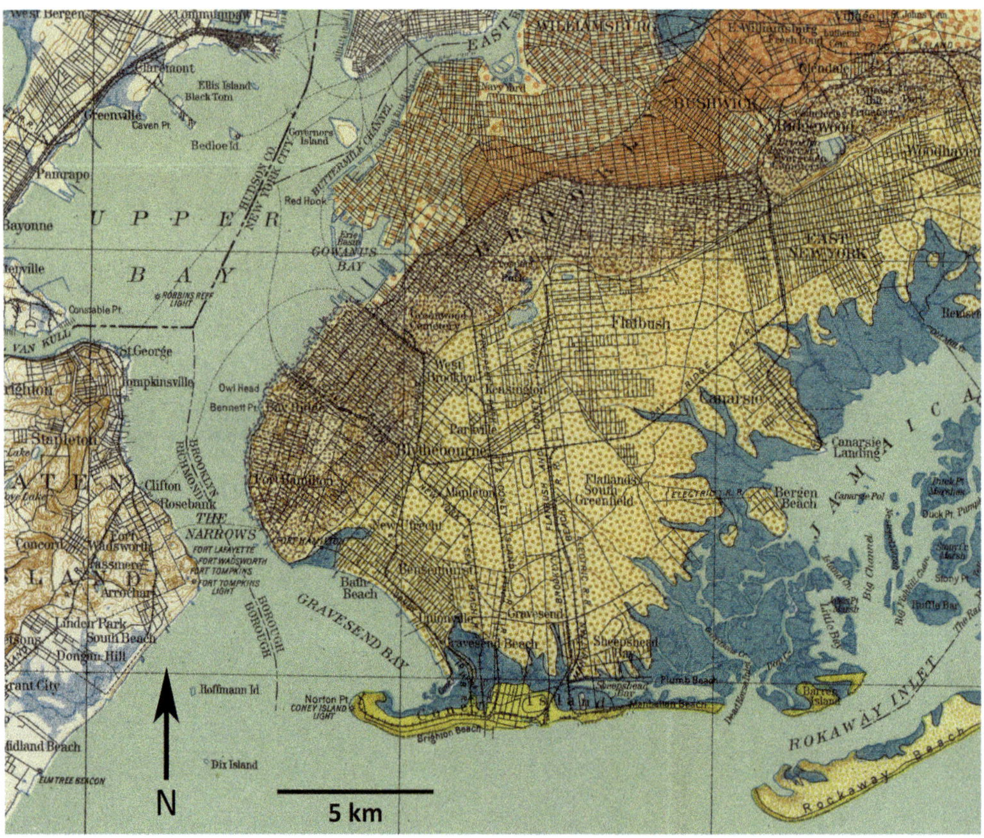

Fig. 6. Geological map of the Narrows, New York City (cf. Fig. 5). The Harbor Hill moraine (coloured brown) extends from the Narrows to the NE, across the word 'BROOKLYN'. Outwash and ground moraine occur on either side of the Harbor Hill moraine (variously coloured buff and red-brown to the NW, and yellow to the SE of the Harbor Hill). Harbor Hill moraine (indicated by brown topographical contours) extends to the SW across the Narrows on Staten Island. From Fuller (1914).

Fig. 7. Aerial view of the Verrazano-Narrows Bridge looking NW at the Narrows, with Fort Thompkins above and Fort Richmond below, on Staten Island (cf. Fig. 5).

the last North American Pleistocene glacial advance (Moss 2013). The Harbor Hill Moraine can be traced from northern New Jersey, across Staten Island and across Long Island. Of the two terminal moraines on Long Island, the Ronkonkoma and the Harbor Hill, the Harbor Hill is the younger and lies to the NW of the Ronkonkoma (Moss 2013).

The Narrows represents a breach in the Harbor Hill Moraine. Several proglacial lakes lay north of this moraine, notably Glacial Lake Flushing to the NE, Glacial Lake Passaic and Glacial Lake Hackensack to the NW, and Glacial Lake Albany almost due north. North of Glacial Lake Albany was Glacial Lake Vermont, the geographical predecessor of present-day Lake Champlain. Lying to the west of Glacial Lake Vermont was Glacial Lake Iroquois, separated from it by the Adirondack Upland. Lake Iroquois occupied the basin southwards of the Laurentide Ice Sheet, west of the Adirondack Upland and north of the Appalachian Plateau. Present-day Lake Ontario occupies the central area of this glacial lake and is about one-third of its size (Donnelly *et al.* 2005; Franzi 2016). As the Laurentide Ice Sheet (which covered much of Canada and the NE USA throughout the Quaternary) retreated, the ice dam between Lake Iroquois and Lake Vermont broke. Dated at approximately 13.35 ka BP (Donnelly *et al.* 2005), the resulting flood dropped the level of Lake Iroquois by about 120 m, moved down Lake Vermont and Lake Albany, burst through the Harbor Hill Moraine at the Narrows, and cut through the moraine and underlying Cretaceous sediment at the Narrows to a depth exceeding 100 m. Concurrently, the flood eroded the coastal plain to the south of the Narrows, forming the 150 km-long Hudson Shelf Valley that lies on the present-day continental shelf (Thieler *et al.* 2007).

Fort Lafayette was situated on a shoal in the Narrows up against the Brooklyn shore. This shoal overlies a ledge known as Hendricks's Reef, composed of Hartland Formation, a Cambrian–Ordovician metamorphic rock unit comprising schist, gneiss and amphibolite with pegmatite intrusions (Nikolaou 2004). This competent metamorphic rock was able to support the weight of the masonry structure. The fort was begun in 1815 as a Second System fort named Fort Diamond and was renamed Fort Lafayette in 1823 to coincide with the return visit of the

Marquis de La Fayette, a hero of the Revolutionary War, to the USA in 1824. It was built of brick with sandstone from Belleville, New Jersey (about 20 km NW of New York City) and granite from Maine, some 600 km to the NE (Julien 1884; Merrill 1889). The sandstone, a reddish arkosic rock known as brownstone, was quarried from Joyce's Quarry (Shaw 1884) and belongs to the Late Triassic–Early Jurassic Passaic Formation (Schweitzer 2015*b*). Fort Lafayette was a water-level fort with three-tiered square brick masonry. Two of its four sides were angled to the channel to allow maximum firepower (Weaver 2001). During the Civil War, like a number of the coastal forts located in the north, Fort Lafayette served as a prison for Confederate prisoners of war and political prisoners. After a fire in 1868, the fort was rebuilt and used for storing munitions during both world wars (Roberts 1988). The fort was torn down in 1960 and the site was used for the eastern support of the Verrazano-Narrows Bridge connecting Staten Island with Brooklyn across the Narrows (Weaver 2001).

The Third System Fort Hamilton in Brooklyn was built on the Harbor Hill Moraine above Fort Lafayette as a dual-purpose structure: to prevent a land-based siege of Fort Lafayette that could be initiated on the high ground of the same moraine on Long Island; and to add long-range fire power from barbette-mounted guns to cover the Narrows (Weaver 2001). Construction of this fort, made of granite from Maine (Julien 1884; Merrill 1889), began in 1825 and it was garrisoned from 1831 (King 1893). Built as a trapezoidal fort with a seacoast front, it had a tier of casemated guns with barbette guns on the top (Weaver 2001). During the Civil War, Fort Hamilton served as a prison. In the 1890s the seacoast front was removed to provide Endicott gun battery emplacements for disappearing guns (guns mounted on a carriage that enabled them to be mechanically rotated or lowered out of enemy view when not in their firing position). The fort was active through both world wars, serving the Coast Artillery, serving as a mobilization centre and storing ammunition from ships entering the Brooklyn Navy Yard (Williford & Polaski 2004).

On the other side of the Narrows on Staten Island (Fig. 5), Third System Fort Richmond and Fort Tompkins were built on the sites of earlier dilapidated Second System forts of the same names. Fort Richmond was built on a point jutting out into the channel at water level (Fig. 7). Utilizing granite from Dix Island, Maine (Julien 1884; Merrill 1889), the fort's construction began in 1847 and was completed in 1864 (Weaver 2001). The next year it was given the name Fort Wadsworth to honour Union Army Brevet Major General James Wadsworth, mortally wounded at the Civil War Battle of the Wilderness in Virginia in 1864. Fort Richmond was built as a four-sided, truncated hexagon with four tiers of guns: three as casemates and the fourth a barbette on top. The three sides that faced the Narrows were about 87 m long and the landwards side with a gorge (ditch) in front was about 137 m long. The fort's 116 guns were able to command a full field of fire out into the Narrows. Landwards defences were minimal, with Fort Tompkins being built on the high ground behind to provide for the defence of Fort Richmond (Weaver 2001). Beginning in the 1890s, Endicott-era disappearing guns were added (Anon 2015*c*).

Third System Fort Tompkins was situated above Fort Richmond on the Harbor Hill Moraine to provide land defence for the latter, as noted above (Fig. 7). Again built of granite from Maine (Julien 1884; Merrill 1889), construction began in 1847 (Weaver 2001), but most of the fort was built between 1859 and 1876 (Anon 2015*c*). It was built as an irregular five-sided structure with two levels of casemates, but without embrasures (an opening in a battlement through which rifles and/or cannon are fired). The longest side faced the Narrows with seacoast guns mounted on the barbette. The other four sides were more subject to land attack and, as such, their defences included counterscarp walls, loopholes for small-arms fire and howitzer positions. In addition to its defensive role, Fort Tompkins provided barracks for personnel stationed at both Fort Tompkins and Fort Richmond (Weaver 2001). During the Civil War, various US regular infantry and artillery units, along with New York volunteer units, garrisoned the fort (Black 1983). In 1902, the entire military post, including Fort Tompkins, was designated Fort Wadsworth, with the original Fort Richmond renamed Battery Weed. It was named after Union Brigadier General Stephen H. Weed, who was killed at the Civil War Battle of Gettysburg in 1863 (Anon 2015*c*). Beginning in the early twentieth century and lasting through World War I, the Coast Artillery manned the guns of the entire post of Fort Wadsworth. By 1924, it became an infantry post; anti-aircraft facilities were added after World War II (Anon 2006).

Fort Pulaski on Cockspur Island

Fort Pulaski, near the mouth of the Savannah River (downstream from Savannah), Georgia (Fig. 1), was built on the sand and saltmarsh mud of Cockspur Island (Fig. 8). The substrate was found to be too unstable to support the original design of two tiers of casemates, so it was redesigned for one tier in 1829 with the fort largely completed by 1847 (Weaver 2001). The fort's primary mission was to control access to the city of Savannah through the Savannah River.

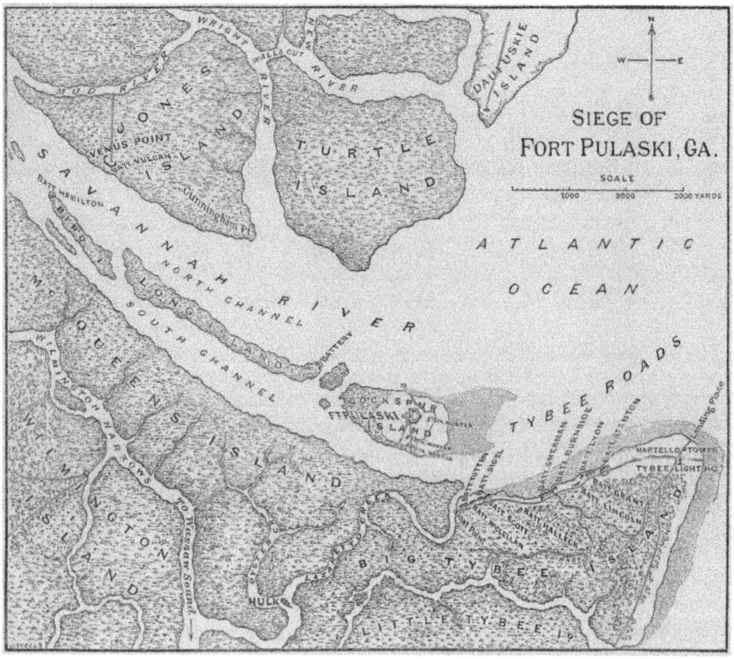

Fig. 8. Map of the mouth of the Savannah River showing Fort Pulaski, the extent of Cockspur Island and Federal battery sites on Tybee Island. From Gillmore (1887).

Cockspur Island had been an important site for fortification since English colonial times. Fort George, a palisaded log blockhouse, was begun in 1761 to prevent attacks by the Spanish. The First System Fort Greene was built in 1794 but in 1804 it was totally destroyed by a hurricane (Kaufmann & Kaufmann 2004).

Cockspur Island is formed by Holocene fluvial sediments brought to the mouth of the Savannah River. It was originally a smaller elevated area surrounded by saltmarsh about 1.6 km long and 1.2 km wide (Meader & Brinkley 2003). Prior to construction of Fort Pulaski, dredge spoil was added to increase the island's size (Howell & Alexander 2004). The island has continued to evolve through natural and anthropogenic agents. The US Army Corps of Engineers constructed jetties at the mouth of the Savannah River in the 1880s and 1890s. The last jetty, built from 1894 to 1896, extended from the NE tip of Cockspur Island outwards into the Atlantic Ocean. This modified the movement of sediment and caused the island's east side to enlarge. Dredging of the river, probably combined with the effect of jetty construction, resulted in Cockspur Island and Long Island becoming one. Continued dredging and deposition of spoils on Oyster Bed Island on the other side of the North Channel of the Savannah River from Cockspur helped protect Cockspur from erosion (Meader & Brinkley 2003; Alexander 2008).

Fort Pulaski was constructed as a five-sided, truncated hexagon, similar to Fort Sumter in Charleston Harbor where in 1861 the first shots were fired in the Civil War. It was orientated such that two sides faced the river, two sides faced the river and marsh, and the longer (gorge) wall faced the longer part of the island. A moat completely surrounded the fort (Fig. 9). Built approximately 1 km from the river, it was designed to withstand cannon fire with a range of less than 1 km (Weaver 2001). The fort was constructed primarily of local bricks made at the Hermitage Plantation near Savannah, along with harder ones for the embrasures, arches and walls facing the parade ground that were from Baltimore, Maryland, and Alexandria, Virginia. Granite shipped from New York and sandstone (brownstone) from Connecticut over 1000 km to the NNE were also used in the construction (Lattimore 1954).

On the eve of the Civil War, Fort Pulaski was not garrisoned by Federal troops. On 3 January 1861, prior to the secession of Georgia from the Union in March 1861, Georgia troops took possession of the fort. The state forces built up the fort's armament by remounting 20 of the 32-pounder naval cannon and procuring other guns. The peak Confederate strength at Fort Pulaski was 48 guns. By the summer

Fig. 9. Fort Pulaski showing its moat and Civil War damage inflicted by long-range rifled cannon.

of 1861, the Federal plan for defeating the Confederacy included a naval blockade of southern ports and the capture of the coastal islands. As such, Confederate positions on nearby Tybee Island were abandoned on 10 November 1861 (Lattimore 1954). Confederate General Robert E. Lee visited Fort Pulaski in late November 1861 as commander of the Military District of South Carolina, Georgia and Florida. Standing on the parapet with the Confederate officer in charge, Colonel Charles H. Olmstead, Lee pointed to the shore of Tybee Island, approximately 1.55 km away, and remarked, 'Colonel, they will make it pretty warm for you here with shells, but they cannot breach your walls at that distance' (Lattimore 1954, p. 28).

However, in February 1862, Union General Sherman's Chief Engineer, Captain Quincy A. Gillmore, took command of the Federal troops on Tybee Island and was ordered to prepare for the bombardment of Fort Pulaski. Gillmore believed the newly developed, longer-range rifled Parrott and James cannon would be able to inflict damage on Fort Pulaski that smoothbore guns could not. Gillmore's forces erected 11 batteries containing 36 pieces of artillery, including 10- and 13-inch mortars, 8- and 10-inch smoothbore Columbiads, 30-pounder Parrott rifles, and 48-, 64- and 84-pounder James rifles. The bombardment opened on the morning of 10 April 1862 and Olmstead surrendered the fort at 2:30 in the afternoon of 11 April after rifled cannon had opened holes in the walls of the fort that exposed the north magazine where 40 000 lbs of black powder were stored. The captured garrison was imprisoned on Governors Island in New York City (Lattimore 1954).

The potential destruction of Fort Pulaski by rifled cannon fire rendered the whole premise of coastal defence by masonry forts obsolete. Overnight, the principle of Third System of fortification was thrown into question and its usefulness was restricted. After capture by Federal forces, Fort Pulaski became a Federal prison. Later, the fort was largely neglected because other sites, such as Tybee Island that directly faced the coast, were more suited to modern long-range artillery. To protect the North Channel of the Savannah River during the Spanish–American War in 1898, a battery was built on the north shore of Cockspur Island (Meader & Brinkley 2003).

Fort Clinch on Amelia Island

Construction of the Third System Fort Clinch on the northern end of Amelia Island, Florida, *c.* 200 km to the south of Savannah (Fig. 1), began in 1847. Amelia Island is a barrier island that lies south of Cumberland Sound between Georgia and Florida (Fig. 10). Like most of the Georgia barrier islands that lie to the north, the core of Amelia Island consists of sediments related to the late Pleistocene Silver Bluff barrier island system formed when sea level was slightly higher than it is today (Henry 1971; Henry *et al.* 1996). It has proved difficult to date the Silver Bluff sediments: current best estimates are 40–50 ka BP (Henry 2005). These barrier islands originated when sea level rose after the last Pleistocene glaciation, flooding the lower areas behind the earlier deposited beach ridges (Hoyt 1967). Amelia Island has been overlain and extended by Holocene sediments (White 1970), and Fort Clinch was constructed on those accreted at its northern end.

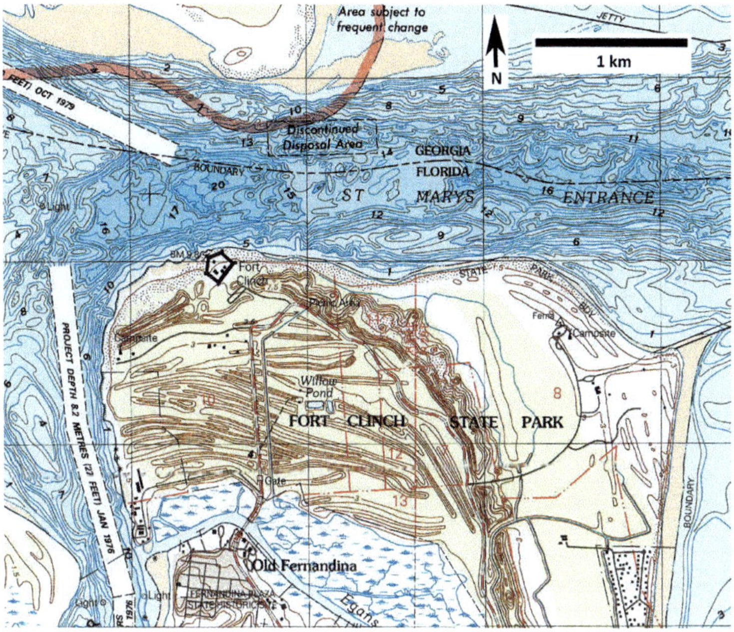

Fig. 10. Portion of the Fernandina Beach topographical map showing Fort Clinch and Holocene beach ridges on the northern end of Amelia Island, Florida. From USGS (1994).

Fort Clinch defended the harbour at the mouth of the St Marys River and the port of Fernandina. It was built as a five-sided, truncated hexagon with a detached scarp and a single barbette tier of 70 guns. The detached scarp consisted of a brick masonry Carnot (loop-holed) wall and a chemin de ronde (raised protected walkway) behind it, providing a primary line of defence for a land-based attack. Only two American forts shared this feature, Fort Clinch and Fort Gaines in Mobile Bay, Alabama (see below). Fort Clinch was predominantly built of bricks like Fort Pulaski, but from a variety of sources. Brown bricks came from brickyards in Savannah and St Marys, whereas red bricks were procured from New York, Pennsylvania and Maine. Additional material included granite from New Hampshire (some 2000 km to the NNW) for staircases, gun platforms, windows, sills and lentils. Bluestone from Pennsylvania and New York was used for flooring and coverings for the freshwater cistern (Ofeldt 2017). This stone, a fine-grained feldspathic sandstone from Upper Devonian Catskill facies units (Albanese & Kelly 1991), was historically quarried in the Hudson Valley, in southern and central New York state, and in northern Pennsylvania.

At the beginning of the Civil War, Fort Clinch had not yet been completed, was unoccupied and had no guns mounted, making it easy to be seized by Confederate forces. As Federal forces began occupying neighbouring Georgia barrier islands, Confederate General Robert E. Lee ordered the evacuation of the fort and Amelia Island in early 1862. Union control of the fort was re-established in March 1862. It was briefly garrisoned during the Spanish–American War of 1898, but was again abandoned after that threat was over (Cox 2013).

Defences of Pensacola Bay

Pensacola Bay, Florida, on the Gulf Coast (Figs 1 & 11) has been defended since it was Spanish territory. In 1698 Fort San Carlos de Austria was constructed, near the later Fort San Carlos de Barrancas (now Fort Barrancas), to protect the Spanish settlement of Pensacola. The French captured the fort and town in 1719 during the 1717–20 War of the Quadruple Alliance, in which an alliance of Britain, France, Austria and the Dutch Republic fought against King Phillip V of Spain's attempt to retake territory in Italy and, perhaps, claim the French throne. Terms of the treaty that ended the war gave Pensacola back to the Spanish in 1722. All of Florida was ceded to Britain under the Treaty of Paris in 1763 that ended the Seven Years War (a conflict that spanned five continents), and Pensacola became the capital of the Province of West Florida. The British erected various forts during their tenure, including the Royal Navy Redoubt erected on the

Fig. 11. Map of Pensacola Bay showing Fort McRee, Fort Pickens and Fort Barrancas. The Water Battery (Baterio de San Antonio) lies on the shore directly south of Fort Barrancas. From Gilman (1887).

high ground of the Red Cliffs, approximately where the present-day Fort Barrancas is situated. In 1781, Spain captured Pensacola while allied with France during the American Revolutionary War. All of Florida was returned to Spain under the articles of the 1783 Treaty of Paris that gave the USA its independence from Britain (Coleman & Coleman 1982).

In 1797, the Spanish built Bateria de San Antonio, a water battery, below Fort Barrancas, facing Pensacola Bay. This brick and masonry structure was designed to mount cannon that would fire at water level. The Royal Navy Redoubt was rebuilt as San Carlos de Barrancas in 1798. Like the Royal Navy Redoubt, the log and sand San Carlos de Barrancas was sited on the Red Cliffs, known to the Spanish as the Barrancas because of the steepness of the cliff face.

Spain ceded Florida to the USA in 1819, and in 1821 Pensacola and its forts were transferred to the USA (Coleman & Coleman 1982). When in 1825 the US Navy made Pensacola Bay the site of the Gulf Coast Navy Yard, further fortifications were needed. As such, four forts were constructed to defend the bay. Fort Pickens on Santa Rosa Island and Fort McRee on Foster's Bank (since connected to Perdido Key as a consequence of longshore sedimentary drift) defended the entrance to the bay (Fig. 11). Fort Barrancas and the Advanced Redoubt on the peninsula just to the west of the Navy Yard defended the bay itself and the land approaches.

The Third System Fort Pickens and Fort McRee were built on Holocene barrier islands. Core data provide evidence that Santa Rosa Island and Perdido Key originated through shoal-bar aggradation in response to several longshore drift cells (Dean *et al.* 1995). Fort Pickens was the only Third System fort designed by two different engineers: plans by Joseph Totten were modified by Simon Bernard. It was begun in 1829 and finished in 1834 as a five-sided fort with two long, angled seacoast fronts, and a single tier of casemates and barbette (Fig. 12) (Weaver 2001). Fort Pickens and Fort McRee primarily used strong Pensacola brick made from high-quality local clay (Bearss 1983). The clay occurs in the Pliocene Citronelle Formation (Rupert 1993). In addition, the lime for Fort Pickens was shipped from Thomaston, Maine. Lead sheeting for the gutters, drains and for waterproofing the casemate arches came from mines in Illinois. The granite that was used for main entrances, stairs, gun carriage traverses, coping, fireplace mantels, sills and lentils came from New York, and was shipped from the wharves at Sing Sing in Westchester County (Bearss 1983). Because the granite was shipped from the Sing Sing wharves, it is very likely that it was locally quarried. The Proterozoic Fordham Gneiss underlies

Fig. 12. Fort Pickens (cf. Fig. 11) from ground level.

this area (Reynolds 1988) so the 'granite' was probably the Fordham Gneiss. Fort McRee was begun in 1838 and completed in 1840 on Foster's Bank (Coleman & Coleman 1982). Because Foster's Bank was so narrow, the fort was built as a long rectangle with rounded ends and was angled in the middle (Weaver 2001). It had two tiers of casemates with a barbette and was designed to mount 128 guns (Coleman & Coleman 1982).

In 1839 work began on the Third System Fort Barrancas, built on the site of the Spanish San Carlos de Barrancas. It was completed in 1844. Because it was designed partially for defence of the bay and partially for defence from attack overland, Fort Barrancas was four-sided, with two sides directed towards the bay and the other two directed landwards. This brick and masonry fort mounted 14 barbette guns for seacoast defence, along with rifle galleries and howitzer positions to defend the fort from landwards attack (Weaver 2001). Bateria de San Antonio was modernized at the same time and renamed the Water Battery. Because Fort Barrancas and the Water Battery were vulnerable to overland attack, the Advanced Redoubt ('Redoubt' in Fig. 11) was begun in 1845 landwards of Fort Barrancas on high ground on the north side of the peninsula. It was built as a brick and masonry fort with granite from New York (Ogden 1997). Fort Barrancas and the Advanced Redoubt lie on a cliff-bounded terrace between 9 and 11 m above sea level. MacNeil (1950) correlated this with the Late Pleistocene Pamlico marine terrace of the SE Atlantic coast. He found this surface well developed at approximately 8–11 m above sea level in NW Florida (Anon 1977). Henry *et al.* (1996) tentatively dated the Pamlico in Georgia to the mid-Sangamonian: the last North American interglacial (Table 1). More recent work by Otvos (2001) considers the use of the term Pamlico for deposits in NW Florida to be invalid because of the difficulty of correlating between the SE Atlantic coast and the Gulf coast. The sediments which form the high ground beneath Fort Barrancas and the Advanced Redoubt have been ascribed to the Prairie Formation, defined in southern Louisiana, of Sangamonian age (Otvos 2001).

When Florida seceded from the Union in January 1861, just before the start of the Civil War, Federal troops in Pensacola abandoned the Navy Yard, along with Fort Barrancas and Fort McRee. They withdrew to Fort Pickens, deciding that Santa Rosa Island, upon which the fort is located, could be defended. At the same time, Alabama state militia occupied the Navy Yard and Fort Barrancas, while Florida state militia occupied Fort McRee. Union reinforcements to Fort Pickens arrived in April and by autumn they numbered about 2000. Artillery duels between Fort Pickens and Fort McRee followed in November 1861 and January 1862. Although Fort Pickens was not severely damaged, Fort McRee was riddled by cannon fire (Coleman & Coleman 1982). The bloody battle of Shiloh in Tennessee and the Confederate loss of New Orleans, both in April 1862, were followed by the Union Navy's operations off Mobile Bay in preparation for taking the city of Mobile. When it became obvious to the Confederate forces in May 1862 that their position could not be held, they were ordered to evacuate Pensacola and its forts (Bearss 1983). With the Federal occupation, most of the troops at Fort Pickens were moved to the mainland. Fort Pickens became a military prison and was garrisoned by a much smaller force. It remained garrisoned until 1867. Fort McRee was so badly damaged that it was not repaired and was almost completely destroyed by a hurricane in September 1906. Fort Barrancas maintained a few guns and stored powder for coast artillery installations until 1947 (Coleman & Coleman 1982).

Defences of Mobile Bay

The defences of Mobile Bay (Fig. 1) were considered to be of primary importance when the plan for the Third System of fortification was formulated because the Bay provided a safe anchorage on the Gulf of Mexico, it connected through an inland waterway to New Orleans and it gave access to interior rivers. The narrow opening to Mobile Bay between Mobile Point on the eastern side and

Dauphin Island on the west made it logical to erect two forts, one on either side of the entrance (Fig. 13). Fort Morgan on Mobile Point was constructed at the western end of the 21 km-long Morgan Peninsula barrier spit and Fort Gaines was constructed on the eastern end of Dauphin Island (Weaver 2001).

Geologically, the Mobile Bay area reflects coastal sedimentary processes. As Morgan Peninsula prograded westwards from longshore drift, it constricted the Bay entrance (Kopaska-Merkel & Rindsberg 2005). Dauphin Island is a Holocene barrier island that formed around a Pleistocene core; sand from the large Mobile Bay ebb-tidal delta and the east–west sand-shoal belt west of the delta facilitated growth of the island (Otvos 2001).

Fort Morgan was one of the earliest Third System forts; its construction commenced in 1819 and it was completed in 1834. Construction used bricks made locally in Mobile and granite shipped from New York (Hyatt 2017). Like a number of the early Third System forts, it was designed as a pentagon with large bastions. A brick and masonry fort, it had a single tier of casemates with a barbette for seacoast cannon (Fig. 14), along with extensive surrounding outworks for land defence (Weaver 2001).

Construction of Fort Gaines on Dauphin Island began in 1853 as a five-sided, truncated hexagon

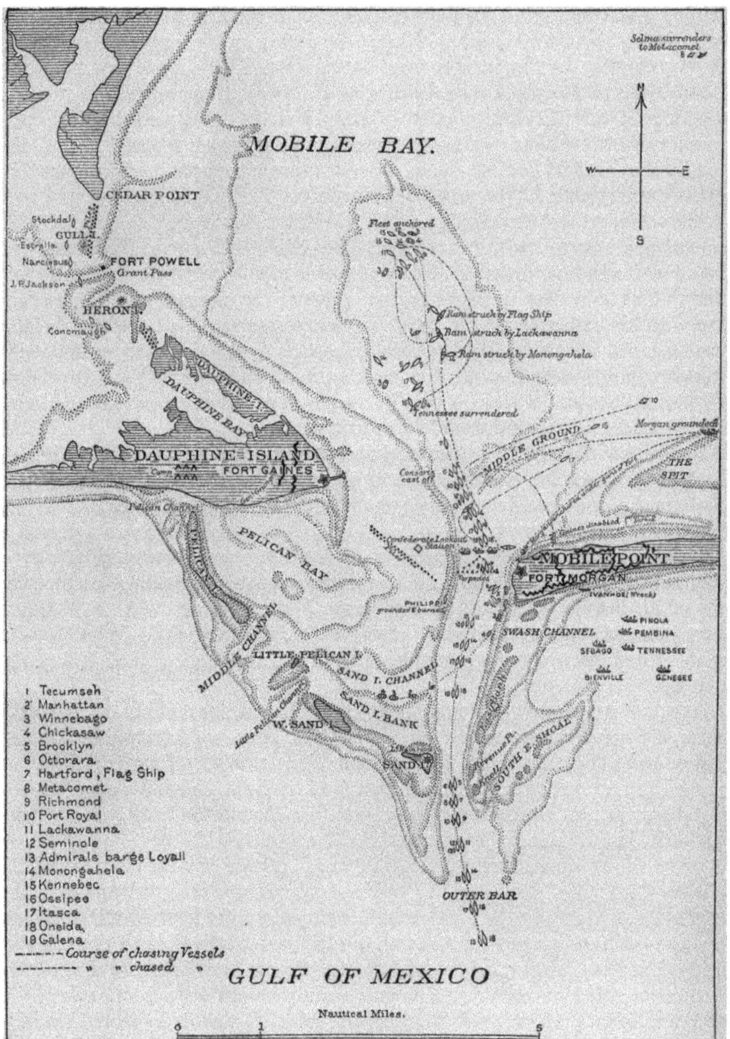

Fig. 13. Map showing Fort Gaines and Fort Morgan at the entrance to Mobile Bay during the Battle of Mobile Bay. From Kinney (1888).

Fig. 14. Fort Morgan (cf. Fig. 13). From Kinney (1888).

made of brick and masonry, but it was not completed until 1861. The original plan was for 10 guns to be mounted 'en barbette' on each of the five sides and for four howitzers to be mounted on each bastion (Anon 2012). Totten's plans for the fort used the latest in French fort design, incorporating a detached scarp consisting of the perimeter Carnot wall combined with the chemin de ronde behind it, providing protection to the troops. As noted above, only one other Third System fort, Fort Clinch, made use of this design (Weaver 2001).

In January 1861, with the coming of the Civil War, Fort Morgan and Fort Gaines were seized by Alabama state troops. The entrance to Mobile Bay was soon mined, leaving a channel for blockade-running ships. With the fall of New Orleans to Union forces in 1862, Mobile was the Confederacy's largest port on the Gulf of Mexico and was extremely important for exporting cotton and importing supplies by means of blockade runners (Mattei 2016).

On 5 August 1864, the Battle of Mobile Bay began when a Union Navy fleet began to enter the bay between the two forts, close to Fort Morgan (Fig. 13). The Union fleet had 199 guns, compared to the 26 guns of Fort Gaines and the 46 of Fort Morgan. At the same time, a Union Army force that had been landed on Dauphin Island to the west of the fort was assaulting Fort Gaines, forcing its surrender on 7 August 1864. From 9 August until 23 August, Union forces laid siege to Fort Morgan. After an intense bombardment on 22 August that seriously damaged the fort, Fort Morgan was forced to surrender the following day (Mattei 2016).

Following the Civil War, the forts underwent changes. In 1895, the Endicott System of new batteries began to be constructed at Fort Morgan. Units of Coast Artillery were stationed at Fort Morgan, creating the largest permanent military base in Alabama in the early years of the twentieth century. Although the fort was closed in 1923, its coastal defence role was renewed from 1941 to 1944 during World War II. In July 1944 Fort Morgan was closed. Fort Gaines began to receive new armament of the Endicott System in 1898, including disappearing guns, and, like Fort Morgan, was garrisoned by Coast Artillery units during World War I. It was declared obsolete in 1923 and sold to the state of Alabama in 1927. In World War II, Fort Gaines was reactivated when it became a camp for the Alabama National Guard, and the US Coast Guard used the fort as a base for patrolling against enemy submarines (Hyatt 2017).

Fort Massachusetts on Ship Island

Ship Island, Mississippi, is a Holocene barrier island in the Gulf of Mexico approximately 17.7 km off the coast on the south side of Mississippi Sound: a shallow body of water lying between the mainland and the Mississippi–Alabama barrier island chain. The island is adjacent to the main ship channel leading into New Orleans (Fig. 1) and shelters a deep-water anchorage (Fig. 15). As such, it was considered an important site for fortification. During the war of 1812, the British used Ship Island to stage a major attack on New Orleans in January 1815 (Weaver 2001).

Geologically, the development of the Mississippi–Alabama barrier island chain was predicated upon several factors. Cores taken on Horn and Ship islands (both in the top right in Fig. 15) show that the sand at the base is different to sand higher in the core. The lower sand is pale orange to light yellowish brown, possibly from dune oxidation. This is likely to indicate subaerial exposure of a Pleistocene core for these islands. The availability of sand from east to west longshore drift, the large Mobile Bay ebb-tidal delta and the sand shoals west of the delta provide sand to the islands. Cores taken in Mississippi Sound show that Petit Bois, Horn, Ship and Cat islands were created by shoals forming over sandy-muddy sediments. The shoals then built upwards to intertidal range (Otvos 2001).

Ship Island was breached by a hurricane in 1947 and permanently divided into two parts by Hurricane Camille in 1969, creating West Ship Island and East Ship Island. Schmid (2003) analysed the erosion of East Ship Island. Prior to the close passage of Hurricane Georges in September 1998, East Ship Island had been relatively stable but during the hurricane the island eroded very rapidly. Over the next several years, however, it recovered most of its sand. Schmid (2001) found that West Ship Island, the site of the fort, had been similarly stable before Hurricane Georges and largely recovered sand it had lost during it. West Ship has a higher elevation and greater sand source than East Ship Island; east–west longshore drift could be providing sand from East Ship to West Ship.

The construction of the fort on Ship Island (Fig. 16) began in 1859 on the island's western tip, where it could overlook Mississippi Sound, and defend both the ship channel and the deep-water anchorage. The fort was built in a D-shape with a single tier of very tall casemates. The initial construction prior to the Civil War utilized brick from New Orleans. During the time that New Orleans was Confederate territory (1861–62) and Ship Island was occupied by Union forces, the bricks were shipped from Maine. It was the only Third System fort to have a circular scarp. The seawards circular portion guarded the channel and anchorage, whereas the landwards-facing, straight gorge defended the fort

Fig. 15. Map of the lower Mississippi River with Ship Island in the NE quadrant. From Porter (1887).

Fig. 16. Fort Massachusetts on Ship Island from the water (cf. Fig. 15).

from a land-based attack emanating from the eastern side of Ship Island. Designed for 37 guns in the casemates and mounted on the barbette, the fort was one of the smallest of the Third System forts (Kanze 2001; Weaver 2001).

In January 1861 Mississippi state troops seized the fort. It was still incomplete; at the time of its occupation by Mississippi forces, its scarp was only 2.7 m high in places, only a few embrasures were completed and no guns were mounted. In July 1861, Confederate forces occupied Ship Island and mounted several cannon that they had brought from the mainland in the fort. In September 1861, the Confederate military determined that the island was indefensible, given the state of the fort, and ordered its evacuation. Union forces from the USS *Massachusetts* promptly occupied Ship Island and soon began to build up the fort's defences and bring in more guns. In early 1862, Ship Island became the staging area for the US Navy's capture of New Orleans in April. Once New Orleans was captured, Ship Island became a prison, first for political prisoners from New Orleans, then for Federal Army prisoners, and later for Confederate soldiers (Bearss 1984). Although never officially named, in about 1862 the fort on Ship Island became known as Fort Massachusetts after USS *Massachusetts*, the ship that brought the first of the fort's guns and landed the Federal forces that reoccupied it (Kanze 2001; Weaver 2001). The fort was finally finished in 1866 after the Civil War ended. It was abandoned in 1870 but in 1873, due to tensions with Spain, additional heavy artillery was mounted, consisting of two 15-inch Rodman guns and two 100-pounder Parrott rifles. The crisis ended and soon the fort was again abandoned (Kanze 2001).

Concluding remarks

The position and use of these Third System forts can thus be directly related to the coastal geomorphology and Quaternary geological history of the coastal regions in which they were constructed. The forts were built to defend major harbours, shipping and military resources such as navy installations. In Boston and New York, the coastal forts were positioned on glacial features including drumlin islands (Boston Harbor) and end moraines (New York City). Fort Pulaski, built to defend the approach to the Savannah River and the city of Savannah in Georgia, lies on a built-up alluvial island in the Savannah River. A number of the forts in the southern USA, including Forts Clinch, Pickens, McRee, Morgan, Gaines and Massachusetts, were built on barrier islands, many of which had Pleistocene cores. On the mainland, Fort Barrancas and the Advanced Redoubt of the Pensacola defences were constructed on a marine terrace related to a Pleistocene sea-level highstand.

These coastal forts were built of materials available both locally and at a distance. Many of the forts in the NE USA were built of stone that was quarried in Maine, New Hampshire and New York. Forts in the southern USA were largely constructed from brick, both local and transported by ship. Geologically, the brick clay was dug from clay pits located on the Cenozoic coastal plain. Stone that was shipped by water over long distances from New England and New York augmented the brick used for these southern forts. The ease of transporting materials by water over great distances made the use of distantly sourced materials economic.

The Third System of Fortification was never designed for a civil war. Attack was largely anticipated from the sea by an extra-continental invading force. As such, the history of these forts during the Civil War from 1861 to 1865 illustrates some of the unanticipated ways in which they were defended, attacked and/or used as prisons.

Military advances made in the Civil War had a profound effect upon the forts. Early in the war (April 1862), the bombardment and capture of Fort

Pulaski through its partial destruction by rifled cannon sounded the death knell for masonry forts. By the 1890s, long-range naval guns meant that cities and harbours could be attacked at a distance not foreseen by the original designers. This resulted in long-range guns being mounted for coastal defence. The new weapons were frequently sited within concrete batteries built extraneously to the original Third System forts, with the forts commonly being used as barracks or storage facilities. With the advent of the Cold War, these coastal defence positions were obsolete.

Some of the original 42 forts have been destroyed, but many have been preserved and are currently maintained as museums. Of the forts discussed, Fort Independence is a Massachusetts state park, and Fort Warren on Georges Island is maintained by the Massachusetts Department of Conservation and Recreation in the Boston Harbor Islands National and State Park. In New York City, although many of the structures of Fort Hamilton in Brooklyn have been removed, what is left is part of the still active US Army Fort Hamilton Garrison. Fort Wadsworth, which includes Fort Richmond and Fort Tompkins, is part of the Gateway National Recreation Area. Fort Pulaski is a National Monument in Georgia. Fort Clinch is a Florida state park. Around Pensacola, Fort Barrancas, the Advanced Redoubt of Fort Barrancas, Battery San Antonio and Fort Pickens are part of Gulf Islands National Seashore. Of the Mobile Bay defences, Fort Morgan and Fort Gaines have been preserved by the State of Alabama. Fort Massachusetts is also part of Gulf Islands National Seashore. Several Third System forts are threatened by continuing coastal erosion and degradation, notably by hurricanes. All are now militarily redundant, but they still represent a substantial and highly visible aspect of the military history of the USA.

Acknowledgements I am grateful to Kathryn Langmyer Henderson for her help with the figures and to the reviewers and editors of the manuscript for their useful comments and suggestions.

Funding Oxford College of Emory University provided partial funding for site visits.

References

Albanese, J.R. & Kelly, W.M. 1991. History, economy, and geology of the bluestone industry in New York State. *In*: Ebert, J.R. (ed.) *63rd Annual Meeting Field Trip Guidebook*. New York State Geological Association, Oswego, NY, 191–203.

Alexander, C. 2008. *Final Report: Rates and Processes of Shoreline Change at Ft. Pulaski National Monument*. Applied Coastal Research Laboratory, Georgia Southern University, Statesboro, GA.

Anon undated. *Governors Island (Now Part of Logan Airport)*. Friends of the Boston Harbor Islands, Hingham MA, http://www.fbhi.org/governors-island.html

Anon 1878. The Rockport Granite quarries. *Scientific American*, **39**, 314.

Anon 1902. The explosion at Governor's Island. *The Sacred Heart Review*, **28**, 3.

Anon 1977. *Proposed Highway Connecting West Gate of Pensacola Naval Air Station to Interstate 10, Escambia County, Florida: Environmental Impact Statement, State Project 48000–1508*. Florida Department of Transportation, Tallahassee, FL.

Anon 1999. *Coquina – The Rock That Saved St. Augustine, Fort Matanzas*. National Park Service, Washington, DC.

Anon 2006. *Fort Wadsworth*. New York State Military Museum and Veterans Research Center of New York State Division of Military and Naval Affairs, Saratoga Springs, NY.

Anon 2012. *Fort Gaines, Dauphin Island, Alabama History and Tour Guide*. Fort Gaines Historic Site. Alabama Tourism Department, Montgomery, AL.

Anon 2015*a*. *Georges Island, Boston Harbor Islands*. National Park Service, Boston, MA, https://www.nps.gov/boha/learn/historyculture/facts-geor.htm

Anon 2015*b*. *Georges Island, Boston Harbor Islands*. Boston Harbor Islands Partnership, Boston, MA, http://www.bostonharborislands.org/georges-island

Anon 2015*c*. *Detailed History of Fort Wadsworth*. Gateway National Recreation Area, NY, NJ. National Park Service, Boston, MA, https://www.nps.gov/gate/learn/historyculture/ftwad.htm

Anon 2016*a*. *Castle Island, South Boston*. CelebrateBoston.com, Campton, NH, http://www.celebrateboston.com/castle-island.htm

Anon 2016*b*. *Fort Hancock and Sandy Hook Proving Ground National Historic Landmark*. National Park Service, Boston, MA, https://www.nps.gov/gate/learn/historyculture/hancock.htm

Bearss, E.C. 1983. *Historic structure report: Fort Pickens historical data section 1821–1895, Gulf Islands National Seashore, Florida–Mississippi*. Denver Service Center, National Park Service, Denver, CO.

Bearss, E.C. 1984. *Historic Resource Study: Ship Island, Harrison County, Mississippi, Gulf Islands National Seashore, Florida/Mississippi*. Denver Service Center, National Park Service, Denver, CO.

Bearss, E.C. & Paige, J.C. 1983. *Historic Structure Report for Castillo de San Marcos National Monument, St. Johns County, Florida*. Denver Service Center, National Park Service, Denver, CO.

Black, F.R. 1983. *Historic Resource Study: A History of Fort Wadsworth, New York Harbor*. Cultural Resource Study No. 7. National Park Service, Boston, MA.

Burrows, E.G. & Wallace, M. 1999. *Gotham: A History of New York City to 1898*. Oxford University Press, New York.

Butler, G.W. 2001. *The Guns of Boston Harbor (From the Bay Colony Through the Present)*. Butler, Boston, MA.

Chartrand, R. 2012. *Forts of the War of 1812*. Osprey, Oxford.

Coleman, J.C. & Coleman, I.S. 1982. *Pensacola Fortifications, 1698–1980: Guardians on the Gulf.* Pensacola Historical Society, Pensacola, FL.

Cox, D. 2013. *Amelia Island's Historic Fort, Fort Clinch State Park, Fernandina Beach, Florida*, http://www.exploresouthernhistory.com/fortclinch.html

Dean, R.G., Otay, E. & Work, P.A. 1995. *Perdido Key Beach Nourishment Project: A Synthesis of Findings and Recommendations for Future Nourishments.* University of Florida, Gainesville, FL.

Donnelly, J.P., Driscoll, N.W., Uchupi, E., Keigwin, L.D., Schwab, W.C., Thieler, E.R. & Swift, S.A. 2005. Catastrophic meltwater discharge down the Hudson Valley: a potential trigger for the Intra-Allerød cold period. *Geology*, **33**, 89–92.

Franzi, D. 2016. The sandstone pavement Pine Barrens of northeastern New York: A legacy of fire and ice. *Vignettes – Key Concepts in Geomorphology*, https://serc.carleton.edu/vignettes/collection/25562.html

Fuller, M.L. 1914. *The Geology of Long Island, New York.* United States Geological Survey, Professional Papers, **82**.

Gibbard, P.L., Boreham, S., Cohen, K.M. & Moscariello, A. 2011. *Global Chronostratigraphical Correlation Table for the Last 2.7 Million Years.* Subcommission on Quaternary Stratigraphy, Department of Geography, University of Cambridge, Cambridge.

Gillmore, Q.A. 1887. Siege and capture of Fort Pulaski. *In*: Johnson, R.U. & Buel, C.C. (eds) *Battles and Leaders of the Civil War, Volume 2.* Century, New York, 1–12.

Gilman, J.H. 1887. With Slemmer in Pensacola Harbor. *In*: Johnson, R.U. & Buel, C.C. (eds) *Battles and Leaders of the Civil War, Volume 1.* Century, New York, 26–32.

Henry, V.J. 1971. *Geological History and Development of Amelia Island, Nassau County, Florida.* A Preliminary Ecological Inventory, A report to the Amelia Island Company, Fernandina Beach. Proposed by D.A. Wallace, I.H. McHarg, W.H. Roberts and T.A. Todd, Inc. by J. McCormick and Associates, Philadelphia, PA.

Henry, V.J. 2005. *The Georgia Coast in the New Georgia Encyclopedia*, http://www.georgiaencyclopedia.org/articles/science-medicine/geology-georgia-coast

Henry, V.J., Alexander, C. & Crawford, J. 1996. *Guidebook to the Mesotidal Barrier, Back Barrier, and Inlet Environments of Sapelo Island, Georgia. International Association of Sedimentologists conference guidebook, Tidalites '96.* Georgia Southern University Applied Research Laboratory, Savannah, GA.

Howell, S.M. & Alexander, C.R. 2004. Shoreline change at Fort Pulaski, Georgia: a remote sensing approach. *Geological Society of America Abstracts with Programs*, **36**, 47.

Hoyt, 1967. Barrier island formation. *Geological Society of America Bulletin*, **78**, 77–78.

Humphrey, J., Hager, J. & Woodhouse, D. 1993. A geohistorical boat tour of Boston Harbor. *In*: Cheney, J.T. & Hepburn, J.C. (eds) *Field Trip Guidebook for the Northeastern United States: 1993 Boston Geological Society of America Annual Meeting, Volume 2.* Department of Geology and Geography, University of Massachusetts, Amherst, MA, Contributions, **67**, W1–W16.

Hyatt, W. 2017. *Fort Morgan State Historic Site*, http://www.fort-morgan.org/history/

Johnson, B. 2005. *Fort Independence: Object of the Month.* Massachusetts Historical Society, Boston, MA, https://www.masshist.org/object-of-the-month/objects/fort-independence-2005-06-01

Julien, A.A. 1884. Stone construction in cities, New York City and environs. *In*: Gannett, H. (ed.) *Report on the Building Stones of the United States and Statistics of the Quarry Industry for 1880.* Final Report of the Tenth Census. Department of the Interior, Census Office, Government Printing Office, Washington, DC, 313–335.

Kales, D. 2007. *The Boston Harbor Islands: A History of Urban Wilderness.* Arcadia, Charleston, SC.

Kanze, E. 2001. *The Fort Massachusetts Story.* Eastern National, Fort Washington, PA.

Kaufmann, J.E. & Kaufmann, H.W. 2004. *Fortress America: The Forts That Defended America 1600 to the Present.* Da Capo, Cambridge, MA.

King, M. 1893. *King's Handbook of New York City 1893.* Moses King, Boston, MA (1972 reprint 2nd edn, Blom, New York).

Kinney, J.C. 1888. Farragut at Mobile Bay. *In*: Johnson, R.U. & Buel, C.C. (eds) *Battles and Leaders of the Civil War, Volume 4.* Century, New York, 379–400.

Kopaska-Merkel, D.C. & Rindsberg, A.K. 2005. *Sand-quality Characteristics of Alabama Beach Sediment, Environmental Conditions, and Comparison to Offshore Sand Resources.* Open-File Report 0508. Geological Survey of Alabama, Tuscaloosa, AL.

Kopera, J.P. 2011. *The Bedrock Geology of Boston Harbor and Outer Harbor Islands.* National Park Service, Boston, MA, https://www.nps.gov/boha/learn/management/research-abstracts-2011-kopera.htm

Lattimore, R.B. 1954. *Fort Pulaski National Monument, Georgia.* National Park Service, Washington, DC.

Lewis, E.R. 1970. *Seacoast Fortifications of the United States. An Introductory History.* Smithsonian Institution Press, Washington, DC (1979 reprint edn, Presidio Press, Novato, CA).

Luedtke, B.E. & Rosen, P.S. 1993. Archaeological geology on Long Island, Boston Harbor. *In*: Cheney, J.T. & Hepburn, J.C. (eds) *Field Trip Guidebook for the Northeastern United States: 1993 Boston, Geological Society of America Annual Meeting, Volume 2.* Department of Geology and Geography, University of Massachusetts, Amherst, MA, Contributions, **67**, T1–T15.

MacNeil, F.S. 1950. *Pleistocene Shorelines in Florida and Georgia.* United States Geological Survey, Professional Papers, **221-F**.

Mattei, E. 2016. *A Tale of Two Forts on Mobile Bay: Fort Gaines and Fort Morgan.* The Army Historical Foundation, Arlington, VA, https://armyhistory.org/a-tale-of-two-forts/

Meader, J.F. & Brinkley, C. 2003. *Fort Pulaski National Monument: Administrative history.* Southeast Region National Park Service, Cultural Resources Division, Atlanta, GA.

Merrill, G.P. 1889. The collection of building and ornamental stones in the U.S. National Museum: a handbook and catalogue. *In*: Goode, G.B. (ed.) *Annual Report of the Board of Regents of the Smithsonian Institution for 1886.* Government Printing Office, Washington, DC, 277–648.

Moss, C.J. 2013. Evidence of glacial readvances during recession from the Last Glacial Maximum from the Citifield Stadium site, Queens, New York City, NY. *In*: Hanson, G.N. (chair) *20th Annual Conference on Geology of Long Island and Metropolitan New York*, 13 April 2013, State University of New York at Stony Brook, NY, Long Island Geologists Program with abstracts, http://pbisotopes.ess.sunysb.edu/lig/Conferences/abstracts13/Moss.pdf

Newman, W.A., Mickelson, D.M., Berg, R.C., Rendigs, R.D., Oldale, R.N. & Bailey, R.H. 1993. Pleistocene geology of the Boston Basin and its adjacent surroundings. *In*: Cheney, J.T. & Hepburn, J.C. (eds) *Field Trip Guidebook for the Northeastern United States: 1993 Boston, Geological Society of America Annual Meeting, Volume 2*. Department of Geology and Geography, University of Massachusetts, Amherst, MA, Contributions, **67**, U1–U24.

Nikolaou, S. 2004. Local geology of New York City and its effect on seismic ground motions. *In*: *Fifth International Conference on Case Histories in Geotechnical Engineering*. University of Missouri–Rolla, Rolla, MO, http://scholarsmine.mst.edu/icchge/5icchge/session00g/2/

Ofeldt, F.A. 2017. *Fort Clinch*. Arcadia, Charleston, SC.

Ogden, D.P. 1997. *The Fort Barrancas Story*. Eastern National, Fort Washington, PA.

Otvos, E.G. 2001. *Mississippi Coast: Stratigraphy and Quaternary Evolution in the Northern Gulf Coastal Plain Framework. Stratigraphic and Paleontologic Studies of the Neogene and Quaternary*. United States Geological Survey, Open-File Report 01-415-H.

Parkman, A. 1978. *Army Engineers in New England: the Military and Civil Work of the Corps of Engineers in New England, 1775–1975*. US Army Corps of Engineers, New England Division, Waltham, MA.

Porter, D.D. 1887. The opening of the lower Mississippi. *In*: Johnson, R.U. & Buel, C.C. (eds) *Battles and Leaders of the Civil War, Volume 2*. Century, New York, 22–55.

Reynolds, R.J. 1988. *Hydrogeology of the Croton-Ossining area, Westchester County, New York*. United States Geological Survey, Water-Resources Investigations Report 87-4159.

Roberts, R.B. 1988. *Encyclopedia of Historic Forts: The Military, Pioneer, and Trading Posts of the United States*. Macmillan, New York.

Rupert, F.R. 1993. *The Geomorphology and Geology of Escambia County, Florida*. Florida Geological Survey, Tallahassee, FL.

Schmid, K. 2001. *West Ship Island Evolution, Morphology, and Hurricane Response – 1995 to 2000*. Open-File Report 133. Mississippi Department of Environmental Quality, Office of Geology, Jackson, MS.

Schmid, K. 2003. *East Ship Island Evolution, Morphology, and Hurricane Response – 1994 to 2001*. Open-File Report 134. Mississippi Department of Environmental Quality, Office of Geology, Jackson, MS.

Schorow, S. 2008. *East of Boston: Notes from the Harbor Islands*. Arcadia, Charleston, SC.

Schweitzer, P. 2015*a*. *Anastasia Formation. Mineral Resources On-Line Spatial Data*. United States Geological Survey (USGS), Reston, VA, https://mrdata.usgs.gov/geology/state/sgmc-unit.php?unit=FLPSa%3B0

Schweitzer, P. 2015*b*. *Passaic Formation. Mineral Resources On-Line Spatial Data*. United States Geological Survey (USGS), Reston, VA, https://mrdata.usgs.gov/geology/state/sgmc-unit.php?unit=NJJTRp;2

Seasholes, N.S. 2003. *Gaining Ground: A History of Landmaking in Boston*. MIT Press, Cambridge, MA.

Shaw, W.H. 1884. *History of Essex and Hudson Counties, New Jersey, Volume 1*. Everts & Peck, Philadelphia, PA.

Smith, E.B. 1923. *Governors Island: Its Military History Under Three Flags 1637–1922*. Valentine's Manual, New York.

Thieler, E.R., Butman, B., Schwab, W.C., Allison, M.A., Driscoll, N.W., Donnelley, J.P. & Uchupi, E. 2007. A catastrophic flood event and the formation of the Hudson Shelf Valley. *Palaeogeography, Palaeoclimatology, Palaeoecology*, **246**, 120–136.

USGS 1994. *Fernandina Beach, FL-GA 7.5-minute Topographic Quadrangle Map*. United States Geological Survey (USGS), Reston, VA.

Weaver, J.R., II 2001. *A Legacy of Brick and Stone: American Coastal Defense Forts of the Third System, 1816–1867*. Redoubt Press, McLean, VA.

White, W.A. 1970. *The Geomorphology of the Florida Peninsula*. Florida Geological Survey (FGS) Bulletin, **51**. Florida Department of Natural Resources, Bureau of Geology, Tallahassee, FL.

Williford, G. & Polaski, L. 2004. *Long Island's Military History*. Arcadia, Charleston, SC.

The northern Atlantic Wall: German engineering geology work in Norway during World War II

HERMANN HÄUSLER

Department for Environmental Geosciences, University of Vienna, Althanstrasse 14, 1090 Vienna, Austria

0000-0003-2900-6222

hermann.haeusler@univie.ac.at

Abstract: German forces occupied Norway from June 1940 to secure the strategically important iron ore railway from Lulea in Sweden to Narvik in Norway and to improve the German base for war against the UK. Increasing in number to a peak in 1943, by 1945 more than 60 German and Austrian military geoscientists had supported the army, air force and paramilitary construction agency, Organisation Todt, from 1942, mostly assisting fortification works along the Norwegian coast that formed part of the Atlantic Wall. In total, six military geology units (Wehrgeologenstellen) were deployed to support Army Headquarters Norway: WG18 and WG33 were assigned to the 17th Fortress Engineers in Oslo; WG22 to the 16th Fortress Engineers at Trondheim; WG3 and WG31 to the 15th Fortress Engineers at Narvik; and WG27, not for construction works, but to support the headquarters of Dietl's Lapland Army, operationally deployed SW of Murmansk. Engineering work at strategic locations also included airfields, U-boat pens, roads and railways. Military geologists contributed to tasks related both to the Precambrian/Paleozoic bedrock and to its locally thick Quaternary cover, notably pioneering applications of rock mechanics and soil mechanics for major excavations, and locating sources of raw materials and drinking water.

An understanding of the military situation in northern and eastern Europe early in World War II can be conveniently gained from online sources and from Tschudi (1960) and Kinder & Hilgemann (2000). The non-aggression pact of 23 August 1939 between Germany and the Union of Soviet Socialist Republics (referred to here as Russia) regulated their spheres of interest in eastern Europe. It was followed by attacks on Poland, a German attack eastwards on 1 September 1939 and a Russian attack westwards on 17 September 1939. A German–Russian frontier treaty signed on 28 September 1939 led to an economic treaty between Germany and Russia that frustrated a British naval blockade intended to deny supply routes to Germany. After bombing the Finnish capital of Helsinki at the end of November 1939, Russian troops invaded Finland and reached its Karelian defence line in February 1940. Russo-Finnish hostilities ceased on 12 March 1940. On 9 April 1940, German forces invaded Denmark and Norway, in part to secure the strategically important iron ore railway route from Lulea in Sweden to Narvik in Norway (Fig. 1) and in part to improve Germany's economic base for war against the UK. On 28 May 1940, Allied troops recaptured Narvik, but were forced to evacuate at the beginning of June. On 7 June 1940 the King of Norway and his government departed for exile in the UK. German troops occupied the whole of the country and from December 1940 their Army Headquarters Norway controlled forces deployed throughout Norway to Finland.

In June 1941 the German army in Norway took part in Operation Barbarossa, the invasion of Russia (https://en.wikipedia.org/wiki/Army_of_Norway_(Wehrmacht)). Finland joined Germany at this time to avenge its earlier defeat by the Russians. German forces were given military responsibility over northern Finland and Army Headquarters Norway was tasked to seize the Russian port of Murmansk (Fig. 1) and the strategically important Murmansk railway. An additional German regional headquarters was created in Finland (Befehlsstelle Finnland) and, in January 1942, this generated the Lapland Army.

In December 1941 a small team of military geologists was deployed to support Army Headquarters Norway. It was redeployed in February 1942 to support Army Headquarters Lapland and, from November 1942, to support the 20th Mountain Army. Between 1941 and the end of hostilities in May 1945, another five military geology teams were deployed to assist Fortress Engineers constructing elements of the defensive Atlantic Wall along Norway's coast (Häusler 2000). Smaller numbers of geologists were employed to guide works units of the German air force (the Luftwaffe) and the German paramilitary construction agency Organisation Todt.

From: ROSE, E. P. F., EHLEN, J. & LAWRENCE, U. L. (eds) 2019. *Military Aspects of Geology: Fortification, Excavation and Terrain Evaluation*. Geological Society, London, Special Publications, **473**, 83–108.
First published online March 6, 2018, https://doi.org/10.1144/SP473.4

Fig. 1. Relief map showing distribution of numbered Fortress Engineer Commands (FestPiKdr = Festungspionier-Kommandeur), Fortress Engineer Staffs (FestPiStab = Festungspionierstab) and military geology units (WG = Wehrgeologenstelle) in Norway and Finland during 1941–45, with current national boundaries and major localities mentioned in the text. GebAOK 20 = GebirgsArmeeOberKommando 20 = High Command of the 20th Mountain Army.

In total, German armed forces had made use of more than 60 geoscientists in Norway by the end of the war, more than double the number of geologists employed by the British armed forces worldwide (Rose 2011).

Geology of Norway

A current understanding of the geology of Norway (Fig. 2) may be summarized from the textbook of Ramberg *et al.* (2008). While the supercontinent of Pangaea was assembling from earlier continental units in the southern hemisphere, the Iapetus Ocean developed between the North American and European–African cratons in the north. Squeezing of these oceanic deposits between 'old' Europe and North America about 490–390 myr ago generated the Caledonian orogenic belt. Southern components of the Caledonian nappe system overthrust the Fennoscandian Shield (in what is at present a southeasterly direction), while the northern components overthrust the North American continent (in a westerly direction) to form the present day Appalachian Mountains of the USA.

A major part of Norway therefore consists of a Precambrian crystalline basement and overlying Lower Paleozoic sedimentary and volcanic rocks,

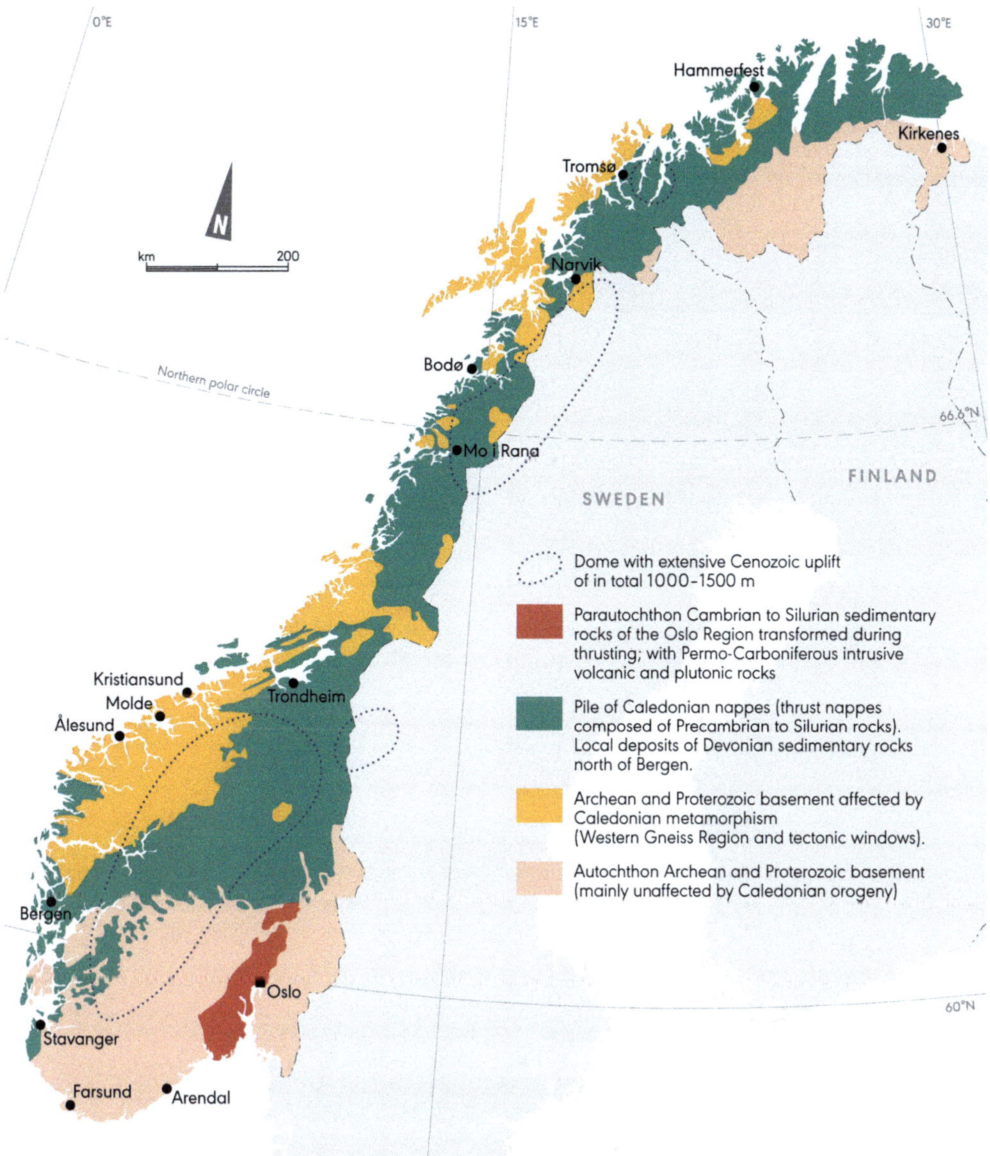

Fig. 2. Geological map of Norway, simplified from Ramberg *et al.* (2008) and www.eurare.eu/countries/norway.html

all *in situ*, but influenced, in part, by the Caledonian Orogeny. These rocks, in turn, are overlain by a lower thrust nappe of Late Proterozoic 'Sparagmites' (mostly massive feldspathic sandstones and greywackes) with an upper nappe of up-thrust crystalline rocks of the Precambrian basement. The younger successions consist of sedimentary rocks of Devonian age. The Oslo rift region contains sedimentary rocks of Cambrian to Silurian age, as well as Permo-Carboniferous extrusive and plutonic rocks.

Details of the offshore geology known at the present time were not known during World War II. However, it was evident that the Norwegian coast had been eroded into a hard rock terrain dominated by cliffs, with relatively rare sandy beaches, and that the fortification of its harbours, fjords and islands and associated tunnelling works would potentially involve engineering geologists with appropriate hard rock expertise.

Features of the Quaternary geology were also important. The surface of Norway had been shaped during many phases of glaciation and deglaciation during the Pleistocene. During the Last Glacial Maximum, about 20 ka ago, glaciers filled the Gulf of Bothnia between Sweden and Finland, as well as much of the Baltic Sea and the North Sea, and extended towards Middle Europe. The land rebounded during the rapid deglaciation of the 3000 m thick Scandinavian ice sheet, with a rate of uplift calculated at 9 mm a^{-1} in the Gulf of Bothnia. In the innermost parts of the Oslo Fjord (Fig. 1), the land has risen by 360 mm in the last century and by about the same amount further north in the Trondheim area (Fig. 1). By contrast, uplift has probably ceased in the outer coastal districts of western and northern Norway; to the south, in Denmark and the Netherlands, the land is actually sinking, which is inferred to be a consequence of isostatic forces beneath Scandinavia. The emergent seafloor became subject to erosion by rivers and streams. Hence the fjord bottom sediments became the foundation of an undulating agricultural landscape in which ravines were eroded into clay flats as the land was uplifted, their steep sides subject to landslides.

Most Quaternary marine clays now exposed in Norway were deposited during the retreat of the major ice sheets and close to their terminal moraines. The thickness of the marine clays can exceed 150 m. The most extensive marine clay deposits are found in the lowlands of eastern Norway and along the western coastal region north of Trondheim. Under particular circumstances, 'quick clays' can form within the marine clays, usually within zones where ion-deficient freshwater has gradually replaced the original saline pore waters. When quick clay loses its support, its internal structure collapses and it starts to flow. In the presence of excess water, the viscosity of the clay decreases instantly and the clay flows as a soup-like liquid, with a landslide rapidly propagating backwards. The process of liquefaction of these post-glacial marine clays by natural or artificial shocks, and in reverse their consolidation, was a regionally significant thixotropic effect. Notably, it affected villages and construction works in coastal areas near Trondheim, posing a particular hazard during Allied air strikes. In addition to the engineering problems associated with construction work on a hard rock terrain, military geologists therefore became involved in many projects associated with quick clays.

Military geologists were required to use their knowledge of geology and to make use of geological maps at a variety of scales to provide soil and subsoil information about the distribution of moraines, peat beds and post-glacial marine clays; sources of raw materials; the suitability of bedrock formations for foundations and for tunnelling; and supplies of potable water, both in coastal regions and on islands.

German and Austrian military geologists in Norway

Graduates from universities or institutes of technology who had experience, or at least an interest, in applied geology were recruited or conscripted to serve as military geologists (Wehrgeologen). Their educational backgrounds differed widely; some were geologists as such, others palaeontologists, mineralogists, petrologists or geotechnical engineers. Some were able to benefit from short training courses provided by the armed forces, but most learned their skills from one or more of the widely available military geological textbooks (e.g. von Bülow *et al.* 1938; Kranz 1938) and training on the job.

Military geologists mostly served on attachment to the army, the air force or the Organisation Todt (Häusler 1995*a*, *b*; Häusler & Willig 2000). Although these were administratively distinct services, geologists could be transferred from one to another. Essentially, they served in uniform, but as government officials (Militärverwaltungsbeamte) to provide technical advice, rather than as officers to take military command decisions, even if they had served as normal officers or non-commissioned officers previous to appointment.

In the army, geologists appointed to the equivalent of commissioned rank were by 1940 usually assigned the status of Technischer Kriegsverwaltungsrat (TKVR) or Technical War Administration Officer. They wore a uniform like that of combat troops, with the national emblem on the breast (Rose & Willig 2002, fig. 3), but with distinctive shoulder straps. Those under the age of 35 years wore insignia equivalent to the rank of captain, whereas those over the age of 35 years were equivalent to the rank of major. According to Hellmut Grabert (pers. comm. 1985), they wore the letters KV to indicate their Kriegsverwaltung (War Administration) assignment.

In the air force, geologists were also appointed as officials rather than officers and assigned to one of three sub-categories: (1) reserve officials (Beamte des Beurlaubtenstandes or BdB, also referred to as BdR); (2) officials in recalled status 'at the disposal of a branch of the armed forces' (Beamte zur Verfügung or BzV); and (3) officials appointed for the duration of the war (Beamte auf Kriegsdauer or BaK). They were effectively civil servants, but served in uniform and with ranks equivalent to air force/army officers, e.g. Regierungsbaurat (government construction administrative officer = major) or Regierungsbauassessor (government construction officer = captain). Geologists could be transferred from army to air force units or vice versa because their duties were similar, sometimes retaining their former rank, and seemingly older army geologists in relatively static 'fortress' assignments might be appointed with these ranks rather than TKVR status.

The Organisation Todt, named from its founder, the engineer Dr Fritz Todt, was a government construction agency founded in 1933 to develop the motorway network in Germany and Austria. In 1938 it took over construction of the Westwall, the zone of fortifications along Germany's western boundary. Tasked in 1942 to assist with the construction of the Atlantic Wall, the series of intermittent coastal fortifications that marked the western boundary of German-occupied Europe (https://en.wikipedia.org/wiki/Atlantic_Wall), it developed into a paramilitary construction agency with a workforce at its peak of *c.* 1.5 million. Most, or at least the (minority) German personnel, wore a uniform (commonly khaki or olive green) and the Germans had the right to bear arms and resist enemy action. The German government came to classify them as militia. A swastika brassard was worn on the left arm with the abbreviation 'Org. Todt' above it (Absolon 1971). The Organisation Todt had its own system of rank names and corresponding uniform insignia, but, in addition, a relatively small group of Beamte (government officials), including geologists, assigned mostly to technical and administrative supervision and control. During the winter of 1944, the German air force had been so degraded by Allied action that all its geologists were transferred to the Organisation Todt.

Duplicate copies of reports prepared by geologists serving with the army or air force were routinely sent to their headquarters in Berlin. Some were relatively long documents named specifically as 'reports' (Berichte), but most were shorter 'expert opinions' (Gutachten) or 'leaflets' (Merkblätter) for more widespread use, or documents given a title, but not a specified category. Towards the end of the war, an archive of this material was stored underground in a former potash mine at Heringen in central Germany to keep it safe from Allied aerial bombardment. This Heringen Collection was captured by American troops and transported to the USA, although it was mostly, if not entirely, returned to Germany in later years (Hadden 2008; Willig 2009). Documents from this collection form the basis for this paper. The author has not been able to locate any reports compiled by Organisation Todt geologists for Norway, either within the range of archives searched or from personal approaches to surviving Organisation Todt geologists in the 1990s. It is presumed that either the advisory system was different and paperwork was not a priority, or the subarctic climatic conditions were generally unfavourable to its compilation.

Use of military geologists from the German army and air force to facilitate construction of part of the Atlantic Wall has so far been described in English only for the Channel Islands of Jersey, Guernsey and Alderney (Rose *et al.* 2002; Rose 2005*a*, *b*, *c*, 2007, 2008; Robins & Rose 2005; Rose & Willig 2009, 2012, 2013, 2014; Robins *et al.* 2012). By coincidence, a geotechnical engineer, Leopold Müller, who served in the Organisation Todt as a supervisor during the construction of the artillery casemates and protective bunkers that proliferated on the islands, was later deployed as a geoscientist to northern Norway for similar fortification work (Müller-Salzburg 1985).

A glossary of English/German technical and military terms is provided as Appendix A to summarize the translations adopted in this paper. Place names have British spellings in current use where these differ from those in the German documents.

Army geologists

A military geological service had been developed within the German army during World War I, but was disbanded at the close of hostilities (Rose *et al.* 2000). The service was effectively reformed as World War II loomed (Häusler & Willig 2000), and by April 1941 had developed to include military geological units, effectively teams (Wehrgeologenstellen) numbered 1–25 (Häusler 1995*a*, *b*, 2000). The number of units increased to 32 in October 1941, to 35 in September 1943 and to 40 in November 1943. Each team typically consisted of two officers or officials of officer status (TKVR), three non-commissioned officers as technical assistants (sometimes including men who were graduate geologists, but not yet deemed suitable for commissioned rank) and four other ranks for less skilled duties (e.g. drilling assistant, driver, typist). It had two vehicles of its own to ensure that it was highly mobile, one with a trailer for an exploratory drilling rig and other equipment.

A Technical Military Geology Group based within the Army's Engineer and Fortress Branch in Berlin guided team deployments. This Group was led until 15 December 1941 by the veteran military geologist Ernst Kraus. Following his retirement and the Group's re-organization, it became the Military Geology Group within the Army General Office under the leadership of Walter von Seidlitz. In 1942, von Seidlitz was styled Chief Army Geologist and made responsible to the Inspector of Fortresses at the High Command of the Army in Berlin.

The first military geology unit to be deployed in Norway, WG27, was deployed at this time of change in command in December 1941. It was assigned to Befehlsstelle Finnland of the Army Headquarters Norway (see Figs 1 & 3), attached to its Engineer Branch as in other German armies operationally deployed elsewhere in Europe. WG27 was led from

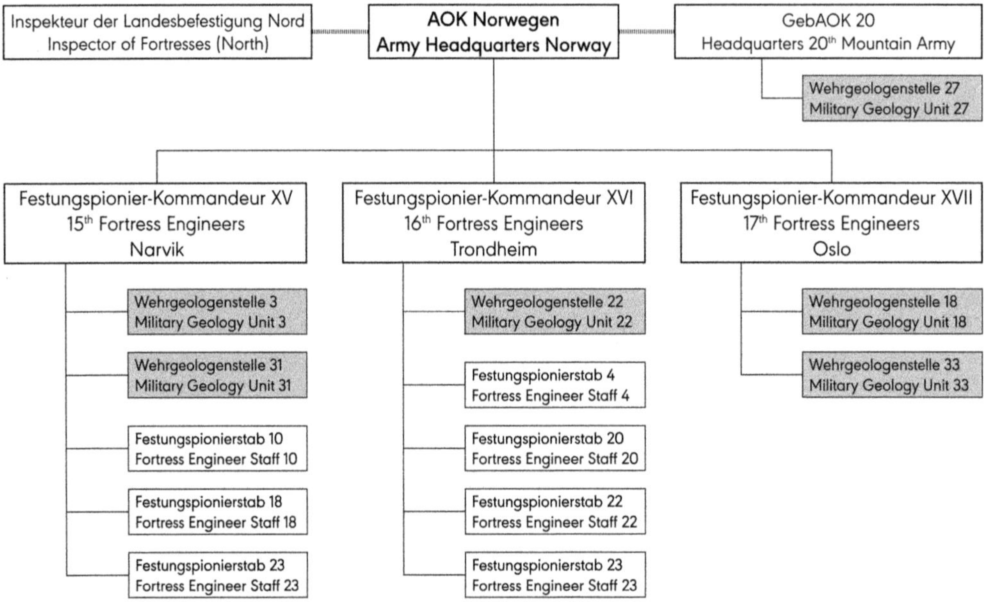

Fig. 3. Chain of command diagram (with German titles and English translation) for 1943, showing assignments of five numbered military geology units plus associated Fortress Engineer Staffs to the three Fortress Engineer Commands directed by Army Headquarters Norway, and its links both to the Headquarters 20th Mountain Army (with an assigned military geology unit) and to the Inspector of Fortresses (North).

January 1942 until 1943 by Dr Hans-Rudolf von Gaertner and thereafter until the end of the war by Dr Georg Knetsch. Its main tasks were to provide assessments of cross-country trafficability and for the construction of log-based roads in marshy soils, water supply to troops both north and south of the Polar Circle during the winter months, and fluctuations of groundwater levels in northern Karelia (on the boundary between Finland and Russia) and on the Kola Peninsula to the NW of Murmansk. One 'expert opinion' was written on the potential for creating landslides by bombing, compiled in March 1942 when the German General Dietl was engaging Russian troops south of Varanger Fjord, the easternmost fjord in Norway.

Five other units were deployed later (Häusler 2000), assigned to support units of Fortress Engineers. WG3 and WG18 were deployed in February 1942, WG31 in September 1942, WG22 in November 1942 and WG33 in October 1943 (Figs 1 & 3). These assignments followed a decision issued by the Supreme Command of the Armed Forces in Berlin to fortify the coasts of German-occupied territories, a decision soon developed for western Europe by implementing the construction of the Atlantic Wall (Rolf 1983). The large number of reports compiled by the military geological teams is indicative of the considerable effort put into Atlantic Wall fortification in their areas of deployment. Precise locations cannot always be identified from their reports, but the report titles generally indicate the broad geographical area to which they relate.

WG3 and WG31 were deployed to the northern region of the Norwegian coast to assist 15th Fortress Engineers. With headquarters in Narvik, this had command of three subordinate Fortress Engineer Staffs. Sometimes it is clear from reports that part of a military geology team could be deployed from the unit's long-term base to form a temporary outstation (Aussenstelle) at another location. Thus WG3 established several such outstations distant from Narvik during its period of deployment, notably at Gravdal (on an island promontory *c.* 160 km to the west), Harstad (*c.* 120 km to the NW), Alta (nearly 300 km to the NE) and Hammerfest (*c.* 500 km to the NE). The furthest outstation was established by WG31 at Kirkenes, close to the Russian border (Fig. 1).

WG22 was deployed to the middle region of the coast to assist 16th Fortress Engineers. From its headquarters in Trondheim, it was assigned to tasks ranging in location from Molde in the south via Kristiansund to Bodø in the north, a region in which four Fortress Engineer Staffs were in operation.

WG18 and WG33 were deployed to the southern region to support 17th Fortress Engineers from a headquarters based in Oslo. WG18 generated temporary outstations at Bergen, Sandvies/Stavanger,

Arendal and Farsund; WG33 generated an outpost at Haugesund.

The names of officers of commissioned status who served with these units can be gleaned from their signatures (normally only recording status and surname), which authorized the reports they compiled and issued. The present author interviewed many of these men when they were still alive in the 1980s and was thereby able to further clarify some names and assignments. In total, at least 38 named officers/officials are known to have served with the German army's six military geology units deployed in Norway between 1941 and 1945 (Table 1).

A set of numbered leaflets compiled by WG18 and WG27 (Table 2) serves to illustrate some of the diversity of tasks undertaken by army geologists in Norway, the scope ranging from water supply and earthworks to flooding in tunnels.

One leaflet deals with an unusual topic: magnetic declination (item 10 in Table 2). In Norway, the declination is highly influenced by the deposits of iron ore common in the country and by widespread ferruginous igneous rocks (such as gabbros). The report contains a map of Norway with isolines of eastern and western declination, revealing that negative declination (in a westerly direction) ranged by up to 10° in the south of the country. The map proved useful for air force navigation, the orientation of troops in the field and the correction of compass readings taken during geological surveys.

In addition to the tasks that generated leaflets, WG18's mainstream engineering geological role can be illustrated from a map at scale of 1 : 250 000 and an accompanying report for the Oslo fjord region (Fig. 4). This was completed in June 1944 (as expert opinion number 263) for 17th Fortress Engineers and issued under the signature of the unit's leader, Professor Dr Konrad Richter, serving in the rank of Regierungsbaurat (= major). The map classified rock formations in an area of 135 × 180 km according to their suitability for use in fortification. It did this by summarizing information for eight properties: (1) the suitability of the rock for the excavation of underground facilities; (2) the potential differential wear on drilling bits; (3) the suitability of rock for use in the construction of obstacles; (4) the suitability of rock for use as broken stone, such as to armour bunkers against explosive bombardment; (5) the suitability of rock for use as building stone; (6) the suitability of rock for use as concrete aggregate; and (7) the suitability of rock for use as road metal; and (8) the electrical resistivity of the rocks, helpful in predicting the potential interference with radio beams. The key to the map distinguishes 11 rock formations, the youngest at the top and the oldest at the bottom, with the rock strength indicated by colour intensity; the darkest tones are used for the strongest rocks. These are illustrated in Figure 5, which, for simplicity, omits formations 5–8. The columns summarize information for each of the eight properties per formation, with a final column for remarks.

Relatively weak rocks are represented by formations 1–4. Formation 1 consists of sands and gravels formed by terminal moraines of the last ice age; formation 2 consists of glacial and post-glacial clay and sand. Both of these are categorized as 'bad' for tunnelling and the remarks warn of the danger of slope failure in excavations where the clay is more than a few metres thick, unconsolidated and water-saturated. Only the boulders in the moraines provide material useful for the construction of obstacles. Formation 3 consists of Permian conglomerate; formation 4 consists of sandstone and quartzite of Late Silurian age. These are categorized as of 'medium' suitability for tunnelling, with abrasion of the drilling bits likely to be 'high'. They were 'unsuitable' for use either in the construction of obstacles or as broken rock for bunker armouring (because of schistosity and cleavage), but their strength made them both suitable for roadstone. Both had a high electrical resistivity. The 'remarks' for quartzite warned of the danger of silicosis to miners when tunnelling.

Stronger rocks are represented by formations 9–11. Formation 9 consists of essexite, here explained as a basaltic rock of Permian age (but a rock that would more explicitly be defined as a nepheline gabbro); formation 10 granite; formation 11 gneiss and ancient altered granite. These formations were all categorized as 'good' for tunnelling, but the abrasion of drilling bits is likely to be 'high'. They are mostly, if not entirely, 'good' for the other purposes tabulated, but there is a risk of silicosis to tunnellers in granite and gneiss. Electrical resistivity is generally high in granites and gneiss, but low in basaltic rock.

The map provides an excellent example of terrain evaluation for potential engineering works by geologists of the German army for, effectively, the final year of the war, when mapping techniques had become well established.

Air force geologists

Norway was invaded both by sea and by air. Airborne troops seized the two main airports at Oslo and Stavanger and supplies and reinforcements were later delivered to German troops by air as well as sea. The German air force thus had a major role in Norway during invasion and subsequent occupation. An Aerial District Command Norway was quickly established, with headquarters at Trondheim, and Professor Dr Ernst Hermann Ackermann was deployed a few weeks after the initial attack

Table 1. *German and Austrian geoscientists assigned as military geologists (Wehrgeologen) in Norway during World War II*

Family name	First name	Nationality	Service years	Assignment
Abels		German	1942	WG3
Ackermann	Ernst Hermann	German	1940–45	Luftwaffe
Bachmayer	Friedrich	Austrian	1942–44	Luftwaffe
Beck	Hans	German	1942–44	Luftwaffe/Organisation Todt
Birther	Walter	German	1942–43	WG22
Birzer	Friedrich	German	1941–43	WG3, WG18, WG31
Breddin	Hans	German	1942–43	WG22
Dette	Horst	German	1944	Organisation Todt
Eder		German	1943–44	WG33
Eigenfeld	Rolf	German	1941–42	WG3, WG27
Felser	Karloskar	Austrian	1942	WG27
Gaertner	Hans-Rudolf, von	German	1940–42	Luftwaffe, WG3, WG27
Gallwitz	Hans	German	1942	WG18
Groschopf	Paul	German	1942	WG3 or WG31?
Guenther	Ekke W.	German	1941–45	Luftwaffe/Organisation Todt
Haber	G.	German	1940–42	Luftwaffe, WG18
Habetha	Ernst	German	1943	Luftwaffe
Hardieck		German	1942	WG3
Heissel	Werner	Austrian	1943–44	Organisation Todt
Helmcke	Hans	German	1943–44	Luftwaffe/Organisation Todt
Hirsch		German	1943	WG3
Hohl	Rudolf	German	1942–43	WG3
Isert		German	1944	WG22
Kahler	Franz	Austrian	1942–43	Organisation Todt
Kattinger		German	1943	WG3
Kieslinger	Alois	Austrian	1942–45	Organisation Todt
Knetsch	Georg	German	1942–45	WG27
Kobold		German	1944	WG22
Koritnig	Sigmund	German	1943	WG3
Köster		German	1945	WG18
Kralik	Bruno	German	1942	WG22
Krasser	Leo	Austrian	1943	WG3
Ladurner	Josef	Austrian	1942–44	Luftwaffe/Organisation Todt
Lüdemann	Fritz	German	1943?	Luftwaffe
Matthes	Horst	German	1943?	WG31
Meixner	Heinz	Austrian	1942–45	Luftwaffe/Organisation Todt
Mertin	Hans	German	1941	WG3 or WG31?
Metz	Karl	Austrian	1943–44	Organisation Todt
Mixius	Friedrich Karl	German	1943–45	WG27
Murban	Karl	Austrian	1943	WG31, WG22
Müller	Leopold	Austrian	1943?	Organisation Todt
Nietsch	Helmuth	German	1944	Forschungsstaffel zbV
Nöring	Friedrich Karl	German	1942–44	WG18
Pankuin	Wilhelm	German	1943	Luftwaffe
Plewe	Ernst	German	1943	WG18
Protzen		German	1943–44	WG33
Purkert	Richard	Austrian	1943–44	WG27, WG33
Rein	Ulrich	German	1942–45	WG27
Reithofer	Otto	Austrian	1941–44	Luftwaffe
Rest		German	1943	WG33
Richter	Konrad	German	1940–44	Luftwaffe, WG18
Richter	Max	German	1941–44	Luftwaffe/Organisation Todt
Schadler	Josef	Austrian	1942–43	Organisation Todt
Schauberger	Othmar	Austrian	1941–42	Luftwaffe
Schmidt	W.	German	1944	WG18
Schulz	Günther	German	1941	WG18
Seelig		German	1945	Luftwaffe/Organisation Todt

(*Continued*)

Table 1. (*Continued*)

Family name	First name	Nationality	Service years	Assignment
Taschenmacher		German	1944	WG3
Treibs	Walter	German	1941–44	Luftwaffe
Waldmann	Leo	Austrian	1942–43	WG3
Wieseneder	Hans	Austrian	1941	WG3
Zanoskar	Walter	Austrian	1941?	Organisation Todt
Zapfe	Helmuth	Austrian	1941–44	Luftwaffe, WG18, Organisation Todt

After Häusler (2015)

to serve as the senior geologist in its construction department (Ackermann pers. com. 1985). He was required to attend daily briefings and there provide advice, including the prediction of dates at which the snow cover would melt completely at specific localities to guide the choice of potential landing sites free of snow. His department was also initially responsible for (1) planning the extension of runways at existing small airports to facilitate use by large transport or combat aircraft and (2) reconnaissance for new airfields by a team consisting of an aviation officer, a surveyor and a military geologist. By the end of the war, the Luftwaffe had made use of at least 19 geologists in Norway (Table 1).

As the German occupation of Norway developed, geological investigations were undertaken at the main airports controlled by Aerial District Command Norway: Oslo (Fig. 1), Kristiansand (*c.* 250 km to the SW), Bergen and Trondheim (Fig. 1). Additional tasks related mostly to new airfields and included:

- planning the extension of existing runways, or the length and alignment of new runways, and the source of aggregates required for their construction;
- investigating potential groundwater resources for the supply of potable water;
- designing measures to protect potable water supplies from contamination arising from accommodation barracks or aircraft hangars;
- locating gravel pits for the production of concrete, bearing in mind that because of the common presence of humic acids or sulphur-bearing minerals such as pyrite, the upper 3–5 m in most gravel deposits was not suitable for use in concrete production;
- site investigation for the construction of underground storage facilities, aircraft hangars, bridges on supply routes and wharves for ports at which supplies were delivered (e.g. at Andalsnes, near Molde, Fig. 1);
- advice on the construction of the Nordland Line, a railway system between Trondheim and Bodø, with regard to air force implications;
- exploratory borehole drilling and geotechnical laboratory studies to investigate the cause of landslides and the liquefaction of quick clays (e.g. the mass movement associated with a burst water pipe at Lade (Fig. 6) and with overweight in Hommelvik, both localities near Trondheim);

Table 2. *Numbered leaflets (Merkblätter) compiled by military geologists of the German army's WG18 and WG27, assigned to support the Inspector of Fortresses North in Norway*

No.	Title	Translation
1	Wasserversorgung und Erdarbeiten in Norwegen	Water supply and earthworks in Norway
2	Beton-Zuschlagstoffe	Concrete aggregates
3	Untersuchung von Trinkwasser	Investigation for potable water
4	Moor als Baugrund in Norwegen	Swamp land as construction sites in Norway
5	Minierarbeiten in Norwegen	Tunnelling work in Norway
6	Beton-Fundamente in Moorboden	Concrete foundations in marshy soils
7	Auswahl und Einbau von Zerscheller Gestein	Selection and use of broken rock (for bunker armouring)
8	Sperrmöglichkeiten in Gebeiten mit weichen Bodenarten	Excavation in regions with weak subsoils
9	Bekämpfung von Frostschäden im Boden	Protection from frost damage in soil
10	Die magnetische Missweisung in Norwegen	Magnetic declination in Norway
11	Bearbeitbarkeit der Felsarten an der Norwegischen Küste	Workability of rock formations at the Norwegian coast
12	Wasser im Stollenbau	Water in tunnel construction

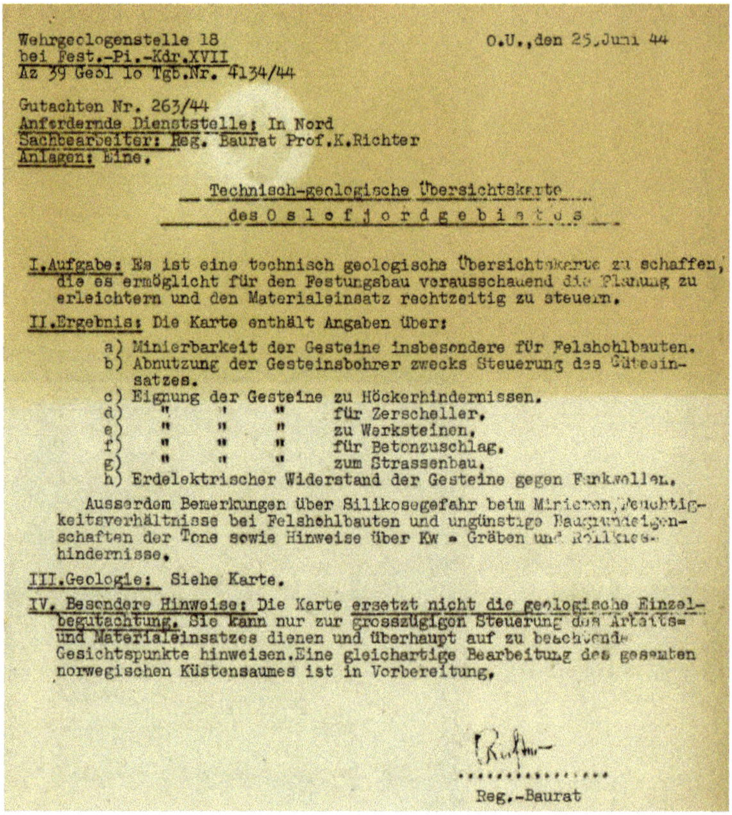

Wehrgeologenstelle 18 O.U., den 25.Juni 44
bei Fest.-Pi.-Kdr.XVII
Az 39 Geol 1o Tgb.Nr. 4134/44

Gutachten Nr. 263/44
Anfordernde Dienststelle: In Nord
Sachbearbeiter: Reg. Baurat Prof.K.Richter
Anlagen: Eine.

Technisch-geologische Übersichtskarte
des Oslofjordgebietes

I.Aufgabe: Es ist eine technisch geologische Übersichtskarte zu schaffen, die es ermöglicht für den Festungsbau vorausschauend die Planung zu erleichtern und den Materialeinsatz rechtzeitig zu steuern.

II.Ergebnis: Die Karte enthält Angaben über:

a) Minierbarkeit der Gesteine insbesondere für Felshohlbauten.
b) Abnutzung der Gesteinsbohrer zwecks Steuerung des Geräteeinsatzes.
c) Eignung der Gesteine zu Höckerhindernissen.
d) " " " für Zerscheller.
e) " " " zu Werksteinen.
f) " " " für Betonzuschlag.
g) " " " zum Strassenbau.
h) Erdelektrischer Widerstand der Gesteine gegen Funkwellen.

Ausserdem Bemerkungen über Silikosegefahr beim Minieren, Feuchtigkeitsverhältnisse bei Felshohlbauten und ungünstige Baugrundeigenschaften der Tone sowie Hinweise über Kw - Gräben und Rollkieshindernisse.

III.Geologie: Siehe Karte.

IV. Besondere Hinweise: Die Karte ersetzt nicht die geologische Einzelbegutachtung. Sie kann nur zur grosszügigen Steuerung des Arbeits- und Materialeinsatzes dienen und überhaupt auf zu beachtende Gesichtspunkte hinweisen. Eine gleichartige Bearbeitung des gesamten norwegischen Küstensaumes ist in Vorbereitung.

....
Reg.-Baurat

Fig. 4. Front page of 'expert opinion' (Gutachten) number 263, dated 25 June 1944, compiled by Professor K. Richter of the military geology unit WG18, to accompany an engineering geological map of the Oslo Fjord region (author's collection).

- investigation of thixotropy as a decisive factor triggering sudden flows of post-glacial clays (see Ackermann 1948);
- prevention of landslides by making use of Norwegian construction reports, notably for localities SW of Trondheim;
- evaluation of subsurface stability and the potential for liquefaction of quick clays through vibrations triggered naturally or by aerial bombing.

Despite the wide range of these tasks, they seemingly generated fewer technical reports than those known by contemporary geologists serving with the German army (cf. Table 2). Two numbered leaflets consisted of:

(1) a general geotechnical guideline for the sampling of undisturbed weak rocks on construction sites, extended by a geotechnical guideline for frost heave experiments on soils and weak rocks potentially subject to freezing;
(2) a reconnaissance for gravel to facilitate the production of concrete to be used in the construction of an airfield on the Ørland Peninsula NW of Trondheim. This included a geological sketch map at scale of 1 : 600 000 based on published literature together with ground reconnaissance by motorboat and by flights using a small aeroplane (a Fieseler Storch). It identified two potential gravel pits, one with reserves estimated at <10 000 m^3 in volume, but the other (at Orkanger, SW of Trondheim) with reserves likely to exceed 1 000 000 m^3.

From other reports, it appears that engineering geological investigations undertaken by air force geologists consisted of both geotechnical studies of strong rocks for the construction or excavation of fortifications and also soil mechanics studies for airfield construction. The following two examples serve to illustrate the type of reports generated.

Guidelines for military geology reports on the excavation of caverns in rock

The cover page for this report (Fig. 7) shows in its top left-hand corner that it was issued by a geologist

lfd. Nr.	Gesteinsart.	Minierbarkeit.	Abwalzung der Gesteinsbohrer	Eignung zu Höckerhindernissen	Eignung für Zerscheller	Eignung zu Werksteinen	Eignung zu Betonzuschlag.	Eignung zum Straßenbau	Erdelektr. Widerstand gegen Funkwellen.		Bemerkungen.
1.	Kiesig-sandige Endmoräne der letzten Eiszeit.	schlecht.		Findlingsblöcke meist zu klein.	Findlingsblöcke meist geeignet.		Meist gut, oberste 2.0 m Humussäure haltig, gelegtl. Schwefelkieshaltig.	gut.	mäßig bis stark.		Stellenweise zu Rollkieshindernissen geeignet.
2.	Ton und Sand der Eiszeit und Nacheiszeit.	schlecht bis sehr schlecht.					Tone ungeeignet sonst wie bei 1.	Tone ungeeignet sonst gut.	wechselnd.		Große Rutschgefahr der Tone, Tone in mehreren Metern Tiefe meist breiig, Möglichkeit von Schlammgräben.
3.	Permische Konglomerate.	mittelmäßig zeitweise naß	stark.	ungeeignet.	ungeeignet.	ungeeignet.	meist ungeeignet.	brauchbar.	stark.		
4.	Sandstein u. Quarzit des oberen Silurs.	mittelmäßig.	mäßig bis stark.	ungeeignet.	ungeeignet.	zu Fliesen.	zu plattig.	brauchbar.	stark.		Silicosegefahr beim Minieren.
9.	Essexit-basaltähnliches Gestein der Permzeit.	Gut aber sehr hart.	sehr stark	meist zu spaltenreich sonst gut falls nicht blasig.	gut.	ungeeignet.	meist sehr gut.	sehr gut.	gering.		
10.	Smaalensgranite.	sehr gut.	stark.	gut.	gut.	gut.	gut.	sehr gut.	stark.		Silicosegefahr beim Minieren
11.	Gneis u. alter gepreßter Granit.	meist gut.	stark bis sehr stark.	wechselnd, meist brauchbar.	wechselnd, meist brauchbar.	meist ungeeignet.	meist gut gelegentlich schwefelkieshaltig.	meist gut bis sehr gut.	meist stark, bei viel dunklen Anteilen geringer.		Silicosegefahr beim Minieren.

Fig. 5. Extracts from the key to an engineering geological map at scale of 1 : 250 000 of the Oslo Fjord region to which Figure 4 refers (author's collection).

(Geol.) based within the senior construction staff for fortification (Befestigungs-Oberbaustab) of the air force (d.Lw. = der Luftwaffe). The title follows, centred and underlined, followed by references to earlier reports relevant to the topic. These consist of preliminary guidelines for underground excavation in rock issued for the German army by the Inspector of Fortresses (North) in 1942; 'construction instructions' numbered 5 and 9 issued subsequently by the Luftwaffe fortification construction staff; a military geology leaflet (A0–A4) issued by the geologist at the Aerial District Command Norway; and leaflets numbered 5, 11 and 12 issued by the geological section based at the army's Inspectorate of Fortresses (North) (leaflets listed by those numbers in Table 2 herein). The report was evidently written after 27 June 1943 (the date shown for 'construction instructions' number five); it indicates that both the German army and the Luftwaffe had staffs that included geologists as such and that there could be cross-references between technical reports compiled by these two separate branches of the German armed forces.

The title page continues by listing four things that must be known to a military geologist before preparing a technical report on cavern construction: (1) the purpose for which the facility was to be excavated; (2) profiles for the access routes and the chambers to be utilized; (3) tactical requirements determining its position and the number of entrances; and (4) the workforce available.

The title page then specifies that such military geological reports should routinely include sections dealing with:

(1) Site investigation.
(2) Appraisals of construction design.
(3) Advice during tunnelling.

The guidelines further include pioneering information on rock mechanics (Häusler 2015), including a classification of rock types according to their strength and suitability for underground excavation (Table 3). The different classes of rock type required different thicknesses of *in situ* rock (ranging from 3 to 9 m) to remain above the excavated caverns if they were to be safe from damage by aerial bombing.

Although not explicitly stated in the guideline text, it can be assumed that the values recorded for uniaxial compressive strength were determined

Fig. 6. Damage in the Lade district, NE Trondheim, caused by mass movement of quick clays. Photograph taken by the Organisation Todt's Area Control Staff for Middle Norway (cf. Figs 9 and 10), 2 August 1944, now in author's collection (see Ackermann 1948).

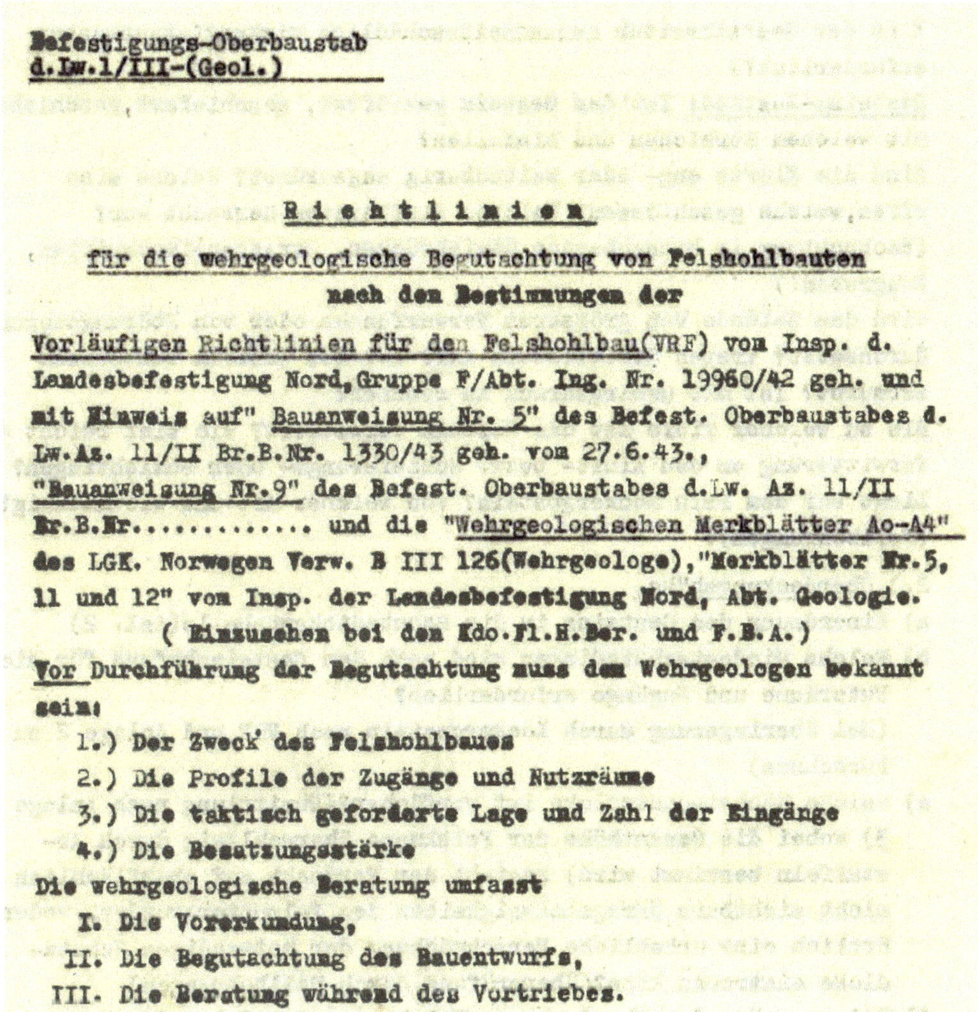

Befestigungs-Oberbaustab
d.Lw.1/III-(Geol.)

Richtlinien

für die wehrgeologische Begutachtung von Felshohlbauten
nach den Bestimmungen der

Vorläufigen Richtlinien für den Felshohlbau(VRF) vom Insp. d. Landesbefestigung Nord,Gruppe F/Abt. Ing. Nr. 19960/42 geh. und mit Hinweis auf" Bauanweisung Nr. 5" des Befest. Oberbaustabes d. Lw.Az. 11/II Br.B.Nr. 1330/43 geh. vom 27.6.43.,
"Bauanweisung Nr.9" des Befest. Oberbaustabes d.Lw. Az. 11/II Br.B.Nr............. und die "Wehrgeologischen Merkblätter A0-A4" des LGK. Norwegen Verw. B III 126(Wehrgeologe),"Merkblätter Nr.5, 11 und 12" vom Insp. der Landesbefestigung Nord, Abt. Geologie.
(Einzusehen bei den Kdo.Fl.H.Ber. und F.B.A.)

Vor Durchführung der Begutachtung muss dem Wehrgeologen bekannt sein:

1.) Der Zweck des Felshohlbaues
2.) Die Profile der Zugänge und Nutzräume
3.) Die taktisch geforderte Lage und Zahl der Eingänge
4.) Die Besatzungsstärke

Die wehrgeologische Beratung umfasst

I. Die Vorerkundung,
II. Die Begutachtung des Bauentwurfs,
III. Die Beratung während des Vortriebes.

Fig. 7. Part of the front page of a report compiled by an air force geologist providing guidelines for military geological expert opinions on the excavation of caverns in rock, based on references up to 27 June 1943 (author's collection).

from laboratory tests on the various rock types listed. The guideline text amplifies the data summarized in the table with regard to 'suitability for mining' and notes that the data relate to fresh, unweathered rocks. Therefore, depending on the rock fabric as observed in the field, a particular rock type might be moved up or down by one class in the classification. A rock that showed moderate jointing and slight weathering needed the minimum 'protection thickness' to be increased by one class; a rock that showed strong (especially vertical) jointing or a high degree of weathering (and so decomposition) required the minimum thickness to be increased by two or three classes. The 'protection thickness' shown in the table referred specifically to that required above chambers and caverns (such as those used for storage) of relatively large dimensions. Access tunnels of smaller dimensions could have the protection thickness reduced by up to 50% according to diameter, but not to less than 3 m. These guidelines were to provide the template for future reports guiding the excavation of underground facilities in Norway by the German air force.

Explanation to accompany a military geological map of the Ørland peninsula

On 10 December 1941, about two years earlier than the guidelines described in the preceding section, a 'brief' (Kurze) map description (Fig. 8) was issued

Table 3. *Classification of rock types in Norway according to their strength and suitability for underground excavation*

Class	Rock quality designation	Uniaxial compressive strength (kg cm^{-2})	Rock type	Protection thickness (m)	Suitability for mining	
					Drilling	Blasting
0	Very weak	<250	Boulder clay (dry) Shale	9.00	Very easy to drill (hand auger)	Breaking-in shots using low blasting explosives
1a	Weak	250–700	Marly limestone Clayey sandstone Dolomite (sandy)	8.50	Easy to drill (hand auger; rotating rock drilling)	
1b	Weak and intact		Slate Phyllite Mica schist	8.00		
2a	Medium strong	700–1500	Dolomite, marble Quartz sandstone Quartzitic breccia Gneiss, rich in mica Granite, rich in mica; Pegmatite	7.00	Moderately hard to medium difficult to drill (light hammerdrill)	Breaking-in shots using high blasting explosives (dynamite or blasting gelatine safety explosives) Perimeter shots using low blasting explosives
2b	Medium strong and intact		Serpentinite Amphibole schist Greenschist Gabbro, coarse-grained	6.00	Medium hard to difficult to drill (light to heavy hammerdrill)	
3a	Strong	1500–2200	Siliceous limestone Gneiss, poor in mica and feldspar Quartzite Greywacke Granite, fine-grained			
3b	Strong and intact		Syenite, fine-grained Siliceous schist Amphibolite, fine-grained Diorite Gabbro, fine-grained	5.00	Difficult to very difficult to drill (heavy hammerdrill; eventually use of sintered carbide drilling rod)	Breaking-in shots and perimeter shots using high blasting explosives
4	Very strong and partly intact	2000–2600	Hornfels Eclogite Porphyrite Diabase	4.00	Very difficult to drill	
5a	Extremely strong	>2600	Basalt	3.00		Breaking-in shots using blasting gelatine; perimeter shots using dynamite
5b	Extremely strong and intact		Quartz porphyry			

From guidelines issued for the German air force in late 1943; geotechnical translation after Bucksch (1998).

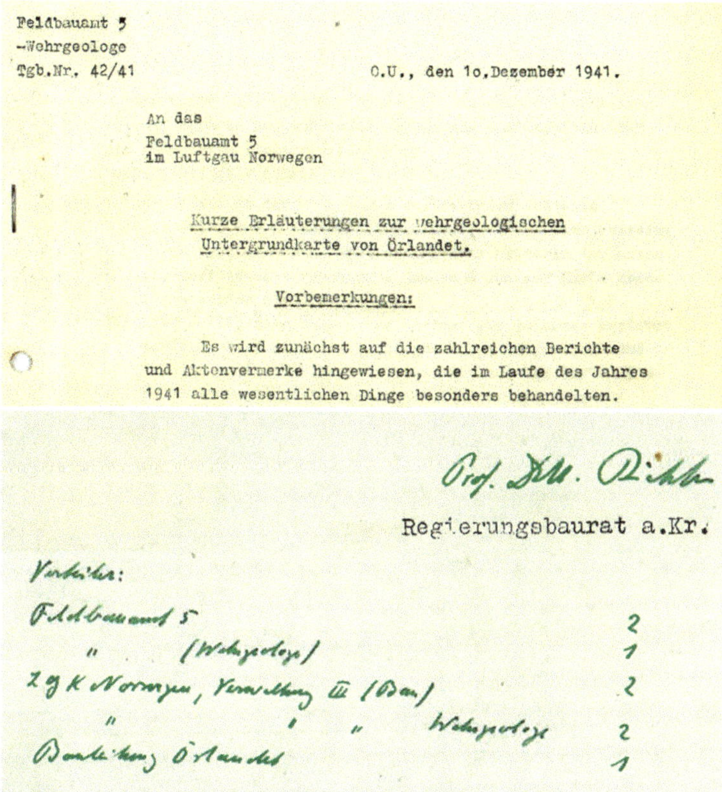
Feldbauamt 5
-Wehrgeologe
Tgb.Nr. 42/41 O.U., den 1o.Dezember 1941.

An das
Feldbauamt 5
im Luftgau Norwegen

Kurze Erläuterungen zur Wehrgeologischen Untergrundkarte von Örlandet.

Vorbemerkungen:

Es wird zunächst auf die zahlreichen Berichte und Aktenvermerke hingewiesen, die im Laufe des Jahres 1941 alle wesentlichen Dinge besonders behandelten.

Prof. Dr. M. Richter
Regierungsbaurat a.Kr.

Verteiler:
Feldbauamt 5 2
" (Wehrgeologe) 1
Lg K Norwegen, Verwaltung III (Bau) 2
" " " Wehrgeologe 2
Bauleitung Örlandet 1

Fig. 8. Extracts from the first and last pages of expert opinion number 42, dated 10 December 1941, compiled by geologist Professor Dr Max Richter serving with the air force's Field Works Office No. 5, a brief explanation to accompany a military geological bedrock map of the Ørland Peninsula, NW of Trondheim (author's collection).

under the signature of Professor Dr Max Richter, then serving in the rank of Regierungsbaurat auf Kriegsdauer (= major) as the military geologist (Wehrgeologe) assigned (probably temporarily, for this specific task) to the 5th Field Works (or Construction) Office (Feldbauamt) of the Luftwaffe's Aerial District Command Norway. The preliminary remarks (Vorbemerkungen) indicate that numerous reports and file notes were generated earlier in 1941, providing data from which the map and its description were compiled. Handwritten annotation shows that eight copies of the 'explanation' typescript were produced, their distribution including two to the Field Works Office requiring the report, one to the author and two to a military geologist higher in the air force chain of command in Norway. The number of copies and their distribution locally and up the chain of command was fairly typical for German military geological reports in general.

The report describes a geological map of the Ørland (Ørlandet) Peninsula, compiled at a scale of 1 : 20 000, essentially from a pre-war civilian map sheet published in Norway at a scale of 1 : 100 000, but with its topographic detail enhanced by aerial photography. The bedrock geology consisted of granite, gneisses and amphibolite schists, plus weakly metamorphosed sandstones and breccias of Late Silurian age. A partial cover of superficial deposits included glacial boulder clay and post-glacial blue clay to a thickness of several metres, and coastal terraces with a cover of sand and gravel ranging from 2 m thick in the north of the peninsula to 6 m in the south, widely surfaced by peat. Peat was commonly 10–20 cm in thickness, locally >1 m and up to a maximum of 2 m thick.

The surface and subsurface geology was interpreted in terms of its relevance to construction works, sources of raw materials and use of groundwater. Granite occupied the northern upland part of the peninsula. This had an even, medium-grained texture and was low in accessory minerals and pyrite. Accordingly, it was judged suitable for use as paving stone, pitched stone work, crushed rock and concrete aggregate – with about 10 000 000 m^3 available.

Elsewhere, gneisses were also in general very strong and could also be used as hardcore or crushed rock. Silurian sedimentary rocks cropped out along the southern coast and consisted of silicified sandstone, breccias, conglomerates and argillaceous slates. Except for the slates, these rocks were strong enough to be used as crushed rock and hardcore and were quarried for building stone.

The Quaternary blue clay was typically overlain by loam and peat. It was firm when dry, but slippery when wet. Investigations, guided by soil mechanics, revealed a maximum load-bearing capacity of 1.5 kg cm^{-2}. Consequently, for a hangar of 8 m^2 floor area with foundations 1.5 m deep, a potential subsidence of 5 cm was predicted for the final building. The boulder clay was slightly firmer than the blue clay, with a load-bearing capacity up to 2.5 kg cm^{-2}. The loam above the clay, with a particle size ranging from sand to clay grade, was typically water-saturated and highly susceptible to frost, creating heave ranging between 2 and 5 cm. Coastal sand on the northern shore of the peninsula was rich in mollusc shells and contained 98% calcium carbonate, whereas on the southern shore it was richer in gravel and free of carbonate due to the action of humic acids. The high content of pyrite and humic acids prevented the use of the coastal sand and gravel for concrete aggregate, but not their use as a filling material for drainage trenches. Gravel suitable for concrete was located near the village of Orkanger, *c* 40 km to the south of the region, SW of Trondheim. The large quantities of peat in the region were deemed a potential source of heating material.

Usable groundwater was assessed as rare on the peninsula. Coastal sand and gravel, thin in the north but thickening to the south, constituted the primary aquifer suitable for exploitation by shallow wells. The water was suitable for drinking if filtered. Potable water was also obtained from springs in the hills forming the western side of the peninsula. Surface water could be used in concrete production because its ion content was suitably low. However, concrete foundations in peaty soil were subject to attack by highly acidic pore waters (40 mg l^{-1}, well above the maximum of 8 mg l^{-1} then permitted by building regulations). Such attack could be avoided by the use of carbonate-free cement, by a protective coat of paint or by installing appropriate drainage.

The map and report thus illustrate many of the aspects of geology deemed to be of interest to German air force geologists from an early stage in the occupation of Norway.

Geologists of the Organisation Todt

The development of the Organisation Todt as a paramilitary construction agency can be determined from a handbook published in English in the final year of the War (Supreme Headquarters Allied Expeditionary Force, Counter-Intelligence Sub-Division 1945), accounts by Dorsch (1950, 1998) and particularly by Absolon (1971), and documents studied at the German Federal Archives. During the war it became responsible for the construction of all the major German military works across Europe, led by its founder the engineer Dr Fritz Todt until his death in an aeroplane accident in February 1942. Thereafter, for the rest of the war, control was vested in the Reichminister for Armament and War Production, Albert Speer.

As the Organisation Todt developed into an organization of *c.* 1.5 million, perhaps 2 million workers, it developed its own command structure (Fig. 9), intended to operate both during the war and afterwards. Under Speer's overall command, a central office (Organisation Todt Zentrale) was directed by the engineer Xaver Dorsch. Organisation Todt Zentrale had responsibility over several specialist sections, notably that led by the Organisation Todt Engineer-in-Chief, Leo Casagrande. This section, in turn, had technical responsibility for other specialist sections and for the main regional executive and operational units, the Einsatzgruppen (Area Control Staffs: Army Group Level). Each Einsatzgruppe functioned in a region effectively equivalent to that of an Army Group, with component Einsatz units (Area Control Staffs: Army Level). By June 1944, eight Einsatzgruppen were based in Germany, one (Einsatzgruppe Wiking) in Norway, one in SE Europe and another in Italy. The Norwegian Einsatzgruppe Wiking was composed of four Einsätze, one each for the polar regions, northern Norway, middle Norway and Finland. Politically, the Organisation Todt as a whole was placed under control of the Nazi Party's SS organization, with an SS liaison officer in every major echelon.

The planning and development of fortifications in any country, particularly with regard to their position and time plan for construction, was ultimately the responsibility of the Inspector of Fortresses, a general within the Engineer Branch, based at the High Command of the Army in Berlin. However, responsibilities were delegated to regional commanders of Fortress Engineers as appropriate. The Organisation Todt worked in close liaison with, and was directed by, the engineers.

Each Einsatz was further divided into a number of component regions, whose work was directed by an Oberbauleitung, the basic operational sector of any large Organisation Todt region of activity, a Construction Sector and Administrative Headquarters that functioned as a principal works management office. Work within an Oberbauleitung sub-region was directed by Bauleitungen (works management offices) and/or Abschnittsbauleitungen (section

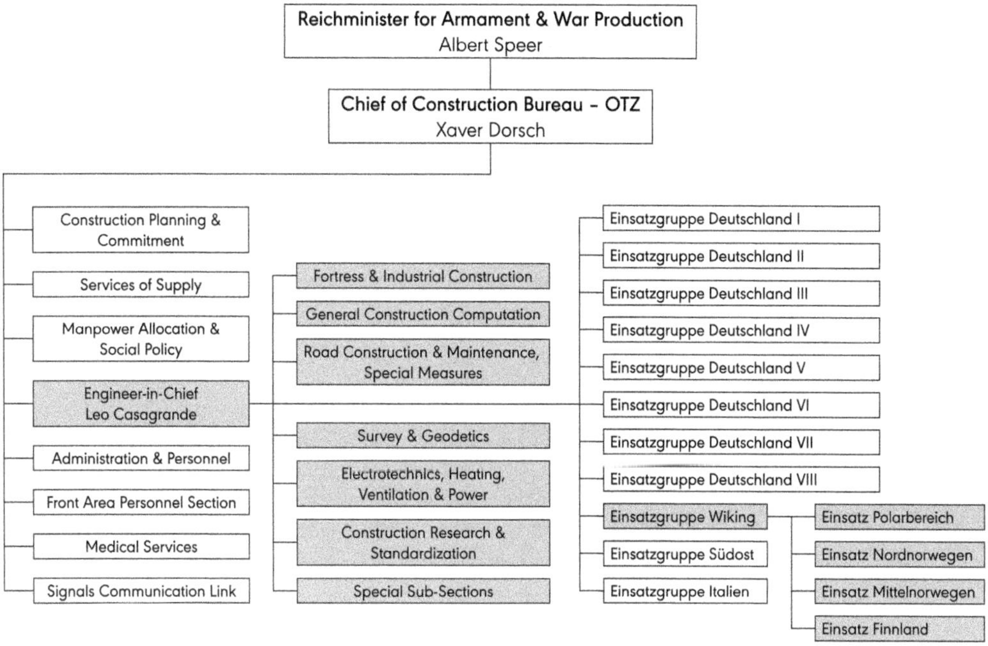

Fig. 9. Administrative structure of the Organisation Todt as at June 1944 (after Supreme Headquarters Allied Expeditionary Force, Counter-Intelligence Sub-Division 1945), with grey stipple indicating the Engineer-in-Chief (Professor Dr-Ing. Leo Casagrande) at its central office (OTZ) in Berlin, his areas of responsibility within the subordinate Area Control Staffs (Einsatzgruppen), such as that for Norway (Einsatzgruppe Wiking), and the four sub-areas within Wiking.

works offices) or sometimes Leitstelle (control centres), all of which had control of varying numbers of Baustellen (construction sites). By May 1943 such units were numerous and widespread in Norway, indicating that major Organisation Todt construction work was in progress (Fig. 10).

The range of work involving the Organisation Todt was considerable. It undertook the construction of the National Trunk Road (Reichstrasse 50), which connected the southern part of Norway to the north and continued to the Finnish border; the extension of the Trondheim–Narvik railway system; the construction of power stations; and the construction of facilities to produce heavy water for the German nuclear programme. It was also involved in mining for metal ores, for nickel in Finland and for iron at Kirkenes near the Russian border (Fig. 1). However, a major effort focused on construction work associated with the Atlantic Wall, which began in the spring of 1942. In Norway, this consisted mainly of numerous fortifications, defended ports and coastal batteries. Fortifications, in turn, involved the construction of reinforced concrete bunkers to a standard design at key points along the coast and on selected offshore islands.

According to Rolf (1983), by April 1944 the Organisation Todt in Norway had a workforce of *c.* 69 000 employees and labourers (the latter including 13 000 prisoners of war and >1000 convicts). The workforce increased to *c.* 80 000 personnel by the middle of 1944, of whom *c.* 50% were engaged in coastal fortification works. Overall, in 1944, rock excavation in Norway by the Organisation Todt had a total volume of 161 000 m^3 and its construction works consumed 207 000 m^3 of reinforced concrete.

Häusler (2015) identified eight military geologists who served exclusively with the Organisation Todt in Norway between 1940 and 1944 (Table 1), seven from Austria: Werner Heissel, Franz Kahler, Alois Kieslinger, Karl Metz, Leopold Müller, Joseph Schadler and Walter Zanoskar (in alphabetical order). Three of these are illustrated in Figure 11. In 1944, when Luftwaffe geologists were transferred to the Organisation Todt, the total number of geologists then serving in Norway with the Organisation Todt increased to 23 (11 from Germany and 12 from Austria).

Alois Kieslinger used his professional experience in Norway to generate three publications post-war (Kieslinger 1953, 1958, 1960). His paper on remnant tension and the relief of tension (Kieslinger 1958) is perhaps particularly significant in inferring that the vertical movement and exfoliation of granite

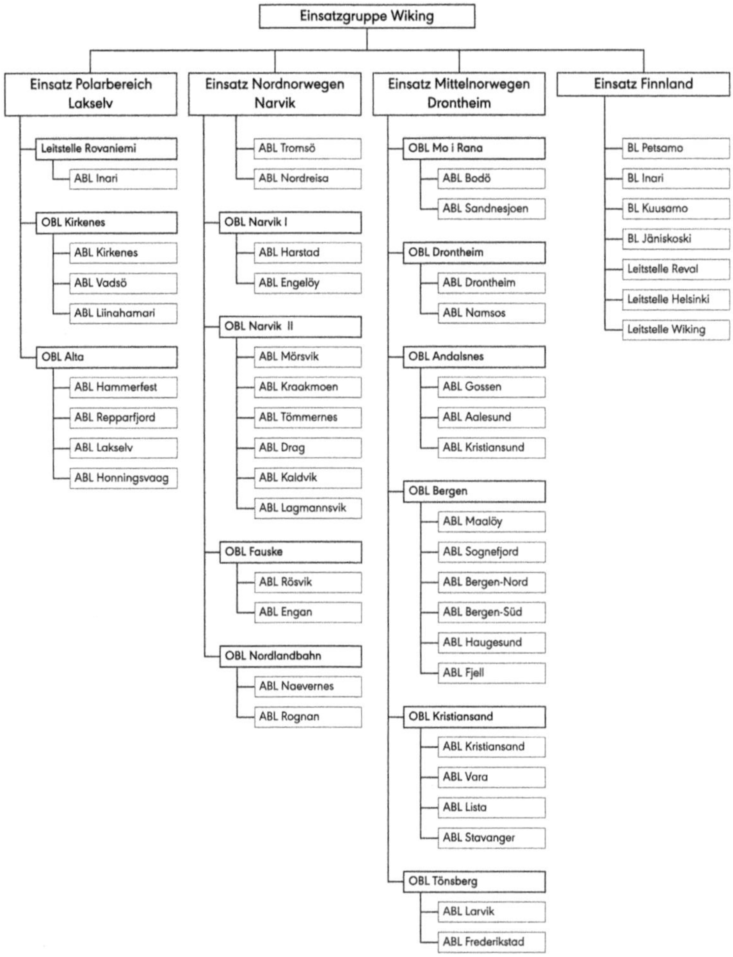

Fig. 10. Administrative structure of the Organisation Todt within Norway (Einsatzgruppe Wiking, cf. Fig. 9) as at May 1943, indicating towns where sub-area control staffs were based (e.g. German Drontheim = Norwegian Trondheim, in Middle Norway), together with control centres (Leitstellen), principal construction regions (OBL = Oberbauleitung), works management offices (BL = Bauleitung) and section works offices (ABL = Abschnittsbauleitung).

domes in northern Norway were the result not of post-glacial rebound, but near-surface stress release in the granite (cf. Fig. 12).

Diaries and maps compiled by Joseph Schadler prove that he worked with the Organisation Todt in Norway from at least 1942 to 1943. One example of his work dates from November 1942: a geological map at scale of 1 : 10 000 for the region near Øysand, adjoining the Gulosen Fjord, *c.* 20 km SW of Trondheim (Figs 1 & 13). The bedrock is mapped in terms of 14 units, shown in cross-section to be overlain by Quaternary sediments and by more recent deltaic deposits of the River Gaula (Häusler 2015). Schadler's documents make no mention of the purpose for his map, but it is known from other sources (e.g. https://en.wikipedia.org/wiki/Nordstern) that the Germans planned to construct a new harbour of strategic importance in this area and to construct a new metropolis (named Nordstern) to house a population of *c.* 250 000–300 000 inhabitants. Planning for this dated back to May 1941, but construction only began in 1943.

A second example of Schadler's work dates from April 1943: a geological map at scale of 1 : 16 000 of the 46.5 km^2 island of Gossa (Gossen), NW of Molde (Fig. 1). Mapping was completed in only five days. Broadly similar to Øysand, the bedrock consisted of strongly folded red biotite gneiss, amphibolite, amphibolite gneiss and garnet amphibolite, seen in many coastal outcrops to be intruded by

Fig. 11. Photograph, taken in 1943 or 1944 outside a barracks in Norway, showing members of the Organisation Todt, of whom three can be identified as military geologists Alois Kieslinger, Franz Kahler and Karl Metz (the last two wearing a brassard on the left upper arm with the Organisation Todt symbol); after Häusler (2015), reproduced courtesy of the Kieslinger estate.

pegmatite. The basinal areas of the central island contained boulder clay of Quaternary age overlain by post-glacial gravels and coarse to fine dune sand. Beach sands fringed the coast on terraces at heights of 25–50 m, overlain by peat from 0.5 to 2.5 m in thickness. This map was required to help

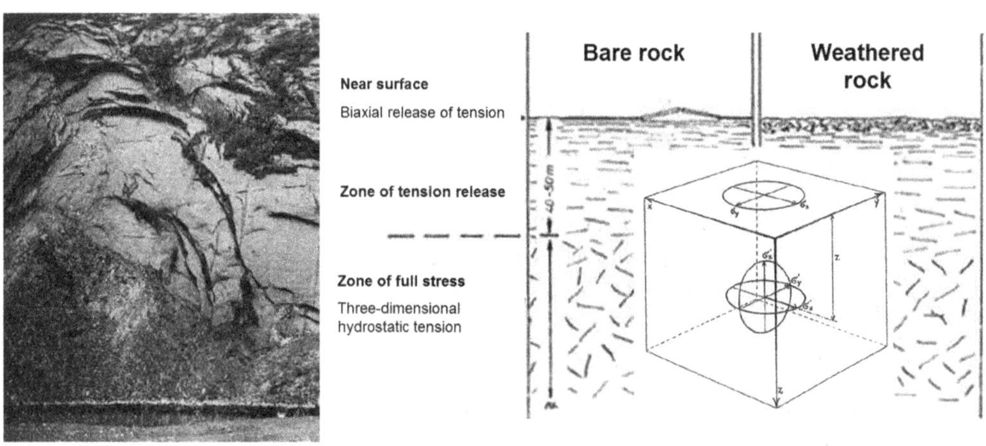

Fig. 12. Exfoliation of granite attributed to near-surface stress ('tension') release, photographed at Hellemobotten near Narvik, Norway, by Alois Kieslinger in 1942 (see Kieslinger 1958), with diagram indicating change from triaxial stress in zone of full stress (below 50 m depth) to biaxial stress nearer the ground surface; after Kieslinger (1960) and Häusler (2015).

Fig. 13. Geological map of Øysand adjacent to Gulosen Fjord, 20 km SW of Trondheim, original at scale of 1 : 10 000, with key to (igneous or metamorphic) bedrock and cross-section south to north, completed by Josef Schadler on 30 November 1942; after Häusler (2015) (author's collection).

plan the fortification of the island; remnants of German fortification can still be seen there (www.bunkerpictures.nl).

Additional to the fortification associated with the Atlantic Wall, work planned in 1940 and begun by the Organisation Todt on a massive scale in 1941 included bunkers to serve as bases for submarines (U-boats) at both Bergen and Trondheim (Fig. 1) (https://uboat.net, https://en.wikipedia.org/wiki/Dora_I). At Bergen, the bunker, when complete, consisted of seven pens: three dry docks plus three wet docks, plus a seventh pen for storing fuel, torpedoes and other essentials. The bunker itself had a roof 6 m thick and occupied an area of 131 m × 143 m. It could house up to nine U-boats. At Trondheim, the Organisation Todt had charge of the construction of two bunkers, named Dora 1 and Dora 2, needed to provide repair facilities for submarines outside Germany itself and forming part of the largest German naval base in northern Europe. Dora 1 was completed by 1943; at 47 m long and 32 m wide it had five pens and space to hold seven U-boats. The concrete roof was 3.5 m thick and reinforced with steel. Construction of Dora 2 only began in 1942 and was never completed. If finished, it would have been 51 m long by 31 m wide, with four pens, and been capable of holding six U-boats. Excavation at Trondheim involved a volume of *c.* 270 000 m^3 of material, a massive achievement at the time.

Dora 1 (Fig. 14) was excavated through soil into subsoil that consisted of uniform, very silty clay down to a depth of *c.* 100 m, a sequence of post-glacial lacustrine sediments. In the undisturbed state the material was weak and in a worked state it became almost liquid. Excavation of this material proved difficult and the problem was reported to the Organisation Todt's Engineer-in-Chief in Berlin,

Fig. 14. Aerial photograph of excavation for a U-boat bunker at Trondheim, probably taken in 1941, showing the entire site surrounded by two rows of steel-sheet piling facilitating electro-osmotic stabilization during excavation; from Casagrande (1952).

Dr Leo Casagrande. It was he who was to develop an improved technique for the electro-osmotic stabilization of soils (Casagrande 1952).

The site lay immediately adjacent to the open sea and so it was decided to surround it entirely by two rows of sheet piling with embedded lengths of 20 m. In addition, two berms were provided to add safety to the excavation. Despite these major protective measures, which were calculated to be adequate defence against any likely construction problems, the sheet piling failed in several places. In one area a considerable length of the sheet pile wall moved horizontally towards the excavation without tilting. A further problem was that excavation could be made to only half of its intended depth of 15 m. When a depth of *c.* 8 m was reached, no deeper excavation proved possible and the ground heaved and filled within a few seconds. It proved impossible to operate excavation plant inside the site as originally planned.

These problems compelled the contractor to accept the recommendation of Leo Casagrande that electro-osmotic stabilization be applied. For that purpose, two rows of electrodes were driven to a depth of 30 m. The cathodes consisted of slotted 8-inch (200 mm) diameter pipes covered with a fine copper mesh and spaced 10 m apart. Common gas pipes were used for the anodes, sited halfway between the cathodes. The soil was found to contain a high concentration of salts and therefore the intended potential of 90 V had to be decreased to 40 V so as not to exceed the capacity of the generator. Two days after the potential was applied to the electrodes, it became possible to continue the excavation to full depth without difficulty and to operate the excavation plant on site as intended (Casagrande 1952).

Leo and Arthur Casagrande: pioneers of soil mechanics on opposite sides

Leo Casagrande (1903–90) was born at Haidenschaft in Slovenia, a town north of Trieste in NE Italy, in a German-speaking part of what was then the Austrian Empire (Tedesko & Peck 1992; https://de.wikipedia.org/wiki/Leo_Casagrande). He was educated at school in Austria, followed by five and a half years of engineering courses at the Institute of Technology in Vienna. He then began his professional career in Germany as a structural engineer in Bavaria, but in 1930 moved to the USA as a research assistant at the Massachusetts Institute of Technology to work with Professor Karl von Terzaghi (renowned as the father of soil mechanics). In 1932 he returned to the Institute of Technology in Vienna to assist Terzaghi, who had been appointed as its first professor of soil mechanics. However, in 1933 he accepted a teaching appointment at the Technological University of Berlin to organize a soil mechanics institute. In 1934 he was asked to head the Soil Mechanics Division of the Office of the Inspector General for German Highways and so began a phase in his career that led eventually to his appointment in 1943 as the Engineer-in-Chief at Organisation Todt headquarters. During the period 1934–45 he solved many problems in foundation and earthworks engineering and in soil stabilization. He developed the principles and practice of electro-osmotic dewatering and stabilization of soils and used it extensively. He rapidly became recognized as one of the most knowledgeable soil mechanics engineers in Europe, experienced both in research and in common-sense practical applications. From 1940 until the end of the war he additionally held a teaching post at the Institute of Technology in Braunschweig, which entitled him to be addressed as Professor.

Arthur Casagrande (1902–81) was Leo's elder brother (Casagrande 1982; https://en.wikipedia.org/wiki/Arthur_Casagrande). His early career was nearly identical to that of Leo, but after graduating from the Institute of Technology in Vienna in 1924 (the year of his father's death), he remained there briefly as a professorial assistant before moving to the USA in 1926 in search of more lucrative employment to support the Casagrande family. He worked from 1926 to 1932 as a research assistant with the US Bureau of Public Roads, assigned to the Massachusetts Institute of Technology, where he assisted Terzaghi in numerous research projects. In 1932 he moved to Harvard University, where he was appointed to a newly created chair of soil mechanics and foundation engineering in 1946. At Harvard he established a school of postgraduate teaching and research that grew from 12 students in 1932 to more than 80 after World War II. During the war, between 1942 and 1944, he trained 400 army officers in the soil mechanics aspects of airfield construction through a series of intensive four-week programmes at the request of the US Army Corps of Engineers. He had assisted the Corps with projects from 1935 onwards and developed the Unified Soil Classification System adopted by the Corps in 1942. Thus while Leo was making a significant soil mechanics contribution to the German engineering war effort, his brother was making a significant soil mechanics contribution to that of the Allies.

After the war the brothers came together again. Leo, as a scientist of particular distinction, was invited by the British to join their Building Research Station in Hertfordshire, then the leading organization in the UK developing the still relatively new field of soil mechanics. He duly worked there from 1946 to 1950. In 1950 he moved to the USA to join his brother Arthur, who had been promoted to professor of soil mechanics and foundation engineering at Harvard University in 1946 (a chair he held until retirement in 1973). Having shared teaching of these subjects, Leo was himself appointed a professor at Harvard in 1956 until his retirement in 1972. The two brothers developed an impressive academic record, Arthur notably becoming President of the International Society for Soil Mechanics and Foundation Engineering from 1961 to 1965 (a body renamed the International Society for Soil Mechanics and Geotechnical Engineering in 1998) and receiving many academic honours. From 1969, in addition to their academic duties, the brothers set up Casagrande Consultants, a practice based in Massachusetts. The pioneering scientific achievements for which they became internationally well-known were, in part, based on their experiences providing technical support for opposite sides in World War II.

Concluding remarks

At least 63 geologists, 44 of German and 19 of Austrian nationality (Table 1), were deployed as such in Norway or adjacent regions between 1940 and 1945 to support units of the German army, Luftwaffe or the paramilitary construction agency Organisation Todt, except for one assigned to an inter-service unit for terrain evaluation (the Forschungsstaffel). This number includes all of those employed specifically as military geologists (Wehrgeologen) and so includes men (they were all men) whose academic training was primarily in geology, palaeontology, petrology or geotechnical engineering, because all would have taken at least foundation courses in basic geology. Some served only briefly in the region, on short assignments, others remained for longer terms. The number in contemporaneous employment increased from 1940 to a peak in 1943, when the last of the army's teams (WG33) was deployed. By then, most had served on assignment to the army, fewer to the Luftwaffe, and a relatively small number to the Organisation Todt (Fig. 15). Geological priorities changed in 1944 when Dietl's army in Lapland required greater operational support and when all Luftwaffe geologists were reassigned to the Organisation Todt. In consequence, in the final year of the war 38 military geologists worked on assignment to the army, 23 to the Organisation Todt (Häusler 2015).

Austrian geologists and engineers assigned to the Organisation Todt for tasks in Norway helped to pioneer the relatively new discipline of soil mechanics. The Austrian-born Arthur Casagrande convened an International Conference on Soil Mechanics and Foundation Engineering at Harvard University in the USA in 1936 that was to lay the foundation for the present day International Society for Soil Mechanics and Geotechnical Engineering formed post-war. His efforts at Harvard were supported from 1950 by his younger brother Leo, who had served as Engineer-in-Chief at the Organisation Todt headquarters in Berlin during the war, but whose expertise was applied to some major tasks in Norway, notably the construction of the massive U-boat bunker at Trondheim.

Austrian geologists and engineers also used their experience gained in Norway during wartime to develop the discipline of rock mechanics. Leopold Müller, who had served on assignment to the

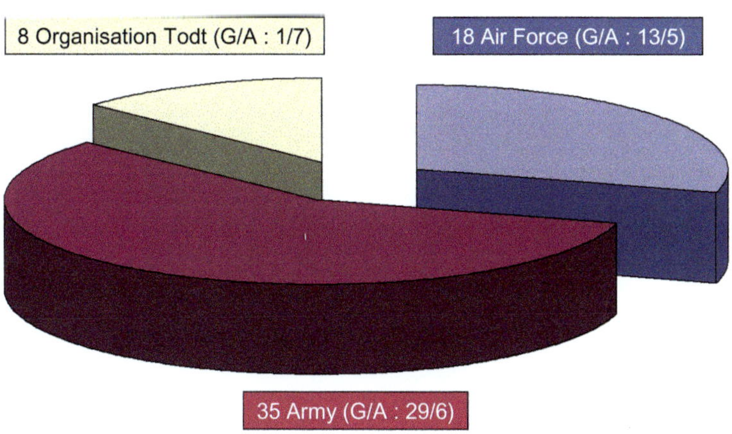

Fig. 15. Pie chart illustrating how of 61 German (G) or Austrian (A) military geoscientists were deployed in Norway between 1940 and 1944 (Table 1, excluding Nietsch, assigned to the inter-service Forschungsstaffel, and Seelig, who arrived in 1945), most (35 (57%)) served under army command, a smaller number (18 (30%)) for the air force and fewest (8 (13%)), mostly Austrians, initially for the Organisation Todt (although late in 1944 all air force geoscientists were transferred to the Organisation Todt).

Organisation Todt (Table 1), formed an Austrian Working Group for Geomechanics after the war, fostering an interdisciplinary interest from professionals working in geology, engineering geology, geophysics, structural geology, mining and construction engineering (Häusler 2015). This group of a dozen research-active geoscientists, usually meeting at Salzburg in Austria, became known as the Salzburg Circle. Müller (1971) was later to record that some of the stimulus for creation of the Circle came from wartime construction work and tunnelling in strong rocks north of Narvik in Norway. Müller was to develop an Austrian School of Rock Mechanics, from which developed the annual Salzburg Colloquium on Geomechanics, a series that attracted international participation and continues to the present day, now hosted by the Austrian Society for Geomechanics. In 1962 the Salzburg Circle developed into the now well-established International Society for Rock Mechanics, a society whose foundation owed much to Müller, who acted as its president until 1966.

Between 1957 and 1965, the Salzburg Circle led by Leopold Müller (later to be named Müller-Salzburg) worked jointly with Ladislaus von Rabcewicz and Franz Pacher to develop the New Austrian Tunnelling Method, also known as the sequential excavation method, a method of modern tunnel design and construction (Häusler 2015). The technique, which takes advantage of the inherent geological strength available in the surrounding rock mass to stabilize the tunnel, has been credited with helping to revolutionize the modern tunnelling industry, another legacy, in part, of experience gained in wartime Norway.

A yet further legacy worthy of note is that of the experience passed on to students by the many geologists and geotechnical engineers recruited or conscripted from university staffs who returned to their teaching careers after the war. Notable among these, in alphabetical order, were Ernst Hermann Ackermann, Friedrich Bachmayer, Friedrich Birzer, Leo Casagrande, Karloskar Felser, Hans-Rudolf von Gaertner, Werner Heissel, Rudolf Hohl, Franz Kahler, Alois Kieslinger, Georg Knetsch, Leo Maria Krasser, Josef Ladurner, Heinz Meixner, Karl Metz, Leopold Müller-Salzburg, Friedrich Karl Nöring, Ladislaus von Rabcewicz, Andreas Thurner, Hans Wieseneder and Helmuth Zapfe (Häusler 2015).

Acknowledgements Thanks for access to unpublished documents are due to: the Bundesarchiv (German Federal Archives) at Freiburg im Breisgau; the Bundesarchiv at Kornelimünster/Aachen; the Geological Survey of Austria, Vienna (Alois Kieslinger estate); the Institute of Engineering Geology, Vienna University of Technology (Leopold Müller-Salzburg estate); and the Norwegian Geotechnical Institute, Oslo (Karl von Terzaghi estate). The author is grateful to P.H. Banham and T. Davenport for reviewing the paper as first submitted, and also to E.P.F. Rose for his significant help to improve it.

Appendix A

Glossary of German/English military and geotechnical terms used in the text, with some standard abbreviations.

German	English
Abschnittsbauleitung (ABL)	Organisation Todt section works office
Aussenstelle	Outstation
Armee Ober Kommando (AOK)	Army High Command (for a numbered army)
Beamter	Government official
Beton	Concrete
Bauleitung (BL)	Organisation Todt works management office
Baustelle	Construction site
Befehlsstelle	Command centre, headquarters
Befestigung	Fortification
Bericht	Report
Einbruchschuss	Breaking-in shot
Einsatz	Organisation Todt area control staff, single army level
Einsatzgruppe	Organisation Todt area control staff, Army Group Level
Feldbauamt	Air force field works (or construction) office
Felshohlbau	Rock cavern construction
Festigkeit	Strength
Festigkeitseigenschaft	Rock quality designation
Festungspionier-Kommandeur	Fortress Engineer command

(Continued)

German	English
Festungspionierstab	Fortress Engineer staff
Forschungsstaffel zbV	Staff for special (terrain evaluation) duties
Gebirge	Mountain
Geschiebelehm	Boulder clay
Gutachten	Expert opinion; a short technical report
Inspekteur der Festungen (InFest)	Inspector of Fortresses (in Berlin)
Inspekteur der Landesbefestigung	Inspector of Fortresses (regional)
Leitstelle	Organisation Todt building control centre
Luftgaukommando	Aerial district command
Luftwaffe (Lw)	Air force
Merkblatt	Leaflet
Oberbauleitung (OBL)	Organisation Todt construction sector and administrative headquarters
Oberbaustab	Senior construction staff
Oberkommando der Wehrmacht (OKW)	Supreme command of the armed forces
Oberkommando des Heeres (OKH)	High command of the Army (in Berlin)
Regierungsbauassessor	Government construction officer = captain
Regierungsbaurat	Government construction administrative officer = major
Sprengstoff	Blasting explosive
Stollen	Tunnel or gallery
Technischer Kriegsverwaltungsrat (TKVR)	Technical war administration officer = major or captain
Tragfähigkeit	Load-bearing capacity
Wehrgeologe	Military geologist
Wehrgeologenstelle (WG)	Military geology unit, centre or team
Wehrmacht	Armed forces
Zerschellergestein	Broken rock, for external armouring of bunkers
Zuschlagstoffe	Aggregate

From Marshall (1944), Bucksch (1998) or Häusler & Willig (2000).

References

ABSOLON, R. 1971. *Die Dienst- und Rechtsverhältnisse der Angehörigen der Organisation Todt (OT)*. Sammlung wehrrechtlicher Gutachten und Vorschriften, **9**. Bundesarchiv, Kornelimünster, 80–98.

ACKERMANN, E. 1948. Thixotropie und Fliesseigenschaften feinkörniger Böden. *Geologische Rundschau*, **36**, 10–29.

BUCKSCH, H. 1998. *Wörterbuch GeoTechnik* [Dictionary of Geotechnical Engineering]. Vol. **II**. Springer, Berlin.

CASAGRANDE, D.R. 1982. Memorial to Arthur Casagrande 1902–1981. *Geological Society of America Memorials*, **13**, 1–4.

CASAGRANDE, L. 1952. Electro-osmotic stabilization of soils. *Journal of the Boston Society of Civil Engineers*, **39**, 51–83.

DORSCH, X. 1950. *Zusammenfassende Darstellung der Geschichte, Organisation und Funktion der OT sowie deren Rolle in der Kriegsvorbereitung und Führung*. Bestand Bundesarchiv, R50I/1a.

DORSCH, X. 1998. Die Organisation Todt. *In*: SINGER, H. (ed.) *Veröffentlichungen deutschen Quellenmaterials zum Zweiten Weltkrieg, Abteilung III: Quellen zur Geschichte der Organisation Todt*. Vols **1 and 2**. Biblio, Osnabrück, 437–610.

HADDEN, R.L. 2008. The Heringen collection of the US Geological Survey Library, Reston, Virginia. *Earth Sciences History*, **27**, 242–265.

HÄUSLER, H. 1995*a*. Die Wehrgeologie im Rahmen der Deutschen Wehrmacht und Kriegswirtschaft. Teil 1: Entwicklung und Organisation. *Informationen des Militärischen Geo-Dienstes, Vienna*, **47**, 1–155.

HÄUSLER, H. 1995*b*. Die Wehrgeologie im Rahmen der Deutschen Wehrmacht und Kriegswirtschaft. Teil 2: Verzeichnis der Wehrgeologen. *Informationen des Militärischen Geo-Dienstes, Vienna*, **48**, 1–119.

HÄUSLER, H. 2000. Deployment and role of military geology teams in the German army 1941–45. *In*: ROSE, E.P.F. & NATHANAIL, C.P. (eds) *Geology and Warfare: Examples of the Influence of Terrain and Geologists on Military Operations*. Geological Society, London, 159–175.

HÄUSLER, H. 2015. Von der Wehrgeologie in Norwegen 1940–45 zum 'Salzburger Kreis' der Geomechanik. *Berichte der Geologischen Bundesanstalt*, **113**, 56–84.

HÄUSLER, H. & WILLIG, D. 2000. Development of military geology in the German Wehrmacht 1939–45. *In*: ROSE, E.P.F. & NATHANAIL, C.P. (eds) *Geology and Warfare: Examples of the Influence of Terrain and Geologists on Military Operations*. Geological Society, London, 141–158.

KIESLINGER, A. 1953. Karsterscheinungen im hohen Norden. *In*: KÜPPER, H., EXNER, C. & GRUBINGER, H. (eds) *Skizzen zum Antlitz der Erde*. Hollinek, Vienna, 371–375.

KIESLINGER, A. 1958. Restspannung und Entspannung im Gestein. *Geologie und Bauwesen*, **24**, 95–112.

Kieslinger, A. 1960. Gesteinsspannungen und ihre technischen Auswirkungen. *Zeitschrift der deutschen geologischen Gesellschaft*, **112**, 164–170.

Kinder, H. & Hilgemann, W. 2000. *dtV-Atlas Weltgeschichte*. Vol. **2**. Deutscher Taschenbuch, Munich.

Kranz, W. 1938. *Technische Wehrgeologie. Wegweiser für Soldaten, Geologen, Techniker, Ärzte, Chemiker und andere Fachleute*. Jänecke, Leipzig.

Marshall, G.C. 1944. TM 30-506. War Department Technical Manual German-English Military Dictionary, www.allworldwars.com/German-English-Military-Dictionary.html

Müller, L. 1971. Die mechanischen Eigenschaften der geologischen Körper. *Carinthia II*, **28** (Festschrift Kahler), 177–191.

Müller-Salzburg, L. 1985. Latente Spannungen in der Skandinavischen Schieferhülle. *Felsbau*, **3**, 34–38.

Ramberg, I.B., Bryhni, I., Nøttvedt, A. & Rangnes, K. (eds) 2008. *The Making of Land – Geology of Norway*. Norsk Geologisk Forening, Trondheim.

Robins, N.S. & Rose, E.P.F. 2005. Hydrogeological investigation in the Channel Islands: the important role of German military geologists during World War II. *Quarterly Journal of Engineering Geology and Hydrogeology*, **38**, 351–362, https://doi.org/10.1144/1470-9236/05-016

Robins, N.S., Rose, E.P.F. & Cheney, C.S. 2012. Basement hydrogeology and fortification of the Channel Islands: legacies of British and German military engineering. *In*: Rose, E.P.F. & Mather, J.D. (eds) *Military Aspects of Hydrogeology*. Geological Society, London, Special Publications, **362**, 203–222, https://doi.org/10.1144/SP362.11

Rolf, R. 1983. *Der Atlantikwall – Perlschnur aus Stahlbeton*. AMA, Beetsterzwaag.

Rose, E.P.F. 2005*a*. Work by German military geologists on the British Channel Islands during the Second World War. Part 1: pioneering studies by Walther Klüpfel (Jersey and Alderney), Walter Wetzel (Guernsey and Alderney), and Friedrich Röhrer (Guernsey). *Channel Islands Occupation Review*, **33**, 93–120.

Rose, E.P.F. 2005*b*. Specialist maps of the Channel Islands prepared by German military geologists during the Second World War: German expertise deployed on British terrain. *The Cartographic Journal*, **42**, 111–136.

Rose, E.P.F. 2005*c*. The first hydrogeological and geological maps of Jersey, Channel Islands: work by Walther Klüpfel in 1942 and Richard Nelson c.1828. *Proceedings of the Geologists' Association*, **116**, 107–116.

Rose, E.P.F. 2007. Work by German military geologists on the British Channel Islands during the Second World War. Part 2: Bernhard Beschoren, Dieter Hoenes, and the role of *Wehrgeologenstelle 4* on Guernsey and Alderney. *Channel Islands Occupation Review*, **35**, 93–114.

Rose, E.P.F. 2008. German military geologists and fortification of the British Channel Islands during World War II. *In*: Nathanail, C.P., Abrahart, R.J. & Bradshaw, R.P. (eds) *Military Geography and Geology: History and Technology*. Land Quality Press, Nottingham, 61–72.

Rose, E.P.F. 2011. Credit due to the few: British field force geologists of World War II. *In*: Häusler, H. & Mang, R. (eds) *International Handbook Military Geography. Vol. 2. Proceedings of the 8th International Conference on Military Geosciences*, Vienna, Austria, June 15–19, 2009. Truppendienst, Ministry of Defence, Vienna, 429–442.

Rose, E.P.F. & Willig, D. 2002. British contrasted with German military geologists and the Battle of France, 1940: no comparison? *Royal Engineers Journal*, **116**, 154–160.

Rose, E.P.F. & Willig, D. 2009. Work by German military geologists on the British Channel Islands during the Second World War. Part 3. Reports with contributions by Walther Klüpfel and Rolf Thienhaus now preserved at the Training and Education Centre of the Bundeswehr Geoinformation Office, Fürstenfeldbruck. *Channel Islands Occupation Review*, **37**, 105–118.

Rose, E.P.F. & Willig, D. 2012. Work by German military geologists on the British Channel Islands during the Second World War. Part 4: site investigation on Guernsey for airfield extension and heavy anti-aircraft battery locations by *Luftwaffe* geologist Hans Schneider. *Channel Islands Occupation Review*, **40**, 188–212.

Rose, E.P.F. & Willig, D. 2013. Work by German military geologists on the British Channel Islands during the Second World War. Part 5: work by Luftwaffe geologist Professor K.G. Schmidt and Hilfsgeologe Dr. K. Diebel, for tunnelling (in general and on Jersey) and water supply (on Alderney). *Channel Islands Occupation Review*, **41**, 78–101.

Rose, E.P.F. & Willig, D. 2014. Work by German military geologists on the British Channel Islands during the Second World War. Part 6: work by the Luftwaffe geologist Franz Schulte on Jersey and Guernsey. *Channel Islands Occupation Review*, **42**, 152–172.

Rose, E.P.F., Häusler, H. & Willig, D. 2000. Comparison of British and German military applications of geology in world war. *In*: Rose, E.P.F. & Nathanail, C.P. (eds) *Geology and Warfare: Examples of the Influence of Terrain and Geologists on Military Operations*. Geological Society, London, 107–140.

Rose, E.P.F., Ginns, W.M. & Renouf, J.T. 2002. Fortification of island terrain: second World War German military engineering on the Channel island of Jersey, a classic area of British geology. *In*: Doyle, P. & Bennett, M.R. (eds) *Fields of Battle: Terrain in Military History*. Kluwer Academic, Dordrecht, 265–309.

Supreme Headquarters Allied Expeditionary Force, Counter-Intelligence Sub-Division 1945. *Handbook of the Organisation Todt (OT)*. **MIRS/MR-OT/5/45**. Military Intelligence Research Section, London.

Tedesko, A. & Peck, R.B. 1992. Leo Casagrande 1903–1990. *Memorial Tributes: National Academy of Engineering*, **5**, 41–45.

Tschudi, E. 1960. *Die militärischen Operationen in Europa 1939–1945. Karte 1:5,000,000*. Kümmerly & Frey, Bern.

von Bülow, K., Kranz, W. & Sonne, E. 1938. *Wehrgeologie*. Quelle & Meyer, Leipzig.

Willig, D. 2009. Die Odyssee des Wehrgeologenarchivs als Teil der Heringen Collection: Versuch einer Rekonstruktion der Vorgänger von März 1945 bis heute. *Geoinformationsdienst der Bundeswehr Schriftenreihe*, **4/2009**, 1–31.

Trench construction and engineering geology on the Western Front, 1914–18

PETER DOYLE

Department of Earth Sciences, University College London, Gower Street, London WCIE 6BT, UK

doyle268@btinternet.com

Abstract: Trench warfare became associated with the First World War from late 1914. Where possible, British and German trenches were laid out by military engineers in line with the most recent military manuals. The effectiveness of individual trenches was to a large extent controlled by the nature of the ground conditions. Engineering geology had a major part to play in slope stability (because the ideal for trenches was to maintain slopes in an over-steepened condition relative to normal angle of repose) and drainage (because excess surface and ground waters not only weakened trench slopes, but also created untenable conditions for the troops occupying their positions). Recent archaeological investigations in Belgium and northern France provided opportunities to examine slope engineering and drainage solutions for trenches in Paleogene clay–silt sediments and Cretaceous chalk. In Flanders, the failure of slope engineering early in the war led to the creation of A-frames to support slopes at the required batter, with the provision of drainage channels beneath duckboard walkways. In northern France, where frost-shattered chalk was close to the surface, drainage and slope support was less of a problem, although trenches cut through thick Quaternary deposits similarly required imaginative solutions.

Trench warfare became associated with the First World War from November 1914. At this point, the Allied and German armies had become locked in a series of battles that saw them attempt to 'turn the flanks' in a phase of the conflict that became known as the 'race for the sea'. As both sets of armies approached the coast, they began to construct trench lines that would remain largely static – although with some periodic movements in the wake of offensives – until 1918.

Although associated with the Great War, the adoption of trench lines in late 1914 owed much to at least two and a half centuries of the application and development of the art of the military siege. Nevertheless, in 1914 the establishment of trench lines was effectively to produce two parallel fortresses that were exceptionally difficult to destroy (see Saunders 2010; Doyle 2017). These lines were in place on the Western Front from the end of the mobile phase in 1914 through to the opening of a new phase of open warfare in the spring and summer of 1918. Establishing and maintaining these trenches required much engineering skill. As excavations, trenches were inevitably subject to geological controls and the engineering geology of these military fortifications became a major consideration in ensuring that they were fit for purpose in static warfare, as first discussed in some detail by American military geologists Brooks (1920) and Johnson (1921) in the immediate post-war interval.

Trench warfare came to encapsulate not only the simple excavation of trenches for protection and defence, but also the development and construction of underground shelters (dug-outs) and concrete surface shelters (pill-boxes), together with subways and tunnels for military use. Some work examining the geological aspects of these has been completed (e.g. Doyle *et al.* 2002, 2006; Rose & Rosenbaum 2011; Doyle 2012, 2015, 2017) and is ongoing. Relatively little attention has been given to trenches themselves, however. As such, the aim of this paper is to examine the development of trenches and trench warfare in relation to geology within the northern sector of the Western Front (Fig. 1), as here it is possible to contrast the geological conditions of the chalk country of Artois and Picardy with that of the Flanders clay plain, with their attendant problems. Its focus, in particular, is the engineering geology challenges faced by the armies. Reference is made to the contemporary military manuals provided to British and, to a certain extent, German armies. In addition, consideration of the resulting trench lines and their geology is made both through access to existing archive resources, but also through the examination of archaeological excavations in this region carried out over the past 25 years. These excavations clearly show the way in which military engineers struggled to adapt their experience to the long-term occupation of the trench lines that were put in place from late 1914 to spring 1918. (see Doyle *et al.* 2001, 2006; Barton *et al.* 2004; De Meyer & Pypе 2007; Brown & Osgood 2009; Dewilde & Saunders 2009; Saunders 2011; Verdegem *et al.* 2013; Doyle 2015, 2017).

From: Rose, E. P. F., Ehlen, J. & Lawrence, U. L. (eds) 2019. *Military Aspects of Geology: Fortification, Excavation and Terrain Evaluation*. Geological Society, London, Special Publications, **473**, 109–130.
First published online March 6, 2018, https://doi.org/10.1144/SP473.6

Fig. 1. Trace of the Western Front in 1915 showing the location of British and French offensives of that year. The area of interest in this study is from Nieuport at the Belgian coast to Albert and the River Somme. Image: public domain.

Geology and trench warfare

The application of geology in war dates back a number of centuries (see Rose 2014), but arguably it was the work of German military fortification engineer Hauptmann (later Major) Walter Kranz that defined the principles of Militärgeologie (Kranz 1913; see also Häusler 2003), although Kranz's ideas were taken seriously only when war broke out in Europe (Brooks 1920, p. 91). A review of Kranz's work in 1915 laid down the basic principles of Militärgeologie:

> Geology is practical and necessary: to prove the stability of parapets and trenches and the stability of dugouts; to identify the speed of digging excavations; to identify water supplies; to assist in rain water and waste water removal; to supply building materials; and to identify mineral raw materials
>
> (Anon 1915, p. 94, translated).

These principles were built upon and expanded in a wartime lecture, Kriegs-Geologie, delivered by Wilhelm Salomon of Heidelberg University in 1915 and published in the same year (Salomon 1915).

Trenches and dug-outs clearly figure in the definitions of the subject by Kranz (1913) and Salomon (1915), as does the need to deal with the waste waters that inevitably accompanied their excavations, both from groundwater build-up, surface water flow and precipitation. Resolving some of the difficulties presented by the trench war formed much of the work of Allied and German military geologists; these activities have been ably described elsewhere (e.g. King 1919; Brooks 1920; Institution of Royal Engineers 1922; Rose & Rosenbaum 1993; Häusler 2000; Rose *et al.* 2000; Willig & Häusler 2012; Rose 2014). This paper examines the evidence of this work and the geological challenges presented to them and to the military engineers of the day.

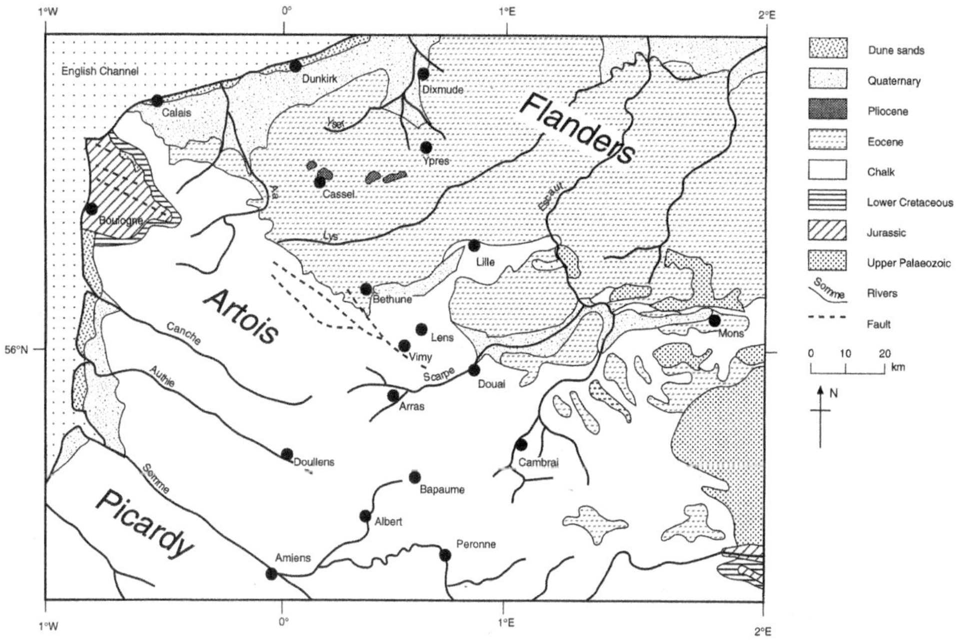

Fig. 2. Geology of the northern sector of the Western Front. Six broad belts of terrain can be recognized from the coast southwards: the coastal dune belt; the Polder Plain; the clay plain of Ypres; the sand ridges to the east and SE of Ypres; the coal belt of Lens-Bethune; and the chalk of Artois and Picardy. Based on Doyle (1998, fig. 6).

Geology of Northern France and Flanders

The geology of Flanders–Picardy is not complex (Figs 2 & 3), but it was to have a major role in influencing the outcome of the war and in the prosecution of trench warfare. As noted in 1917 by Sir Aubrey Strahan, the Director of the Geological Survey of Great Britain, direct geological comparison could be made between the region of France and Flanders, and that of southern England, with particular regard

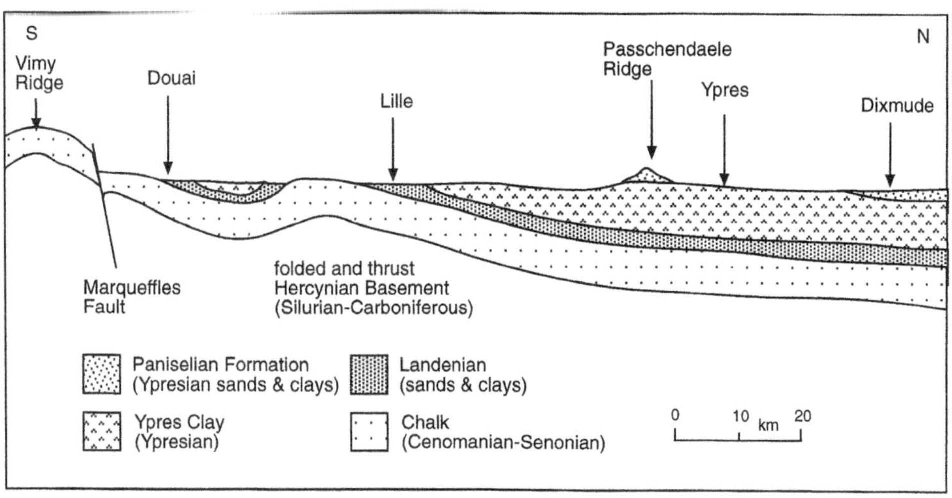

Fig. 3. Sketch cross-section of the geology of the northern sector of the Western Front. The section runs from Arras to Douai, Lille to Ypres, Ypres to Dixmude and the coast (see Fig. 2). Based on Doyle (1998, fig. 7).

to the chalk, the Thanetian sands and the clay of Flanders, so readily associated with the London Clay (Strahan 1917, p. 70).

Military geographer Douglas Johnson studied the region and, for convenience, divided it into broad belts (Johnson 1918, 1921). These belts have been discussed in detail by Johnson (1921) and used by others as a shorthand for the geology (e.g. Doyle & Bennett 1997; Doyle 1998, 2017; Doyle *et al.* 2000; Barton *et al.* 2004). In simple terms, from the Belgian coast to the Somme, these fall into at least six zones or belts: the coastal dune belt; the Polder Plain; the clay plain; the sand ridges; the coal belt; and the chalk upland (Figs 2 & 3).

(1) The coastal dune belt consists of a 2 km wide strip of Holocene coastal dunes that fringe the North Sea. Each of the dunes is made up of loose sands, although they are stabilized by marram grass and other hardy vegetation typical of the sand hills along the North Sea coast (De Moor & Heyse 1978; Heyse 2015).

(2) North of the Belgian city of Ypres, commencing for the most part at the village of Dixmude and passing north to Nieuport, is the coastal plain of Flanders, 'the Polders', an area regularly flooded during the Holocene that has been engineered to provide access to agriculture. The Polder Plain is at best just 5 m above sea-level (De Moor & Heyse 1978; Heyse 2015), but farther south the clay upland provides a barrier to inundation by the sea. Across the coastal plain runs the Yser, joined and linked to the complex of artificial canals and drainage ditches between them. Not surprisingly, the Polder Plain is composed of a sequence of sands, silts and organic-rich deposits, all of which are conspicuously water-bearing. (see Doyle 1998, 2017; Baeteman 1999).

(3) From Ypres southwards, Flanders consists of an extensive, flat plain composed of thick deposits of clay, overlain by soils and other Quaternary deposits (De Moor & Heyse 1978; Heyse 2015). The plain itself is mostly made up of clays once traditionally called the 'Ypres Clay' – now the Kortrijk Formation (see Steurbaut & Nolf 1986; Steurbaut 1987; Nolf & Steurbaut 1990; King 1990; Laga *et al.* 2001; De Geyter *et al.* 2006; see also Rose & Rosenbaum 2011). This thick blanket of clay underlying the city of Ypres and spreading its influence across the region consists of heavy clays of variable depth, reaching up to 130 m thick, but reducing to *c.* 50 m in the west. The clay is blanketed by Quaternary soils to a greater or lesser extent (see Heyse 2015), with considerable thicknesses of alluvium in the river valleys.

(4) Rising slowly up from the clay plain is a low range of hills that curve to the east of the city of Ypres. These low hills are part of a relict ridge system that has been sculpted by streams flowing onto the clay plain. This ridge had been occupied by the Germans from 1914, providing them with direct observation of their enemies on the plain. The low ridge system is composed of what contemporary geologists called the 'Paniselian sand complex' of Eocene–Pliocene age, named after Mont Panisel near Mons, where it was first described in detail. The Paniselian sediments overlie the Ypres Clay and consist of a stratigraphic sequence of sands, sandy clays and clays of variable composition and extent (see Laga *et al.* 2001; De Geyter *et al.* 2006). Today, the Paniselian is defined in modern stratigraphic terms as belonging to the Gent Formation and lower levels of the Tielt Formation (see Steurbaut & Nolf 1986; Steurbaut 1987; King 1990; Nolf & Steurbaut 1990).

(5) South of the Messines Ridge, the ground descends from the Paniselian hills, first to the valley of the Douve and then, farther south, to the Lys. These rivers flow over an extension of the clay plain, with chalk deposits at depth. Farther south into French Flanders, Carboniferous coal-rich deposits are found closer to the surface due to the Marqueffles fault system. This is most obviously associated with the Vimy Ridge, a fault scarp that now divides Flanders from Artois and Picardy. Below the surface, the geology is complex, with multiple thrust slices associated with the Variscan Orogeny (see Delattre *et al.* 1973).

(6) Beyond Flanders, the Artois region consists of a broad anticline, an extension of the Weald. This chalk upland has, in contrast with the buried Lens Coalfield, a relatively simple, flat-lying structure. The chalk of the Somme and Artois is broadly equivalent to that of southern England and, like it, consists of three divisions: a lower, more clay-rich part (les marnes crayeuses), a middle, flint-bearing level (la craie grise à gros silex cornus, overlain by craie blanche à silex) and an upper, pure white, flint-poor part (craies blanche) (see Delattre *et al.* 1973, pp. 25–26; Bureau de Recherches Géologiques et Minières 1982). The chalk is mostly frost-shattered in its upper part and other evidence of the Quaternary is provided by the soil layers that lie above the white chalk. Typically there are four separate units (clay with flints, loess, loam and alluvium), although not all will be present at any given

location (see Johnson 1921). The loess and loam are often classified together on French maps as Limons de Plateaux – a fine covering on the upland areas, but sometimes reaching a thickness of up to 10 m. It forms a cap to the hilltops and slopes and helps to fill the valleys.

Trench warfare

The primary purpose of trenches as used in World War I was to hold the enemy until an assault could be deployed and to protect their occupants (General Staff 1908, 1917*a*, *b*, *c*; see also Saunders 2010; Doyle 2017). However, as was experienced during the war, not all trenches were equal in value and effectiveness and, to a greater or lesser degree, this was a product of geology and terrain. Trenches could suffer enfilade fire – the opportunity for an enemy to fire along their length – which was in part due to the method of construction and the position relative to topography. They could be exposed to observation – again a function of slope and position – and their occupants again be vulnerable for this reason. And from a purely geological perspective, trenches could be perennially wet, flooded or flooding; they could be subject to collapse due to weakness in the rock mass or through over-steepened and/or poorly engineered slopes, and they could be exceptionally difficult to dig, depending on the soil cover (see Doyle 2017 for discussion).

Although trench warfare was becoming a much more professional business by 1916, it is still important to understand that the situation and condition of Allied trench systems was often greatly influenced by their General Staff's offensive policies and a general unwillingness to yield ground, in its early days at least. For the Allied General Staff there was a constant fear that trench warfare would engender a defeatist attitude, an 'unhealthy' dependence on shelter or safety, with

> an insidious tendency to lapse into a passive and lethargic attitude, against which officers of all ranks have to be on their guard, and the fostering of the offensive spirit, under such unfavourable conditions, calls for incessant attention
>
> (General Staff 1916, p. 8).

The Germans, on the other hand, had gained ground early in the war and were able to retain their position through the maintenance of an overall defensive attitude. There was no squeamishness about maintaining strong positions. Their field manual, *Stellungsbau*, of 1916, translated by the British General Staff in 1917, made clear the principal considerations – and the commitment to attrition:

> Field positions when constructed afford considerable advantages to the defence. The important points to be borne in mind by the *defence* in a *war of positions* [includes]: utilization of ground so that conditions favourable for combat are obtained, while they are made unfavourable to the enemy
>
> (General Staff 1917*c*, pp. 3–4).

As a result, deep shelters – dug-outs – designed to protect troops from direct shellfire were mostly a feature of German defensive positions, while Allied positions had mostly small excavations in trench walls or, later, cut-and-cover shelters using corrugated iron and sandbags (see discussion in Brooks 1920; Johnson 1921; Doyle *et al.* 2001, 2006; Barton *et al.* 2004). This meant, in many cases, that Allied lines were poorly situated with respect to topography and ground conditions, whereas German policy allowed for strategic withdrawal to carefully prepared and suitably located positions (see Doyle 2017 for review).

For the most part, similar trench systems were used by the main protagonists on the Western Front, directed by their high commands, but influenced by experience on the ground. For example, a British manual published in 1916 gives a list of the typical components to give the maximum confidence of 'defence in depth'. (Lake 1916, p. 47; Fig. 4).

(1) Obstacles (generally barbed wire) in front of the first line trench, concealed if possible from artillery observation.
(2) Listening posts, look-out posts, machine guns.
(3) Fire trenches, recessed and traversed.
(4) Communication trenches to the rear, linking up the whole system.
(5) Shelters and dug-outs. These should be immediately behind the first line fire trenches, with easy communication to them.
(6) Support trenches (traversed) from 25 to 100 yards in rear of fire trenches.
(7) Dressing stations, kitchen, etc., branching from communication trenches.
(8) Second line trenches. Fire trenches, machine guns, etc., similar to organization of first line.
(9) Supporting points, behind the second line, well defended by parties of 20–40 men, serve to hold up enemy assault on first and second line; such points should be entirely surrounded by barbed wire.

In ideal circumstances, trenches were to be planned, laid out and traced across ground to take in natural characteristics and use them to advantage, as well as avoiding the difficulties presented by the local geological conditions. As the British *Field Defences* manual of 1908 wryly commented:

> The ideal site for trenches is one from which the best fire effect can be obtained, in combination with complete

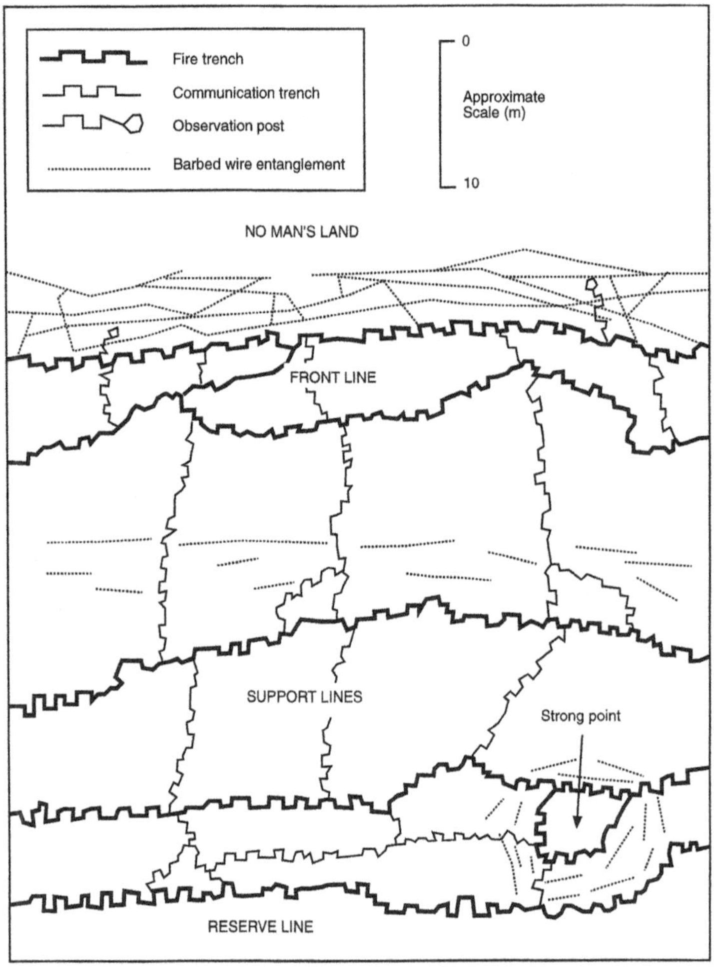

Fig. 4. Idealized trench system as developed on the Western Front in 1915–17. Based on Doyle (1998, fig. 3).

> concealment of the trenches ... As such positions will rarely be found, the best compromise must be sought for
>
> (General Staff 1908, p. 48).

For the German High Command, committed to ensuring their trench lines were the strongest, the use of natural features as strong points was the most important factor.

> Great use has been made by the Germans of natural strong points, such as villages, farms, and woods. The normal procedure now, when taking up a new position, is to fix on a general line of natural strong points, and to prepare these for defence first and then to join them up by fire trenches, without much regard to the field of fire of the latter
>
> (General Staff 1917*a*, p. 17).

Trench types

In their simplest sense, the trenches of the Great War were linear excavations of variable depth that were mostly open to the sky, but were sometimes, rarely, roofed for concealment purposes (see Brooks 1920; Doyle 1998; Doyle *et al.* 2001). Despite their simplicity, the function of trenches varied and, as the war progressed with no absolute sign of a break in the deadlock, more and more types were developed. The pace of change was so rapid that the third edition of *The Royal Engineers Field Service Pocket-Book* of 1916 contained only blank pages for field defences, accompanied by the statement: 'Owing to the constant changes taking place in Field Defences it has been decided ... to omit this section which will be rewritten after the war ...'

(Johnston 1916, p. vii). However, in the main there were two consistent types: fire trenches, which formed the front lines, and communication trenches, which joined them.

Fire trenches (i.e. fighting trenches) were divided into a regular pattern of fire bays (facing the front and occupied by soldiers on guard) and traverses (which linked the bays) (Fig. 4). This system meant that no soldier could walk in a straight line for long without having to switch back on himself. Such movement was intended to limit the effects of shellfire exploding in the trenches or the possibility of enfilade rifle and machine gun fire along the length of a trench.

British and German fire trenches were alike in this respect. French versions varied; some had a more leisurely, curved, zigzag layout, but they were all calculated to reduce the impact of explosions and enfilade fire. French military engineering was the inspiration for much trench warfare. Thus the spoil removed in digging a trench was used to form a parapet (a mound of earth in front of the trench on the enemy side, intended to stop bullets) and a parados (a slightly higher mound at the rear, which would interrupt the movement of bullets, reduce the impact of exploding shells and prevent soldiers' heads from being silhouetted against the skyline). Firesteps were constructed to allow the infantry to fire over or through the parapet (General Staff 1908).

Communication trenches varied in length. Running from the rear areas and connecting all the forward trenches up to the front line, they offered protection for supply and troop movements from the rear (Fig. 4). They were usually dug in a zigzag or wavy pattern and in Flanders, where the geological conditions meant that revetment – officially defined as 'any artificial material used for retaining earth at a steeper angle that it would normally assume' (General Staff 1908, p. 39) – was essential, communication trenches had similar dimensions to a fire trench. In the coastal strip at Nieuport, communication trenches of French construction were close boarded and roofed by timber, and referred to as boyaux couverts (Doyle *et al.* 2001).

The front line trenches of the opposing armies were separated by a belt of contested ground known as No Man's Land, usually extensively pitted by shell holes and mine craters. In some cases, only mine craters lay between the two lines and guidance on capturing and strengthening craters was given in all the manuals (War Office 1921; General Staff 1917*a*). Behind the front line, *c.* 10–30 m, were the support and reserve lines, trenches that were still 'organized for fire', with the intention of holding back the enemy if a breakthrough of the frontline was achieved. Farther back were the reserve lines. The whole system could be encapsulated within a zone of 50–150 m width and access from the rear areas to each of the trenches in turn was through communication trenches, which traversed the ground between the lines of fire trenches. In some cases, particularly where the enemy controlled the high ground, tunnels or subways were constructed to allow the safe passage of troops to the front from the rear areas. This became a feature of the trench war in 1917 (see Doyle 2017 for discussion).

Trench construction

Trench construction was usually carried out by the infantry (General Staff 1914). Fatigue parties and later labour battalions had mostly to dig their trenches rapidly under the cover of darkness, generally with men spaced between two to three paces apart. For the Germans, there were explicit instructions that the construction of trenches was to be entrusted to those men who would, at least at first, be occupying them, again under the control of engineers (General Staff 1917*c*, p. 4).

The speed of digging was directly influenced by the ground conditions, prevailing weather and the nature of the troops. The rate of digging was generally taken to be 30 ft^3 (*c.* 10 m^3) per hour, increasing by 50% where the soils were 'very easy' and decreasing by 30% where the soils were 'very hard' (General Staff 1908, p. 14). For the most part, the geology of the battle zone in northern France and Flanders consists of variable thicknesses of Quaternary soils overlying soft sediments, clays, silts and sands. Where trenches were to be dug into chalk ground in Artois and Picardy, ground conditions could vary. In some cases, Quaternary soils capped chalk spurs; in other cases, this had been removed, leaving a capping of bare chalk. In most situations, however, this chalk was extensively shattered by freeze–thaw and therefore relatively easy to work.

With, in most cases, the Germans dictating the line of the opposing trenches, more often than not the trace of the British defences was unfavourable, but it is hard to generalize. For the British at least there was little chance to reference the nature of the local geology. For the Germans the ideal was to have two types of geologist to advise: 'the trained ones who are of use in the field before and after the battle' and 'the officer on the front during battle' (Anon 1915, p. 95).

Position relative to slope was an important factor to be considered in the construction of trenches. Basic principles for entrenchment were laid down, which emphasized the theoretical aspects of position in relation to forward and reverse slopes, valleys and spurs, and topographical height (Pressey 1919; War Office 1921, 1925). These aspects were considered important to allow the effective observation of opposing trench positions, to prevent enemy observation of forward and reserve trenches, and to provide necessary supportive arcs of fire for small arms. Ideally, the trench lines were to be designed to contour the hills and valleys, particularly important in providing 'mutual enfilade' fire in valleys, the assaulting troops being attacked from both sides of the valley and on the slopes of spurs by the defenders (Pressey 1919; War Office 1921, pls 14, 15).

In all cases, the positioning relative to slope was to maximize observation of the enemy to control artillery and machine gun fire. In the Ypres area, the British trench lines were positioned either at the foot of, or on the reverse slopes of, the Paniselian (Gent Formation) ridge complex overlooking the Ypres clay plain. This made artillery observation difficult and provided ample opportunity for accurate offensive fire from the German artillery. In the rolling chalk upland of the Somme, the British positions were more variably located and more able to follow the guidelines laid down by the official regulations (see Doyle 2017 for discussion). But again, for the most part, the German lines were more formidably sited (see Willig 2011).

Slopes and revetment

As open excavations, trenches were subject to the normal considerations of slope engineering, as the 'required stability of rock slopes will vary depending on the type of project and the consequence of failure' and 'there is usually little flexibility to adjust the orientation of the slope to suit the geological conditions encountered in the excavation' (Wyllie & Mah 2005, pp. 1–2). In military trenches, more often than not constructed under fire and with little regard to best practice, the consequences of failure would bring obvious problems for those men seeking shelter and, for the Allies in particular, there were few options available.

In the ideal situation, each fire trench was *c.* 2 m deep and 0.6 m wide at the bottom, usually widening to 2 m at its top, but depths varied according to topography and the depth to permanent water saturation. The trench width at floor level was *c.* 0.8 m and at the top, given the appropriate regulation 'batter' or slope, up to 3 m (Fig. 5). In official terminology, the trench sides were known as 'slopes' and were routinely kept at over-steepened angles relative to the normal angle of repose for soft sediments. The British wartime manual *Field Defences*, first published in 1908, was clear in its insistence that, to be effective, the 'interior slope should be as steep as possible in order to increase the protection from projectiles falling at a steep angle' (General Staff 1908, p. 55). The same manual determined that:

> … in almost any type of soil, it should be possible, with sods, lumps of earth, or other means, to keep the slope up to 3/1 for a moderate height; but unless great care be exercised, it will seldom be possible to build a slope steeper than 2/1
>
> (General Staff 1908, p. 55; Fig. 5).

These gradients suggest slopes should be at least 65–70°, rather than the angle of *c.* 35° more likely to be achieved for clay soils, without some form of engineered stabilization methods. These were known as revetments, defined as 'any artificial material used for retaining earth at a steeper slope than that which it would naturally assume' (General Staff 1908, p. 20).

In the case of trenches, with the requirement to maintain steep slopes, some form of revetment to maintain the steepened slope was therefore essential, particularly in the soft sediments encountered in Flanders from the coast to the Marqueffles Fault Zone. Although steep slopes in clay are possible for short periods, especially during dry weather, the absorption of water will soon lead to slope failure, particularly in a clay with a high percentage of montmorillonite with its propensity to swell (Verrett 1980). The periodic failure of slopes was, to a certain extent, to be expected given their purpose; this is not usually an acceptable aspect of slope engineering within civilian projects (Wyllie & Mah 2005, p. 4). What is required is the establishment of a factor of safety, the limit equilibrium of the slope, with stability achieved where the factor of safety is greater than unity. Given that trenches were cut through a variety of materials, were subject to changes in pore pressure in permeable sediments, especially in the vadose zone above the level of water saturation, and suffered from the violence of explosions, this was not always an easy proposition. In addition, it has to be recognized that military trenches were intended from the outset only to be a 'phase' of the military campaign as directed by the General Staff and that detailed examination of their engineering geological aspects was singularly lacking in the early stages of the war. This would change as it was realized that trenches had become a major component of the architecture of the war.

With trench side slopes cut at a higher angle than the soils could naturally support, there was a real need for an engineered solution to prevent collapse and revetment was carried out with whatever was available, sometimes wattle, often corrugated sheeting and expanded metal, sometimes chicken wire, or even timber boards if they could be salvaged (Fig. 6). 'Brushwood' and 'well-built sods' were also recommended and used (General Staff 1908, 1917*a*, *b*, p. 12). Timber was often used to hold these materials in place, as were angle iron posts anchored deep into the earth; layers of bonded sandbags strengthened the whole. The military manuals laid down the parameters.

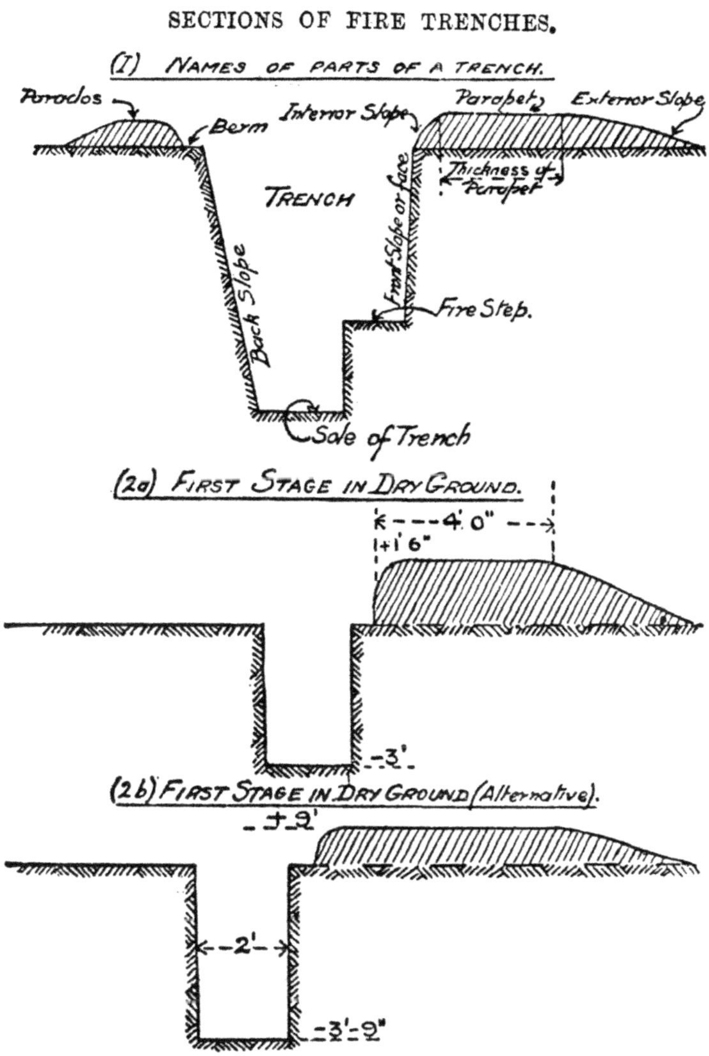

Fig. 5. Sections of a typical fire trench showing the stages of development and over-steepened slopes or trench sides. Reproduced from General Staff (1908).

> The side of trenches which have to be occupied for a long time, and particularly in wet weather on a damp site, must be reveted. Hurdles or rabbit netting held up by stout stakes securely wired to short pickets firmly anchored in the parapet or parados, form a useful type of revetment for this purpose. Sandbags are not so suitable. In the winter in Flanders some really solid form of revetment, such as planks or timber, or expanded metal sheets, is necessary
>
> (General Staff 1917*a*, p. 12).

In addition, by 1917, to assist with the construction of trenches in 'soft ground', 'trench frames' were recommended in British use (Fig. 7). The intention of the frames was to provide engineered support for over-steepened and often water-saturated slopes. The idea was that they would not only help retain the trench profile, but would also allow a firmer anchoring of the revetment materials, as well as a means of creating a walkway elevated above the trench bottoms. The idea for the frames was published in the December 1916 revision to the earlier British manual *Notes for Infantry Officers on Trench Warfare*, drawn up in March of that year. Such frames were an inverted 'A', the cross of which provided an elevated foundation for a duckboard track, with the main line of the 'A' providing a support for both the trench revetment and the sandbag walls of a breastwork (General Staff 1917*a*, fig. 13; War Office

Fig. 6. Well-built German trench with brushwood revetment, developed in friable soils *c.* 1915–16. Photograph taken in the summer months. Image: public domain.

1921, plate 69). Major Buckingham Royal Engineers, Assistant Inspector of Mines, was impressed with them when he encountered the frames in a communication trench in 1917.

> I was very struck with this trench which is the most elaborate trench I have seen. It was made in accordance with recent E in C's [Engineer in Chief] plates with expanded metal revetment held in place by [inverted A] frames about 3′ or 4′ above duck board level & then a flat ledge, and unreveted sides above
>
> (Diary Major W.E. Buckingham RE, National Archives file WO158/140).

The development of such structures indicates the need to deal with the unsupported and over-steepened slopes that were such a component of the trench war in 1914–15, with new engineering solutions widely adopted by 1917.

Drainage

The provision of adequate drainage was obviously of great significance in the construction and maintenance of trenches. In extreme cases, particularly where thin soil cappings sit on clay and in other areas where groundwater was close to surface, borrow pits were dug on either side of the trench to supply the extra earth needed to build up a sufficient height to protect the troops. In some cases sandbag breastworks were constructed where the ground conditions prevented even the most rudimentary trench lines. Captain F.C. Hitchcock of the Leinster Regiment described his ideal trenches in the clay ground near Armentières, in 1915.

> Our trenches appeared to be very formidable; they were duckboarded, and the parapets and paradoses were completely reveted with sand-bags. The parapets were 6 feet high, and the wooden fire steps being 1 1/2 feet in height, gave a fire position of 4 1/2 feet. Owing to the low-lying nature of the terrain the trenches were breastworks
>
> (Hitchcock 1937, p. 22).

Such breastworks (Fig. 8), known as high command trenches, in contrast with the normal low command

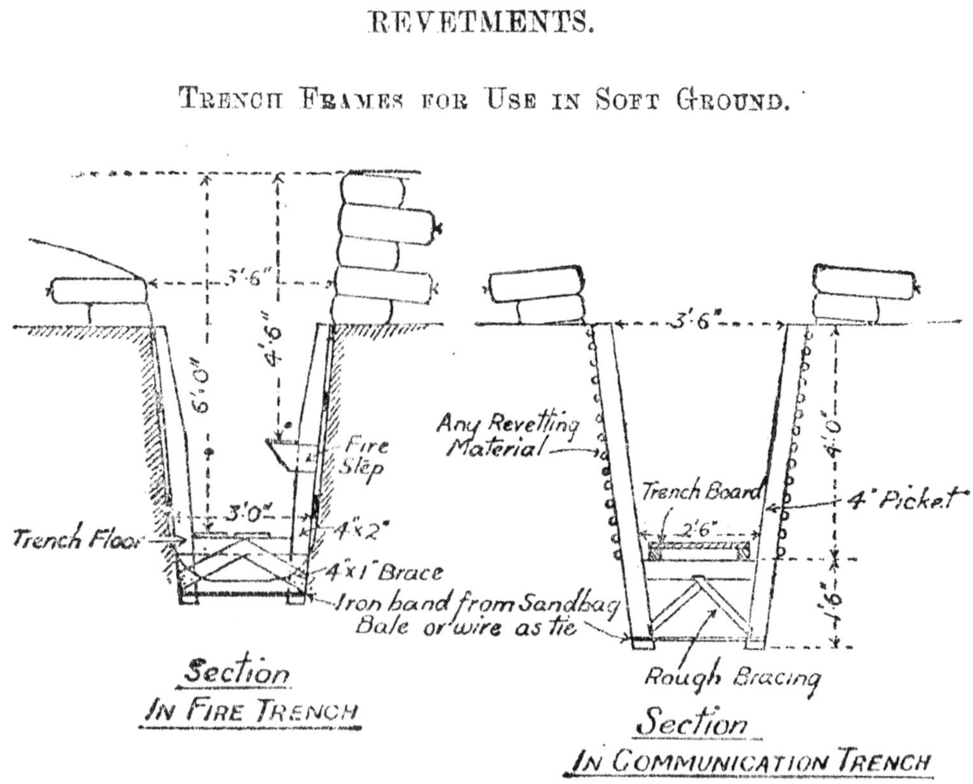

Fig. 7. Trench or A-frames introduced to British use in 1916–17. The inverted 'A' was intended to support the revetment and provide a foundation for duckboard tracks. Reproduced from General Staff (1917*a*).

types (General Staff 1908), were common in the Ypres Salient due to the usual waterlogged nature of the Quaternary soils or Paniselian sands that overlie the Kortrijk Formation. Digging down further simply created a void, one lined with impervious clay, that would soon be filled with water and which would require pumping against both groundwater and rainfall (Fig. 9).

Ensuring that the trench was adequately drained and floored was clearly essential. On both sides, trenches were floored with wooden duckboards, which were built up to allow drainage beneath – in

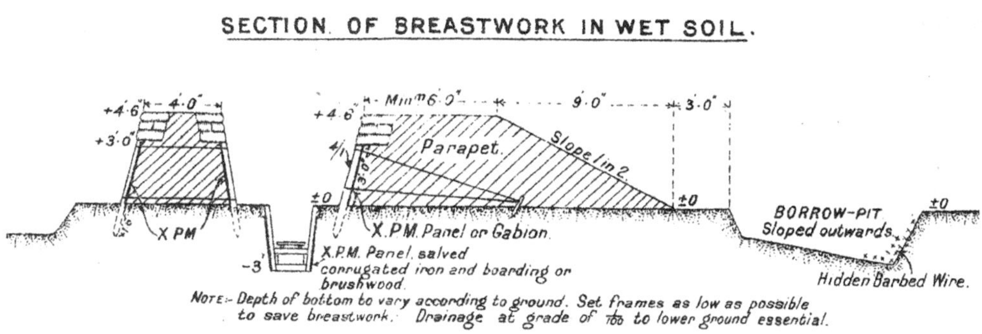

Fig. 8. In wet ground, usually associated with thin soils on clay, drainage was a problem and an insufficient depth could be developed without superior drainage conditions. In such cases, breastworks were constructed. Reproduced from General Staff (1917*a*).

Fig. 9. Problems of drainage in wet trenches in 1914–15. Pumping was required before more sustainable solutions could be achieved. Image: public domain.

fact, it was common for successive levels of duckboards to be laid one on top of another to combat the difficult conditions encountered. In rare cases, when trench lines snaked through destroyed villages and houses, bricks and rubble were used. Although the men hoped that waters would subside inside a flooded trench, this factor was at the mercy of ground conditions. As the British post-war *Manual of Field Works (All Arms)* was quick to point out, 'sumps or soakage pits should not be relied upon unless natural drainage is possible … unless the sump reaches a permeable stratum, it must be pumped or baled out …' (War Office 1921, p. 65; 1925, p. 73).

With the possibility of troops standing in cold water in undrained or waterlogged trenches came the probability of 'trench feet', a condition that directly affected the effectiveness of the fighting forces. Field Marshal Sir John French, commanding the British Expeditionary Force, estimated casualties from trench feet to be somewhere in the region of 20 000 men in the winter of 1914–15 (French 1919, pp. 288–289). As such, by 1916, the advice of the *Notes for Infantry Officers on Trench Warfare* (Lake 1916) was that drains of 'adequate capacity' should be engineered into the floor of the trench as it was being built, sturdy and boxed in to prevent collapse.

As had been suggested by Walter Kranz, the drainage of field fortifications was intended to present a major task for engineers and their geological advisers. The German manual, *Stellungsbau*, noted sagely, 'If the water level in the country is high, special measures to deal with it may be necessary' (General Staff 1917*c*, p. 6). In this respect, the German army became expert in the drainage of trenches as the war progressed. This expertise was recognized by their enemies, the British:

> Great attention is paid to drainage of trenches, as on the success or non-success of the measures taken may depend whether this position can or cannot be held in the wet season. It is laid down that the drainage must be done on a definite plan which must be carried out in good time. Drainage engineers and geologists are to be consulted, and use made of existing maps and plans. Wherever possible, the drainage water is to be led in the direction of the enemy, pipes being put through the parapet for this purpose
>
> (General Staff 1917*c*, p. 6).

Case histories

The examination of trenches in archaeological investigations in France and Belgium over the past 25 years provides a means of testing the application of contemporary British and German military manuals in practice and of examining the details of military engineering geology on the ground, particularly with regard to slope stability and drainage. Three excavations are discussed here, two in the vicinity of the Belgian city of Ypres (Ieper) in Flanders, and one in Picardy, France on the Somme battlefield.

Boesinghe, Belgium

The entrenchments at Boesinghe consist of both trenches and deep-mined dug-outs. The trench/dug-out system is located close to the Yser Canal to the north of Ypres, close to the end of the British line (Fig. 10).

These British trenches were constructed in late 1915, after the German offensive of spring 1915. At this stage in the war the system of trenches was constructed with standard front, support and reserve lines, connected by communication trenches. In 1992 the area was being cleared as part of the development for an industrial park north of the prosperous city of Ypres. During these works, a trench system was uncovered with an associated deep-mined dug-out; this had a maximum depth of *c.* 10 m and is thought to have been constructed by the 173rd Tunnelling Company (Royal Engineers). This dug-out served as the headquarters of the 13th and 16th Royal Welsh Fusiliers in the Third Battle of Ypres (Saunders 2011). Development proceeded while local amateur archaeologists excavated the site, which was associated with both the Second and Third Battles of Ypres in 1915 and, subsequently, in 1917. Part of the site is now preserved in concrete (Saunders 2011).

The trench system exposed by the 1992 excavations was shallow; it had been developed in a thin layer (up to 2.2 m thick, according to the contemporary geological maps) of the Quaternary soils, sands and silty clays that overlie the Ypres Clay plain. Situated east of the Yser Canal, cut deeply into the Ypres Clay, the British line had the canal to the rear, its embankment raised up from the clay forming a significant position for the construction of dug-outs and other shelters. The trench line lies roughly on the 15 m contour line and faced a German line that sits close to the 20 m contour, the westwards extension of the Pilkem Ridge. The difference in heights also represents the differences in geology: the Germans

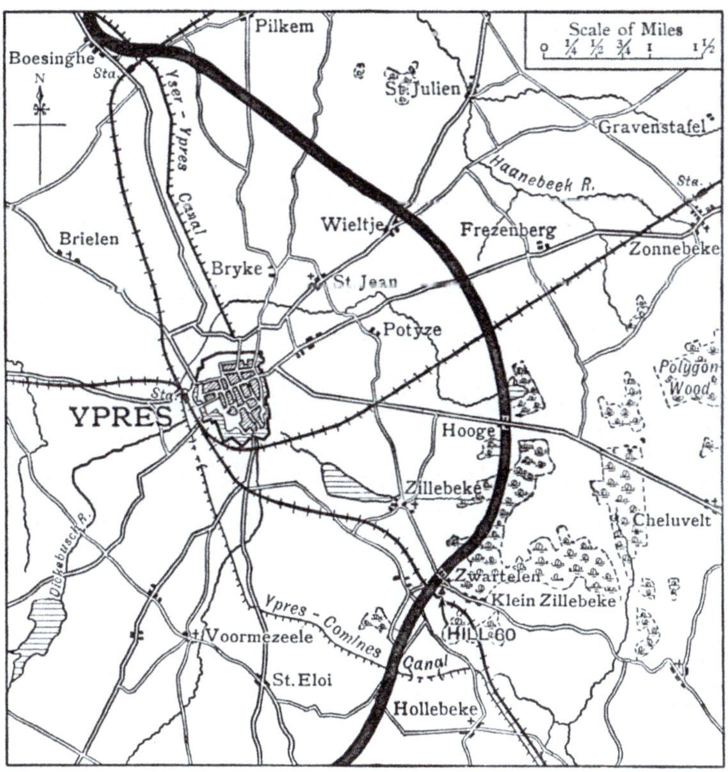

Fig. 10. Location of the Boesinghe trench lines, close to the low ground of the Ypres–Yser Canal. The arc of the trench lines is from 1915. Messines is just off the image, south of Hill 60. Image: public domain.

squarely on the sands that sit on the clay (Kortrijk Formation), the British at its feather edge. Comparison with the most modern geological map, published by the Belgian Geological Survey (1999), shows the correspondence of the old front line with the junction between the Ypres Clay (Kortrijk Formation) and the Paniselian sands (Tielt Formation).

Here, a German strong point, Caesar's Nose, pushed westwards towards the British line using a low-lying spur as its point of reference. This work had, in fact, been part of a contested system of trenches, with both sides effectively sharing the line in 1915 (MacGreal 2011). The 6th West Yorkshire Regiment arrived in the line in July 1916 and found it to be dry, although these conditions were not to last.

> When the battalion first took over the line, the trenches had not fallen in. The precautions which began to be adopted on a big scale in 1916 to preserve trenches from collapsing in the winter months (such as A frames and other systems of reveting) had not been taken in the Salient. The Division [49th] arrived too late to organize draining and reveting on a sufficiently elaborate basis. The result was disastrous. When the Autumn rains began in August, the trenches disappeared, or became canals … The line could only be held by a system of detached posts, where men were prisoned till nightfall, up to the knees in water day and night … Tremendous efforts were made throughout the Division to combat the most serious danger of water-logged trenches – 'trench feet'…
>
> (Tempest 1921, p. 50).

The trench system had to be considerably improved, in line with the advice provided by *Notes on Trench Warfare for Infantry Officers* and, in particular, the use of A-frames as illustrated in its revised diagrams (General Staff 1917*a*, fig. 13; see Fig. 7). By mid-1916, these A-frames were in use. Lieutenant Grover of the King's Shropshire Light Infantry recalled their function in this sector:

> I spend a lot of time in the Salient, where the water level is just below the surface, and when it rains the whole thing is just one field of mud, so all the defences have to be built up. Which means you not only have to build them up with sandbags, you have to revet them very strongly. We had things called 'A Frames', in the form of an inverted A with a trace across to allow a drain below, and that was the inside of the trench, these were fitted in and were reveted inside
>
> (MacGreal 2011, pp. 109–110).

Fig. 11. Trench A-frames exposed in archaeological investigations near Ypres (Ieper). (**a**) Boesinghe, showing the cross of the 'A', with duckboard track sitting on it. The trench cuts through Quaternary soils and is floored by clay. (**b**) Forward Cottage near Wieltje (Fig. 10) showing the engineered slopes maintained in an over-steepened state, with corrugated iron sheeting in position, and once again floored by clay. Photographs by P. Doyle.

These trenches were therefore constructed using breastworks and the archaeological work demonstrated the presence of the inverted A-frame structures, intended to support the revetments and breastworks of sand bags and to ensure that there was an engineered solution to the maintenance of over-steepened slopes (Figs 7 & 11a, b). Examining the site carefully, it is possible to see that the trenches themselves were about 1 m in depth to the duckboard surface, with up to another metre beneath this to the level of water saturation. This is consistent with a position for the trench just overlying the impervious Ypres Clay. Other, more simply constructed, trenches, earlier versions, were also found in this area. In at least one part of the trench the presence of several successive layers of duckboarding suggests attempts to raise the trench floor level above the saturated ground before the A-frames were installed.

Messines, Belgium

If the trenches at Boesinghe were typical of British trench architecture of 1915–17, comparison with German lines would be instructive. German front line trenches at Messines at the southern extremity of the Ypres Salient (south of St Eloi) were excavated in 2012–13 as part of a major replacement of water mains in the region (Verdegem *et al.* 2013).

The excavation site uncovered a German trench system located on the slopes of the Messines Ridge, just above the valley of the Douve, surrounding the town of Messines itself. Messines sits on a spur of the main ridge system east of Ypres and forms part of a wider plateau, with Wytschaete to the NW and Messines to the south (Fig. 12). This plateau had seen hard fighting in 1914 and had stabilized in its position in 1915, becoming a dominant fortress consisting of a system of strong points

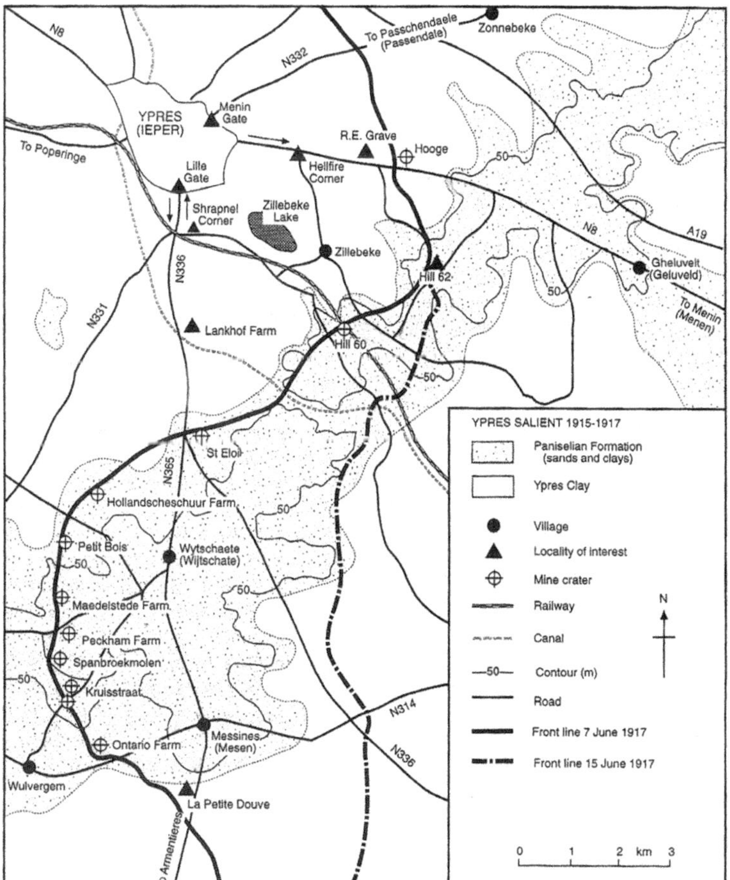

Fig. 12. Map of the Ypres (Ieper) Salient, 1915–17, showing the location of Messines and the British front line at the start and end of the Battle of Messines in 1917. Based on Doyle (1998, fig. 13) with permission of the Geologists' Association.

designed to break up any attempt to take the ridge top by frontal assault. With the Germans occupying the high ground, with every spur and building built into the line as a fortress (in line with the doctrine laid down in the German manual, *Stellungsbau*) (General Staff 1917*c*, p. 3), the British were forced to build trenches that were effectively 10 m lower, facing the forward slopes of the ridge (Doyle *et al.* 2002).

Capping the ridge top at Wytschaete are the driest sands, known to the British (not surprisingly) as the Wytschaete Sands (Fig. 13; see Rose & Rosenbaum 2011, tables 2 and 3). A recent map (Belgian Geological Survey 1999) has this as the Gent Formation, situated just to the west of the village. Messines is built on the Tielt Formation, the sands and clays of what was called the Paniselian. For the most part, however, the rest of the ridge is composed of the various levels of the Paniselian, overlying, as always, the Ypres Clay. Here are found the waterlogged Kemmel Sands, sandwiched between the clay-rich layers, and there were even wetter 'alluvial' soils, that thicken to considerable depth in the valley of the Douve at the foot of the ridge (see Rose & Rosenbaum 2011).

The trench system was constructed as a strong point in the German front line. The excavation trenches cut by the archaeologists exposed, in turn, the details of the German Great War trenches, incredibly well preserved. Here was Eckert Graben (known as Uhlan Support by the British) a second line position organized as a fire trench; a machine gun position; and a communication trench, Blauer Graben. To the south of Messines, on the slopes facing the Douve, was exposed Emil Graben (Uhlan Avenue), a line to the rear of the second line (Verdegem *et al.* 2013, p. 104). The German lines here almost exactly follow the junction of the Kemmel Sands between the 50 and 60 m contours. Although the centre of Messines sits on drier ground, the trenches constructed in front of the village sit squarely within the outcrop of the water-rich Kemmel Sands and

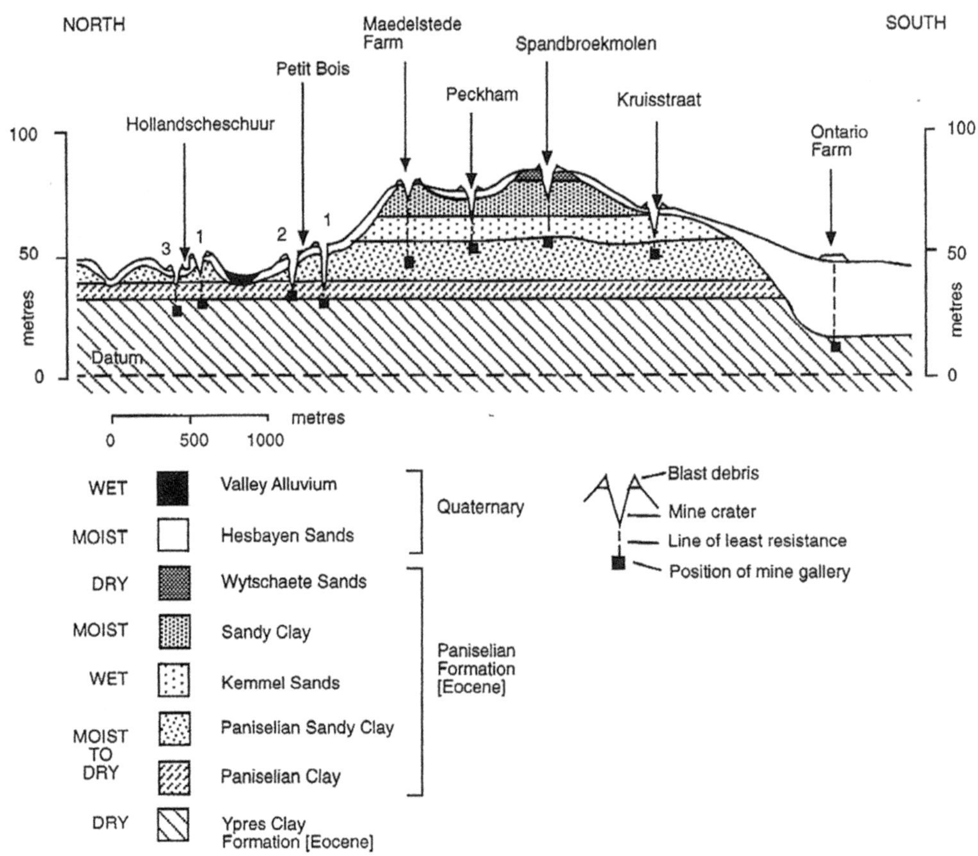

Fig. 13. Geological section north–south through the Wytschaete–Messines ridge showing the location of the Kemmel Sands towards the top of the ridge and the use of the lower levels of the Paniselian for the construction of the British mine galleries used in the Battle of Messines, in June 1917. Based on Doyle (1998, fig. 22).

once again there was the threat of flooding from the ever-present impervious clay beneath, as seen in the contemporary geological maps constructed by the British in 1917 (*1:10 000 Map Geological, Ploegsteert, 28SW4*, May 1918, The National Archives file WO297/2475; see Rose & Rosenbaum 2011). Not only that, but the valley of the Douve was filled with water-rich alluvium, sediments that had been built up by the flow and overflow of the river in a subsiding valley to an extent that it was *c.* 30 m thick at its deepest (see Fig. 13; beneath the location known as Ontario Farm). This wet ground caused problems for the Tunnelling Companies that were trying to undermine the German defences here (see Barton *et al.* 2004; Willig 2011; Doyle 2012). It also meant that, as the British and German trench lines breasted the ridge and moved down into the valley on to Ploegsteert and beyond, there was increasingly difficult, wet and waterlogged conditions for both sides (Brown & Osgood 2009).

Evidence of periodic flooding from the sands is provided by successive layers and levels of timber flooring and duckboarding. This is not surprising given the geological situation here. On top of the typical long trench boards are timbers laid so that they resemble duckboards of the British type. It is probable that these trenches were drained using the German system, on the basis that, wherever possible, the fall of drainage should be towards the enemy. This approach was facilitated by the simple fact that the German lines lay on the slope above the British.

Exposed in the archaeological investigations at Messines were narrow trenches that were boarded throughout and that were reveted by a variety of means (Fig. 14). This included the rescue of doors and other timbers from damaged and destroyed buildings in the village. Timber used in trenches in this way was potentially dangerous – it could provide splinters that would add to the problems of men tightly packed in the line. Also *in situ* is the brushwood hurdling commonly seen in German trenches. This type of revetment was the preferred means of protection, but was difficult to replace in the shell-blasted landscape of Flanders. The manual *Stellungsbau* was clear on this point.

> The sides of trenches must not be reveted with any material that may make traffic in the trench impossible or even difficult after bombardment. Planks and timber should not be used if possible. Hurdles are not so objectionable. The best reveting material is sods or thin loose brushwood
>
> (General Staff 1917*c*, p. 15).

Here and there the revetment shows the impact of British shelling, with extreme disturbance to the timbers of the German line. Such revetment was held in place by vertical members, some of them cut branches rather than shaped timber, and the trenches are uniformly cut through the water-rich silts and sands of the Paniselian.

The German trench system also included a concrete shelter with room for six or seven men, located at the termination of Eckert Graben (Fig. 15). This was cast *in situ* and must have been a response to the damp conditions. The water-rich sands, into which the trenches were dug, meant that concrete was the solution to providing shell-proof shelters, rather than deep dug-outs. This concrete shelter

Fig. 14. German frontline positions at Messines. Situated close to the Kemmel Sands, there was constant flooding, evidenced by successive levels of duckboarding. Here, inadequate revetment is provided by timbers salvaged from destroyed buildings. Photograph by P. Doyle.

Fig. 15. German solution to flooding and inadequate revetment in the frontline trenches at Messines: a concrete shelter cast *in situ*. Photograph by P. Doyle.

gave access to the second line through a reveted and planked trench and was fully equipped with an internal pump. At its entrance was a carefully prepared rifle rack and a recess with a hinged lid that contained German stick grenades, found as they had been left by the last German occupants in June 1917.

Four shallow mined tunnels were also uncovered in the excavations (Verdegem *et al.* 2013, p. 150). At a depth of just 2–3 m, the protection given to them from howitzer shells was likely to have been limited head cover, although it appears that there is a rubble 'burster course' of building waste, suggesting that the tunnels were constructed as cut-and-cover. Each tunnel was constructed using a 'mining case' timber revetment, supported by bridle joints and pegs. Most likely the purpose of these tunnels was for storage and concealment, rather than protection from artillery bombardment. The tunnels have no inclines – just simple adit entrances from the trench line to the south of Messines. The tunnels sit between the 45 and 50 m contours in Paniselian sandy clays that underlie the difficult and waterlogged Kemmel Sands (geological detail from 1:10 000 map, Geological Ploegsteert 28 SW4, The National Archives file WO297/2475) – a factor that suggests that, in winter at least, these tunnels had to be drained effectively, either by drains running downslope or by pumping.

La Boisselle, Picardy, France

In Picardy, the opportunity to observe excavated trenches has been provided by research at La Boisselle on the Somme battlefield (Fig. 16). Here, in 2011–12, one of the last remaining pieces of original extant battle terrain on the Somme battlefield was excavated by the La Boisselle Study Group (see www.laboisselleproject.com for details). Situated in the village of La Boisselle, the terrain captures a snapshot of the Allied (British) and German front line trenches dug into chalk from the period 1914–16. Here, at a farm known to the French as Ilôt and the Germans as Granathof, the French stopped the German advance on the Somme on 14 September 1914. With the village firmly held by the Germans, this part of the front line was exploited by underground warfare, with numerous attempts to dislodge the enemy using mines, commencing in December 1914 and continuing until the British took over the front line in August 1915. From this point on the British took over the tunnelling activity, deepening the system from *c.* 12 m to 24–30 m below ground. These mines, forming craters at the surface, formed part of No Man's Land – just 45 m apart – up to the opening days of the Battle of the Somme.

The nature of these very trenches was recorded by Charles Douie, a British officer who served in them in 1916.

> I learned something of the reputation of the La Boisselle trenches. They were among the most notorious in the British lines. For a considerable distance the opposing lines were divided only by the breadth of the mine craters: the British posts lay in the lips of the craters protected by thin layers of sandbags and within bombing distance of the German posts; the approaches to the posts were shallow and waterlogged trenches

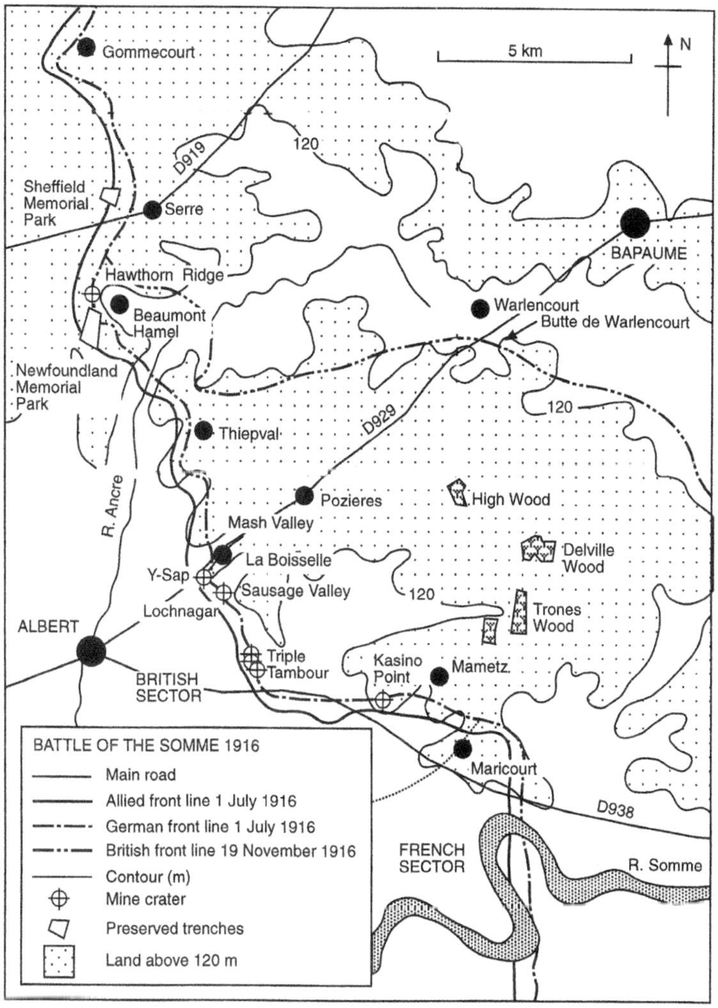

Fig. 16. Map of the Somme Battlefield showing the location of La Boisselle. The chalk is exposed at the surface here, although much of the Somme is blanketed by Quaternary deposits. Reproduced from Doyle (1998) with permission of the Geologists' Association.

> below the level of the German lines, and therefore under continuous observation and accurate fire by snipers
>
> (Douie 1929, pp. 87–88).

With No Man's Land so challenging, and with the need to maintain the offensive, mining activity here was extensive – and had been so since late 1914 (Simon Jones, pers. comm. 2014). La Boisselle was captured by the British on 4 July 1916 and the notorious trench system and its mines were left behind.

The trenches excavated by the La Boisselle Study Group expose a small system facing the mine craters and supporting a number of adits that lead underground (War Diary, 53rd Infantry Brigade, Headquarters, The National Archives file WO95/ 2033). As described by Charles Douie, the trenches were relatively shallow here, dug directly into the chalk and therefore directly visible from the air. Here, there are only relatively thin surface soils compared with other parts of the Somme (see Bureau de Recherches Géologiques et Minières 1982), although the chalk is extensively shattered (Fig. 17). No doubt some of that shattering could be attributed to the effects of the almost continuous bombardment, but it is more likely that this is a by-product of Quaternary freeze–thaw action and the development of permafrost, which naturally destroyed the integrity of the chalk strata. Lying above this layer is a thin level of soil, consisting of clay or loess

Fig. 17. Scone Street, one of the frontline trenches cut in chalk at La Boisselle. There is direct access to the chalk, with a limited Quaternary cover. Trenches are more capable of sustaining steep slopes, although the frost-shattered chalk often required chicken wire netting support. Photograph by P. Doyle.

with an admixture of chalk blocks, attributed to backfill and mine spoil by the La Boisselle Study Group. It is likely that these trenches were capable of sustaining relatively steep and therefore regulation slopes, although the frost-shattered chalk would no doubt have required some stabilizing revetment from expanded metal or chicken wire.

Scone Street, one of the main trenches excavated, has access to extensive underground workings. With weakened chalk for much of its trench slopes, this trench would have required revetment and there are remains of timber there to support this inference. The level of weak chalk is variable and, at the mid-part and base of the trenches, there is much stronger, robust chalk. Flint levels are evident *in situ*; these would have provided an extra hazard, the brittle flints providing sharp shards if hit by explosives. The presence of the flints shows that the chalk here is from the second of the three main chalk units within the region, of la craie blanche à silex (see Bureau de Recherches Géologiques et Minières 1982).

Given that the clay-rich chalk is at depth, the Tunnelling Companies here found that the water level was at least 100 ft beneath the surface (Simon Jones, pers. comm. 2014), typical of the chalks at an upper level, at least in summer, with ground waters percolating downwards through the various fractures and levels that characterize the blockwork nature of this pure lime rock. It was for this reason that the tunnellers of two specialist Royal Engineer Tunnelling Companies, the 179th and 185th, were able to work below ground to lay their offensive mines. This contrasts with the depiction of 'waterlogged trenches' by Charles Douie, but his experiences surely belong to winter conditions when the water saturation levels were highest, as well as to local conditions, where the surface soils prevented water drainage.

Conclusions

From the outset of the war, trench fortifications as deployed by both sides had only ever been intended to be a temporary solution to a military necessity: holding the enemy in a static position until an offensive could be launched, while maintaining the safety of the troops garrisoning them. Although trenches were constructed according to the specifications of the field manuals developed for the use of military engineers, in many cases these specifications were undermined by the nature of the geological conditions. Thus, although it was expected that trench slopes would be maintained at over-steepened angles, this was not achievable without sufficient slope engineering and adequate revetment to support the slopes, together with drainage solutions that would reduce the relative ineffectiveness of pumping. Trench systems from 1914 to 1916 show the construction of breastworks, multiple layers of duckboard tracks and extemporized revetments (as in the German lines exposed at Messines). It took a more carefully engineered approach, with the development of A-frames, to maintain adequate trench positions in the clay ground at Ypres (Ieper). That these frames achieved their aim is shown by their continued existence below ground, evidenced by their presence in archaeological investigations near

Boesinghe and Wieltje, Ypres. On the Somme, where trenches were developed in the thick Quaternary soils capping the chalk, similar solutions were necessary, although it was possible for trenches to be excavated into the frost-shattered chalk directly, as at La Boisselle, with less need for extensive revetment. These examples illustrate the continuous struggles the armies in France and Flanders had in constructing adequate trench lines and the development of innovative engineering solutions to the weak ground, over-steepened slopes and the persistence of flooding.

Acknowledgements I am grateful to many colleagues working in the archaeology of the Western Front for their valuable input, including Peter Barton, Matthieu De Meyer, Simon Jones and Simon Verdegem.

References

ANON 1915. Geologie im Kriege. *Geologische Rundschau*, **6**, 94–95.

BAETEMAN, C. 1999. The Holocene depositional history of the Ijzer palaeovalley (Western Belgian coastal plain) with reference to the factors controlling the formation of intercalated peat beds. *Geologica Belgica*, **2**, 39–72.

BARTON, P., DOYLE, P. & VANDERWALLE, J. 2004. *Beneath Flanders Fields. The Underground War 1914–18.* Spellmount, Staplehurst.

BELGIAN GEOLOGICAL SURVEY 1999. *Kaartblad (27-28-36), Proven–Ieper–Ploegsteert, 1:50 000.* GSB, Brussels.

BROOKS, A.H. 1920. *The Use of Geology on the Western Front.* US Geological Survey Professional Papers, **128-D**, 85–124.

BROWN, M. & OSGOOD, R. 2009. *Digging up Plugstreet.* Haynes, Yeovil.

BUCKINGHAM, W.E. 1917. Manuscript Diary. The National Archives, London, File TNA WO158/140.

BUREAU DE RECHERCHES GÉOLOGIQUES ET MINIÈRES 1982. *Carte Géologique de la France a 1/50,000, Albert XXIV-8.* BRGM, Paris.

DE GEYTER, G., DE MAN, E., HERMAN, J., MOORKENS, T., STEURBAUT, E. & VANDENBERGHE, N. 2006. Disused Paleogene regional stages from Belgium: Montian, Heersian, Landenian, Paniselian, Bruxellian, Laekenian, Ledian, Wemmelian and Tongrian. *Geologica Belgica*, **9**, 203–213.

DE MEYER, M. & PYPE, P. 2007. Scars of the Great War (Western Flanders, Belgium). *In*: SCOTT, D., BABITS, L.E. & HAECKER, C. (eds) *Fields of Conflict.* Praeger, Westport, **2**, 359–382.

DE MOOR, G. & HEYSE, I. 1978. Dépots quaternaires et géomorphologie dans le nord-Ouest de la Flandre. *Bulletin de la Société belge de Géologie*, **87**, 37–47.

DELATTRE, C., MERIAUX, E. & WATERLOT, M. 1973. *Région du Nord, Flandre, Artois, Boulonnais, Picardie.* Guides Géologiques Régionaux. Masson, Paris.

DEWILDE, M. & SAUNDERS, N.J. 2009. Archaeology of the Great War. The Flemish experience. *In*: SAUNDERS, N.J. & CORNISH, P. (eds) *Contested Objects. Material Memories of the Great War.* Routledge, London, 251–265.

DOUIE, C. 1929. *The Weary Road. The Recollections of a Subaltern of Infantry.* John Murray, London.

DOYLE, P. 1998. *Geology of the Western Front, 1914–18.* Geologists' Association Guide, **61**. Geologists' Association, London.

DOYLE, P. 2012. Examples of the influence of groundwater on British military mining in Flanders, 1914–17. *In*: ROSE, E.P.F. & MATHER, J.D. (eds) *Military Aspects of Hydrogeology.* Geological Society, London, Special Publications, **362**, 73–83, https://doi.org/10.1144/SP362.5

DOYLE, P. 2015. Examples of the geo-archaeology of trench warfare in Flanders. *In*: WILLIG, D., ROSE, E.P.F., HEYSE, I., ALLERSMEIER, C. & DOYLE, P. (eds) *Militärhistorisch-kriegsgeologischer Reiseführer zum Wytschaete-Bogen (Messines Ridge) bei Ypern (Belgien). Geoinformationsdiesnt der Bundeswehr, Schriftenreihe*, **4**, 193–202.

DOYLE, P. 2017. *Disputed Earth: Military Geology and the Western Front, 1914–18.* Uniform Press, London.

DOYLE, P. & BENNETT, M.R. 1997. Military geography: terrain evaluation and the British Western Front, 1914–18. *Geographical Journal*, **163**, 1–24.

DOYLE, P., BENNETT, M.R. & COCKS, F.M. 2000. Geology and warfare on the British sector of the Western Front, 1914–18. *In*: ROSE, E.P.F. & NATHANAIL, C.P. (eds) *Geology and Warfare: Examples of the Influence of Terrain and Geologists on Military Operations.* Geological Society, London, 179–235.

DOYLE, P., BOSTYN, F., BARTON, P. & VANDEWALLE, J. 2001. The underground war 1914–18: the geology of the Beecham dugout, Passchendaele, Belgium. *Proceedings of the Geologists' Association*, **112**, 263–274.

DOYLE, P., BARTON, P., ROSENBAUM, M.S., VANDEWALLE, J. & JACOBS, K. 2002. Geo-environmental implications of military mining in Flanders, Belgium, 1914–18. *Environmental Geology*, **43**, 57–71.

DOYLE, P., BARTON, P. & VANDEWALLE, J. 2006. Archaeology of a Great War dugout: Beecham Farm, Passchendaele, Belgium. *In*: POLLARD, T. & BANKS, I. (eds) *Past Tense: Studies in the Archaeology of Conflict.* Brill, Leiden, 45–66.

FRENCH, J. 1919. *1914.* Constable & Co, London.

GENERAL STAFF, WAR OFFICE 1908. *Military Engineering (Part 1) Field Defences.* HMSO, London.

GENERAL STAFF, WAR OFFICE 1914. *Infantry Training.* HMSO, London.

GENERAL STAFF, WAR OFFICE 1916. *Notes for Infantry Officers on Trench Warfare.* HMSO, London.

GENERAL STAFF, WAR OFFICE 1917*a*. *Notes for Infantry Officers on Trench Warfare.* Revised Diagrams, December 1916. HMSO, London.

GENERAL STAFF, WAR OFFICE 1917*b*. *Summary of Recent Information Regarding the German Army and its Methods.* HMSO, London.

GENERAL STAFF, WAR OFFICE 1917*c*. *The Construction of Field Positions (Stellungsbau)* [translated from German by the General Staff, May 1917]. HMSO, London, 3–4.

GEOGRAPHICAL SECTION, GENERAL STAFF 1918. *1:10,000 Map, Geological, Ploegsteert 28W4.* Geographical Section, General Staff, London. The National Archives file TNA WO297/2475.

Häusler, H. 2000. Die Österreichische und Deutsche Kriegsgeologie 1914–18. *Informationen des Militärischen Geo-Dienstes*, **75**, 1–161.

Häusler, H. 2003. Dr. Walter Kranz (1873–53): Der erste Militärgeologe des 20. Jahrhunderts. *MILGEO: Organ des Militärischen Geowesens des Österreichischen Bundesheeres*, **12**, 1–80.

Heyse, I. 2015. Geomorphology and geology of 'Flanders Fields' (the Ypres to Coastal Plain Region of West Flanders, Belgium) and their impact on military tunnelling and defensive flooding in the 1914–18 world war. *In*: Willig, D., Rose, E.P.F., Heyse, I., Allersmeier, C. & Doyle, P. (eds) *Militärhistorisch-kriegsgeologischer Reiseführer zum Wytschaete-Bogen (Messines Ridge) bei Ypern (Belgien). Geoinformationsdienst der Bundeswehr, Schriftenreihe*, **4**, 51–70.

Hitchcock, F.C. 1937. *'Stand To' A Diary of the Trenches 1915–18.* Hurst & Blackett, London.

Institution of Royal Engineers 1922. *The Work of the Royal Engineers in the European War, 1914–19: Geological Work on the Western Front.* Institution of Royal Engineers, Chatham.

Johnson, D.W. 1918. *Topography and Strategy in the War.* Constable, London.

Johnson, D.W. 1921. *Battlefields of the World War: Western and Southern Fronts.* American Geographical Research Series, **3**. Oxford University Press, New York.

Johnston, W.J. (ed.) 1916. *The Royal Engineers Field Service Pocket-Book.* Institution of Royal Engineers, Chatham.

King, C. 1990. Stratigraphy of the Ieper Formation and Argile de Flandres (Early Eocene) in western Belgium and northern France. *Bulletin de la Société Belge de Géologie*, **97**, 349–372.

King, W.B.R. 1919. Geological work on the Western Front. *Geographical Journal*, **54**, 201–221.

Kranz, W. 1913. Militärgeologie. *Kriegstechnische Zeitschrift*, **16**, 464–471.

Lake, B.C. 1916. *Knowledge for War. Every Officer's Handbook for the Front.* Harrison, London.

Laga, P., Louwye, S. & Geets, S. 2001. Paleogene and Neogene lithostatigraphic units (Belgium). *Geologica Belgica*, **4**, 135–152.

MacGreal, S. 2011. *Boesinghe.* Pen & Sword, Barnsley.

Nolf, D. & Steurbaut, E. 1990. Stratigraphie de l'éocène en Flandre occidentale et dans les regions limitrophes. *Bulletin d'Information des Géologiques du Bassin de Paris*, **27**, 9–36.

Pressey, H.A.S. 1919. Notes on trench war. *Royal Engineers Journal*, **29**, 297–315.

Rose, E.P.F. 2014. Military geosciences before the twenty-first century. *In*: Harmon, R.S., Baker, S.E. & McDonald, E.V. (eds) *Military Geosciences in the Twenty-first Century.* Reviews in Engineering Geology, **XXII**. Geological Society of America, Boulder, CO, 19–26.

Rose, E.P.F. & Rosenbaum, M.S. 1993. British military geologists: the formative years to the end of the First World War. *Proceedings of the Geologists' Association*, **104**, 41–50.

Rose, E.P.F. & Rosenbaum, M.S. 2011. British geological maps that guided excavation of military dug-outs in Belgium during World War 1. *Quarterly Journal of Engineering Geology and Hydrogeology*, **44**, 293–306, https://doi.org/10.1144/1470-9236/08-111

Rose, E.P.F., Häusler, H. & Willig, D. 2000. Comparison of British and German applications of geology in world war. *In*: Rose, E.P.F. & Nathanail, C.P. (eds) *Geology and Warfare: Examples of the Influence of Terrain and Geologists on Military Operations.* Geological Society, London, 108–140.

Salomon, W. 1915. *Kriegs-Geologie.* Carl Winters Universitätbuchhandlung, Heidelberg.

Saunders, A. 2010. *Trench Warfare 1850–1950.* Pen & Sword, Barnsley.

Saunders, N.J. 2011. *Killing Time. Archaeology and the First World War.* The History Press, Stroud.

Steurbaut, E. 1987. The Ypresian in the Belgian Basin. *Bulletin de la Société Belge de Géologie*, **96**, 339–351.

Steurbaut, E. & Nolf, D. 1986. Revision of the Ypresian stratigraphy of Belgium and northwestern France. *Mededelingenvan de Werkgroep voor Tertiarie en Kwartaire Geologie*, **23**, 115–172.

Strahan, A. 1917. Geology at the seat of war. *Geological Magazine*, **64**, 68–74.

Tempest, E.V. 1921. *History of the Sixth Battalion West Yorkshire Regiment. Vol. 1, 1/6th Battalion.* Percy Lund, Humphries, Bradford.

Verdegem, S., Billemont, J. & Genbrugge, S. 2013. *Acheo Rapport 28. Mesen – Collector Aquafin.* AEDE, Ghent.

Verrett, D.D. 1980. Etude statistique des résultats d'essais géotechniques réalises en laboratoire sur l'Argile des Flandres (nord de la France). *Bulletin of the International Association of Engineering Geology*, **22**, 253–255.

War Diary, 53rd Infantry Brigade 1916. Manuscript War Diary, The National Archives, London, File WO95/2033.

War Office 1921. *Manual of Field Works (All Arms).* HMSO, London.

War Office 1925. *Manual of Field Works (All Arms).* HMSO, London.

Willig, D. 2011. Mining warfare in the Wytschaete Ridge 1914–17 – advantages and disadvantages of high ground emplacements. *In*: Häusler, H. & Mang, R. (eds) *International Handbook Military Geography.* Vol. 2. Truppendienst, Vienna, 510–524.

Willig, D. & Häusler, H. 2012. Aspects of military hydrogeology and groundwater development by Germany and its allies in World War I. *In*: Rose, E.P.F. & Mather, J.D. (eds) *Military Aspects of Hydrogeology.* Geological Society, London, Special Publications, **362**, 85–103, https://doi.org/10.1144/SP362.6

Wyllie, D.C. & Mah, C.H. 2005. *Rock Slope Engineering, Civil and Mining.* 4th edn. Spon, London.

German military geology and military mining on the Eastern Front in World War I

DIERK WILLIG

Bundeswehr Geoinformation Centre BGIC, Frauenbergerstr. 250, 53879 Euskirchen, Germany

0000-0002-5279-7555

DierkWillig@bundeswehr.org

Abstract: At its peak, the Eastern Front encompassed the entire frontier between the Russian Empire and Romania on the Allied side and the Austro-Hungarian Empire, Bulgaria, the (Turkish) Ottoman Empire and the German Empire on the other: a distance of *c.* 1500–1700 km. Mobile warfare alternated with periods of static trench and siege warfare, when 'mines' (tunnels charged with explosives detonated to breach the overhead fortifications) and 'dugouts' (to protect troops from artillery or aerial bombardment) were constructed in at least 50 localities between 1915 and 1917, from Mitau in present-day Latvia to Rarancze in Moldova. Terrain ranged from plateaux floored by Cenozoic sediments in the north to mountains with more complex, older, stronger rocks in the south. Despite lessons learnt from the Russo-Japanese war of 1904–05, both sides began the conflict unprepared for tunnelling warfare. However, German and Austro-Hungarian forces soon developed units of military geologists whose duties included guidance of tunnelling projects. Eight teams of German military geologists and four of Austro-Hungarian, in total over 60 men who can be named, are known to have served on the Front, as did newly formed and equipped specialist engineer mining battalions from Prussia, Bavaria and Austria-Hungary.

Supplementary material: Examples of equipment schedules are available at https://doi.org/10.6084/m9.figshare.c.4280300.

The impact of the terrain on the conduct of military operations in mobile warfare is generally obvious and has long been recognized in military practice and science (cf. von Clausewitz 1937). Moreover, in trench warfare or attempts to breach fortified positions, the battlefield extends underground from the land surface through linear elements such as trenches and tunnel systems. During the Great War of 1914–18, once the front lines had stabilized, soldiers in trenches were exposed to the devastating effects of artillery and mortar bombardment without the freedom to move to safer positions. They had to seek shelter in hardened bunkers or underground. Yet the constant increase in ground penetration, achieved by progressive improvements during the war in artillery and mortar ammunition, forced the construction of tunnelled 'dugouts' at increasingly greater depths. This developing use of underground terrain stimulated an awareness that geology had significant military applications.

The trench war with its battles of attrition that started in late 1914 along the 740-km Western Front, across Belgium and northern France, has been described extensively in English literature (e.g. Grieve & Newman 1936; Westwell 2010; Doyle 2018 and references therein). Less well known in English is the conflict on the Eastern Front, ultimately *c.* 1500–1700 km long and so of far greater extent (cf. Churchill (1931) *The Unknown War: The Eastern Front*). This began in the summer of 1914, and at its peak encompassed the entire frontier between the Russian Empire and Romania on the Allied side and the Central Powers – the Austro-Hungarian Empire, Bulgaria, the (Turkish) Ottoman Empire and the German Empire – on the other. On this front, phases of static warfare and sieges of forts (most famously Przemyśl) alternated with offensive mobile operations (Anon. 2005, 2017; Westwell 2010).

It was only in the phases of trench war and during sieges that mine warfare was feasible: the excavation of tunnels, beneath the enemy's strongly held front line or fortifications, that were charged with explosive and detonated prior to infantry assault. Kranz (1934) noted that such warfare took place at about 50 locations along the Eastern Front, although without naming these localities. This article confirms a similar number of sites, which are listed by name and coordinates in Table 1 and mostly shown on Figure 1. Tunnelling operations were mostly concentrated in the south, and so largely performed by Russian forces and their opposing Austro-Hungarian (and later German) sappers/miners, to whom Turkish combat engineers at times gave support. A smaller

From: Rose, E. P. F., Ehlen, J. & Lawrence, U. L. (eds) 2019. *Military Aspects of Geology: Fortification, Excavation and Terrain Evaluation*. Geological Society, London, Special Publications, **473**, 131–150.
First published online November 1, 2018, https://doi.org/10.1144/SP473.14

Table 1. *Localities at which tunnelling can be proved to have taken place on the Eastern Front, with contemporary German place name and date where known, and coordinates in decimal degrees*

Location	Date	Longitude	Latitude
Annenburg*		23.72	56.65
Baranowitschi		26.01	53.13
Baruny		25.66	54.35
Batkow*		24.89	49.80
Brzezany, Zlota Lipa	1916	24.91	49.46
Burkanow		25.36	49.28
Cebrow		25.49	49.57
Chwojna/Kowel		24.83	51.19
Cisna–Balinogrod road*	March 1915	22.26	49.29
Czernowitz, Rarancze		25.94	48.29
Dolzok	January 1916	26.04	48.46
Dünaburg		26.53	55.88
Dzikie Lany		24.91	49.46
Hladki	May 1916	25.52	49.64
Kalnischki		25.05	54.88
Kisielin (Kyssylyn)	February 1917	24.82	50.86
Kolpytiv, Kremalz		24.72	50.93
Konin	March 1917	18.28	52.19
Kopzewitschi near Schitomir*	May 1916	28.42	52.19
Koziowka		22.49	51.45
Koziuwka (Lemberg)	April 1915	23.50	49.07
Kozlow*		25.29	49.56
Kranzem	August–September 1917	24.61	58.86
Kuskowo-Glinki		19.69	52.56
Labusy-Darewo	July 1916–March 1917	26.27	53.11
Lipnica Gora	March 1917	21.05	49.85
Luck	February 1917	25.08	50.75
Lysonia Höhe		24.99	49.43
Manilowa Höhe	March 1915	22.26	49.29
Mitau		23.72	56.65
Mogily, Mogila	April 1915	20.11	52.06
Moskiska		23.14	49.79
Narajowka		24.91	49.46
Novo Georgievsk		20.68	52.44
Oginski Kanal		25.44	52.54
Oglenda/Kot Höhe 137	March 1915	20.95	53.09
Patolenka Schanzen (Przasnysz)*		20.95	53.09
Petrikow-Kopzewitschi	May 1916	28.42	52.19
Popielcha Höhe		24.91	49.46
Przasnysz-Czarzaste*	Februay–March 1915	20.95	53.09
Puzniki Tlumacz	March 1915	21.22	49.56
Rawa, Höhe 10	April 1915	20.24	51.77
Sapanow		25.13	51.05
Sokolow, Stryj		24.02	49.18
Smorgon	June–July 1916	26.38	54.47
Stanislau		24.72	48.92
Swiniuchy	February 1917	24.75	50.63
Tarnopol	April 1916	25.49	49.57
Toboly		24.91	49.46
Toutry	July 1916	25.49	49.57
Turcina Höhe, Stubno		23.14	49.79
Usiscko		26.04	48.46
Wertepy		25.49	49.57
Wladimir-Wolinski		24.82	50.87
Worobijowka*		25.52	49.64
Zlodzow, Obydra Höhe Wertepy	May 1917	24.89	49.80
Zlota Gora		25.16	49.71
Zwynyn (Zwinin)	April 1915	23.50	49.07

See Figure 1, except for localities marked*.

Fig. 1. Map showing locations/regions along the Eastern Front in World War I where mine warfare could be proved: contemporary German name, as listed in Table 1, highlighted in red with location indicated by red arrow. Names highlighted in blue are for other significant localities: see text. Kurland, western Latvia; Litauen, Lithuania; Polen, Poland; Wolhynien and Podolien, parts of NW Ukraine; Ungarn, Hungary; Galizien, Galicia, a crown land of Austria-Hungary straddling the border of Poland and the Ukraine; Karpathen, Carpathian Mountains; Bukowina, a region now split between Romania and Ukraine. Solid black lines from west (left) to east (right) show the frontline as at 1 May 1915, mid-July, mid-August, and mid-September 1915. Based on contemporary German maps by Baumgarten-Crusius (1919).

number of examples is known from the north, where the ground was contested by Russian and German forces.

Poor communications and an inadequate road and rail infrastructure across the wide front hampered the rapid closing of breaches in the front line torn by military offensives. This inhibited trench warfare and facilitated traditional mobile cavalry-led operations. However, the autumn and winter of 1914–15 saw trench wars develop near Warsaw and Lodz in east-central Poland, and in the Carpathian Mountains and elsewhere in southern Poland and territories nearby. At the end of September 1915, another trench war flared up, along the line Riga–Daugavpils–Pinsk–Smarhon'–Czernivtsi–Lutsk–Ternopil (see Fig. 1) (Baumgarten-Crusius 1919). Warfare was to continue intermittently until early 1918.

The preconditions for mine warfare comprised a relatively short distance between the trenches of the warring parties, technical engineering and tunnelling expertise amongst their troops, and a suitable underground (determined by geology). Once one side had started mining, the opposing forces needed to develop appropriate countermeasures such as camouflets (tunnels charged with explosive detonated to collapse an opponent's tunnel).

Information sources for the Central Powers

Few sources still exist in Germany for primary documents recording military mining and other tunnelling activities during World War I on the Eastern Front. Post-war, all German states stored their military archive documents within the Prussian army archives, in Potsdam (Willig 2015). However, during World War II, in April 1945, most of this 1914–20 material went up in flames during an Allied air raid on Berlin and Potsdam nearby. The surviving material is now held within the Military Archive (*Militärarchiv*): part of the German Federal Archive (*Bundesarchiv*) in Freiburg-im-Breisgau. Additionally, the states of Bavaria, Württemberg and Saxony had separate archives. Extensive duplicate documents relating to military units from these three states are preserved in their provincial archives, although most of the material of the Saxon *Kriegsarchiv* (War Archive) was either also destroyed during World War II air-raids (in February 1944) or removed by Soviet troops at the end of the war. Austro-Hungarian military records are held by the Austrian State Archives (in their *Kriegsarchiv* section) in Vienna: no documents on mining warfare could be found in the Hungarian War History Archive, in Budapest. Military-historical research is consequently reliant to a significant extent on the regimental histories and memoirs of veterans written after 1918.

The main information sources used to generate this article therefore comprise:

- German Federal Archive (Military Division), Freiburg-im-Breisgau;
- Baden-Württemberg State Archives (Central State Archives Department), Stuttgart;
- General State Archives, Karlsruhe;
- Bavarian Central State Archives (Division IV War Archives), Munich;
- Austrian State Archives (War Archive), Vienna;
- Bundeswehr Geoinformation Centre, Euskirchen;
- Heringen Collection, within the Bundeswehr Geoinformation Centre, Euskirchen;
- diaries and histories of military units;
- family histories online;
- field postcards (author's collection);
- contemporary photographs (author's collection).

Russian experience of mine warfare 1904–05

By 1914, the Russians already possessed significant experience of 'modern' mine warfare, gained from the Russo-Japanese War barely ten years earlier. That war had begun with the Battle of Port Arthur, a surprise attack by a fleet of Japanese destroyers on the Russian fleet anchored at Port Arthur in Manchuria, NE China, on 8–9 February 1904. The subsequent Siege of Port Arthur (1 August 1904 to 2 February 1905: see Kranz 1909; de Fonclare 1928) was the longest and most violent battle of the war – Port Arthur being widely regarded at that time as one of the most strongly fortified positions in the world – and provides a good example of the use of 'mines' early in the twentieth century. The siege became a model for subsequent trench wars. Eventually, it was the war fought underground that forced the besieged Russian troops to surrender.

Russian troops had withdrawn to Port Arthur on 30 July 1904, within a perimeter protected by a line of hills with concrete forts, machine-gun emplacements and connecting trenches. They were besieged by a Japanese army of about 150 000 men, supported initially by artillery of 474 guns. The Japanese succeeded in neutralizing the Russian artillery by their own guns within two weeks, but the concrete walls of the forts were generally resistant to bombardment. Moreover, during Japanese assaults on the Russian trenches and fortress, murderous infantry and artillery fire caused heavy losses among both attackers and defenders. The Japanese therefore abandoned frontal assaults and began a protracted siege, developing the techniques of sapping (driving trenches forward from their positions so as to provide cover for troops prior to assault) and mining (tunnelling under fortifications so as to breach their defensive line by detonation of explosives). By the end of August the siege operation had progressed so far that the Japanese troops were no more than 100 paces from the Russian forward defence installations of the fortress (the Waterworks and Temple Redoubts) and 300 paces in front of its caponier no. 3: a structure that allowed defensive fire across a ditch in front of the fort (von Tettau 1912). By mid-September the Japanese had dug some 8 km of trenches. They attacked and captured the Waterworks Redoubt on 19 September and the Temple Redoubt soon afterwards.

Mining had been employed at the Waterworks Redoubt, but a more significant case history is provided by the attack on the Russian 'Fort II' (Fig. 2) (von Tettau 1912; Ortenburg 2005). To prevent the Japanese from undermining it, the commander of a Russian engineer company, Lieutenant-Colonel Zherebtsov, initiated countermining to blow up tunnels being constructed by the enemy. The Japanese had extended a tunnel ('c' on Fig. 2) from the edge of the glacis (the artificial slope surrounding the fort) which continued as a mine adit towards the western corner of the fort's parapet ('C' on Fig. 2). The Russians learned of this work and attempted a minor attack in order to assess it, but this failed. They therefore dug countermines from the shoulder points of the parapet to create a defensive tunnelling work that has been described in detail in the diary of the officer in charge of the Russian tunnel warfare, engineer Lieutenant-Colonel Rashevsky (von Tettau 1912). Initially, progress was through rock and thus slow but gathered speed through the made ground of the glacis area. The plan was to dig two 10 m-long countermines, connected by a gallery, at shallow depth (occasionally only 1 m under the earth surface). However, they feared that the Japanese might destroy these countermines by means of demolition shafts from above and when, after only a short time of tunnelling, the enemy was heard apparently a short distance ahead, the Russians felt compelled to detonate their camouflet prematurely, on 27 October. The oversized charge not only collapsed the enemy tunnel but also created a deep crater ('a' on Fig. 2) that damaged the corner of the fort – although according to von Tettau (1912), Rashevsky recorded that the charge consisted of only 8 lbs (3.6 kg) of gunpowder. Subsequently the Japanese extended their tunnel forward to this crater to create an opening from which they could rake the area with machine-gun fire, and after further detonations, the Russians had to abandon the parapet of the fort in this region.

Another oversized Russian camouflet detonated on 17 November, creating a crater that uncovered their own mine adit ('b' on Fig. 2) without causing damage to the enemy. A Japanese mine charge was successfully detonated within the fort on 26 November (Fig. 2, at the end of the right-hand tunnel shown in green). The Russians were forced to withdraw from the parapet, which fell into the hands of the Japanese. The Russians heard the enemy continue tunnelling but did not attempt to construct camouflets.

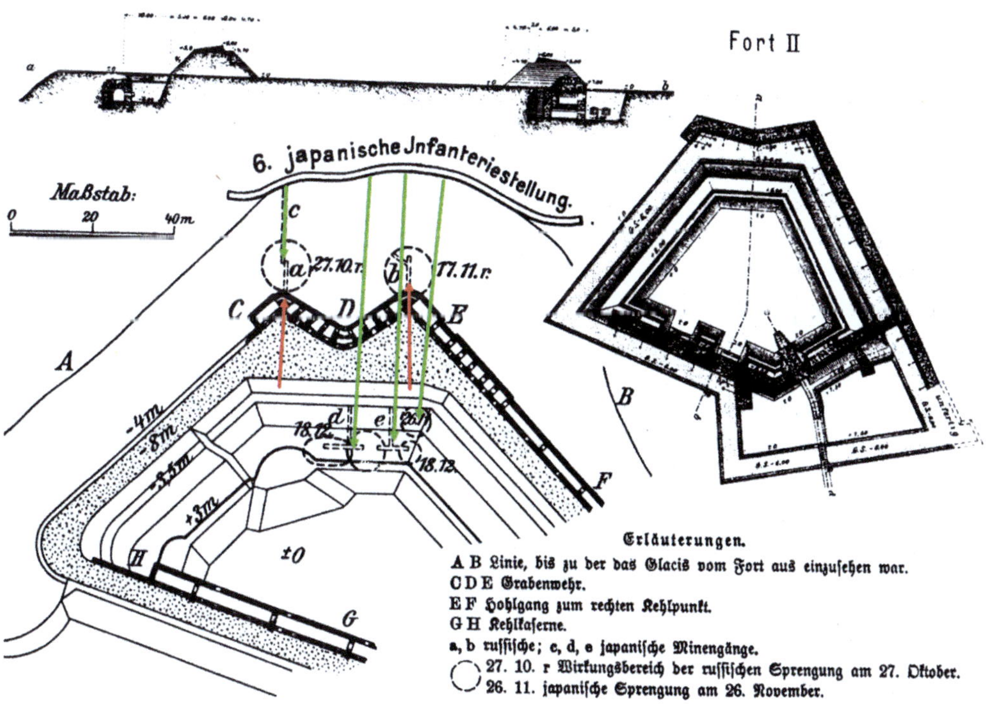

Fig. 2. Fort II of Port Arthur (Manchuria, China) in 1904, during the Russo-Japanese War. Green lines show four tunnels driven forward from the Japanese 6th Infantry position, red lines two countermines driven by the Russians. Circles indicate the sites of explosives detonated: a Russian demolition (a) of 27 October which destroyed attack tunnel 'c'; another demolition (b) of 17 November which was ineffective; a Japanese charge detonated within the fort on 26 November, and three more on 18 December, that forced the Russians to abandon the fort. After Kranz (1909), von Tettau (1912) and Ortenburg (2005); adapted by kind permission of Mönch Verlagsgesellschaft mbH.

On 18 December, detonation of three charges (emplaced by means of the adits shown as 'd' and 'e' on Fig. 2) tore a large breach in the wall of the fort, enabling the Japanese to capture it entirely. Fighting took place in close proximity, by means of small-arms fire accompanied by throwing of hand grenades. It was thus a prelude for the later static battles in World War I, as noted by Ortenburg (2005).

On the 31 December, Fort III was also seized by the Japanese after mine detonations had generated a breach. According to von Tettau (1912), the charges employed in the two adits consisted of a total of 3000 lbs (1360 kg) of dynamite and 2400 lbs (1089 kg) of gunpowder.

Tunnelling was also employed against field positions. Before the Russo-Japanese War, mining had been used only in the sieges of fortresses. However, it was adopted in open terrain at the Shen (the present-day Hun) river during trench warfare there, that extended from October 1904 to February 1905. Thus, whilst the Russians tunnelled towards a Japanese bridgehead in front of a railway bridge over the Shen in Mukden, the Japanese built two defensive mines which became a site of conflict with close-quarter weapons (Kranz 1936, p. 73). Both warring parties also tunnelled near the town of Linschinpu. The Russians attacked by means of an adit in mid-January 1905. Kranz (1936) recorded that the tunnelling speed reached 7.5 m per day on average. When the adit came upon a Japanese countermine after approximately 150 m, the Russians detonated 900 kg of gunpowder, on 28 February 1905 at 10 p.m. Their subsequent infantry assault was repulsed by the Japanese with heavy casualties. Eventually, the Russians had to withdraw, abandoning several mine shafts under construction.

Despite the relatively recent experience gained from these different aspects of mine warfare in the Russo-Japanese War, it seemingly did not provide a particular advantage for the Russians in the subsequent trench conflicts of World War I; although mine warfare was deemed to be a subject of military importance in Russia both during World War I and afterwards (Antulajev 1930). Indeed, it was employed on a few occasions in World War II, notably the destruction of a German stronghold on 10 November 1942 when Russian combat engineers undermined the building with a 63 m-long adit and blew it up with three tons of TNT (Afalašev & Badanin 1969).

Development of a German military geological organization

In Europe, the unexpected revival of mine warfare early in World War I did not initially stimulate an appropriate response. 'The European armies were therefore neither tactically nor technically prepared for extensive and sustained tunnel warfare' (translated from Kranz 1936, p. 73). Austro-Hungarian, German and, later, Turkish troops were drawn into mine warfare without the necessary training and equipment. There was a lack of military regulations covering mining and demolition procedures, of trained combat engineers, of the equipment and tools necessary for working underground, and of materials to shore up shafts and adits. The German Army possessed only one relevant military manual: a pre-war demolition regulation (Anon. 1911).

The German Army also began World War I without the benefit of organized military geological advice. At that time, neither geologists at universities and other academic institutes nor those at the regional geological survey departments within Germany had been called upon to contribute to military matters. Philipp (1919) noted that state institutions had in no way made any preparations in peacetime to meet predictable military requirements (see also Willig 1999; Häusler 2000; Rose *et al.* 2000; Willig & Häusler 2012). Until the outbreak of hostilities, geologists at the universities had not developed research into potential military applications of geology, certainly not for a war underground.

However, with the onset of the trench war on the Western Front in the winter of 1914, the need for geological advice in aspects of underground excavation and of groundwater development became evident. Shortly before the war, the visionary combat engineer officer and geologist Walter Kranz had advocated the need for a military geological organization to support the German Army (Kranz 1913). He had also recognized that lessons could be learnt from the Russo-Japanese War and especially from the Siege of Port Arthur on the importance of mine warfare in contesting fortifications (Kranz 1909). Kranz, the father of twentieth century German military geology (Häusler 2003), promoted the establishment of a military geological organization by means of newspaper articles, petitions, position papers and presentations given to the military. The military value of geology was also demonstrated by other geologists who volunteered for service on the Western Front early in the war. 'In addition to Kranz, who started working as a war geologist for the Strasbourg [Fortress] Government in November 1914, Prof. Dr Passarge was employed in the area of the 4th Army on the Western Front in the winter of 1914/15' (translated from Willig 1999, p. 13). From May 1915, Professor Dr Hans Philipp developed and led a group of six geological units with a total staff of 25 officers and other ranks to support operations by 121 Infantry Division in the Priesterwald (Bois-le-Prêtre) region.

A major step towards the establishment of a military geological organization was made on 21 May

1915. On that day, the Chief of the General Staff of the Field Army ordered that 'the generals of the Engineer Corps at army headquarters should be informed that they could consult geologists, and that these should be made available to them on request' (translated from Kranz 1927, p. 9). The next milestone was a decree by the Chief of the General Staff of the Field Army on 6 September 1916 that initiated the integration of military geology into the Military Survey branch of the German Army. A non-geologist, Major Siegfried Boelcke, chief of the Military Survey branch, therefore also became the head of a new military geological organization. His senior personal assistant, however, was a geologist: Lieutenant (of the Reserve Army) Julius Wilser – who later used his experience to generate an early post-war textbook on military applications of geology (Wilser 1921). Each of 28 Survey departments (*Vermessungsabteilungen*) was assigned a small team of geologists which in turn provided support to the headquarters of army corps or divisions. The geological department of the *Preußische Landesaufnahme* (Prussian Land Survey Office) served as a co-ordinating body for the military geological organization and was responsible for both the provision of geological maps and literature and the training of junior military geologists (*Kriegsgeologen*, literally 'war geologists').

However, affiliation with Military Survey created organizational problems, and the Chief of the General Staff of the Field Army decided on 13 March 1918 that 'in peacetime and wartime military geology will be a responsibility of the Engineer Corps' (translated from Kranz 1927, p. 13). According to Kraus (1941), one of the last organizational changes to be effected was the establishment of geologist units with the Survey departments in Lithuania and the Baltic region. Ultimately, German military geologists were deployed on both the Western and Eastern Fronts, in the Balkans and in Palestine. However, their re-assignment from Survey to Engineer command did not come into effect before hostilities ended, on 11 November 1918.

Of the 28 Survey departments with attached military geological centres or teams, eight were assigned at some time for operational service on the Eastern Front, numbers 4, 5, 9, 10, 12, 18, 20 and 22. No documents have been traced that reveal names of geologists who served with department 4, and the records for other departments may be incomplete. However, they are sufficient to reveal the identity of 60 individuals who together contributed geological support to German forces on the Eastern Front through these units (Table 2) (Uhl serving at different times in both departments 9 and 20).

It is noteworthy that few of these were commissioned as officers, and of those, none is known to have held a rank higher than that of lieutenant. There were in addition at least five German military geologists who served as such on the Eastern Front whose work cannot be related to that of Survey departments: Hans Cloos (who assisted prospecting for nickel in the German province of Silesia, for the Silesian Army); a geologist named von Czernowitz; O. Grupe who served in the province of Kurland in western Latvia; E. Henning who was deployed in the Carpathian Mountains in 1916; and Leopold Sommermeier who was deployed in the winter of 1916 to Romania (Häusler 2000). Willig (1999) and Häusler (2000) together list the names of over 200 geologists known to have served as such with the German Army as a whole in World War I, but Table 2 amplifies this knowledge with many additional names.

Additionally, four War Survey departments (*Kriegsvermessungsabteilungen*) of the Austro-Hungarian Imperial and Royal Army are known to have served on the Eastern Front and to have been supported by military geologists: numbers 2 (in Ukraine), 3 (with 3rd Army Corps, in Galicia and Ukraine), 6 (in Ukraine and Crimea) and 14 (whose precise area of assignment has not been ascertained) (Häusler 2000). The number and names of these geologists are not known (only that of Dr Jan Nowak, separately assigned as a geologist lieutenant to Krosno in Poland: Häusler (2000, p. 108)), but in total, the number of military geologists deployed to support the Central Powers on the Eastern Front was clearly considerably greater than the two British geologists and nine American geologists known to have supported Allied forces on the Western Front (Brooks 1920).

Roles of German military geologists

The tasks undertaken by the military geological unit led by Lieutenant Carl Mordziol of Survey department 20 are presumed to be fairly representative of these units as a whole. In effectively the final year of the war on the Eastern Front (where a cease-fire was declared in December 1917) his unit had completed at least 215 tasks (Mordziol 1917*a*, *b*, *c*). Most of these related to water supply (Fig. 3), a smaller but still highly significant proportion to supplies of raw materials (primarily sand or aggregate) or to problems of site drainage, and smaller proportions still to field fortifications, mining work and road construction.

For German military geological units in general, Bärtling (1917), noted that geologists had been used in trench warfare to give advice that improved the living conditions of the troops and also strengthened the resistance of field fortifications to bombardment or assault. He listed as specific examples of primary geologist tasks:

Table 2. *Military geologists and their assistant personnel (*) assigned to Military Survey departments* (Vermessungsabteilungen) *of the German armies on the Eastern Front*

Department	Location	Name	Rank or status if known
5	Army Group Woyrsch	Frank, Dr W.	
		Philipp, Prof. Dr H.	
		Reitz	
9 (Bavarian)	Latvia	Doss, Prof. Dr K.*B.*	Chief geologist
		Beck	Civil servant
		Berz, *K.C.*	Civil servant
		Burhenne, *H.*	Civil servant
		Edel*	Corporal, assistant geologist
		Fischer, *H.*	Corporal
		Götze*	2nd lieutenant, assistant geologist
		Herbig	Lance-corporal
		Holtermann*	Driver
		Knauer, *J.*	Civil servant
		Leuchs, *Dr K.*	2nd lieutenant (R)
		Manhart*	Corporal, assistant geologist
		Marx	Lance-corporal
		Regelmann, *C.*	Kriegsgeologe
		Reissinger*	Soldier
		Sandkühler	Kriegsgeologe
		Schlossmacher, *Dr K.*	Soldier (R)
		Simmrt*	Soldier
		Sinner*	Soldier
		Stille, *Dr H.*	2nd lieutenant
		Uhl	Soldier (R)
		Völker*	Soldier, assistant geologist
		Wolf	2nd lieutenant
10	Army Group Eichhorn	Finckh	Kriegsgeologe
		Gothan, *K.*	Kriegsgeologe
		Lehman	Lance-corporal
		von Linstow, Prof. Dr *C.*	Kriegsgeologe
12	2nd Army	Büchner	Grenadier
		Kittler	Sergeant-major
		Mylius	Civil servant
		Watter	Soldier
18	Estonia; Latvia 8th Army	Gäbert, Dr *C.*	Chief geologist
		Kiekebusch, Dr	
		Klamm	Lieutenant
		Oertel, Dr	
		Rieger	Lance-corporal
		Rüger	
		Scupin, *Dr* H.	
		Stock	Lance-corporal
		Thiem, Dr Ing. G.	
		Uhlig, V.	
20 (Bavarian)	Galicia	Mordziol, Dr C.	Chief geologist
		Banse, E.	Civil servant
		Baur, *C.*	Soldier (R)
		Dürrfeld	
		Kraiss	
		Leier	Sapper
		Lindenbein	Lance-corporal
		Pulfrich	
		Uhl	
		Wegener	Civil servant
		Wohlstedt, *P.*	
22 (Bavarian)	Ukraine; Poland	Gagel, C.	
		Hietz	Lance-corporal
		Kulisch	Soldier (R)
		Lang	2nd lieutenant
		Scheu	Corporal
		Steinert	Lance-corporal

Note: titles and initials shown in italic are derived from Willig (1999) and/or Häusler (2000), not primary documents; R, reserve army.

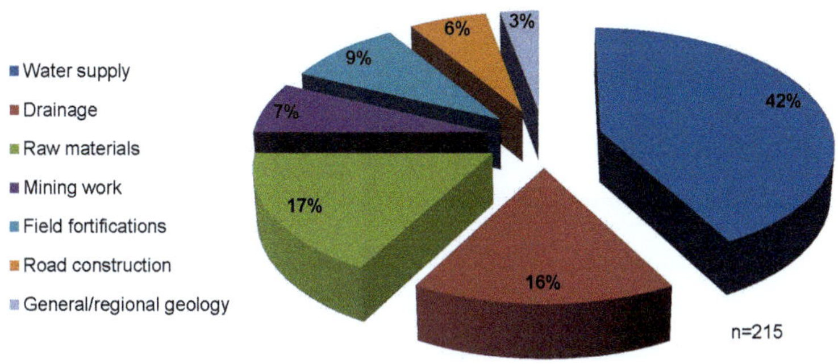

Fig. 3. Main categories of 215 tasks undertaken between January and November 1917 by the geological unit attached to Survey department 20, assigned to the German-Austrian South Army on the Eastern Front; data compiled from Mordziol (1917*a*, *b*, *c*).

- water supply to field positions and fortifications;
- disposal of waste and rainfall water from field positions and fortifications;
- investigation of soil conditions for the construction of rearward positions and bomb-proof dugouts (ground investigation);
- cooperation in mining and sapping work;
- advice on the construction of defensive positions in mountainous regions;
- acquisition of aggregate for concrete and of other building materials;
- advice in the construction and maintenance of railway lines;
- investigation of swamps and similar terrain obstacles with regard to cross-country trafficability;
- fuel supply to troops in sparsely wooded areas;
- supply of the home country with mineral resources;
- recording of geological conditions in the operational area.

Additional tasks included:

- advice in reply to questions on earth telegraphy (*Etel-Arendt* listening stations);
- assessment of cross-country trafficability for armoured vehicles;
- advice in demolitions and determining the modulus of resistance for mine demolitions;
- planning of artificially induced floods to create obstacles to cross-country movement;
- search for hot springs to be used for military baths;
- search for limestone deposits suitable to create quicklime for decontamination purposes;
- investigation of underground quarries with respect to their use for further quarrying, as shelters for troops or civilians, or as threats to the stability of surface buildings;
- epidemic control, building of war cemeteries, latrines and hygienic facilities;
- advice with regard to geological hazards such as landslides (artificially induced or natural), mudslides, rockfalls and avalanches.

German military geologists were thus used for a wide variety of tasks during the war, of which mining work and advice for the construction of underground shelters (dugouts) were a significant although not a predominant part. A wide range of military geologist roles had also been recognized in Austria-Hungary, and by February 1916: see section III *Über die 'Kriegsgeologie'* in Schafarzig (1916).

Prerequisites for mine warfare

Mine warfare in general was subject to several basic constraints. Notably, the front line had to remain static over a long period. During this static battle, there was sufficient time for trenches, dugouts and fortified trench systems to be constructed.

The area between the trench lines of the warring parties was a no-man's-land that was normally reinforced with barbed wire intended to slow down enemy assaults. This zone was exposed to the projectiles of 'mine launcher' mortars and artillery, and to machine guns and other infantry weapons, from both sides. Accordingly, the no-man's-land became a killing zone that could usually be conquered only at the sacrifice of a high number of casualties. The length of the static front line and a frequent shortage of ammunition, which occurred particularly on the German side, often made it impossible to concentrate sufficient artillery fire power to destroy enemy barbed wire and fortifications before launching an assault. If the trench lines of opposing troops were not too far apart from each other, attempts could be made to 'sap' towards the opponent. Such sapping was achieved by open trenches, or by attack tunnels at shallow depth. The final assault on an enemy trench would then be launched from the head of such a trench or tunnel. To facilitate an assault, auger borings were occasionally driven from the head of a sap or tunnel towards and beneath the enemy. These borings were then filled with relatively weak explosive charges and ignited prior to the assault, to breach fortifications and to inflict casualties on the opposing troops.

Given suitable geological conditions and in the absence of groundwater, attack tunnels were commonly driven forward at a depth of 5–10 m from frontline trenches towards exposed strongpoints such as hardened machine-gun positions. Once one side had started tunnelling the other had to follow suit in order not to give advantage to the enemy. The necessary depth was reached from the land surface by constructing either vertical access shafts, shallowly inclined ramps or 'stairwells'. Because the shafts could easily be destroyed by enemy artillery shelling or demolitions, troops were normally concentrated further to the rear, in a second trench line, where tunnelling generally took place at a lower level (20–30 m). The tunnel system thus soon consisted of two storeys. Vertical shafts near the front line were connected by galleries from which both attack and sound intercept adits were dug towards the target.

As digging became deeper, the impact of geological environmental conditions became more obvious. Expert knowledge of miners and military geologists inevitably came into demand in the armed forces on both sides. Moreover, the combat engineers in German forces that were initially employed for the tunnelling work were converted into specialized tunnelling units – combat engineer/miner companies (*Pionier-Mineur-Kompagnien*) in Prussia and miner companies (*Mineur-Kompagnien*) in Bavaria. Although with different names, similar specialized units were created in their allied Austro-Hungarian forces, guided by comparable principles, adopting similar tactics, and provided with similar equipment (Österreichisches Bundesministerium 1930). To this end, miners and appropriate craftsmen (Fig. 4) were recruited and ideally put under the command of an officer with experience as a mining engineer in his pre-war civilian life (cf. Sappeurkompagnie 5/4 1915; Sappeurkompagnie 2/22 1916; von Ardenne 1917; Sappeurkompagnie 2/11 1918; Sappeurkompagnie 2/13 1918; Sappeurkompagnie 2/43 1918; Sappeurkompagnie 3/13 1918; Sappeurkompagnie 5/11 1918*a*, *b*; Sappeurkompagnie 5/13 1918; Hartenstein 1936; Willig 2015). Experienced troops (German combat engineers under the command of German officers) even supported Ottoman Turk forces when trench warfare developed in 1915 near Gallipoli at the Dardanelles, the narrow strait forming part of the continental boundary between Europe and Asia, following the now famous but ultimately disastrous amphibious landings of British, French and ANZAC (Australian and New Zealand Army Corps) troops (Grieve & Newman 1936; Lawrence 1981).

Fig. 4. World War I Austro-Hungarian mining troops (Bosnians with fez headdress) ready to ignite a demolition charge, according to an inscription on the reverse of the photograph. Author's collection.

The influence of geology on mine warfare

The Eastern Front, both at the start of the war in 1914 and when warfare in this region came mostly to an end in 1917, crossed a region of relatively low plateaux underlain predominantly by sedimentary rocks of Cenozoic age in the north and a more mountainous region with older, stronger rocks and a more complex geological structure to the south (Fig. 5).

The underground, i.e. geology, has a particularly decisive influence on tunnelling projects. Additional to the construction of tunnels and dugouts, geology is of paramount importance in the planning and conduct of warfare underground. A lesson learnt during World War I was that the planning of attacks that involve preparatory mining should ideally always be undertaken with geological support. German military geologists came to realize that that support could most effectively be provided for the military leadership by specialist thematic maps. Based on conventional geological maps, these simplified the geology and provided additional information relevant to a specific military purpose, in a form easily understood by engineers and staff officers who were not geologists. Thus, to guide underground excavation, thematic maps were created to show the principal lithologies and variations in the depth to groundwater in the region of proposed military operations (cf. Fig. 6).

In general, underground excavation and so mine warfare was influenced by factors determined by lithology, groundwater, subsurface features, rock fractures and porosity, and the swell/shrink capacity of clays:

Lithology	Progress of mining
	Sound propagation
	Radius of rupture
	Crater depth
	Maximum throw of debris
	Explosive required
	Shock propagation/shock absorption
	Storage capacity for explosion gases
	Stability under load/demand for mine timber (props)
	Camouflage of excavated material
	Splinter effect
	Air pressure impact
Groundwater	Presence of ground water
	Seasonal variations of water table
	Quicksands, running sands
Subsurface features	Existence of vaults
	Occurrence of channels
	Dip of strata
Rock fractures and porosity	Progress of mining
	Retention ability and sudden emission of toxic or suffocating explosion gases
Swell/shrink capacity of clays	Need for time-consuming emplacement of support/shoring with pit props

Of these, progress of mining and sound propagation are notable amongst the factors whose importance can be documented from the Eastern Front.

Progress of mining

Rock strength, largely dependent on lithology and fractures and porosity within the rock, was an important influence on mining progress. Based on experience in general during World War I, Kraus (1941, p. 19) distinguished five categories of military engineering in which estimates of rock strength had particular relevance:

(1) during rock excavation (workability by digging, drilling or blasting);
(2) during slope/tunnel stabilization and loading on construction sites (admissible load, compressive strength, rock pressure, and sliding tendency being factors to calculate);
(3) under shelling (resistance to bullets and bombs, and splinter effect, being significant);
(4) during applications involving electrical sound and shock waves; and
(5) during changing weather conditions (e.g. the effect of waterlogging or frost).

For underground excavation, 'workability' came to be classified into five categories according to the tools necessary:

(1) Digging ground. 'Soil' that could easily be dug by hand with a spade; estimated rate *c.* 1 m^3 per 1 man/hour.
(2) Pickaxe ground. Compacted soil interspersed with rocky slabs that required use of a pickaxe; estimated rate *c.* 1 m^3 per 1.5 man/hours.
(3) 'Breakable' ground. Soft stone alternating with stronger rock that required use of a crowbar in addition to a pickaxe, estimated rate 1 m^3 per 2.5–3.5 man/hours.
(4) Easy demolition ground. Medium-strength rock with planar features (jointing, cleavage or fissures) contributing to weakness that required use of a crowbar following detonation of explosive charges, estimated rate 1 m^3 per 3–5 man/hours.
(5) Difficult demolition ground. Strong rock lacking in planar features (or at least features whose orientation contributed to weakness) that required a greater quantity of explosives and effort, estimated rate 1 m^3 per 5–10 man/hours.

Such principles were established from experience on the Eastern Front (see, for example, Figs 7 and 8) as well as the Western and other fronts.

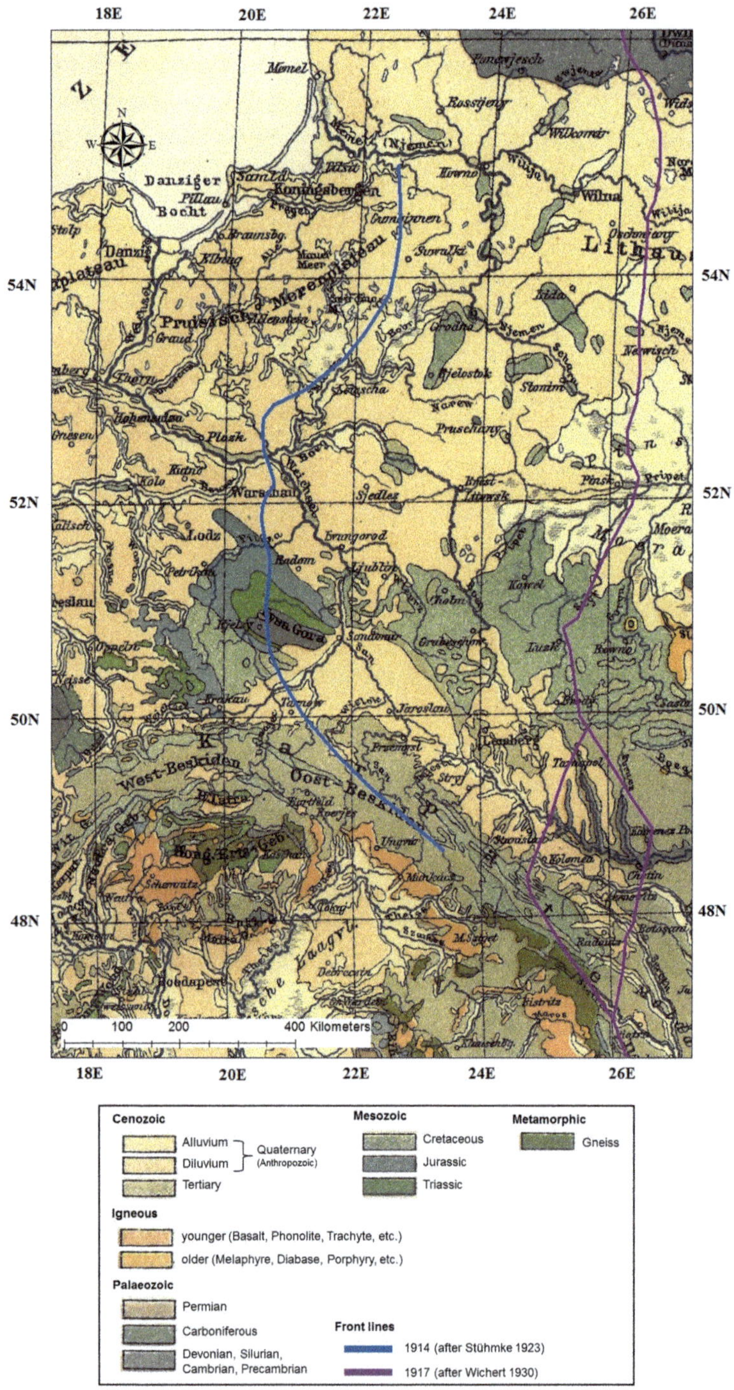

Fig. 5. Geological map of the Eastern Front (cf. Fig. 1). Part of a geological map from Zondervan (1907), with addition of front lines as at 1914 (after Stühmke 1923) and 1917 (after Wichert 1930) and lines of longitude and latitude at 2° intervals.

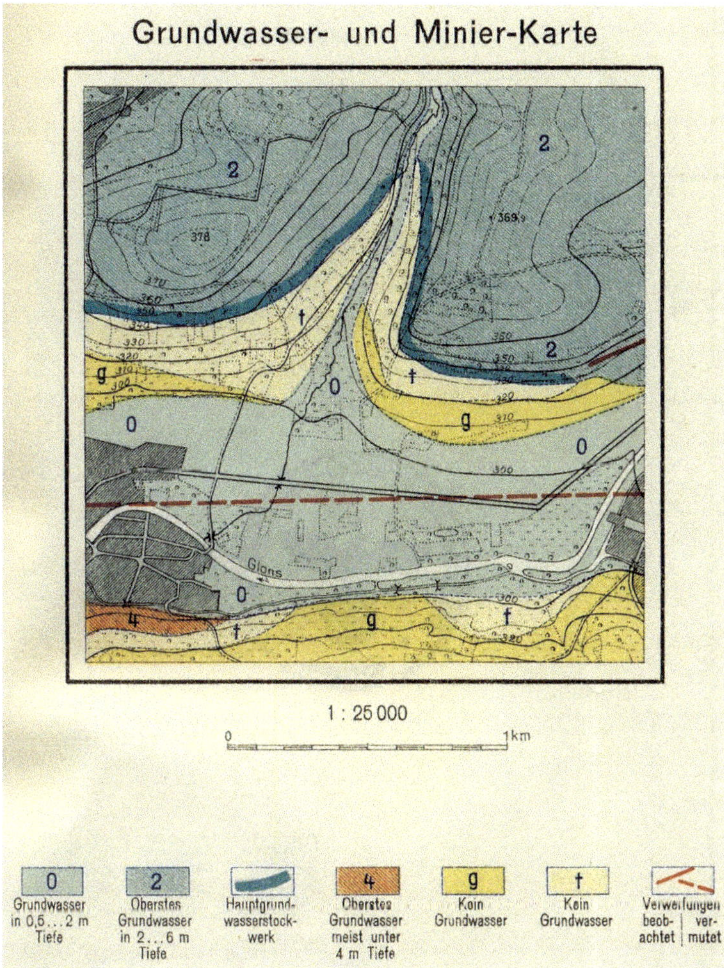

Fig. 6. Simplified groundwater and mining map, one of a set of small thematic maps at scale of 1:25 000 that illustrate different military applications of geology. The key shows, from left to right: groundwater at depth of 0.5–2.0 m; highest groundwater at depth of 2.0–6.0 m; uppermost aquifer; upper groundwater mostly below 4 m in depth; no groundwater; (again) no groundwater; faults/barriers observed or inferred. From von Bülow *et al.* (1938).

Sound propagation

The importance of geology with regard to sound interception of tunnelling activities can also be applied to the Eastern Front as well as the Western. Detection of where and in which phase of the mining process the enemy was working was critical for survival on the battlefield. For this, it was necessary to locate enemy activity underground. Different types of noise were audible over different distances depending on the properties of the rocks through which the sound travelled.

Once this principle had been recognized by military engineers, a systematic sound intercept routine was introduced. First, before sound interception was attempted at a tunnel head, all working noise in that tunnel had to be silenced. Next, a trained sound intercept miner would attempt to locate the direction from which any noise of enemy tunnelling arrived. The source of the noise could be geometrically calculated by taking bearings of noise direction from two or more positions within the observer's tunnel. Experienced sound intercept miners were not only able to determine the direction of tunnelling noise but also to estimate the distance to and the type of enemy tunnelling activities being performed at that time.

The range at which enemy tunnelling activities could be detected (Table 3) varied with the lithology

Fig. 7. Mining in soft sands at Zlota Góra (cf. Fig. 1), at a depth of 25 m, according to an inscription on the reverse of the photograph. To conceal the works from detection by aerial reconnaissance as well as generate material to shore up cut slopes, the white sands shown were quickly used to fill sandbags. Author's collection.

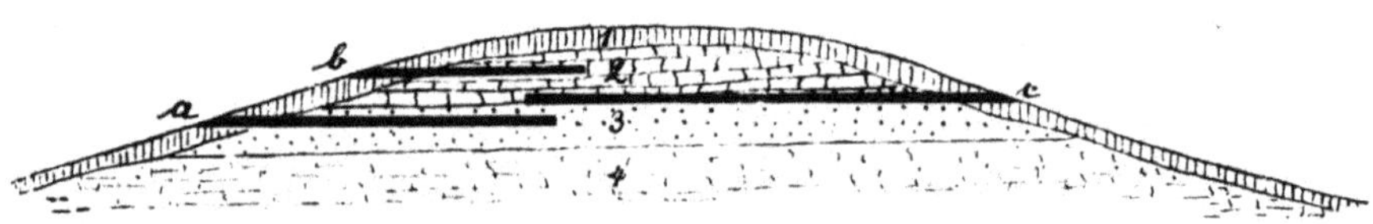

Bild 12. Verschieden günstige Lage von Minenstollen.

1. Löss und Lehm. 2. Harter galizischer Algenkalk. 3. Galizischer Meeressand, leicht bearbeitbar, wenig Miniergeräusche. 4. Podolische Kreide, mittelschwer bearbeitbar, jedoch wenn möglich zu vermeiden. a) Geologisch günstige Lage. b) Ungünstige Lage. c) Anfangs günstig erscheinender, später aber ungünstig werdender Minierstollen.

Fig. 8. Situations varying in suitability for mine tunnels. Key: (1) Loess and loam (Quaternary); (2) Hard Galician algal limestone (Tertiary); (3) Galician marine sand (Tertiary) easily worked, with little noise from mining; (4) Podolian chalk (Upper Cretaceous), can be worked with moderately difficulty, but should be avoided if possible; (a) geologically suitable position; (b) unfavourable position; (c) a mine tunnel situation that seems to be suitable at first, but later becomes unsuitable. From Mordziol (1917*c*).

Table 3. *Empirically determined sound intercept ranges (in metres) for different types of underground noise detected within three major rock types (cf. Fig. 8) by (a) unaided ear, (b) seismomicrophone and (c) geophone*

Type of noise/rock	Chalk			Clay			Sandy loam		
	(a)	(b)	(c)	(a)	(b)	(c)	(a)	(b)	(c)
Digging by pickaxe	41	69	90	37	48	48	15	25	30
Digging by shovel	21	25	36	15	27	33	3	9	9
Walking on wooden flooring	15	21	24	12	18	18	3	6.6	7.5
Falling of tunnelled subsoil	10.5	18	18	9	15	16.5	1.5	4.8	6
Dragging of sandbags	9	16	16	4.5	9	10	1.5	4.5	4.8
Talking	3.6	13.5	15	1.8	4.5	9	1.5	3.6	4.5

Note: data compiled from Ball (1919), Anon. (1922) and Kraus (1941).

of the surrounding bedrock, the type of activity creating the noise, and whether detection was attempted by ear alone or by means of the listening devices newly developed by both sides during the war and utilized on both Western and Eastern Fronts: notably the seismomicrophone (an electric detector of sound but not its source of direction) and the geophone (a device similar in principle to the medical stethoscope) (Ball 1919; Anon. 1922; Kraus 1941). Two listening devices were listed amongst the equipment standard for a Bavarian mining company in December 1915 (see Supplementary material 1), and a geophone system with accessories for a similar Prussian company in 1917 (Supplementary material 2).

As the war went on, procedures became more sophisticated. Advanced tunnel systems with special sound intercept adits were built in which increasing numbers of listening devices were operated. In addition, these devices were interlinked, and intercept centres were established to save time and manpower. However, the role of a sound intercept miner remained hazardous because opposing tunnellers tried to deceive each other by various means, such as 'knocking devices' that were used to make the enemy believe that work was still going on and there was no risk of imminent detonation, whereas in fact a camouflet had been completed and an explosive charge long been primed. Consequently, sound intercept miners or even entire squads of active tunnellers were severely injured or killed by unexpected camouflet detonations.

Typically, detachments of sound intercept miners were commanded by a non-commissioned officer and consisted of 10 to 18 junior ranks. An example of such a unit is that in the 3rd company, Engineer Battalion No. 29, deployed with the German Reserve Infantry Regiment No. 217 and the Austro-Hungarian Infantry Regiment No. 58 forward from Kolpytiv/Kremalz (Fig. 1 and Table 1; *c.* 40 km SE of Wladimir-Wolynski) in February 1917 (Tröbst & Leipold 1932).

German military mining in Ukraine: two examples

The importance of military geological guidance to German mining on the Eastern Front can be illustrated by an example from Mogila Hill (Mogila Höhe of Fig. 1 and Table 1). Dr Carl Mordziol (Table 2) was the head of the military geologist unit attached to Military Survey department 20 when operational on this part of the Eastern Front. He recorded (Mordziol 1938, pp. 97–98, translated here) for an offensive in July 1917 that:

> Concurrently with the offensive of the Russian 8th Army near Stanislau south of the Dniester river valley, the Russian 11th Army initiated their breakthrough north of it along line Tarnopol [present-day Ternopil]–Lemberg [Lviv] [cf. Fig. 1 and Table 1]... In this front sector, the left wing of the German South Army, were two long-embattled hot spots: Zlota Góra (411 m) north of the Zborow [Zboriv] railway station and the Mogila Hill (394 m) [nearby] southwest of the town of Zborow [Zboriv].

The entrances to military tunnels at Zlota Góra are illustrated in Figure 7 and the principles of tunnelling in similar rock nearby at Mogila Hill in Figure 8. Figure 9 illustrates the crater from a mine successfully detonated on this feature. Hills in this region have superficial deposits of Quaternary loess several metres in thickness that cover a near-horizontal bedrock sequence of Tertiary limestone underlain by marine sands resting upon Cretaceous chalk (Wegner 1917; Mordziol 1938). Tunnelling within the sands was achieved by means of an access ramp, allowing an enemy adit to be undermined relatively quickly and without generating detectable noise of this activity, so as to emplace a successful camouflet.

Figure 10 illustrates a second geological feature from western Ukraine. Operational use of military aircraft was one of the major innovations of World War I. It is well known that aircraft engaged in air-

Fig. 9. Mine crater through the loess layer on the surface of Mogila Hill (cf. Fig. 8: Mogila-Höhe of Fig. 1), SW of Zboriv, a town in western Ukraine, historically in the region of Galicia. Author's collection.

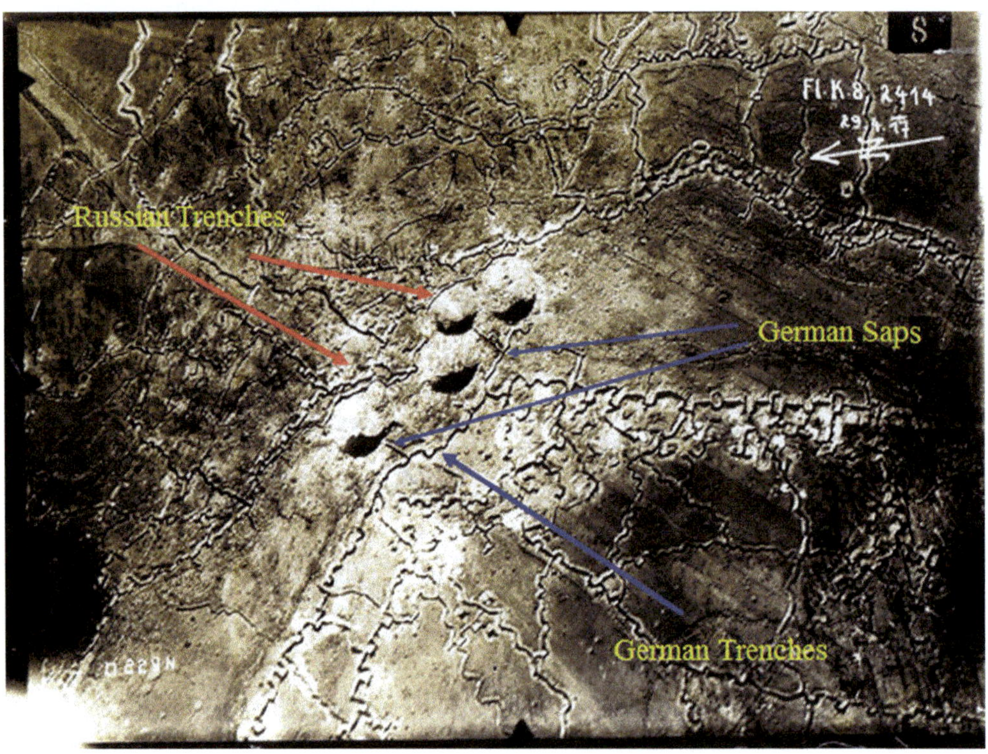

Fig. 10. Four mine craters in the rock of the Zwinin ridge in the outer eastern Carpathian Mountains of eastern Galicia, in present-day Ukraine, SE of Przemyśl in southern Poland (cf. Fig. 1). A Russian trench runs exactly along the crater lip. Saps extend from the German trenches to their side of two craters. It is assumed that the craters were blown by the Germans because it would not have made sense for the Russians to cause explosions so close to their own positions. Author's collection.

to-air combat, and that they also dropped bombs on enemy positions. It is less well known that aerial photographic reconnaissance was born at this time. Photographs taken from aircraft allowed cartographers to compile accurate maps of trench systems on both sides of the front line. The technique was applied on the Eastern Front (e.g. Fig. 10) as on the Western Front and elsewhere and was important for revealing evidence of mining activities from conspicuous heaps of spoil (cf. Fig. 7).

Equipment supplied to German mining units

At the start of the war, German combat engineers did not possess a scale of equipment suitable for mining operations. It was only a demolition regulation of 1911 that provided some guidance (Anon. 1911). At first, German miners could only improvise from equipment and material to hand. However, as the war progressed, scales of equipment were developed for the work underground. Examples of such scales for Bavarian and Prussian tunnelling units are provided online as supplementary information to this article (Supplementary material 1 and 2).

Bavarian Mining Company No. 2 (see Bayerische Mineur-Kompagnie Nr. 2 1916) is representative of such units. It comprised six officers and 258 other ranks. The equipment listed in Supplementary material 1 is that allocated when the company was assigned on 19 January 1916 for service with the 1st Bavarian Reserve Corps. It was transported on two carts, each drawn by a pair of horses (*Kriegsministerium Erlaß* 1915 – War Ministry Ordinance (1915); see Supplementary material 1). The company's other vehicles comprised a field-kitchen cart, a baggage cart and a food cart (each also drawn by a pair of horses) and two bicycles.

For comparison, Supplementary material 2 (2a and 2b) records the equipment and the load plan for the three equipment carts (each drawn by a pair of horses) later mandatory for Prussian mining units (Ingenieur-Komitee 1917*a*, *b*). The equipment includes a mine-surveying device missing from the earlier Bavarian record.

Generators, electric fans to ventilate tunnels during construction, hammer drills to speed excavation, and pumps to drain rain or groundwater, are notably absent from both lists, as are oxygen supplies for breathing apparatus. Breathing apparatus, to facilitate rescue of personnel from tunnels whose air supply either became contaminated by gas or was low in oxygen, is represented by only two items on the Bavarian list for 1915 but nine on the Prussian list for 1917. Consequently, fatalities occurred relatively frequently in the early years among both tunnellers and their would-be rescuers.

Although none of the combatant nations had been prepared for conflict in the lithosphere at the start of the war, both sides created units specialized for tunnelling from 1915 onwards. Yet as a lack of raw materials became increasingly acute in Germany from 1917 as a consequence of the Allied embargo, skilled miners had to be withdrawn from military mining units because they were needed to extract coal and mineral ores for Germany's home industry. Mining companies in the German Army had passed their operational peak.

Concluding remarks

Conflict on the Eastern Front ended more quickly than that on the Western Front (Westwell 2010). A cease-fire between Russia and the Central Powers was agreed on 16 December 1917, but talks to agree an armistice faltered. German forces launched an attack on 17 February 1918 to bring the Russians back to the negotiation table, advancing *c.* 240 km in two days and forcing Russia to accept peace terms 48 hours later. The Treaty of Brest-Litovsk signed on 3 March 1918 brought conflict between the Central Powers and Russia to an end. On the Western Front, hostilities ended only with the armistice between Germany and the Allies on 11 November 1918.

Casualties on the Eastern Front have proved difficult to estimate, and different figures are given in different publications (cf. Anon. 2018). In broad terms *c.* 65 million troops were mobilized between 1914 and 1918 and of these *c.* 42 million served with the various Allies and 22 million with the four Central Powers (Westwell 2010). The Allies recorded *c.* 4.8–5.7 million military dead and 12 million wounded, the Central Powers *c.* 3.1–4.0 million dead and 8.4 million wounded. Civilian deaths, excluding those from an epidemic of influenza, totalled *c.* 6.6 million. Russia suffered the highest numbers of dead (*c.* 1.8 million military plus 1.5 million civilians) and military wounded (*c.* 5 million), all therefore on the Eastern Front. Austria-Hungary suffered 1.1 million dead and 3.6 million wounded, Germany *c.* 2.0 million dead and 4.2 million wounded, on all fronts but large numbers in the east. In comparison, French military deaths in the war came to *c.* 1.4 million in total, for British Imperial Forces (mostly from the UK) *c.* 1.1 million, and for the USA *c.* 53 000. The casualties on the Eastern Front were thus even more appalling than those on the Western Front.

This article demonstrates that geology and geologists were involved in military operations, notably mining operations, on the Eastern Front as well as the Western – aspects of applied geology hitherto documented in few German publications and, for the Eastern Front, scarcely at all in English. The use made of military geologists by German and Austro-Hungarian forces in World War I is strikingly greater than that made by the Allies to support their opposing armies.

Acknowledgements The author is grateful for peer review of this paper by Peter Doyle and Hermann Häusler; to Florian Malm for help in revising several figures; and to Ted Rose for editorial work that significantly improved the paper's final content and appearance.

Funding This research received no specific grant from any funding agency in the public, commercial or not-for-profit sectors.

References

Afalašev, D. & Badanin, B. 1969. Die pioniertechnische Sicherstellung der Stalingrader Verteidigungs-operation. Translation into German (now held by the Bundeswehr Geoinformation Centre, Euskirchen) from *Voenna-istorischeskij Shurnal*, **12**, 26–35.

Anon. 1911. *Sprengvorschrift von 1911 mit den eingearbeiteten Deckblättern Nr. 1-116.* (Spr.V.). Barth Verlag, Berlin [Reprinted 1918].

Anon. 1922. *The Work of the Royal Engineers in the European War, 1914–19: Military Mining*. Institution of Royal Engineers, Chatham.

Anon. 2005. *Berichte aus dem deutschen Grossen Hauptquartier 1914–18.* http://www.stahlgewitter.com/welt krieg/1916_russische_maerzoffensive.htm [last accessed 6 August 2018]

Anon. 2017. *4 Division (Alte Armee).* http://wiki-de.geneal ogy.net/41._division_(alte_armee) [last accessed 6 August 2018]

Anon. 2018. *World War I casualties.* https://en.wikipedia.org/wiki/World_War_I_casualties [last accessed 13 August 2018]

Antulajev, (or Ammulajew), E. 1930. Unterirdischer Minenkrieg [from] Krieg und Technik (Militärtechnisch-wissenschaftliche Zeitung). Translation into German (now held by the Bundeswehr Geoinformation Centre, Euskirchen) from *Vojna i Technika (Vojenno-techničeskij i naúcno-issledowatelskij zhurnal)*, **3**, State Publishing House, Moscow.

Ball, H.S. 1919. The work of the miner on the Western Front, 1915–18. *Transactions of the Institute of Mining and Metallurgy*, **28**, 189–248.

Bärtling, R. 1917. Grundzüge der Kriegsgeologie. *Zeitschrift der Deutschen Geologischen Gesellschaft*, **68**, 70–85.

Baumgarten-Crusius, A. 1919. *Geschichte der Sachsen im Weltkrieg, Vol. 2, von Sachsen in großer Zeit.* Lippold, Leipzig.

Bayerische Mineur-Kompagnie Nr. 2 1916. *Ausrüstungsnachweis für die Bayer. Mineur-Komp. Nr. 2 beim K. 1. bay. Reserve-Korps.- K.M.E. v. 15.12.15, Nr. 117003, Nachweis vom 19.1.16.* Unpublished report in Kriegsarchiv, Munich [Ing. Korps. 1724].

Brooks, A.H. 1920. The use of geology on the Western Front. *United States Geological Survey Professional Paper, 128-D*, 85–124.

Churchill, W.S. 1931. *The Unknown War: The Eastern Front.* Charles Scribner, New York.

Doyle, P. 2018. Trench construction and engineering geology on the Western Front, 1914–18. *In*: Rose, E.P.F., Ehlen, J. & Lawrence, U.L. (eds) (2019) *Military Aspects of Geology: Fortification, Excavation and Terrain Evaluation.* Geological Society, London, Special Publications, **473**, https://doi.org/10.1144/SP473.6

de Fonclare, G. 1928. A travers la Grande Guerre. La Cote 108 pres de Berry-au-Bac et la guerre de mines (1915). *Revue des Questions Historique*, series 3, **13**, 62–82.

Grieve, W.G. & Newman, B. 1936. *Tunnellers. The Story of the Tunnelling Companies, Royal Engineers, During the World War.* Herbert Jenkins, London; reprinted in 2002 by the Naval & Military Press, Uckfield.

Hartenstein, O. 1936. *Das Kurhessische Pionier-Bataillon Nr. 11 im Weltkrieg 1914–18. Aus Deutschlands großer Zeit: Band 75. Heldentaten deutscher Regimenter.* Verlag Bernhard Sporn, Zeulenroda.

Häusler, H. 2000. Die Österreichische und Deutsche Kriegsgeologie 1914–18. *Informationen des Militärischen Geo-Dienstes (MILGEO-INFO)*, **75**, Vienna, 1–161.

Häusler, H. 2003. Dr. Walter Kranz (1873–1953) – Der erste Militärgeologe des 20. Jahrhunderts. *MILGEO*, Bundesministerium für Landesverteidigung, Vienna, **12**, 1–80.

Ingenieur-Komitee 1917*a*. *Ausrüstungs-Nachweis für eine Mineur-Kompagnie, Kgl-Preuß. Ingenieur-Komitee Abtlg. 4 vom 14. Juli 1917.* Unpublished report in Kriegsarchiv, Munich [Ing. Korps 1724].

Ingenieur-Komitee 1917*b*. *Beladeplan des zweispännigen Mineur Gerätewagens, Kgl-Preuß. Ingenieur-Komitee Abtlg. 4 vom 14. Juli 1917.* Unpublished report in Kriegsarchiv, Munich [Ing. Korps 1724].

Kranz, W. 1909. Beitrag zum Studium des Nahangriffs auf Port Arthur, Die Nordostfront. *Beiheft zum Militär-Wochenblatt*, Berlin, **1909**, 213–249.

Kranz, W. 1913. Militärgeologie. *Kriegstechnische Zeitschrift*, Berlin, **16**, 464–471.

Kranz, W. 1927. *Die Geologie im Ingenieur–Baufach.* Enke-Verlag, Stuttgart.

Kranz, W. 1934. Beiträge zur Entwicklung der Kriegsgeologie. (Fortsetzung.) III. Russland. *Geologische Rundschau*, **25**, 194–201.

Kranz, W. 1936. Feldminierkampf und Kriegsgeologie. Monte Cimone – Col di Lana – Mont Sief. *Zeitschrift für das gesamte Schieß- und Sprengstoffwesen*, **31**, 73–75, 118–121.

Kraus, E. 1941. *Angewandte Geologie-Wehrgeologie.* Unpublished preface, written in 1941 but extended in 1968, for a planned textbook; author's collection.

Lawrence, C. 1981. *The Gallipoli Diary of Sergeant Lawrence of the Australian Engineers, 1st A.I.F., 1915.* Melbourne University Press, Melbourne.

Mordziol, C. 1917*a*. *Geologisches Merkblatt der Vermessungsabteilung 20, Geologengruppe Mordziol.* Document in Heringen Collection, Euskirchen.

MORDZIOL, C. 1917*b*. *Erläuterungen zum geologischen Merkblatt der Vermessungs-Abteilung 20*. Document in Heringen Collection, Euskirchen; also filed in the Militärarchiv at Freiburg-im-Breisgau, as Vermessungs-Abteilung 20, MilArchiv H 07- 1/20, fol. 1-305.

MORDZIOL, C. 1917*c*. *Geologie und Stellungsbau im Gebiet des Durchbruchs bei Stanislau 6.-12.7.17 nebst geologischen Bemerkungen über das Durchbruchsgebiet bei Konjuchy*. Document in Heringen Collection, Euskirchen.

MORDZIOL, C. 1938. *Einführung in die Wehrgeologie*. Verlag Otto Salle, Frankfurt.

ORTENBURG, G. 2005. *Heerwesen der Neuzeit: Waffen der Millionenheere, 1871–1914*. Bernard & Graefe Verlag, Bonn.

ÖSTERREICHISCHES BUNDESMINISTERIUM FÜR LANDESVERTEIDIGUNG UND VOM KRIEGSARCHIV (ed.) 1930. *Österreich-Ungarns letzer Krieg 1914–18. Sechster Band. Das Kriegsjahr 1917*. Verlag der Militärwissenschaftlichen Mitteilungen, Vienna.

PHILIPP, H. 1919. Die Entwicklung der Kriegsgeologie. *Technik und Wehrmacht*, Berlin, **22**, 129–134.

ROSE, E.P.F., HÄUSLER, H. & WILLIG, D. 2000. Comparison of British and German military applications of geology in world war. *In*: ROSE, E.P.F. & NATHANAIL, C.P. (eds) *Geology and Warfare: Examples of the influence of terrain and geologists on military operations*. Geological Society, London, 107–140.

SAPPEURKOMPAGNIE 5/4 1915. *Die Sprengung der Höhe 810 auf der Manilova.- Lt Vandoffy, Pion. Bataillon Nr. 4. Manilova, 11.-12. März 1915*. Unpublished report in Austrian State Archives [Abt. Kriegsarchiv Wien AdTK, Kiste 1393].

SAPPEURKOMPAGNIE 2/22 1916. *Minenkrieg bei RARANCZE (Kote 258) 23. März bis 10.06.16, K.u.k.5.J.T.D.-11.K.K.* Unpublished report in Austrian State Archives [Abt. Kriegsarchiv, AdTK, Kiste 1393].

SAPPEURKOMPAGNIE 2/11 1918. *Kampfleistungen der 2./11 früher 7./9. Sappeur.Komp., an Erl. v. 16.Nov.1917. Präs. No 31697, Beibl. 57./17, Feldpost 523 am 1918, I V· Komp Kmdtr.* Unpublished report in Austrian State Archives [Abt. Kriegsarchiv Wien AdTK, Kiste 1393].

SAPPEURKOMPAGNIE 2/13 1918. *Lt. i. d. Res. Stan. Tomic: Abschrift. Beiträge. Auf K.M.Erl. v. 16. Nov. 1917 Präs.Nr. 31607, Bblt.57/1917. G.I.d.S.u.P.Tr. E.Nr. 362/4 v. 20./4.1918. – an K.u.k. Sappeur-Ersatzbaon Osijek (früher Ersatzkomp. Sapp. Baon Nr. 13)*. Unpublished report in Austrian State Archives [Abt. Kriegsarchiv Wien AdTK, Kiste 1393, 21a].

SAPPEURKOMPAGNIE 2/43 1918. *Hptm. Kornicki: Abschrift. Beiträge. Auf K.M.Erl. v. 16. Nov. 1917 Präs.Nr.31607, Bblt.57/1917. G.I.d.S.u.P.Tr. E.Nr. 362/4 v.20./4.1918.- an K.u.k. Sappeur-Ersatzbaon Osijek (früher Ersatzkomp. Sapp. Baon Nr. 13)*. Unpublished report in Austrian State Archives [Abt. Kriegsarchiv, AdTK, Kiste 1393, 1a].

SAPPEURKOMPAGNIE 3/13 1918. *Abschrift. Minenkämpfe Brückenschanze USCISCKO II. 1916, Sapp. Komp. 3./13 Beiträge – Auf K.M.Erl. v. 16. Nov. 1917 Präs.Nr. 31607, Bblt.57/1917. G.I.d.S.u.P.Tr. E.Nr.362/4 v. 20.4.1918. – an K.u.k. Sappeur-Ersatzbaon Osijek (früher Ersatzkomp. Sapp. Baon Nr. 13)*. Unpublished report in Austrian State Archives [Abt Kriegsarchiv Wien AdTK, Kiste 1393].

SAPPEURKOMPAGNIE 5/11 1918*a*. *Oblt.i. d. Res. A. Schreiber: Abschrift. Beiträge. Auf K.M.Erl. v. 16. Nov. 1917 Präs. Nr. 31607, Bblt.57/1917. G.I.d.S.u.P.Tr. E.Nr. 362/4 v. 20.4.1918. – an K.u.k. Sappeur-Ersatzbaon Osijek (früher Ersatzkomp. Sapp. Baon Nr. 13)*. Unpublished report in Austrian State Archives [Abt. Kriegsarchiv Wien AdTK, Kiste 1393, 35a].

SAPPEURKOMPAGNIE 5/11 1918*b*. *K.u.k. SAPPEUR. ERS. BAON OSIJEK, An das K.u.k. Kriegsarchiv (Schlachtfeldführergruppe) in Budapest, am 4.6.1918 – Wien, auf K.M.Erl. v. 16. Nov. 1917. Pr.Nr. 607*. Unpublished report in Austrian State Archives [Abt. Kriegsarchiv Wien AdTK, Kiste 1403].

SAPPEURKOMPAGNIE 5/13 1918. *Lt. i. d. Res. L. Gonda: Abschrift. Beiträge. Auf K.M.Erl. v. 16. Nov. 1917 Präs.Nr. 31607, Bblt.57/1917. G.I.d.S.u.P.Tr. E.Nr. 362/4 v. 20.4.1918. – an K.u.k. Sappeur-Ersatzbaon Osijek (früher Ersatzkomp. Sapp. Baon Nr. 13)*. Unpublished report in Austrian State Archives [Abt. Kriegsarchiv Wien AdTK, Kiste 1393, 23a].

SCHAFARZIG, F. 1916. Eröffnungs-Vortrag anlässlich der am 9 Februar 1916 abgehaltenen Hauptversammlung der Ungarischen Geologischen Gesellschaft. *Földtani Közlöny*, **46**(suppl.), 43–56.

STÜHMKE, R. 1923. Das Infanterie-Regiment 'Kaiser Friedrich, König von Preussen' (7. Württembergisches) Nr. 125 im Weltkrieg 1914–18. *Band 31 In*: FLAISCHLEN, H. (ed.) *Die württembergischen Regimenter im Weltkrieg 1914–18*. Belser, Stuttgart.

TRÖBST, H. & LEIPOLD, A. 1932. *Das Posensche Pionier-Bataillon Nr. 29 und seine Kriegsformationen Reserve-Pionier-Bataillon 33, Pionier-Kompagnie 107, 361 und 401, Pionier-Bel.-Tr. 10 und 29*. Verlag Tradition, Berlin.

VON ARDENNE, A. 1917. *Der Krieg 1914–19 in Wort und Bild, Vol. 2*. Deutsches Verlagshaus Bong & Co, Berlin.

VON BÜLOW, K., KRANZ, W. & SONNE, E. 1938. *Wehrgeologie*. Quelle & Meyer, Leipzig.

VON CLAUSEWITZ, C. (VON COCHENHAUSEN, F. (ed.)) 1937 reprint. *Vom Krieg*. Insel-Verlag, Leipzig.

VON TETTAU, E. 1912. *Der Russisch-Japanische Krieg. Amtliche Darstellung des Russischen Generalstabes, Port Arthur. Deutsche vom russischen Kriegsministerium mit Allerhöchster Genehmigung autorisierte Ausgabe des russischen Generalstabswerkes. Teil 2*. Mittler und Sohn, Berlin.

WAR MINISTRY ORDINANCE 1915. *Ausrüstungsnachweis für die bayr. Mineur-Komp. Nr.2, Königlich Ministerialer Erlass Nr. 117003 v. 15.12.15.- Band 3 Pionier*. Document in Kriegsarchiv, Munich.

WEGNER 1917. *Gutachten für Kdr. Der Pioniere 197 I.D. über das Grundwasser im Minierstollen 1a an der Zlota Gora*. Unpublished report by author, given surname only, in Militärarchiv, Freiburg (file reference PH 14/316 Pionier –Btl 197).

WESTWELL, I. 2010. *World War One: An illustrated history*. Hermes House, London.

WICHERT, P. 1930. *Das Ehrenbuch der Pommern. Die Pommern im Weltkrieg. Ein Volks- und Gedenkbuch*. Vaterländische Verlagsanstalt Oscar Hinderer, Berlin.

WILLIG, D. 1999. Entwicklung der Wehrgeologie: Aufgabenspektrum und Beispiele. 1. Von den Anfängen bis 1918. *Fachliche Mitteilungen*, Amt für Wehrgeophysik, Traben-Trarbach, **225**, 1–116.

WILLIG, D. 2015. German military mining on the Western and Eastern Fronts. *In*: BYLEDBAL, A. (ed.) *Les Taupes de la Grande Guerre: Combats et combattants souterrains*. Artois Presses Université, Arras, 29–54.

WILLIG, D. & HÄUSLER, H. 2012. Aspects of military hydrogeology and groundwater development by Germany and its allies in World War I. *In*: ROSE, E.P.F. & MATHER, J.D. (eds) *Military Aspects of Hydrogeology*. Geological Society, London, Special Publications, **362**, 85–103, https://doi.org/10.1144/SP362.6

WILSER, J. 1921. *Grundriß der angewandten Geologie (unter Berücksichtigung der Kriegserfahrungen für Geologen und Techniker)*. Borntraeger, Berlin.

ZONDERVAN, H. (ed.) 1907. *Geologische kaart van Middeleuropa*. Scale 1: 7 500 000; in *Winkler Prins' Geïllustreerde Encyclopaedie*, 3rd ed., **7**, Epacris-Genie. Elsevier, Amsterdam.

Quarrying Companies Royal Engineers in World War I: a geologically constrained innovation to support British armies on the Western Front

EDWARD P. F. ROSE

Department of Earth Sciences, Royal Holloway, University of London, Egham, Surrey TW20 0EX, UK

0000-0003-4182-6426
ted.rose@earth.oxon.org

Abstract: Quarrying Companies were a new type of unit first raised within the Royal Engineers in World War I. Thirteen served in northern France, on the Western Front: two from late 1916 (198 and 199 Quarrying Companies) and 11 more from 1917 (320–329 and 348 Quarrying Companies). Recruited from Great Britain, Ireland and the Channel Island of Guernsey, each consisted of four officers and 264 soldiers; over 3000 men in total, assisted by *c.* 4000 less skilled labourers. They were used to support the British Expeditionary Force by providing 'stone', mostly from existing quarries near Marquise, NE of the port of Boulogne. There they excavated Devonian and Carboniferous 'limestones' in the Ferques Inlier, relatively strong rocks within a region of wide Mesozoic and Cenozoic outcrop. As the British Expeditionary Force expanded to a peak of five armies and *c.* 1.5 million troops, 'stone' was required for the enhancement or repair of the roads, railways and associated facilities that formed a crucial element of its infrastructure, essential for the efficient movement of soldiers and their copious supplies of food, stores and ammunition. The requirement ceased soon after the end of hostilities and all Quarrying Companies Royal Engineers were disbanded in 1919.

World War I, the Great War of 1914–18, is associated with many military innovations (Holmes 1999). Of these, many were in response to the near-static conditions of trench warfare that followed the stabilization in strength of the opposing forces on the Western Front, across Belgium and part of northern France (Fig. 1). Many were innovations generated by or within the Royal Engineers, as technology new to the armed forces was adapted for military use (Pritchard 1952*a*, *b*). This led to the foundation of many new types of military unit, including:

- Boring Sections Royal Engineers: five sections in 1917, one to each of the five British armies then in operation within the British Expeditionary Force on the Western Front, to supply drinking water from deep boreholes. The deployment of these sections was guided by Lieutenant (later Captain) W.B.R. King, from June 1915 the first British military geologist ever to be assigned to serve as such on a battlefield (Anon 1922*a*; Rose 2009, 2012*a*).
- Tunnelling Companies Royal Engineers: nine companies in 1915, expanded to a total of 25 from 1916 to 1918, together with three companies from Australia, three from Canada and one from New Zealand, initially to tunnel 'mines' for explosive detonation that would breach the enemy's front line fortifications, and later to excavate dug-outs that sheltered troops from aerial and artillery bombardment. These companies were guided from May 1916 by the first British geologist to be assigned to support combat operations as an engineering geologist, Major (later Lieutenant-Colonel) (Sir) T.W. Edgeworth David (Anon 1922*b*; Grieve & Newman 1936; Doyle *et al.* 2000; Rose & Rosenbaum 2011; Rose 2015).
- Quarrying Companies Royal Engineers: two companies in 1916, 11 more in 1917, primarily to provide the crushed stone and aggregate required in large quantities by the British Expeditionary Force for the repair or enhancement of roads and railways and their associated facilities. Their deployment was constrained essentially, if not completely, to the Boulonnais area of northern France by the regional geology and contemporary logistics, although the British military geologists 'Bill' King and Edgeworth David did assess other localities as potential sources of road metal or of sand and aggregate for concrete.

The foundation and role of the Quarrying Companies, hitherto poorly documented, are described here from a geological perspective. Data have been compiled from the following sources: the only Quarrying Company War Diary known to exist, file WO 95/4057 at the National Archives (of the UK) at Kew in London; other files there as cited, from the

From: Rose, E. P. F., Ehlen, J. & Lawrence, U. L. (eds) 2019. *Military Aspects of Geology: Fortification, Excavation and Terrain Evaluation*. Geological Society, London, Special Publications, **473**, 151–171.
First published online March 6, 2018, https://doi.org/10.1144/SP473.1

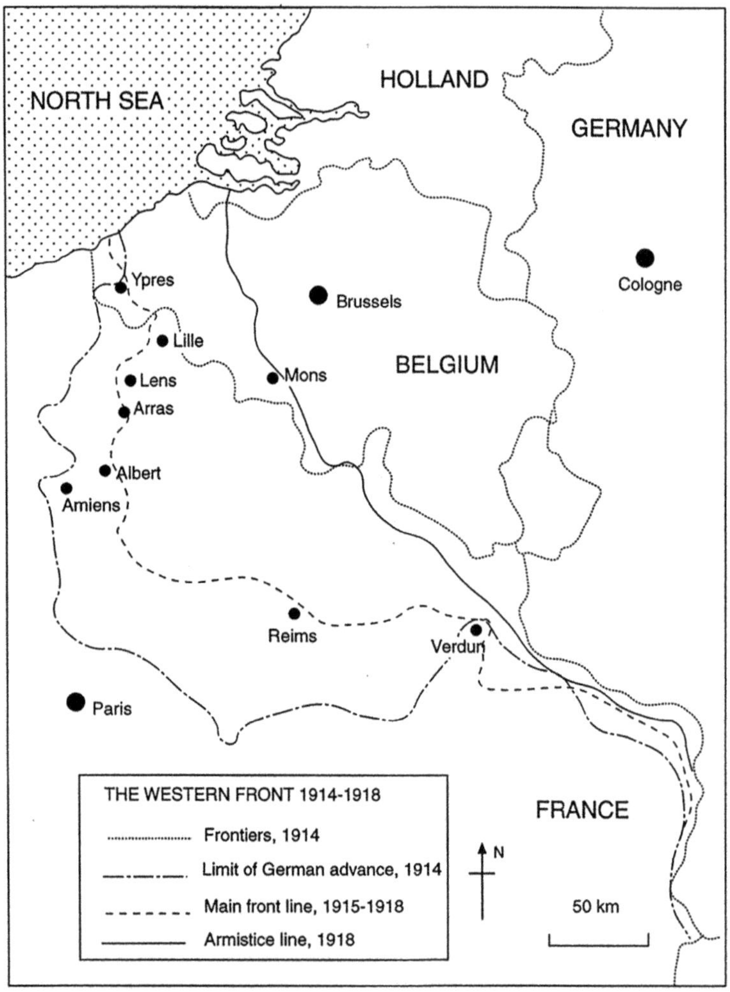

Fig. 1. The Western Front of 1914–18, extending from the North Sea (top left) to the border of France with Switzerland (bottom right). The front line was generally static from late 1914 to early 1918, with only slight adjustments despite the large set piece battles of 1915, 1916 and 1917. Reproduced by permission from Doyle (1998, fig. 2) *Geology of the Western Front, 1914–18*; © 1998 The Geologists' Association, London.

War Office and Ministry of Transport record series; War Diary summaries preserved in the Royal Engineers Museum, Library and Archive (REMLA) at Chatham in Kent; and from other published, archive and web sources as indicated in the text. A work published anonymously in accordance with military practice (Anon 1922*a*) is of particular importance because it is known that the authors were, in fact, the contemporary geologists involved in the work described: T.W. Edgeworth David and W.B.R. King (Branagan 2005). Imperial units of measurement are used for consistency with the historical record, with metric equivalents where deemed necessary.

Background

Following the German invasion of Belgium on 3 August 1914, the UK declared war on Germany on 4 August 1914 and elements of a British Expeditionary Force (BEF) began to arrive in France on 7 August 1914 (Edmonds 1922–25; Holmes 1999; Westwell 2010). The BEF expanded to a force of *c.* 125 000 men and moved forward into southern Belgium on 21 August 1914, fighting its first action the next day. However, its war of movement effectively ended in late 1914 with the First Battle of Ypres (19 October to 22 November 1914), a narrow British victory bought at high cost. Thereafter the

Western Front largely transformed into a zone of static trench warfare and the BEF was progressively expanded, from five divisions initially to 61 divisions at the end of hostilities in November 1918 (including units from Canada, Australia, New Zealand and other parts of the British Empire). At its peak in early 1917, the BEF consisted of five armies and 82 divisions (Pritchard 1952*a*, p. 630), in total more than 1.5 million men, and 0.5 million horses and mules, mostly to move supplies forward from the nearest railheads.

Roads and railways were an important element of the military infrastructure, essential for the efficient movement of people and their copious supplies of food, stores and ammunition. The winter of 1914–15 had caused many kilometres of roads to break up under the weight and volume of military lorry traffic. Few roads in Belgium and northern France had at that time been built to withstand heavy usage. Moreover, lorries at that time were fitted with hard, narrow, solid rubber tyres, which caused rapid wear on roads and allowed little cross-country mobility in dry weather and none at all in wet. Initially, the French had undertaken all responsibility for the upkeep of their roads, but in Belgium local responsibility had been accepted by the BEF from the outset. However, in 1915 the French also asked for assistance and in December called on the British to provide all stone required for the repair or enhancement of the roads in the BEF operational area (Pritchard 1952*a*, pp. 599–601).

Pre-war, most road stone used in this area had been transported by rail from major quarries in Belgium (cf. Rose 2012*b*), which worked strong rocks eminently suitable for the purpose. However, the German advance had seized control of both the quarry areas and their rail distribution network. New sources of supply had to be developed for the BEF, without conflicting with the needs of the French armies or the local population. Initially, the Director of Works for the BEF obtained stone from three principal sources.

(1) Normandy, in NW France: quartzite from Feuguerolles and Val de Maizet quarries in the Calvados district, SW of Caen (according to Anon 1924, p. 21); from a quarry at Necy (in the Orne district nearby), a site handed over to the BEF Directorate of Roads on 1 December 1916 (Anon 1924, p. 214); and from Cherbourg, on the Cotentin Peninsula (according to Lawson 1920). However, stone from Normandy involved long haulage by rail and there was an acute shortage of rail transport available.

(2) The British Isles: slag from Middlesbrough in north Yorkshire (Lawson 1920), and granite from the Channel Island of Guernsey and from Penlee, near Penzance in Cornwall (Anon 1924). By December 1915 the supply of stone averaged 2350 tons/tonnes per day (Pritchard 1952*a*, p. 559). However, these sources involved transport by sea, which became increasingly committed to other priorities and which became vulnerable to attack as the German submarine fleet developed. On 7 December 1915 a ‘road metal’ order was prepared for 5000 tons to be supplied from Penmaenmawr in North Wales by the Penmaenmawr & Welsh Granite Company, 3400 tons from Pwllheli (also in North Wales) by the Pwllheli Granite Company, 3500 tons from Newlyn in Cornwall by the Penlee & St Ives Stone Quarries and 13 100 tons via the Channel Islands of Guernsey, Jersey or Alderney by J. Mowlem & Company. The subsequent estimate of 41–45 ships needed for delivery was deemed to be prohibitive and the quantity for transport was scaled down accordingly (according to correspondence in National Archives file MT 23/522/6).

(3) Quarries in the Boulonnais region, in northern France inland from the port of Boulogne (Fig. 2): none of these quarries had produced stone deemed suitable for use as road metal while stone of better quality was available from Belgium.

During 1916 it became evident that the Boulonnais quarries provided the only viable source of stone near to the BEF area of operations. Initially, stone was bought under contract from civilian operators. As military demand increased, production was enhanced, sometimes by means of additional equipment provided from the UK and/or additional labour – civilians from the UK or elsewhere in Europe, prisoners of war or Labour Battalions of the BEF. Quarrying on a small scale took place under the BEF Director of Works, but the development of a Transportation Directorate and the appointment of a Director of Roads (Brigadier-General (later Sir) H.P. Maybury) in the autumn of 1916 provided catalysts to increased activity. ‘Stone’ was required primarily for the enhancement or repair of roads and railways, in competition with that for hard standings for the many depots in which supplies could be accumulated and then distributed. By the end of 1917, the Director of Roads was maintaining 1900 miles (>3000 km) of roads in army areas and 1200 miles (nearly 2000 km) elsewhere. By 1918 the demands for the maintenance of roads and for the construction of new ones, and especially the metalling of yards at railheads, each equivalent to several miles of road, ‘had grown out of all proportion to the amount of stone available’ (Pritchard 1952*a*, p. 637). To produce stone in adequate quantities, it was resolved in 1916

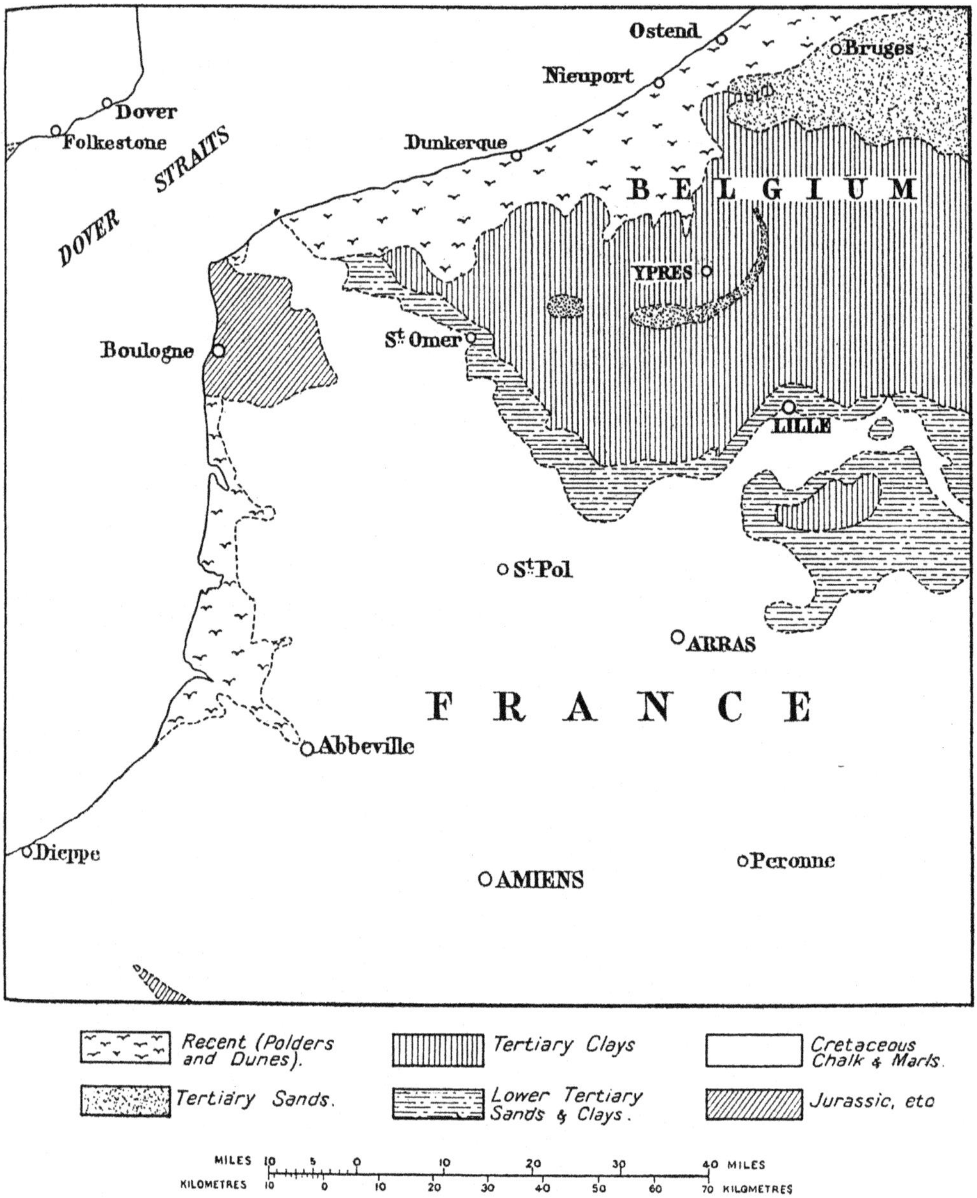

Fig. 2. Near-contemporary geological sketch map of the British-occupied sector of the Western Front. From Anon (1921).

to raise Quarrying Companies Royal Engineers by the direct recruitment of men with the appropriate specialist skills from within the UK extractive industry.

Formation of Quarrying Companies

A war establishment was set in July 1916 as a recruitment target (Walton 2010). Each company was to consist of three officers (a captain and two subalterns: lieutenants or second lieutenants), two warrant officers (a company sergeant major and a company quartermaster sergeant for administration), 28 non-commissioned officers (four sergeants, eight corporals and 16 lance-corporals), five infantry men (one as cook and four as officers' batmen/orderlies), one horse transport driver from the Army Service Corps and 228 sappers. Specialist trades to be included within the sappers consisted of 16 steam engine drivers, four internal combustion engine drivers, four smiths, two bootmakers, two carpenters/joiners, two gas fitters/plumbers, two masons, two platelayers, one tailor, one tinsmith and 197 quarrymen. It was intended that each company would thus be self-sufficient in its operational role. For transport, each company was to be assigned four bicycles, one wagon and two horses or mules. However, when the first Quarrying Company went to France in August 1916, it did so with a revised establishment of four officers and 264 other ranks (REMLA file 70), as did the final company in October 1917, with a wagon, two horses and four bicycles as originally planned (National Archives file WO 95/4057). Additional labour was to be provided when required by local civilians, Army Troops Companies Royal Engineers, Labour Battalions Royal Engineers or prisoners of war. Although the sappers were all enlisted into the Royal Engineers and wore its uniform, they received no specific military training prior to deployment in France apart from (for the final company at least) training in protection from gas attack.

The first two Quarrying Companies were formed in 1916 (Table 1 and Appendix A) at Clipstone in NW Nottinghamshire. Located on land that was to become Clipstone Colliery, Clipstone Camp was opened in 1915, initially for the infantry. A Royal Engineers depot was founded there in March 1916 to recruit men for Tunnelling Companies Royal Engineers, but following an outbreak of 'spotted fever' at the Royal Engineers' Chatham base, it dealt also for a time with recruits intended for Chatham (REMLA file 70). 198 and 199 Quarrying Companies formed at Clipstone and proceeded overseas on 26 and 30 August 1916, respectively, as the depot closed (28 August 1916).

Eleven additional Quarrying Companies were raised in 1917, the first seven (numbers 320–326) in the first six months, at the Road and Quarry Troops Depot Royal Engineers at Buxton in Derbyshire. Used also for training Army Troops Company men, this was known locally as the School for Instruction in Engineering. A series of postcards produced in Buxton during the war shows trainee soldiers of the Royal Engineers building bridges and pontoons across the lake in its Pavilion Gardens. Buxton Cemetery contains the graves of two sappers from Army Troops Companies who died from illness in 1916 and of Corporal Francis Lockard from 323 Quarrying Company, who had enlisted in Scotland (Paisley, near Glasgow), but died in Buxton on 14 April 1917 aged 42 (http://buxtonwarmemorials.mrallsophistory.com/, accessed 1 October 2015).

Company 327 was raised in Ireland, from local people, and two companies, 328 and 329, were later formed in France, at the Transportation Troops Depot Boulogne, from suitably skilled men transferred from other branches of the Army. The final company to be raised, 348, was formed at Aldershot in Hampshire on 12 October 1917 (Appendix A).

Initially, the Quarrying Companies recruited experienced men who had received exemption from military service on medical grounds, or who were above the age limit for recruitment as combat troops. An example is Sapper 'Teddy' Hughes, who had 37 years of service and was 59 years old when he went to France with 198 Quarrying Company (REMLA file 70). Yet the formation of companies 328 and 329 in 1917 by troops drawn from fighting units indicated the high priority given to meeting demand for road stone at this time. Heavy casualties in the BEF that year caused 'a crisis in manpower' (Pritchard 1952*a*, p. 568) and the fittest men working in rear areas were as a rule drafted

Table 1. *Foundation of Quarrying Companies Royal Engineers (from unit War Diaries)*

Company number	Foundation date	Location	Captain commanding on foundation	Principal quarry deployment
198	August 1916	Clipstone	E. Matthews RE	M-R: Vallée Heureuse
199	August 1916	Clipstone	E.J. Stead RE	M-R: Beaulieu
320	January 1917	Buxton	C.S. Darbishire RE	M-R: Haut Banc
321	January 1917	Buxton	H. Brooks RE	M-R: Haut Banc
322	February 1917	Buxton	W.M. Kennedy RE	M-R: Beaulieu, Haut Banc, Banc Noir
323	February 1917	Buxton	A. Seaton RE	M-R: Beaulieu
324	March 1917	Buxton	F. Brooke KORLR	Rebreuve, Baraffle, Marqueffles
325	March 1917	Buxton	T. Gordon RE	M-R: Beaulieu, Vallée Heureuse
326	April 1917	Buxton	E. Marshall RE	M-R: Vallée Heureuse
327	May 1917	Dublin	S.H. Fleming RE	M-R: Haut Banc
328	April 1917	Boulogne	A.L. Mann RE	M-R: Vallée Heureuse
329	April 1917	Boulogne	P.F. Jackson RE	M-R: Beaulieu
348	October 1917	Aldershot	H.J. Herbert RE	Blecquenecques, Macquinghen

M-R, Marquise-Rinxent group of quarries; RE, Royal Engineers.

into fighting units, rather than the reverse. However, stone at times received priority, even over ammunition, for transportation by rail. By late 1917, in total >3000 men were serving with Quarrying Companies Royal Engineers in France, mostly in the Marquise-Rinxent group of quarries; its headquarters used an address stamp with that name.

The officers included men who had significant quarrying leadership experience. Examples are 'Stephen' Darbishire of 320, born in 1886, the son of a respected quarry manager who himself became a quarry manager in the family business on release from the army; and Herbert Brooks of 321, born in 1883, a married man who was 33 years old when he applied for a commission in January 1917, with nine years of experience as a quarry and works manager. Quarrying Company officers can rarely be identified in contemporary photographs, but Darbishire (Fig. 3) is an exception. Second Lieutenants Darbishire and Brooks were promoted together on 16 February 2017 to the rank of temporary captain (*London Gazette Supplement*, 6 March 1917, p. 2257), thereby confirming them as the officers commanding their respective companies.

Other officers included qualified engineers no longer fit for front line service in combat units. An example is Frederick Basil Lawson, born in 1891, who had applied for a commission in the infantry in October 1914, soon after the outbreak of war, but been wounded in the scalp and shoulder by shrapnel in 1915, and shell-shocked by a trench mortar bomb that killed four of his men in 1916 while serving as a temporary captain with the 1st/8th Battalion of the Nottinghamshire and Derbyshire Regiment. A qualified mining engineer, he was seconded for duty with Quarrying Companies Royal Engineers on 28 December 1916, following discharge from hospital in Oxford earlier that year (WO 374/41124). Another example, Frank Brooke, was seconded from the King's Own Royal Lancaster Regiment to command 324 Quarrying Company Royal Engineers (REMLA file 84).

Fig. 3. Captain C.S. Darbishire, Royal Engineers (left), photographed with an unnamed Royal Engineers lieutenant presumed to be under his command, at Haut Banc quarry in the Boulonnais region of France. Part of a photograph reproduced by permission from the Archives and Special Collections, Bangor University, Darbishire Additional Papers.

Others included engineers deemed to be too old for combat commissions. An example of these is Ralph Spencer, born in 1874, and almost 43 years old when he applied for a commission in 1917. Resident in Hounslow, Middlesex, by then he had been the general manager of a road construction company for eight years. The 'certificate of moral character' on his application was signed by a quarry owner resident in Enderby, near Leicester (WO 339/ 94669). Spencer was to serve with 323 Quarrying Company.

Marquise quarries

The original intention was that each company, being self-contained, should move from place to place and work stone easily accessible from the roads requiring repair. However, 'owing entirely to geological reasons [nearly] all the companies were concentrated in one area' (Lawson 1920, pp. 101–102): quarries in the vicinity of Rinxent and Marquise, small towns *c.* 14 and 16 km, respectively, NE of Boulogne. The 'geological reasons' were that relatively strong Upper Paleozoic limestones cropped out as the Ferques Inlier in this region (Ager & Wallace 1967*a*); the BEF area otherwise exposed only relatively weak Mesozoic and Cenozoic sedimentary rocks (Figs 2 & 4), far less suitable for use as road metal or for railway ballast. Other inliers were sparse, smaller and less well served by railways.

The geology of the Boulonnais was relatively well known by that time. The Ferques Inlier had been illustrated (Fig. 5) on sheet 3 Boulogne of the 1:80 000 *Carte géologique détaillée*, surveyed in 1871–72 (Douvillé *et al.* 1875). This distinguished four units within the local Devonian and, consequently, mapped two Devonian limestones as distinct from the succeeding Carboniferous limestone (Table 2), in main outcrop all with an overall dip to the SW. Subsequent literature drew attention to clastic sediments between the limestones and the sequence recognized in wartime was essentially that of six Devonian units as described by Pruvost & Pringle (1924). The most recent medium-scale French government map, sheet 5 Marquise of the 1:50 000 *Carte géologique de la France*, redefines the six units within the 700 m thick succession as formations, subdivided into members in accordance with modern stratigraphic practice (Mansy *et al.* 2007). The two basal formations (of Givetian age) consist of a unit of sandstone, shale and conglomerate, overlain by limestone with thin claystone intercalations. This is succeeded by three formations of Frasnian age, lower and upper formations dominantly of shales separated by a middle formation mostly of dolomite and limestone. The final formation consists of Famennian sandstone and shale. The Carboniferous is of similar (600–700 m) thickness and is largely limestone and dolomite, of Tournaisian–Visean age, now mapped as five formations (Table 2). The stronger rocks in the sequence had been quarried pre-war and the BEF initially secured rights to deploy Quarrying Companies Royal Engineers to six of the existing quarries (Table 3 and Fig. 6).

Three quarries worked the Carboniferous 'limestone':

- Vallée Heureuse (Happy Valley to British soldiers) was the largest (Fig. 6, locality 16), although the limestone it yielded was 'rather soft' (Anon 1924, p. 211). It had supplied stone under civilian contract to meet both British and French needs since at least 1915, but was handed over to the BEF Directorate of Roads on 1 December 1916 (Anon 1924). At that time the quarry face was *c.* 0.75 mile (1200 m) in length and *c.* 120 ft (37 m) high. Sidings within the quarry were already connected by a broad-gauge track with the main railway line between Boulogne and Calais. The face of the quarry was reached by a broad-gauge railway; its west end was served by chain haulage working on an incline and delivering to two jaw crushers. Overburden was removed by chain haulage and by tubs on a Decauville track (ready-made sections of light, narrow-gauge track fastened to steel sleepers that were portable and easily disassembled and re-positioned according to need).
- Haut Banc quarry nearby (Fig. 6, locality 15; Fig. 7) had a length of *c.* 1250 yards (1140 m) and a depth of 100 ft (30 m). It was initially worked in two 'lifts', but a third lift was introduced by the British. The top lift was worked by Decauville haulage, which delivered stone to the existing crusher plant. A broad-gauge track took the stone in the second lift (hand-broken), while the bottom lift was reached and sunk by means of a Blondin (an aerial ropeway, named after the famous tightrope walker Charles Blondin (1824–97)). This delivered stone to two 50 ton h^{-1} crushers (1 ton = 1.016 tonnes).
- Saint André was a small quarry to the east of Haut Banc that had opened just before the war (seemingly Fig. 6, locality 17, and renamed Basse Normandie post-war). The British installed three small 25 ton h^{-1} crushers, fed by chain haulage, to deal with the output. The face was deemed too small to necessitate the installation of larger equipment.

Three quarries worked Devonian 'limestones':

- Beaulieu (otherwise known as Galtier's quarry, having been leased by the landowner M. Lonquéty to Messrs Raymond, Galtier and Brizard) had been opened before the war to produce

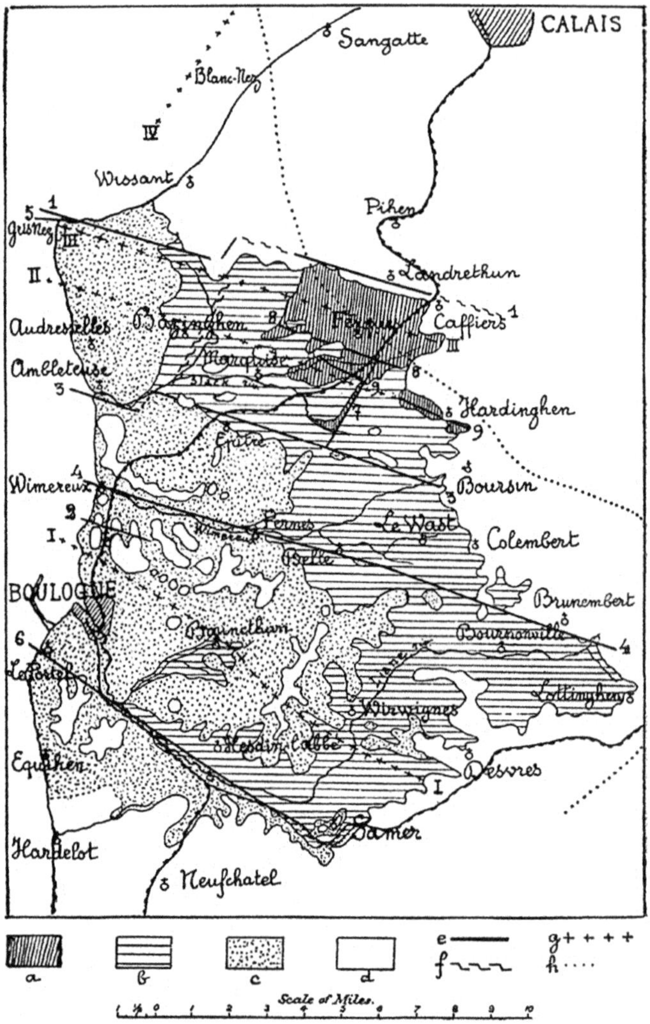

Fig. 4. Geological sketch map of the Boulonnais. Key: a, Paleozoic; b, Middle Jurassic and lower part of the Upper Jurassic; c, Upper Jurassic (Kimmeridgian and Portlandian); d, Cretaceous, Tertiary and Quaternary; e, faults; f, flexures; g, anticlinal axes; and h, eastern limit of the Jurassic deposits. Roman numbers I–IV indicate anticlines, Arabic numbers 1–9 indicate 'post-Cretaceous' faults. Scale: 5 miles = 8 km. From Pruvost & Pringle (1924).

stone for harbour works at Boulogne, mainly as stone blocks, but with a small 25 ton h^{-1} jaw-crusher for the production of aggregate for concrete. The quarry was equipped with railway sidings and connections to the main line. Facilities were greatly enhanced by the British, who worked the quarry to its limit. They installed three air compressors to work pneumatic drills and jack-hammers, two generators, a workshop, three crushers and two chain haulages (overhead Decauville) for delivering stone to the crushers. As the face advanced, the quality of stone rapidly deteriorated and a considerable proportion had to be sent to the tip with the spoil. An increasing amount of the clay underlying the limestone was exposed, causing several large (10 000–20 000 ton) landslides from the limestone. Fortunately, these occurred at night, when work had ceased, so without casualties. Located 'about half-a-mile' from the Banc Noir quarry (Lawson 1920, p. 105) and working a stratigraphically higher bed of limestone, this quarry was later renamed La Parisienne (Fig. 6, location 9), according to Edleston (1940).

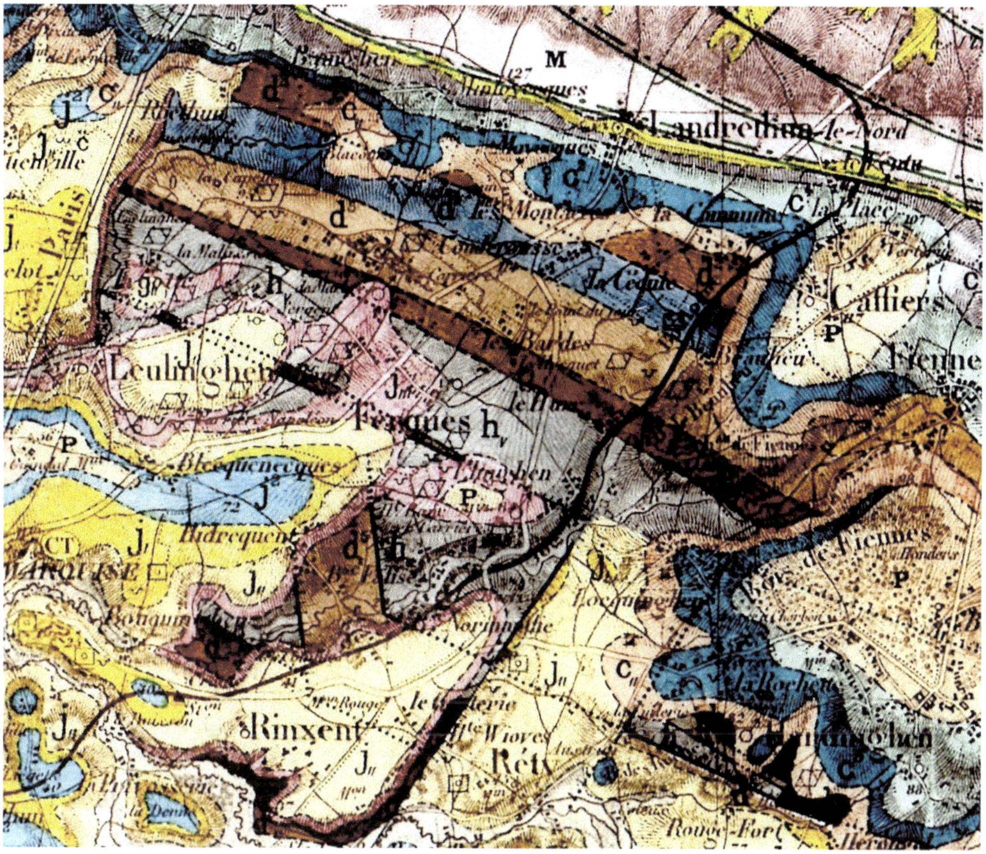

Fig. 5. Part of sheet 3 Boulogne of *Carte géologique détaillée de France*, original scale 1:80 000, published in 1876, showing the Ferques Inller east of Marquise and north of Rinxent. Beneath a Cretaceous unconformity (strata lettered 'c'), four Devonian units with ENE strike (d^2 Poudingue de Caffiers, d^4 Calcaire de Blacourt, d^5 Calcaire de Ferques, d^6 Grès de Fiennes) are succeeded to the SW by Carboniferous limestone (h), unconformably overlain by remnants of a Jurassic cover (J), a sequence complicated to the south by faulting. The main Calais Boulogne railway line crosses the region from near Caffiers in the NE past Rinxent in the SW. Courtesy of the Geological Society of London Library.

- Lonquéty quarry (Longuetty of Lawson 1920) yielded limestone of 'Fair Quality' (Anon 1924, p. 209). Named by the British after its owner, it was 'merely an extension of the same beds as found at Beaulieu, but situated on the other side of the Calais–Boulogne main line' (Lawson 1920, p. 104), so seemingly location 7 on Figure 6. 'No great attempt was made to develop this quarry, as the area worked was so small.' The quarry was *c.* 250 yards (230 m) long by 70 ft (20 m) high. Branch sidings had been laid into the quarry to connect it with the main railway line prior to its handover to the BEF Works Directorate on 1 December 1916.
- Banc Noir quarry (Fig. 6, location 5; Fig. 8) 'just north of Beaulieu quarry, and with rail connection' (Anon 1924, p. 215) worked a lower limestone, the upper beds being almost entirely worked out before handover to the British. A Blondin was installed, the quarry floor lowered by *c.* 25 ft (8 m) and pumps were added to prevent flooding. Ultimately, the old chain haulage was extended to the new beds and used for feeding the existing crusher. 'The large amount of rotten shale overlying the true limestone made the working of this quarry very expensive from the point of view of labour' (Lawson 1920, p. 105). The quality of stone was good, but the face was low.

The use of Quarrying Companies Royal Engineers trebled the output of the quarries at Marquise in 1917 (Pritchard 1952*a*, p. 626), the year in which the strength of the BEF reached its peak. They

Table 2. *Paleozoic succession as distinguished in the Ferques Inlier*

Douvillé *et al.* (1875)	Pruvost & Pringle (1924)	Mansy *et al.* (2007): formations
Calcaire carbonifère	Carboniferous limestone (seven subdivisions)	Calcaires Lunel, Napoléon, Joinville Dolomie à *Siphonodendron martini* Haut Banc: calcaires Dolomie du Hure Calcschistes de la Vallée Heureuse
Grès de Fiennes	Grès blanc de Sainte Godeleine Schistes rouges de Fiennes	Sainte Godeleine: grès micacés Hydrequent: argilites
Calcaire de Ferques	Calcaire de Ferques Schistes de Beaulieu	Ferques: calcaires, dolomies Beaulieu: argilites
Calcaire de Blacourt	Calcaire de Blacourt	Blacourt: calcaires, argiles
Poudingue de Caffiers	Conglomérat de Caffiers	Caffiers: shales, grès, poudingues

continued in production until the end of the war and beyond (Fig. 9), their activities under the overall control of a lieutenant-colonel (Fig. 10). All of the quarries were potentially accessible to a field trip led to the area in 1966 (Ager & Wallace 1967*b*), but far fewer in 2011 (Mistiaen *et al.* 2011).

Seemingly only the War Diary for 348 (WO 95/4057) records that another quarry in this region was also worked: Blecquenecques (the Napoléon quarry, location 19 on Fig. 6; village also named on Fig. 5). This exposes the upper part of the Carboniferous Limestone (Table 2), which has been faulted and folded (Ager & Wallace 1967*b*). The quarry, situated some distance from all the others worked locally by Quarrying Companies Royal Engineers, was among those assessed by Edleston (1940) for potential use during World War II. He noted that it had good road access, but was potentially of little military use because of its remoteness from the railway. Its output in World War I was not registered with the material transported by rail from the other quarries, but used more locally for roadworks. The same inference can be applied to the new quarry developed by 348 at Macquinghen, to the SE of Figures 5 and 6 and therefore the Paleozoic outcrop.

Other sources (Table 3) contributed a relatively small tonnage overall.

- A flint gravel pit at Audrehem, on the north bank of the River Hem in the Pas-de-Calais region, produced flints mixed with a good deal of clay (Anon 1924). There was a narrow-gauge line running past it to Pont d'Ardres, where the material was transferred to broad-gauge trucks. The material made a fair surfacing to a road and bound well, but formed a dirty road. The pit was initially worked by a contractor, but was among those handed over to the BEF Directorate of Roads on 1 December 1916.
- Slag heaps from the Outreau Steel Works, on the River Liane, above Boulogne, formed a very large (mostly pre-war) accumulation, conveniently

Table 3. *Output from the Marquise-Rinxent group of quarries operated by Quarrying Companies Royal Engineers*

Quarry name	Number on Figure 6	Stratum	Pre-war daily output (tons)	October 1916 output (tons)	Final British daily output (tons)
Vallée Heureuse	16	Carboniferous limestone	300	750	3000
Beaulieu (Galtier's)	9	Devonian limestone	350	1000	2500
Haut Banc	15	Carboniferous limestone	300		2000
Lonquéty	7?	Devonian limestone	50	400	350
Banc Noir	5	Devonian limestone	150	60	1000
Saint André	17?	Carboniferous limestone	50	40	400
Other sources		Outcrops, slag, gravel		150	750
Total			1200	2400	10 000

Compiled from Lawson (1920) and Anon (1924). Spelling Longuety in British records corrected to Lonquéty.

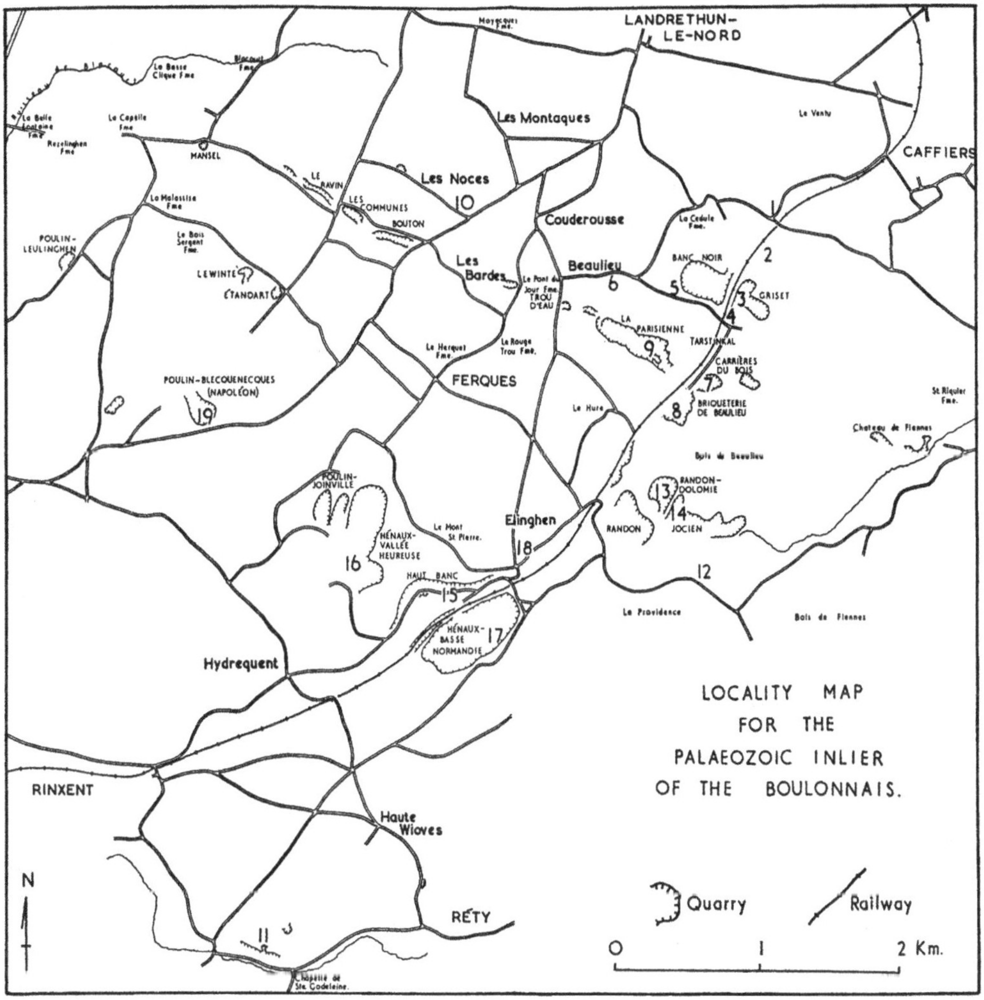

Fig. 6. Map showing the location of quarries in the Paleozoic rocks of the Ferques Inlier accessible in April 1966. Reproduced by permission from Ager & Wallace (1967*b*, fig. 1), *Easter Field Meeting in the Boulonnais, France*; © 1967 The Geologists' Association, London.

close to a railway siding. The Commander Royal Engineers Boulogne arranged with the Steel Works company for the exploitation with his own labour of this slag, of which *c.* 100 000 tons had been extracted by the end of hostilities in November 1918 (Anon 1924).

In addition, some material was provided under contract as before (Anon 1924), notably from Etavaux (near Caen, in Normandy), a quartzite and schist quarry; Incheville, on the banks of the River Bresle, *c.* 37 km NE of Dieppe, a flint gravel deposit; Milly-sur-Thérain (a small commune in Picardy), sand suitable only for concrete; Saint Pierre Halte (near Calais), a beach shingle deposit, only suitable as concrete ballast; and Arques la Bataille (near Dieppe), a flint gravel deposit.

Geological assessment of Quaternary gravels and sands

Although 'Bill' King and Edgeworth David, the two geologists who served as such at BEF General Headquarters, were respectively tasked primarily with guiding the development of water supplies and the excavation of underground facilities (Rose 2015), they were also required to provide guidance on the 'winning of raw materials', both 'road metal' and 'sand and aggregate for concrete' (Anon 1922*a*,

Fig. 7. Near-contemporary postcard showing Haut Banc quarry, near Marquise, when under civilian management, www.wikipasdecalais.fr (accessed 14 August 2015).

Fig. 8. Western part of Banc Noir quarry as recently exploited, showing SSW face on 22 July 2009, with vehicles on upper level serving to indicate scale. © Bruno Mistiaen and reproduced by kind permission.

the Royal Engineers in charge of the marble quarries at Marquise, 25 November 1918. The
the then-current sleeve cuff insignia of a lieutenant-colonel: a crown above a four-pointed Bath
es. The two officers beside him to the right of the photograph clearly have the cuff insignia of
with two stripes. The white-shirted officer to the left of centre has the three stripes indicative of a
colonel and seemingly the badge (crown) of a major. The officer to the far left of the front row is
ptain from his front-rank position in this photograph and centrally on another photograph in the
illustrated). The end officers in the rear rank are seen to be subalterns (lieutenants or second
ir single cuff stripe, and the intervening officers are presumed from their position also to hold
captains and nine subalterns together apparently constitute the officers of three Quarrying
ngineers. Part of the Ministry of Information First World War Official Collection of photographs.
ission: © Imperial War Museum (Q 9706).

k near Péronne (Fig. 2). A dolo-
ong used as a building stone for
es was seen to have been remark-
t to frost. The geologists there-
ended that stone from the local
s tested for use as road metal,
recommendation was seemingly

vel at Vecquemont. On the banks
Somme 11 km east of Amiens
locality was estimated to contain
ic yards (150 000 m^3) of good
0 000 cubic yards of flint gravel.
ef quarries in the basins of the
the Escaut (known as the Scheldt

in Belgium and the Netherlands) and on the plateau near Cambrai were supplied to the Chief Engineer of each British army. Earth augers were used to make test borings to determine the thickness of sands and flint gravels in the vicinity of the River Oise.

(4) Gravel pits in the Authie and Grouches river valleys. Gravel beds were thought, for geological reasons, to exist in the vicinity of Doullens, which might be exploited if German advance towards Amiens denied the pits near that city to the Allies as a source of gravel for concrete. However, the extent of the gravels was not known, nor the thickness of overburden, so test bores were made in the valleys of both

Fig. 9. Catalogued as 'Royal Engineers staff at the marble quarries appears to show the officers and men of one of the Quarrying Com company sergeant major stand at the front: a grey-haired captain to company); the company sergeant major; and three subalterns, presu right of the photograph. Part of the Ministry of Information First W Reproduced by permission: © Imperial War Museum (Q 9714).

pp. 58–61). Their guidance was shaped initially by local knowledge. Early in the war, two Belgian engineer officers attached to the headquarters had compiled a report on road metal in Belgium; information on the availability of road metal in northern France was supplied by the senior French regional government and railway engineers. The chief sources of road stone identified for the BEF in northern France consisted of:

- limestone from the Marquise-Rinxent group of quarries;
- quartzite from Devonian inliers within the Cretaceous chalk, notably Dennebroeucq (worked briefly by 325), Matringhem, 'Bergin' (perhaps Beugin, worked briefly by 198) and Rebreuve and Marqueffles (worked by 324);
- porphyrite from Brittany;
- *pierre de fosse*, a red Carboniferous shale burnt to the consistency of brick through the spontaneous combustion of colliery dumps;
- fli Pa
- sa Te

Of th speci regic not e

(1)

Fig. 10. Officers c central figure wear star, with three stri captains: three star major or lieutenant presumed to be a c same collection (n lieutenants) from t this rank. The three Companies Royal Reproduced by pe

(2) Yellow ch mitic chalk local churc ably resist fore recom quarries w although th not adopte

(3) Sand and g of the rive (Fig. 2), th 200 000 c sand and Lists of c Somme an

Fig. 10. Officers of the Royal Engineers in charge of the marble quarries at Marquise, 25 November 1918. The central figure wears the then-current sleeve cuff insignia of a lieutenant-colonel: a crown above a four-pointed Bath star, with three stripes. The two officers beside him to the right of the photograph clearly have the cuff insignia of captains: three stars, with two stripes. The white-shirted officer to the left of centre has the three stripes indicative of a major or lieutenant-colonel and seemingly the badge (crown) of a major. The officer to the far left of the front row is presumed to be a captain from his front rank position in this photograph and centrally on another photograph in the same collection (not illustrated). The end officers in the rear rank are seen to be subalterns (lieutenants or second lieutenants) from their single cuff stripe, and the intervening officers are presumed from their position also to hold this rank. The three captains and nine subalterns together apparently constitute the officers of three Quarrying Companies Royal Engineers. Part of the Ministry of Information First World War Official Collection of photographs. Reproduced by permission: © Imperial War Museum (Q 9706).

(2) Yellow chalk near Péronne (Fig. 2). A dolomitic chalk long used as a building stone for local churches was seen to have been remarkably resistant to frost. The geologists therefore recommended that stone from the local quarries was tested for use as road metal, although that recommendation was seemingly not adopted.

(3) Sand and gravel at Vecquemont. On the banks of the river Somme 11 km east of Amiens (Fig. 2), this locality was estimated to contain 200 000 cubic yards (150 000 m^3) of good sand and 100 000 cubic yards of flint gravel. Lists of chief quarries in the basins of the Somme and the Escaut (known as the Scheldt in Belgium and the Netherlands) and on the plateau near Cambrai were supplied to the Chief Engineer of each British army. Earth augers were used to make test borings to determine the thickness of sands and flint gravels in the vicinity of the River Oise.

(4) Gravel pits in the Authie and Grouches river valleys. Gravel beds were thought, for geological reasons, to exist in the vicinity of Doullens, which might be exploited if German advance towards Amiens denied the pits near that city to the Allies as a source of gravel for concrete. However, the extent of the gravels was not known, nor the thickness of overburden, so test bores were made in the valleys of both

Fig. 9. Catalogued as 'Royal Engineers staff at the marble quarries at Marquise, 25 November 1918', this photograph appears to show the officers and men of one of the Quarrying Companies. Four officers (see also Fig. 10) and the company sergeant major stand at the front: a grey-haired captain to the left (presumably the officer commanding the company); the company sergeant major; and three subalterns, presumably the junior officers of the company, to the right of the photograph. Part of the Ministry of Information First World War Official Collection of photographs. Reproduced by permission: © Imperial War Museum (Q 9714).

pp. 58–61). Their guidance was shaped initially by local knowledge. Early in the war, two Belgian engineer officers attached to the headquarters had compiled a report on road metal in Belgium; information on the availability of road metal in northern France was supplied by the senior French regional government and railway engineers. The chief sources of road stone identified for the BEF in northern France consisted of:

- limestone from the Marquise-Rinxent group of quarries;
- quartzite from Devonian inliers within the Cretaceous chalk, notably Dennebroeucq (worked briefly by 325), Matringhem, 'Bergin' (perhaps Beugin, worked briefly by 198) and Rebreuve and Marqueffles (worked by 324);
- porphyrite from Brittany;
- *pierre de fosse*, a red Carboniferous shale burnt to the consistency of brick through the spontaneous combustion of colliery dumps;
- flint gravels of Quaternary age (mostly in the Pas-de-Calais region);
- sarsens (occasional loose blocks of well-cemented Tertiary sandstone).

Of these, King and David were not required to report specifically on limestone, quartzite or porphyrite regions. Their studies focused principally, although not exclusively, on Quaternary sands and gravels.

(1) Gravel being worked in pits at Arques and Aire, *c.* 16 km SW of the BEF General Headquarters at St Omer (Fig. 2): a 'remarkable' deposit, of significant, but unknown, extent. Trial pits were sunk to depths of 20–25 ft (6–8 m), revealing an immense development of flint gravels under an overburden of 10–15 ft (3.0–4.5 m) of sandy clay (*limon*). A report was also made on unworked gravel deposits that extended from Wizernes to Wisques, *c.* 6 km SW of St Omer, a resource estimated at *c.* 3 million tons.

rivers. Although each valley was estimated to contain *c.* 500 000 tons of gravel, the average thickness of overburden (8 ft (2.5 m)) made working impracticable.

(5) Sarsens. Widely scattered over the surface of the chalk in northern France, these concretions derived from the former Thanet Sand cover provided a local source of road metal and particularly of *pavé*, cobblestones traditionally used to surface local roads.

(6) *Pierre de fosse*. Special geological reports were prepared in September 1918, when Allied forces were at last making significant advances, showing the exact location and approximate amounts of this material on the British First and Second Army fronts as far as Mons (Fig. 1). The amount available was estimated as at least 50 million tons, although whether this was suitable for road metal as such or merely as fill suitable for use as a sub-base could not be determined from the documents accessed.

The British geologists prepared resource maps to guide military planning, notably a map of Belgium as a whole, based on the road metal report compiled by Belgian engineers early in the war. This showed the positions of major quarries, such as Quenast, Lessines, Soignies, Maffle, Ecaussines, and Tournai. The quartz porphyry intrusions of Lessines and Quenast near Brussels, and the Carboniferous limestone of Soignies between Mons and Brussels, were to be worked by Quarrying Companies Royal Engineers in World War II rather than World War I (Rose 2012*b*). The map also showed sand pits, sand being required for packing under *pavé*, the cobblestones then commonly used for road surfacing in western and central Belgium. A map to show Carboniferous and Devonian 'strong rock' inliers within the Avesnes-Bavay area of northern France was another specially prepared thematic map.

Sand and aggregate for concrete were obtained primarily from the *ballastières* (ballast pits) of the French railways at Aire, Arques and Engoudsent. Lesser amounts were obtained from Quaternary terrace gravels of the River Somme near Flexicourt and from Paleocene 'Thanet Sands' in pits near Authie and at Blaireville, near Arras (Fig. 2). Flint shingle deposits at the mouth of the river Somme, and older beds inland, dune sand near Dunkerque and 'Thanet Sands' near Arques were all assessed as potential sources, but deemed to be unsatisfactory.

Quarry operation

Work in the Marquise quarries has been described in some detail by Lawson (1920), who served periods in command both of 198 and 199 Quarrying Companies. It initially involved the removal of the shale or soil overburden, a labour-intensive task. Prisoners of war were therefore used on a large scale to assist the quarrymen: ten companies of them, each with a working strength of *c.* 400 men, so *c.* 4000 in total. The rock was then drilled and blasted from the quarry face and transported to a crusher to reduce it to a size suitable for use as road metal, railroad ballast or concrete aggregate. After processing, it was transported offsite by rail.

Drilling was carried out by Ingersoll machines. Large holes 20 ft (6 m) deep by 2.5 in (63.5 mm) in diameter were put down for the main blasts, while the blocks which were too large to be broken by jack-hammer were afterwards split by pop shots drilled by jack-hammer. For explosive, blastine and gelignite were chiefly used. The large holes were given a charge of from 15 to 20 lb (6.8–9.1 kg), with pop shots from 2 to 8 ounces (0.057–0.227 kg). The large holes were fired electrically and the pops by fuse and detonator.

Work continued 24 hours a day, seven days a week in a shift system. Initially, there were two shifts per day, 6.00 am to 6.00 pm and 6.00 pm to 6.00 am, both with two-hour meal breaks. However, these hours were found to be excessive when working seven days per week irrespective of weather conditions. When more labour became available, three eight-hour shifts were adopted, with a half-hour meal break. Shift time included walking to the quarry from the base camp and return, so the actual time working was only seven hours. Men were given one day off in seven, but not necessarily a Sunday, to ensure a constant supply of stone for rail transport. In all, 3500 wagons and from 70 to 80 main line locomotives were dedicated to moving stone to the rail heads, where *c.* 1000 lorries were used for unloading.

Output figures vary between sources, but the increase brought by the use of the Quarrying Companies was certainly considerable. An official account (Anon 1924, p. 22) records that, even with additional labour provided by the British, the 'average output from the Marquise quarries in December, 1915, was 410 tons daily', whereas Lawson (1920) recorded the pre-war (1914) daily output at 1200 tons (Table 3). The increased daily output achieved by October 1916 was 2400 tons according to Anon (1924, p. 24), but Pritchard (1952*a*, p. 562) cites only 1900 tons, with 860 tons from the Channel Islands, 420 tons from Caen and 70 tons from Audrehem. At this time, the quarries (apart from Beaulieu) were still largely worked by French and Belgian rather than British labour, before being taken over by the Quarrying Companies Royal Engineers. Edleston (1940) recorded that during the war the total output of road material of all kinds for the British armies in the northern zone of France reached a maximum of *c.* 15 000 tons per day in March 1918 and demand remained at about that rate until the Armistice in November 1918. Of this, *c.* 5000 tons per day were obtained from sources under the French administration, while the remainder was produced at the Marquise group of quarries and a few other locations (gravel pits) in the British area of operations. Walton (2010) quotes a despatch from Field Marshal Sir Douglas Haig as reporting 'over 600 000 tons of material was produced [from the Marquise quarries] in the nine months ending 31st August, 1917', and Lawson (1920, p. 103), from personal experience, reported that a 'final output of 10 000 tons of macadam daily was reached after about 18 months, and was afterwards fully maintained'. However, his figure does not include output from the quarries worked by either 324 or 348. The figure for stone quarried for the BEF is thus likely to have been greater than 10 000 tons per day during most of 1918.

The stone was won at some cost. Although the Marquise area lay well to the rear of the front line, the huge glare from quarry lights when working at night, and the proximity of the quarries to the ports of Calais and Boulogne, attracted enemy bombing on several occasions in the autumn of 1917. Generally this was to little effect, but Lawson (1920, p. 108) records that on one occasion 60 prisoners of war were killed and on another 'twenty quarrymen at work in Beaulieu'. The figures may be exaggerated, but certainly an air raid on 25 September 1917 left at least six dead (Sappers J.H. Pick, H. Slingsby and F. Taylor of 198, A.J. Girdler of 199) or dying (Sappers S. Falla of 321, W.T. Tite of 198) according to grave registers for the nearest war cemetery, Les Baraques Military Cemetery at Sangatte, and more injured (e.g. Lance Corporal C. Lancaster of 321) (according to Walton 2010). The War Diary for 198 confirms five men killed and six wounded from this company on that occasion. Thereafter sentries were posted to warn of potential attack and the lights were extinguished on receipt of a warning, with quarrying continuing by moonlight. Other quarrymen were to die of illness or through accident during their military service, but the death rate from accident was reported by Lawson (1920) to be lower than that in British quarries at the time, a remarkable achievement in view of the pressure and conditions of work, associated with a great deal of unskilled labour. But the war brought other hazards. The War Diary of 328 records that Captain Mann was killed in action and seven other ranks wounded when their unit was deployed forward to help construct defensive earthworks. The War Diary of 348 records that four men (one each of Scottish, Irish, English and Welsh parentage) were honoured for their bravery in extricating a dying pilot from his crashed and burning aeroplane, despite the danger from flames and exploding ammunition.

Concluding remarks

Quarrying Companies were first raised within the Royal Engineers in 1916 for service with the BEF on the Western Front in France: in total, 13 by late 1917. Eleven came under the command of the Headquarters Marquise-Rinxent Group of Quarries (Table 1) within the Line of Communication to the rear of the front line; their War Diaries include pages with that official stamp. A twelfth Quarrying Company, the final one to be raised (348), deployed almost to the same area, albeit under separate command. Yielding on average >10 000 tons of crushed rock per day, principally for road metal and railway ballast, the 'Group' when fully operational, made a significant contribution to the war effort. Geoffrey Walton (pers. comm. 2016) has noted that the output

> is equivalent to 3.5 Mtpa which is more than three times the output in Britain of roadstone in 1913 and equivalent to a quarter of its whole output of limestone for the same year, in all a massive achievement.

Four of the first ten Quarrying Companies to be raised are known to have recruited in specific geographical areas: 198 in Leicestershire and Kent; 320 in North Wales and Lancashire; 321 in the Channel Island of Guernsey; and 327 in Ireland. It seems likely that the other six may also have focused recruitment on quarrymen in specific regions, but the details are not recorded in the unit War Diaries. However, the home details are known for sufficient individuals to be sure that recruitment included England, Wales, Ireland, Scotland and Guernsey. Formed mostly in England, at Clipstone, Buxton or Aldershot, movement in a few weeks (or initially days) of foundation left little, if any, time for preparatory training.

Additional to the output from the Marquise-Rinxent group, some Quarrying Companies contributed more closely to combat operations. 324 Quarrying Company, commanded by an experienced infantry officer on secondment, was assigned to provide closer quarrying support to the British First Army, including road works for the Canadian troops who stormed the supposedly impregnable Vimy Ridge near Arras in 1917. Three Quarrying Companies (325, 328 and 329) went forward from the Marquise-Rinxent area on 24 March 1918 to help construct a line of defences to counter the 'German offensive against the Third and Fifth Armies, which opened on 21st March and halted just short of Amiens on 4th April'. Twenty-seven miles of a rear defence belt were constructed by personnel (>23 000 men) from transportation units. (Pritchard 1952*a*, pp. 634, 646.) It was presumably not a coincidence that 328 and 329 were the only Quarrying Companies to be recruited from line regiments rather than civilian quarrymen: their 'sappers' were trained soldiers, so better able to survive on the battlefield.

Hostilities ceased on the Western Front with the Armistice that came into force on 11 November 1918, although the war there as such did not end until peace treaties were signed during 1919. All the Quarrying Companies returned to England in 1919, where they were disbanded, and their soldiers returned to civilian life. Some returned in February and March via Fovant Camp on Salisbury Plain for dispersal on demobilization (Walton 2010); others served longer, until May at least. They were not among the first servicemen to qualify for demobilization; the Quarrying Companies had only been raised in 1916 and 1917 and men who had joined the army in 1914 and 1915 qualified for earlier release. Moreover, stone was required for reconstruction and route repair in the battle-ravaged terrain, even though hostilities had ended, and military quarrying continued until civilian agencies became fully operational. Thus Sapper William Forbes served with 348 until 7 May 1919, when he was transferred to 328 Quarrying Company and thence to England prior to demobilization (De Santis 2002).

About 3000 men in total had served within the 13 Quarrying Companies, effectively within the last two years of the war. The number was small in relation to the Royal Engineers as a whole; the Corps consisted of *c.* 1328 units and nearly 230 000 men at the end of hostilities (Pritchard 1952*a*, *b*). However, it represents part of the phenomenal growth in size of the Corps during wartime, from the small numbers it contained at the outbreak of war in 1914: 1056 officers and 10 394 other ranks in the Regular Army and Special Reserve, together with 513 officers and 13 127 other ranks in the Territorial Force. In addition, because *c.* 4000 prisoners of war were used to assist the quarrymen, the actual workforce employed on quarrying was >7000 men.

Concentration of effort on the Marquise-Rinxent group of quarries was determined by the geology (rock relatively strong and abundant enough to serve as road metal or railway ballast), location (within convenient transportation distance of the

Table 4. *'Légende technique' on 1:80 000 geological map of Boulogne (Douvillé* et al. *1875)*

Agriculture et Hygiène (agriculture and hygiene)	
Marne (marl)	Exploitation à ciel ouvert (opencast)
Phosphate de chaux (phosphate of lime)	
Sources (springs)	
Eaux superficielles (surface water)	
Nappes d'eau souterraine (groundwater aquifers)	Exploitation par puits (shafts)
Chauffage (combustibles)	
Houille maigre à longue flame (bituminous coal)	Puits de mine/Sondages (mines)
Arts chimiques (economic minerals)	
Pyrites (pyrites)	Exploitation à ciel ouvert (opencast)
Lignites pyriteux (pyritic lignites)	Exploitation par puits (shafts)
Métallurgie (metals)	
Minerals de fer (iron ore)	Exploitation à ciel ouvert/par puits
Arts de construction (construction materials)	Exploitation à ciel ouvert (opencast)
Quartzite (quartzite)	
Grès à ciment calcaires (calcareous sandstone)	
Calcaire lithographique (lithographic limestone)	
Calcaire oolithique grossier (thick oolitic limestone)	
Calcaire oolithique fin (thin oolitic limestone)	
Calcaire schisteux (fissile limestone)	
Limon pour briques (brick clay)	
Glaise (clay for tiles/pottery)	
Pierre à chaux grasse (stone for 'fat'/pure lime)	
Pierre à chaux maigre (stone for 'lean'/poor lime)	
Pierre à chaux hydraulique et à ciment (stone for hydraulic lime and for cement)	
Sables (sands)	
Silex pour empierrement (chert/flint for road metal)	
Calcaire dur pour empierrement (hard limestone for road metal)	
Cailloux et galets (gravels)	
Grès calcaire pour pavés (calcareous sandstone for cobblestones)	
Grès siliceux pour pavés (siliceous sandstone for cobblestones)	
Arts mécaniques et décoratifs (decorative and craft materials)	Exploitation à ciel ouvert (opencast)
Marbres gris et noire (white and black marble)	
Marbre tacheté (mottled marble)	

operational area of the BEF) and good rail access (essential for efficient bulk transportation). The quarries had supplied stone to the BEF under civilian contract until they were transferred to military control to enhance output, so were a known resource that was enhanced as the result of a military decision rather than a new resource developed as a consequence of geological survey. They had, however, not been used as a source for road stone before the war (because far better material was then accessible in Belgium) and none had been equipped for large outputs (Lawson 1920).

British military geologists did, however, compile maps and reports and guide exploratory boring to help the development of sources of sand and gravel for additional military use. Their work was facilitated by local expertise willingly contributed by their French and Belgian allies, and by the use of Belgian 1:40 000 and French 1:80 000 scale geological maps. Consistent with a long tradition in France in 'applied' as much as 'pure' geology (West & Rose 2005), French maps incorporated a 'technical legend' (Table 4) with a marginal key to strata that drew attention to the actual or potential uses of the rock units depicted on the map, and their likely means of exploitation. The maps were thus potentially of more immediate military use than if they had been in the style of the contemporary geological maps of the UK.

Largely at the instigation of 'Bill' King (Rose 2015), British geologists identified the source in Germany of much of the aggregate used by German forces in the construction of their concrete pill-box fortifications in northern France and Belgium (Anon 1922*a*; Sabine 1991; Branagan 2005). Some pebbles could be matched by gravels from the Rhine rather than from Belgium. In addition, angular fragments of basalt that had obviously been generated by quarrying were of Niedermendig-type lavas characteristically found in the Rhine valley near the towns of Linz and Erpel. There was a notable absence of Belgian rocks. The aggregate must therefore have been transported into Belgium and France via the Netherlands, where the transport of such war material should have been prevented by the terms of that country's neutral status. Geological advice formed the basis of a protest duly made by the British government to the Netherlands government early in 1918. British geologists thus not only made a small, but significant, contribution to Royal Engineers quarrying for the Western Front, but ultimately helped to deny illegally transported quarry products to the enemy.

Acknowledgements I am grateful to staff at the Geological Society of London Library, the National Archives and the Royal Engineers Museum, Library and Archive (REMLA), for access to documents in their care; to Lynette Hunter, Archivist at Bangor University, for help to locate items within the Darbishire Papers and Additional Darbishire Papers; to Geoffrey Walton and David Branagan for reviewing the submitted manuscript; and to copyright owners as acknowledged in figure captions for permission to reproduce those illustrations not yet in the public domain.

Appendix A

Formation and deployment of Quarrying Companies Royal Engineers in World War I

198 Quarrying Company was recruited largely from quarries in Leicestershire and Kent (REMLA file 70). It included men from Mountsorrel, a village *c.* 10 km north of Leicester in central England, whose quarry is currently the largest 'granite' quarry in Europe; 29 were photographed together in uniform and were two killed in an air raid on 25 September 1917 (mountsorrelarchive.org/Mountsorrel-Quarrymen-and-Quarry-Companies-RE.pdf, accessed 14 August 2015). The two graves lie in Les Baraques Military Cemetery at Sangatte, north of Boulogne. According to its War Diary, the company formed on 22 August 1916, left England on 24 August (rather than 26 August as recorded in the Clipstone Depot War Diary), arrived at Rouen in France on 26 August and reached Vallée Heureuse near Rinxent on 1 September. It quarried there until deployed from 30 June to 17 September 1917 to Beugin, a village *c.* 48 km SW of Lille (Fig. 1) in the (British) First Army area, returning to Beaulieu, Rinxent on 18 September 1917 for the rest of the war. Its founding captain, Ernest Matthews, was succeeded in command by F.B. Lawson and, in turn, by W.H. Hickman on 9 April 1918 when Lawson was transferred to an appointment under the Director of Roads.

199 Quarrying Company was formed on 22 August 1916, but the source of its recruits is not recorded in its War Diary (REMLA file 70). It left England on 30 August 1916, arrived at Rouen the next day and reached Beaulieu, Rinxent on 9 September 2016. No further movement is recorded in the War Diary. Initially commanded by Captain E.J. Stead, this company was also commanded for a time by F.B. Lawson and by Captain A. Meredith.

320 Quarrying Company was recruited from quarries in North Wales and Lancashire (REMLA file 84), notably Penmaenmawr and Trefor. According to contemporary records, the Welsh quarries made setts and the skills of the sett-makers were needed to help build vital roads in France and Flanders. Founded on 5 January 1917, the company left Buxton for France on 26 February 1917. It was commanded by Charles Stephen Darbishire (1886–1971), youngest son of Colonel Charles H. Darbishire, whose family controlled the Penmaenmawr and Welsh Granite Company, from which the UK government had earlier intended to buy stone for shipment to France. 'Stephen' Darbishire was commissioned as a temporary Second Lieutenant Royal Engineers on 10 January 1917 (*London*

Gazette Supplement, 2 February 1917, p. 1165) and promoted to be a temporary captain on 16 February 1917 (*London Gazette Supplement*, 6 March 1917, p. 2257) as the company formed. The Darbishire Papers and Darbishire Additional Papers held by Bangor University archives (www.archiveswales.org.uk) contain *c.* 100 of his letters, sent to his mother between his arrival in France and completion of service in January 1919. These and associated photographs offer an insight into camp life, employee welfare, quarrying activities, the day-to-day pursuits and views of a Quarrying Company officer during World War I and the effects it had on him and others around him. They indicate that his company was based near Rinxent, at Haut Banc.

321 Quarrying Company formed from 22 January 1917 on the Channel Island of Guernsey (REMLA file 84). The designated officers were all former quarry managers: Captain Herbert Brooks (Officer Commanding) had been with A. & F. Manuelle, Second Lieutenant G.H. Lock with Mowlems and Second Lieutenant Arthur Falla with his own company at Bordeaux (Anon 1990; Walton 2010). Most of the soldiers were mature family men, mainly of local or French origin, who had been living in the north of the island, where most of its quarries were situated (cf. fig. 13 of Rose 2005). The unit left Guernsey on 25 January 1917, trained from 26 January to 15 February 1917 at Buxton in England, arrived via Southampton at Le Havre in France on 17 February 1917, and arrived at Rinxent on 24 February 1917 (REMLA file 84). At Buxton, the unit was commanded by Second Lieutenant W. Middleton; from Buxton and in France up to 14 March 1917 by C.S. Darbishire (of 320) and from 15 March 1917 by Second Lieutenant Lock until the arrival of Captain Brooks on 14 April 1917. Some of its members were recognized at the end of September by a fellow Guernseyman (serving with the Royal Guernsey Light Infantry on its way to the Battle of Cambrai), working near Etaples (Walton 2005) – the port which became the principal depot and transit camp for the BEF – but the War Diary notes only that the company was based at High Cliff (Haut Banc) Camp until moving to Beaulieu Camp on 14 May 1917. It returned to Guernsey early in 1919 and its soldiers slipped quietly back into civilian life. Its last survivor was Percy Renouf, a 22-year-old clerk at Mowlem's quarry when 321 was formed, who had volunteered for the Royal Guernsey Light Infantry, but failed the prerequisite medical examination (Anon 1990). He recalled at the age of 95 that 321 served in France with companies raised in Ireland (presumably 327) and Wales (presumably 320, since also at Haut Banc); that a large number of German prisoners of war were also employed there, under guard; and that some Guernseymen served within 329, employed in the same quarry group.

322 Quarrying Company was formed on 16 February 1917, arrived in France at Le Havre on 17 March 1917 and arrived in the Rinxent area on 30 March 1917 (REMLA file 84). It was accommodated in camps at Beaulieu, Vallée Heureuse and Banc Noir, and worked quarries at Beaulieu, Haut Banc and Banc Noir.

323 Quarrying Company was formed on 20 February 1917, under command of Second Lieutenant Eric Cole until the arrival of Captain Arthur Seaton on 11 May 1917, when the unit moved to France, arriving at Beaulieu Camp on 16 May 1917 (REMLA file 84).

324 Quarrying Company was formed on 1 March 2017, left Buxton on 27 March 1917 and arrived in France via Southampton at Le Havre on 28 March 1917 (REMLA file 84). This Quarrying Company was the first (of two) not to be deployed to the Marquise-Rinxent group of quarries. On 1 April 1917 half of the company was sent to Calais. The other half was assigned to the (British) First Army, attached to Canadian Corps Troops, to work day and night until 8 April 1917 on roads in the forward area at Marqueffles in the Vimy Ridge area NE of Arras (Fig. 1) and quarries to the rear. The two halves rejoined on 28 April 1917, after the Battle of Vimy Ridge (9 12 April 1917), part of the opening phase of the British-led Battle of Arras, to work quarries at Rebreuve (30 km west of Arras) and Baraffles, with detachments prospecting for stone at Marqueffles. On and after 9 October 2017 its attachment was transferred to other Corps within the First Army. This company contained at least one quarryman from Mountsorrel, Sapper Cecil Porter, who died on 1 April 1917 soon after arrival in France. He is buried at Sainte Marie Cemetery, Le Havre (mountsorrelarchive.org/Mountsorrel-Quarrymen-and-Quarry-Companies-RE.pdf, accessed 14 August 2015).

325 Quarrying Company was formed on 9 March 1917 (REMLA file 84). It left England on 1 June 2017 and arrived at Beaulieu Camp in France on 6 June 1917 to work in the Beaulieu quarries, moving to the Vallée Heureuse quarries on 1 August 1917 and back to Beaulieu on 7 November 1917. However, on 24 March 1918 the company was moved from the Marquise-Rinxent area 'to undertake works of a defensive nature near Amiens' (see Fig. 1), from 26 to 29 March 1918 digging trenches in the Cretaceous chalk and making dug-outs as part of the Fifth Army. On 16 April 1918 it moved to Dennebroeucq to take over three quarries being worked by French civilians, to generate macadam for road works in the rear of the First Army, an area reassigned to the Fifth Army on 1 July 1918.

326 Quarrying Company was formed on 10 April 1917, moving to France on 23 May 1917 to quarry stone at Vallée Heureuse (REMLA file 84).

327 Quarrying Company was raised in May 1917 at The Curragh, near Dublin (De Santis 2002), the only Quarrying Company to be formed in Ireland (REMLA file 84). It left Ireland on 31 May 1917 and arrived in France on 2 July 1917. Its War Diary gives its base in June the following year as High Cliff (Haut Banc) Camp, consistent with the record by Percy Renouf of 321 Quarrying Company.

328 Quarrying Company was formed in France on 16 April 1917 and moved to quarries at Vallée Heureuse on 5 May 2017. It included a miner from Barnsley in Yorkshire, posted from the infantry on 17 April 1917 after service in the Mediterranean region with the 14th Yorkshire and Lancashire Battalion (the Barnsley Pals) – the National

Archives make use of his career as a teaching aid (http://nationalarchives.gov.uk/documents/education/allpals.pdf, accessed 14 August 2015). It was moved from the Line of Communication area into an army area on 24 March 1918 and during the period from 28 to 31 March 1918 the officer commanding (Captain Mann) was killed in action. The company returned to the Line of Communication on 12 April 1918 and Second Lieutenant G.W. Mitchell was promoted to command from 1 May 1918 (REMLA file 84).

329 Quarrying Company formed in France on 20 May 1917, by transfer of men from 62 various regiments (REMLA file 84), including some Guernseymen (Walton 2010). It was assigned to Beaulieu quarry initially and continued to quarry in the Rinxent area (the Guernseymen at Noir Banc) until deployed forward on 24 March 1918 into Fifth Army area. It worked on defences in the same region as 325 until 11 April 1918, before returning to the Marquise-Rinxent quarry group on 13 April 1918.

348 Quarrying Company was formed at the Road Troops Depot Royal Engineers, Talavera Barracks, at Aldershot in north Hampshire on 12 October 1917 (WO 95/4057; also REMLA file 84). The company embarked at Southampton on 25 October 1917 for service in France, departed on 27 October 1917 and disembarked at Le Havre on 28 October 1917, with four officers (Captain H.J. Herbert, Second Lieutenants (T.) H. Mutton, A. Sturges and W. Hancock, all Royal Engineers) and 262 of its 264 establishment of other ranks. Herbert's promotion from temporary second lieutenant to temporary captain was dated as from embarkation (*London Gazette Supplement*, 16 November 1917, p. 11 905). On 31 October 1917 the company reached Boulogne, to be placed under the Deputy Assistant Director Roads (Major Evans Royal Engineers), and to be responsible for the upkeep of roads and quarrying for stone in the Boulogne/Calais area, beginning roadwork on 2 November 2017. During December, the company headquarters was established in northern Boulogne at Wimille Mill and a pattern of working in three detachments developed: (1) working on roads within the BEF northern Line of Communication; (2) developing an existing quarry within the Ferques Inlier, at Blecquenecques, to the NW of the quarry group worked by other Quarrying Companies; and (3) developing a new quarry at Macquinghen, near La Capelle and Baincthun, closer to the eastern edge of Boulogne and so within Mesozoic rather than Paleozoic rocks. This pattern of work continued until 31 March 1919, when the War Diary ends, presumably because the company as a whole, or at least its key members, were demobilized the following month. Its officer commanding relinquished his commission, after a period of leave, on 17 June 1919, according to his personal file (WO 339/88296). The War Diary records relatively little apart from weather conditions (generally cold and/or wet during the winter of 1917–18), men reporting sick and occasional warnings of potential air attack; no details of the quarries or their output. Even the end of hostilities on 11 November 1918 is marked only by a terse entry: 'Dull morning, rained afternoon & evening. Armistice signed'.

References

Ager, D.V. & Wallace, P. 1967*a*. The environmental history of the Boulonnais, France. *Proceedings of the Geologists' Association*, **77**, 385–417.

Ager, D.V. & Wallace, P. 1967*b*. Easter field meeting in the Boulonnais, France, 7–13 April 1966. *Proceedings of the Geologists' Association*, **77**, 419–435.

Anon 1921. *The Work of the Royal Engineers in the European War, 1914–19: Water Supply*. Institution of Royal Engineers, Chatham.

Anon 1922*a*. *The Work of the Royal Engineers in the European War, 1914–19: Geological Work on the Western Front*. Institution of Royal Engineers, Chatham.

Anon 1922*b*. *The Work of the Royal Engineers in the European War, 1914–19: Military Mining*. Institution of Royal Engineers, Chatham.

Anon 1924. *The Work of the Royal Engineers in the European War, 1914–19: Work under the Director of Works (France)*. Institution of Royal Engineers, Chatham.

Anon 1990. Guernsey company of Sappers' big war effort. *Guernsey Evening Press and Star*, No. 526, 4 August 1990, 1–2.

Branagan, D. 2005. *T.W. Edgeworth David: a Life: Geologist, Adventurer, and Knight in the Old Brown Hat*. National Library of Australia, Canberra.

De Santis, E. 2002. Sapper William Forbes RE, www.reubique.com/298272.htm [last accessed 7 August 2015].

Douvillé, H., Élie de Beaumont, J.-B. & de Chancourtois, B. 1875. *Sheet 3 Boulogne. Carte géologique détaillée 1:80 000*. Imprimerie nationale, Paris.

Doyle, P. 1998. *Geology of the Western Front, 1914–1918*. Geologists' Association Guides, **61**. Geologists' Association, London.

Doyle, P., Bennett, M.R. & Cocks, F.M. 2000. Geology and warfare on the British sector of the Western Front 1914–18. *In*: Rose, E.P.F. & Nathanail, C.P. (eds) *Geology and Warfare: Examples of the Influence of Terrain and Geologists on Military Operations*. Geological Society, London, Miscellaneous Publications, 179–235.

Edleston, H.G. 1940. *General Report on Marquise Quarries*., Unpublished Report April 1940, within National Archives file MT 39/492.

Edmonds, J.E. 1922–25. *History of the Great War Based on Official Documents. Military Operations: France and Belgium 1914*. **2** vols. Macmillan, London (1915, 2 vols; 1916, 2 vols; 1917, 3 vols; 1918, 5 vols).

Grieve, W.G. & Newman, B. 1936. *Tunnellers: the Story of the Tunnelling Companies, Royal Engineers, During the World War*. Herbert Jenkins, London.

Holmes, R. 1999. *The Western Front*. BBC Worldwide, London.

Lawson, F.B. 1920. The quarrying operations of the British Army in France. *Transactions of the Institution of Mining Engineers*, **59**, 101–108.

Mansy, J.-L., Guennoc, P. *et al.* 2007. *Notice explicative de la feuille Marquise à 1/50 000. Carte géologique de la France*. BRGM Éditions, Orléans.

Mistiaen, B., Brice, D., Hubert, B.L.M. & Loones, C. 2011. Field trip 2: Classical Devonian and Carboniferous sites in the Ferques area, Boulonnais, northern France. *In*: Aretz, M. & Poty, E. (eds) *11th International*

Symposium on Fossil Cnidaria and Porifera – Liège, Belgium, 22–26 August 2011. Kölner Forum für Geologie und Paläontologie, Field Guides, **20**, 51–98.

Pritchard, H.L. (ed.) 1952*a*. *History of the Corps of Royal Engineers. Vol. V. The Home Front, France, Flanders and Italy in the First World War*. Institution of Royal Engineers, Chatham.

Pritchard, H.L. (ed.) 1952*b*. *History of the Corps of Royal Engineers. Vol. VI. Gallipoli, Macedonia, Egypt and Palestine*. Institution of Royal Engineers, Chatham.

Pruvost, P. & Pringle, J. 1924. A synopsis of the geology of the Boulonnais, including a correlation of the Mesozoic rocks with those of England, with report of an excursion 25 August to 1 September 1923. *Proceedings of the Geologists' Association*, **35**, 29–67.

Rose, E.P.F. 2005. Specialist maps of the Channel Islands prepared by German military geologists during the Second World War: German expertise deployed on British terrain. *The Cartographic Journal*, **42**, 111–136.

Rose, E.P.F. 2009. Water-supply maps for the Western Front (Belgium and Northern France) developed by British, German and American military geologists during World War I: pioneering studies in hydrogeology from trench warfare. *The Cartographic Journal*, **46**, 76–103.

Rose, E.P.F. 2012*a*. Groundwater as a military resource: pioneering British military well boring and hydrogeology in World War I. *In*: Rose, E.P.F. & Mather, J.D. (eds) *Military Aspects of Hydrogeology*. Geological Society, London, Special Publications, **362**, 49–72, https://doi.org/10.1144/SP362.4

Rose, E.P.F. 2012*b*. Aggregates pave the way to victory: work of Royal Engineers geologists and Quarrying Companies during World War II, especially for the liberation of Normandy. *In*: Hunger, E. & Walton, G. (eds) *Proceedings of the 16th Extractive Industry Geology Conference*. EIG Conferences, Charlbury, 24–40.

Rose, E.P.F. 2015. 'Abstract from Geology at the Western Front' by T. W. Edgeworth David. *Earth Sciences History*, **34**, 1–22.

Rose, E.P.F. & Rosenbaum, M.S. 2011. British geological maps that guided excavation of military dug-outs in Belgium during World War I. *Quarterly Journal of Engineering Geology and Hydrogeology*, **44**, 293–306, https://doi.org/10.1144/1470-9236/08-111

Sabine, P.A. 1991. Geologists at war: a forensic investigation in the field of war-time diplomacy. *Proceedings of the Geologists' Association*, **102**, 139–143.

Walton, L. 2005. Guernsey's lost generation. *Report and Transactions of the Société Guernesiaise*, **25**, 907–928.

Walton, L. 2010. The Guernsey quarrymen. *Channel Islands Great War Study Group Journal*, **35**, 26–30.

West, G. & Rose, E.P.F. 2005. Discussion of 'The first engineering geological publication in the UK?' by M. G. Culshaw. *Quarterly Journal of Engineering Geology and Hydrogeology*, **38**, 215–219, https://doi.org/10.1144/1470-9236/05-009

Westwell, I. 2010. *World War One: an Illustrated History*. Hermes House, London.

Quarrying Companies Royal Engineers in World War II: contributions to military infrastructure within the UK and to Allied forces during the North African, Italian and NW Europe campaigns

EDWARD P. F. ROSE

Department of Earth Sciences, Royal Holloway, University of London, Egham, Surrey TW20 0EX, UK

0000-0003-4182-6426

ted.rose@earth.oxon.org

Abstract: Eight Quarrying Companies Royal Engineers were raised during World War II, the first four in 1940. Unable to deploy to France as planned, these were used initially for bomb disposal, but from January 1941 companies 851 and 854 (succeeded in 1942 by 857) quarried stone for the construction of two military ports in western Scotland. In early to mid-1943, these companies were sent to support Allied forces in the North African Campaign and, from December 1943, the subsequent Italian Campaign. Other companies initially focused on ports and airfields associated with operations in the Atlantic: Company 853 quarried from 1941 to 1942 in Northern Ireland; Company 855 was deployed in 1941 to Poolewe in NW Scotland, before assignment in 1942–43 to Gibraltar; and companies 125 and 856 quarried from 1942 to 1943 on the Faroe Islands. Four companies were successively employed from 1942 to 1944 quarrying in Oxfordshire to aid the construction of a Central Ordnance Depot. Although three companies were operational in the Mediterranean region, the other five were united in 1944 to form a Quarry Group to support Allied operations in Normandy and the subsequent campaign across northern France, Belgium and into Germany. All eight received some geological guidance; all were disbanded before 1947.

Quarrying Companies of the Royal Engineers were an innovation of World War I. Thirteen were raised for service in France to support the Transportation Units of the British Expeditionary Force on the Western Front by providing crushed rock for road metal and railway ballast, principally from the Marquise group of quarries NE of Boulogne (Rose 2018*a*). Victory accomplished, all were disbanded soon after the close of hostilities.

The process by which Quarrying Companies Royal Engineers were raised anew in World War II has been described by Rose (2016*a*) from documents preserved in the UK's National Archives at Kew (file MT 39/492). Following the outbreak of war in 1939, the UK government's War Office proposed re-forming the Quarrying Companies to support the new British Expeditionary Force, which began deployment to France on 4 September 1939. However, the proposal was deferred when the French refused to assign quarries for British use. Plans changed in late March 1940, when the British Expeditionary Force had been progressively reinforced to the strength of an army, *c.* 400 000 men (Dear & Foot 1995). It was then planned to enlarge it further, by a second army, and to create up to 11 Quarrying Companies to provide 'stone' for roads and airfields. These plans were frustrated by the German attack westwards that began on 10 May 1940 (Dear & Foot 1995). The British Expeditionary Force was soon defeated; it was mostly evacuated between 26 May and 4 June 1940 and, on 22 June 1940, the French government was compelled to sign an armistice with Germany. The British Expeditionary Force, and so the purpose intended for the Quarrying Companies, no longer existed.

A halt was therefore called to their formation, except for four companies whose recruitment was too far advanced to be stopped (Table 1). The target at this time was for four officers and 168 other ranks per company (Fig. 1; Table 2). Of the 116 tradesmen (107 sappers plus nine higher ranks), 76 were quarrymen (rock drillers, getters/rockmen, plus powder men). Officers without previous military service had to be men with quarry management experience and older than 31 years of age; the maximum age for the other ranks was 55 years.

It was realized that each company would need modern equipment, but uncertain whether that equipment should be to maximize production from a central source (as in World War I), or to allow Quarrying Companies to be more mobile than previously, and so be able to produce aggregate closer to where it was actually needed. A scale of equipment was proposed early in the planning process

From: Rose, E. P. F., Ehlen, J. & Lawrence, U. L. (eds) 2019. *Military Aspects of Geology: Fortification, Excavation and Terrain Evaluation*. Geological Society, London, Special Publications, **473**, 173–200.
First published online March 6, 2018, https://doi.org/10.1144/SP473.2

Table 1. *Foundation of Quarrying Companies Royal Engineers 1939–45*

Company	Foundation date	Place of origin	Officer commanding
123	Not formed	Airdrie, Scotland	H.K. Symington
125	1 July 1940	Skipton, Yorkshire	J.E. Weatherill
851	31 July 1940	North Scotland	R.D. Forrester
852	Not formed	Lanark, Scotland	
853	17 July 1940	Cornwall, England	J. Setchell
854	24 July 1940	Central Wales	T.W. Owen
855	22 April 1941	Mostly from Wales	H.W. Morton
856	June 1941	Barton Stacey, Hampshire	E.R. Briggs
857	2 October 1942	From 851, at Garelochhead, Scotland	P.G.J. Gray
858	23 March 1943	Street, Somerset	W.H.R. Evans

After Rose (2012*a*).
For 855, 83 of its earliest members were transferred from 681 General Construction Company Royal Engineers, and although 856 and 858 were founded at specific training camps, personnel were also drawn largely from existing Royal Engineers units rather than directly from the civilian quarrying industry in particular geographical areas. Formation of 123 and 852 was in progress when halted in 1940.

(Table 3), but the War Diaries indicate that there was later flexibility in the items actually supplied.

The first four Quarrying Companies were thus founded during the Battle of Britain, the attempt by the German air force (the Luftwaffe) from mid-June to mid-September 1940 to gain aerial superiority over southern England (Dear & Foot 1995). This period was succeeded by the Blitz, which raged from August 1940 to May 1941, with sustained night bombing attacks against British cities. Consequently, the companies were at first restructured, trained and assigned to deal with the obvious urgent problem of abundant unexploded ordnance in southern England. Reduced in size to units initially of four officers and *c.* 120 other ranks, they began to quarry only at the start of 1941.

Details of Quarrying Company activities can be gleaned from their monthly War Diaries held in the UK's National Archives (Appendix A). There is a wide variation between units in the type and quantity of information recorded. Nevertheless, it is evident that, after their initial role in bomb disposal, the companies contributed significantly to military construction work in nine geographical regions, as well as to miscellaneous more minor assignments within the UK (Fig. 2 and Table 4).

The principal Quarrying Company activities and their links with military geologists are reviewed here to provide the first comprehensive account of these specialist units in the wartime British Army. To achieve overall consistency and balance and to facilitate the concluding analysis, sections on NW Africa and the Central Mediterranean (based entirely on unpublished data) are complemented by similar sections derived from separate more extensive regional accounts: for Scotland (Rose 2014*a*, 2018*c*), Northern Ireland (Rose 2018*c*), Gibraltar (Rose 2017), the Faroe Islands (Rose 2018*c*), England (Rose 2014*b*) and NW Europe (Rose 2012*a*). Contemporary English names are used for geographical localities and imperial units of measurement, for consistency with the historical sources, but with modern and metric equivalents where deemed appropriate (1 ton = 1.016 tonnes). Contemporary working practices are illustrated in various figures, although these would not be considered entirely safe by modern standards.

Scotland: Military Ports 1 and 2

The armistice of 22 June 1940 enabled Luftwaffe squadrons to be based in France. In consequence, ports on the southern and eastern coasts of Great Britain and their access routes came within easy striking distance and were in essence closed to ocean-going ships. As described more fully by Rose (2014*a*, 2018*c*), from which the following paragraphs have been mostly summarized, it was quickly decided to build two new ports and, for reasons of secrecy, that these should be located in remote areas of western Scotland: Military Port No. 1 in Faslane Bay near Garelochhead and Military Port No. 2 bordering the village of Cairnryan *c.* 100 km to the south (Fig. 3).

The geology of Scotland in general was well known (Wilson 1977; Trewin 2002) and geologists were not included in the port reconnaissance team. Geological maps and information would have been readily available from the Geological Survey of Great Britain. Coincidentally, surveys that culminated in the publication of the first official geological map of Scotland in 1836 had been initiated during the Napoleonic Wars and under military (Board of Ordnance) auspices (Rose 1996, 2009).

Military Port No. 1 lay near the southern boundary of the fault-bounded region now known as the Grampian Highland terrane. Even by 1940 this was

Fig. 1. 853 Quarrying Company Royal Engineers, formed in July 1940, photographed at Woodstock in Oxfordshire in June 1943. From the photographic albums of the late Major J. Setchell, Royal Engineers, courtesy of Peter Setchell.

Table 2. *Personnel proposed for each Quarrying Company Royal Engineers at 15 May 1940*

Captains	1
Subalterns (Lieutenants/Second Lieutenants)	3
Company Sergeant Major	1
Company Quarter-Master-Sergeant	1
Military Mechanist (Staff Sergeant)	1
Sergeants	2
Corporals	4
Drivers (including batmen and water duties)	47
Cooks	4
Sappers (from the following trades)	107
Blacksmiths	6
Carpenters and Joiners	2
Clerks	3
Driver Mechanics	6
Draughtsman (Mechanical)	1
Engine Artificers	6
Electrician	1
Fitters	4
Masons	2
Platelayers	2
Tinsmiths and Whitesmiths	1
Engine Hands	8
Quarrymen	76

Data from National Archives file MT 39/492.

recognized as a region dominated by the Dalradian, a complex of metamorphic rocks, commonly mica schists, altered sediments of latest Precambrian (Neoproterozoic) age. Military Port No. 2 lay near the northern boundary of the Southern Uplands terrane, by 1940 known to be a region dominated by highly folded, but only weakly metamorphosed sedimentary rocks (sandstones and mudstones) of Early Paleozoic (Ordovician and Silurian) age.

Port construction was normally a task undertaken by civilian firms, under contract to the military where appropriate, but this was difficult in wartime, especially in remote areas with minimal local labour. It was therefore decided that the two new ports would be built and operated entirely by military personnel. At the peak of construction, each port involved a workforce of *c.* 4000 troops. These included Royal Engineer units of many types, notably from January 1941 the Quarrying Companies to provide the considerable quantities of 'stone' required for construction or the improvement of roads and railways associated with the ports, and use as fill during wharf construction and dockside reclamation of land from the sea.

851 Quarrying Company, recruited initially from quarry workers in Scotland (Table 1), moved to Military Port No. 1 in January 1941 after bomb disposal duties in England (SE London). It was led by Captain D.A. Robertson, then 49 years of age and formerly a partner of Messrs Anderson & Robertson, Quarrymasters, of Newburgh in Fife. He had served as a lieutenant in the Royal Artillery in World War I. The unit worked several quarries near the port and, on 19 November 1941, the quarrymen were treated to 'a lecture on geology', an indication of local interest and expertise in this subject (Appendix A, WO 166/3927). Quarrying continued through 1942 (WO 166/8337), but the year ended with military training in Edinburgh prior to embarkation at Glasgow for Algiers (WO 175/703) and service in the North African Campaign.

854 Quarrying Company, recruited almost exclusively from Blaenau Ffestiniog in Wales (Table 1), also moved to Scotland in January 1941 after bomb disposal duties in England (near London) (Appendix A, WO 166/3929). Based primarily at Military

Table 3. *Key equipment recommended for a Quarrying Company Royal Engineers in 1939 for the production of 25–30 tons per hour of dry macadam*

Compressors, portable: ×2
Piping: 1000 ft × 2 inch, 1000 ft × 1 inch, 500 ft flexible hose
Drills: 4 × 4 inch drifters, six hand-drills various
Steels: 12 complete sets of 1.5 inch diameter hollow round steels
Drill sharpeners: 1 compressed-air operated, 1 oil-fired furnace
Excavators: 2 × 0.50 or 0.75 yard diesel engined on caterpillar tracks
Dump carts: 6 × 2 yard or 3 yard Muir Hill dumpers with balloon tyres
Quarry wagons: 100 × 1 cubic yard side tipping trucks for 24 inch gauge track
Tractors: 2 × diesel engined to suit 24 inch gauge track
Track: several hundred yards of 24 inch gauge, complete with points/crossings
Crushing plant, with subsidiary equipment: 2 × 25 inch by 13 inch jaw-breakers
Power plant: 80/100 horse power diesel engine
Loaders, portable hand conveyor: ×4
Lighting unit, diesel or petrol operated: ×1
Loose tool assortment: including shovels, picks, hammers, crowbars, explosives

Data from National Archives file MT 39/492.

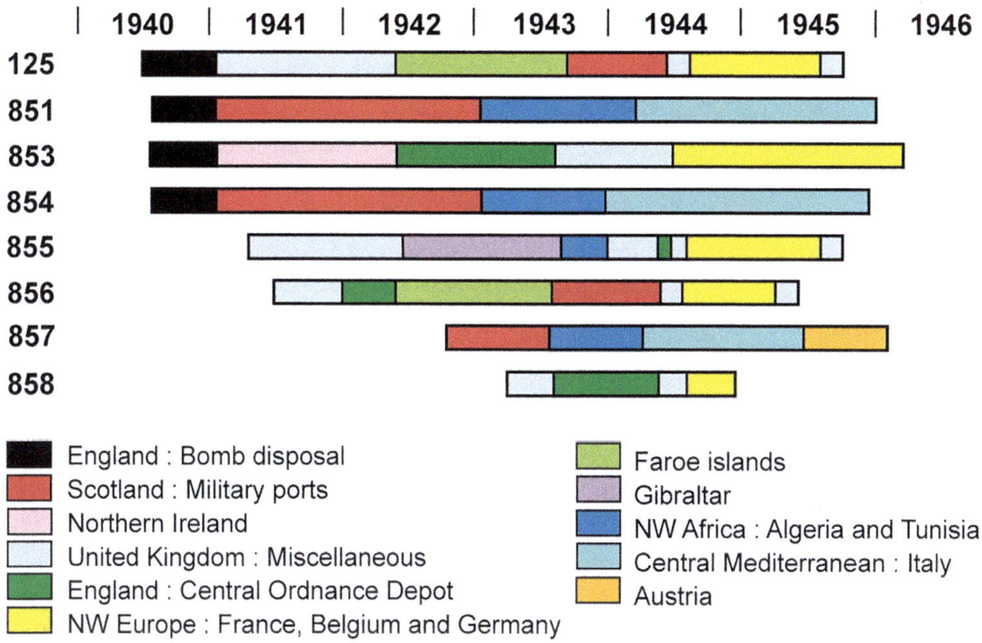

Fig. 2. Deployment of the eight Quarrying Companies Royal Engineers (numbered in left-hand column) during the period 1940–46, in terms of their main projects/regions (see Table 4). After Rose (2016*a*), courtesy of the Institution of Royal Engineers.

Port No. 2, it was led initially by Captain (later Major) T.W. Owen, an ex-serviceman and formerly Manager of the Cwt-y-Bugail Slate Quarries, until his death in hospital. He was succeeded in May 1942 by Captain A.G. Freebairn, formerly Manager at Colzium Quarry and a resident of Stirlingshire in Scotland; his only military experience had been as a corporal in his school cadet force. Quarrying continued through 1942 but, as for 851 Quarrying Company, the year ended with military training and, in January 1943, embarkation for Algiers (WO 175/704).

857 Quarrying Company was formed at Military Port No. 1 from October 1942 (Table 1), with core personnel from the home details of 851 – that is, men deemed by age or medical fitness to be unfit for service overseas (Appendix A, WO 166/8341). It was led by Captain P.G.T. Gray, promoted from 851 like A.G. Freebairn for 854, in May. An appendix to the November 1942 War Diary provides the most detailed record available for work by any of the Quarrying Companies at the ports. Work then consisted of the excavation of rock for the construction of rubble banks on the foreshore of Gare Loch to

Table 4. *Quarrying Company Royal Engineers tasks by region in World War II, with number of companies involved, aggregate of their total time of commitment and estimated total output*

Regional task	Personnel assigned	Output in tons
Scotland: Military Ports 1 and 2	5 companies = 6 years 3 months	500 000?
Central Mediterranean: Italy	3 companies = 4 years 10 months	Not known
NW Europe: France, Belgium and Germany	5 companies = 4 years 5 months	1502 209
NW Africa: Algeria and Tunisia	4 companies = 3 years 1 month	Not known
England: Central Ordnance Depot	4 companies = 2 years 6 months	280 000?
Faroe Islands	2 companies = 2 years 5 months	902 917
Northern Ireland	1 company = 1 year 4 months	136 232
Gibraltar	1 company = 1 year 2 months	2 000 000?
Austria	1 company = 8 months	Not known
UK: miscellaneous	5 companies = 5 years 3 months	150 000+

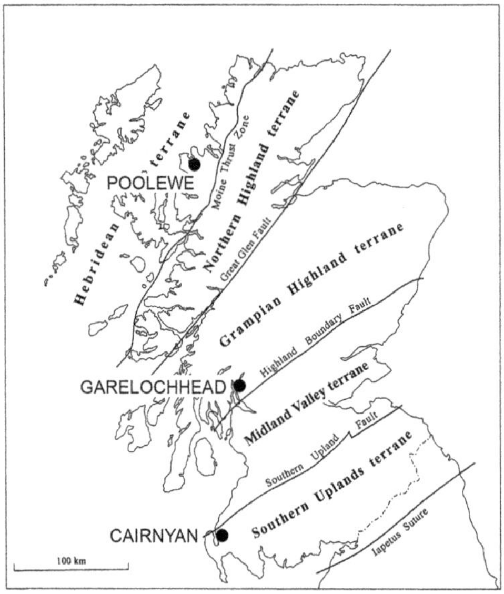

Fig. 3. Map showing positions of Cairnryan, Garelochhead and Poolewe relative to the five geological terranes that constitute present-day Scotland. From Rose (2018*c*), courtesy of Extractive Industry Geology Conferences Ltd.

carry road and rail access to the wharves of the port. Material was excavated from Louden Quarry in the adjacent hillside, working 'an undifferentiated Highland Schist or phyllite, with a 20 ft [6 m] band of the harder Greywacke or schistose-grit about ⅔ of the way along the quarry face' (Fig. 4). Quarrying continued in 1943, interrupted by military training from late March (WO 166/12263). However, in July 1943, 857 Quarrying Company embarked, like 851 and 854 before it, for Algiers.

856 Quarrying Company succeeded 857 in July 1943, on return from the Faroe Islands. One of the later units to form (Table 1), it had been led since August 1941 by Captain E.R. Briggs, one of the original three men nominated as officers for the 125 Quarrying Company founded by the Skipton Rock Company of Yorkshire. During 1943 the unit worked quarries at both Military Ports 1 and 2 (Appendix A, WO 166/12262), continuing until May 1944 (WO 171/1718), although with interruption for military training. In mid-May, 856 Quarrying Company moved to Ravenfield Camp, near Rotherham in Yorkshire, to train for deployment in Normandy.

125 Quarrying Company, recruited initially in Yorkshire (Table 1), had returned to northern England after bomb disposal service in southern England (Essex) for further training prior to mobilization for service in the Middle East. However, plans changed. The unit quarried in Yorkshire from May 1941 to May 1942 and then, until September 1943, in the Faroe Islands. From there it arrived in Scotland at Garelochhead on 4 September 1943 (Appendix A, WO 166/12057), still under the

Fig. 4. Sketch plan showing south end of construction work at Military Port 1 near Garelochhead, with Louden Quarry and areas (lettered A, B, C and D) proposed for filling from quarries adjacent to the bank already made (enclosing an area for proposed hydraulic filling). From Rose (2014*a*) and Appendix A to November Diary within WO 166/8341, by permission of The National Archives.

command of its founding officer, Captain J.E. Wetherill, formerly Secretary and Manager of the Skipton Rock Company and so, in civilian life, a senior colleague of 856's Captain Briggs; he was 36 years of age with 20 years of quarrying experience when commissioned in 1940. However, there was relatively little quarrying work left to do. Two quarries were worked until the following year, yielding stone for pitching, filling and crushed aggregate, but in relatively small quantities (WO 171/1551). The work was handed over to 178 Tunnelling Company Royal Engineers on 31 May 1944, and 125 moved (like 856) to Ravenfield Camp, in preparation for service in Normandy. Work by Quarrying Companies Royal Engineers on the military ports was thus ended.

The total output (Table 4) has been estimated at *c.* 500 000 tons (Rose 2018*c*). The two ports together consumed at least 165 000 tons of stone produced by 854; presumably a similar amount by 851; >80 000 tons produced by 856; presumably a similar amount by 857; and *c.* 7000 tons by 125. The rate of production was determined more by the demand for stone, varying diversions of company personnel for training and other engineering duties, and the prevailing weather conditions during the Scottish winter months, than by other factors: a company in Scotland could achieve an output of at least 4000 tons per week when required. The War Diaries of 857 indicate that high output was facilitated by use of Ruston Bucyrus 19 and 33 (face shovel) excavators, mechanical equipment of then modern design.

Military Port No. 1 was sufficiently complete to be operational from December 1942, when work by 851 and 854 finished. Military Port No. 2 was fully operational from July 1943, when 857 also left for service in North Africa. 856 and 125 helped to enhance the port facilities before they were deployed on operational service in Normandy. On completion, the ports were used primarily for secretive export rather than the import of troops and materiel, notably for the eastern part (Algeria) of the Allied invasion that initiated the North African Campaign.

NW Africa: Algeria and Tunisia

The North African Campaign began with largely American landings in French Morocco and Algeria on 8 November 1942, soon reinforced in the east (near Algiers, Fig. 5) by the British 1st Army (Dear & Foot 1995). It ended with the surrender of German and Italian forces in Tunisia on 13 May 1943. 851 and 854 Quarrying Companies left Scotland in January 1943, to be followed by 857 in July. On arrival in Algeria, the companies assisted in the development of the infrastructure necessary to support troops occupying the eastern region and confining prisoners of war there, and troops staging to campaigns in Sicily (from July) and mainland Italy (from September).

851 Quarrying Company arrived at Algiers on 17 January 1943 and, after military training, split into two parts, one at Philippeville (a major port 340 km to the east, renamed Skikda in 1962) and the other

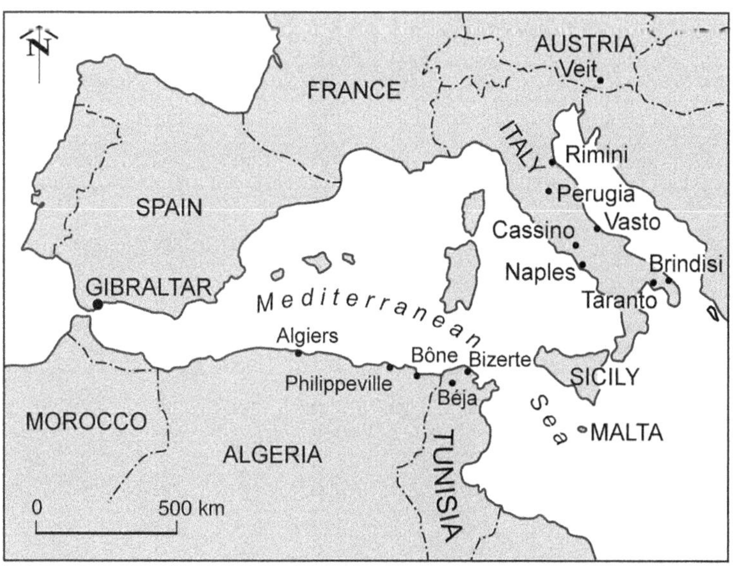

Fig. 5. Locality map for the western Mediterranean region showing the principal localities mentioned in the text.

at the coastal city of Bône (present day Annaba, *c.* 75 km further east), to begin quarrying in mid-February (Appendix A, WO 175/703). By 1 April 1943 there were detachments working two quarries at Bes-Bes (24 km to the south of Bône) and one at Saint-Antoine (*c.* 6 km south of Philippeville), both with men camped under canvas. The half-company at Bône soon operated quarries at 'Daroussa, Zerizer, Paviend, Pentièvre and Barral' and a gravel pit at Mondovi. By 24 April 1943 the 'Commanding Royal Engineer' at Bône was 'quite pleased with quarrying operations', but the officer commanding the company was not: 'bad quarries, poor quality stone, supervision by officers inadequate because of lack of number [only four in total], arab labour poor'. By 2 June 1943 Italian prisoners of war were employed to provide additional labour. Work continued in NE Algeria near Philippeville and Bône for the rest of the year (WO 169/10777) and seven quarries were still in operation near Bône as work drew to a close early in January 1944 (WO 170/1781).

854 Quarrying Company arrived at Algiers on 1 February 1943 (WO 175/704). After due preparation, it deployed to Souk Ahras (*c.* 90 km SSE of Bône), close to the border with Tunisia, beginning work in three quarries on 4 March 1943. In April–May, the unit moved its headquarters to Béja in Tunisia and was similarly busy, reporting in May 1943 that although all demands for stone were being met, it did not have enough men to work all the quarries then in operation. Demand for stone fluctuated thereafter, but from September 1943 there was firm demand to meet a need for airfield and access road construction, until the company embarked on 19 December 1943 at the Tunisian port of Bizerte, the northernmost city in Africa, for service in Italy (WO 169/10778).

857 Quarrying Company arrived in Algiers on 28 July 1943, but its quarrying plant only on 18 August 1943 (WO 169/10779). The unit worked two quarries *c.* 2 km north of Bizerte in Tunisia 'for road construction and other purposes' by October, while waiting for movement instructions. It moved west to Algiers in December 1943, taking over two former civilian quarries from 855 as this unit prepared to return to the UK: at Rivet, *c.* 20 km SE of Algiers (Figs 6 & 7) and Beni Messous, a suburb of the city of Algiers. It worked these quarries until movement

Fig. 6. Two officers with 855 Quarrying Company Royal Engineers at Rivet (near Meftah, SE of Algiers) in late 1943, discussing quarrying of stone for road repairs. From Rose (2016*a*); photo ET/1581, by permission of the Royal Engineers Museum, Library and Archive.

Fig. 7. 855 Quarrying Company Royal Engineers at Rivet (near Meftah, SE of Algiers) in late 1943, sorting and loading stone by hand rather than by means of an excavator. From Rose (2016*a*); photo ET/1586, by permission of the Royal Engineers Museum, Library and Archive.

orders were issued in early February 1944 and embarked at Algiers on 29 February 1944 (WO 170/1783).

855 Quarrying Company had seemingly arrived in Algeria in August 1943 having completed work in Gibraltar, but its War Diary for this period could not be traced in the National Archives (Appendix A). It is known, however, that 855 was at sea on 1 January 1944, *en route* to the UK from the port of Algiers (WO 171/1717), and there is photographic evidence additional to the record in 857's War Diary that it did indeed quarry rather than merely 'train' during the four to five months spent in Algeria (Figs 6 & 7).

Overall output (Table 4) could not be estimated from the information provided by the War Diaries. However, it is likely to have been considerable because the War Diaries indicate that many quarries were in operation and that companies were tasked primarily with quarrying as such rather than other duties. For 851, output is known to have averaged 25 000 m^3 (probably *c.* 50 000 tons) of stone per month for the second half of 1943 and so achieved *c.* 300 000 tons in total for that period alone.

The companies may well have received some military geological guidance during the latter part of their time in the region. No. 2 Boring Section Royal Engineers had arrived at Algiers on 5 March 1943 with an experienced geologist among its three officers: Lieutenant William Archibald Macfadyen MC (1893–1985) (Rose 2012*b*). His first action on arrival was to accompany the officer commanding his section on a visit to the local Geological Survey to obtain geological maps and information. In July he was detached from the section to serve with the Works Directorate at Allied Forces Headquarters, where he achieved the rank of 'temporary major' on 30 December 1943. The Headquarters thus had ready access to his geological expertise as well as to geological survey resources.

Central Mediterranean: Italy and Austria

Allied troops landed in Sicily on 10 July 1943 and, after fierce fighting, the last enemy forces were finally evacuated to the Italian mainland on 17 August 1943 (Dear & Foot 1995). The Italian

Campaign began with the amphibious invasion of southern Italy by the Allies from Sicily on 3 September 1943 and continued, despite the Italian government's announcement of national surrender on 8 September 1943, until German forces in Italy accepted a ceasefire on 2 May 1945.

Geological guidance to the Allied forces was provided from the start by Captain (later Major) John Victor Stephens (1897–1980), a veteran artillery officer of World War I recruited from the staff of the Geological Survey of Great Britain (Rose & Rosenbaum 1993; Rose & Clatworthy 2007*a*; Rose 2012*b*) (Fig. 8). He landed in Sicily on that campaign's D-Day as the Staff Officer (Geologist) for Allied troops. Initially, 'Steve' Stephens provided both 'engineer intelligence' about country occupied by the enemy and 'works' information to guide the development of a military infrastructure in already occupied ground. Later, he was assisted as the campaign progressed by another geologist staff officer: Major 'Bill' Macfadyen from the North African Campaign. Macfadyen was appointed to the Directorate of Works as the campaign progressed and so allowed Stephens to focus on his intelligence role. Their work was facilitated by pre-war geological maps of Italy reprinted by the Geographical Section General Staff of the British Army (Rose & Clatworthy 2007*b*, 2008*b*); from November 1943 by maps

Fig. 8. Captain (later Major) J.V. Stephens, Staff Officer (Geology) for the Allied armies in Italy 1943–45. Reproduced by permission of the British Geological Survey CP17/009.

and reports from the Geological Section of the Inter-Service Topographical Department based at Oxford in the UK (Rose 2016*b*); and by thematic maps and literature compiled by the Military Geology Unit of the United States Geological Survey (Nelson & Rose 2012).

Apart from 855, which moved back to England in January 1944, all Quarrying Companies in North Africa moved to Italy. The war establishment for Quarrying Companies in general changed again at this time: officers commanding as in 1940 were to be ranked as majors, with a captain as second-in-command. In Italy, the companies changed headquarter locations and the quarries they worked to keep pace with Allied troops fighting their way northwards. All companies were still in operation when the war in Europe ended on 8 May 1945.

854 Quarrying Company arrived in the port of Taranto (Fig. 5) on 26 December 1943 (Appendix A, WO 169/10778). It quickly moved north to Avelino, a town *c.* 40 km NE of Naples, to quarry there (18 284 tons in June 1944) until moved further north to Frosinone in August and to Perugia in November (WO 170/1782). Quarrying continued in this region during 1945, by March with detachments at Camucia, Arezzo, Cavi Chelli, Fossombrone, San Qurico, Castiglione dale Vale and Perugia (WO 170/5304). It remained in Italy to quarry after the end of hostilities, until disbanded at Foligno, a town 40 km SE of Perugia, on 5 December 1945.

857 Quarrying Company arrived, mostly via Naples, in March 1944, taking over quarries at Vairano (Fig. 9) and Tora, towns *c.* 60 km north of Naples (WO 170/1783). On 17 April 1944 work also began at Porchio quarry, 7 km south of Cassino. Output from all quarries reached a new high daily total on 5 May 1944 (960 tons) as Allied preparations were made for the Fourth (final) Battle of Monte Cassino, waged from 11 to 18 May 1944. The unit moved its headquarters north to Acquasparta in June and to Perugia in July to operate different quarries. An Adriatic offensive began in August 1944, with significant troop movements that required 'gravel' for route deviations and repair and the consequent continued movement of the headquarters, eventually to Rimini in NE Italy in October. A 'soft limestone' quarry was operated nearby at Ponte Verucchio, with rock crushing at various sites in the region nearby. Quarrying locations changed during early 1945 and the unit headquarters moved 50 km north to Forli at the end of March 1945 (WO 170/5308). It moved further north intermittently in the following months and into Austria in July 1945. Initially based at Veit, it subsequently worked quarries at several locations, finally disbanding at Haimberg on 31 January 1946 (WO 170/8153).

851 Quarrying Company also arrived in Italy via Naples in March 1944 (WO 170/1781). On 15

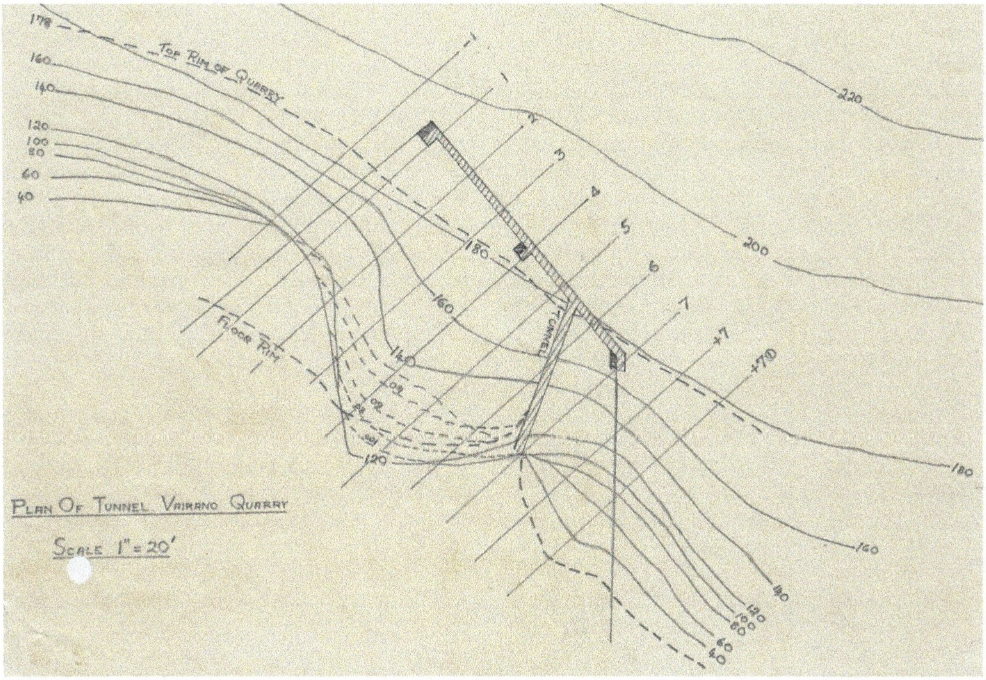

Fig. 9. Sketch plan of Vairano Quarry at the SW corner of Monte Sant'Angelo, worked in 1944 by 857 Quarrying Company Royal Engineers *c.* 500 m from the Vairano railhead. About '400–500 feet across', the quarry exposed 'a massive limestone, close grained and homogenous [with] no defined bedding planes, but three definite sets of cleavages [joints] are apparent … and divide the stone into very large blocks'. War Diaries recording quarrying work in Italy do not normally contain plans of quarries or notes on their geology, but this quarry, because of work involving the tunnel shown, is an exception. Reproduced by permission from The National Archives file WO 170/1783, from a report annexed to the War Diary for April.

March 1944 it assembled in the transit camp at Brindisi, the first time the whole company had been together since February 1943. Quarrying began at Mesagne nearby, to supply 400 tons per day of chippings for roads. Other quarries in the region were added and, in May, arrangements were made to supply stone for the construction of American airfields. In June the headquarters moved north to Lebda, near the east coast town of Vasto, to work quarries at Lebda and Poggio. The posting of one of his subalterns to a mine warfare course in September led to an exasperated note in the War Diary from the officer commanding: 'Can't afford this. My second experienced section officer to go on Field Coy [i.e. Company] work. Takes months to train a quarry officer.' By November practically all of the output from the Poggio quarry was destined for American use, but in December the company moved to a new base near Naples. A day off for everyone was allowed on 25 March 1945, the first for a long time (WO 170/5303). The company moved south in May 1945 to a region near Taranto to produce stone for airfield and prisoner of war camp construction, and north again to near Naples in September 1945. It was still quarrying there (for airfield construction) at the close of 1945, when its War Diaries come to an end.

As for these companies in North Africa, the total output cannot be estimated from data in the War Diaries (Table 4). However, it is again likely to have been considerable. 851 was tasked with producing 400 tons per day in March 1944; 857 achieved 960 tons per day in May; 854 achieved 18 284 tons (609 tons per day) in June. However, periods of intensive quarrying were punctuated by periods when both men and equipment were moving to new locations and so not producing stone.

Northern Ireland

As described more fully by Rose (2018*c*), 853 Quarrying Company arrived in Northern Ireland in late January 1941. It had been raised from the quarrying industry in Cornwall (Table 1) and was led from foundation by Captain (later Major/Lieutenant-Colonel) Joseph Setchell, formerly General Manager of the Old Delabole Slate Company. His supporting officers in the early years were J.G. McLeod and R.D. McLeod of Messrs Freeman & McLeod Ltd, Granite Quarry Owners, of Penrhyn (James being the Quarry Manager and Robert the Company Secretary and Estimator of the family firm). Like the other Quarrying Companies Royal Engineers founded in July 1940, 853 was assigned after initial training to a role in bomb disposal. Between the end of September 1940 and early January 1941 it had cleared *c.* 500 bombs within the southwestern counties of England, working mostly in or near the port of Bristol (Appendix A, WO 166/3928).

Military construction work in Northern Ireland during the war was largely defensive or in support of operations focused on the Battle of the Atlantic (Rose 2018*c*). Notably, between June and December 1940, seven new airfield sites were approved and 15 were soon constructed.

The geology of Northern Ireland was well known by 1940 (Herries Davies 1995; Mitchell 2004). A Geological Survey of Ireland had been founded in 1826 under military (Board of Ordnance) auspices, led by Captain J.W. Pringle and four other Royal Engineers officers (Rose 1996, 1999). Suspended in 1828, the Survey was reactivated in the 1830s under the command of Pringle's protégé Captain J.E. Portlock. Transferred to civilian control with the Geological Survey of Great Britain in 1845, it was led for the first year by another Royal Engineers officer, Captain Henry James. A Geological Survey of Northern Ireland remained part of the 'British' Geological Survey following the partition of Ireland in 1921.

February and March 1941 marked a period of adjustment to 853's new role, but by April the unit was in long-term operation in McBriar's Quarry near the village of Carrowdore on the Ards Peninsula of County Down and Morrow's Quarry at White Mountain in County Antrim (Fig. 10). The bedrock on the Ards Peninsula is predominantly greywacke sandstone, siltstone and mudstone of Silurian age, deformed, but affected by only low-grade metamorphism and equivalent to the Southern Uplands terrane of Scotland. It was therefore similar to the rock being quarried contemporaneously in Scotland for Military Port No. 2 at Cairnryan. By contrast, Figures 11 and 12 show that the quarry in the White Mountain region was excavated in very different rock, seemingly within a thick sequence of basaltic lavas, so part of the sequence now known as the Antrim Lava Group, Paleocene in age and part of the North Atlantic Igneous Province (Mitchell 2004).

Between April 1941 and May 1942, 853 Quarrying Company produced 136 232 tons of stone from these two areas (Table 4), on average nearly 10 000 tons per month (Rose 2018*c*, table 2), principally by hand rather than by mechanical excavators (Fig. 11). The company then returned to England between 22 May and 3 June 1942 to quarry in Oxfordshire.

Scotland and Gibraltar

855 Quarrying Company was founded in late April 1941 (Table 1) in England (Appendix A, WO 166/3930). Its 13-strong core staff was extended by the

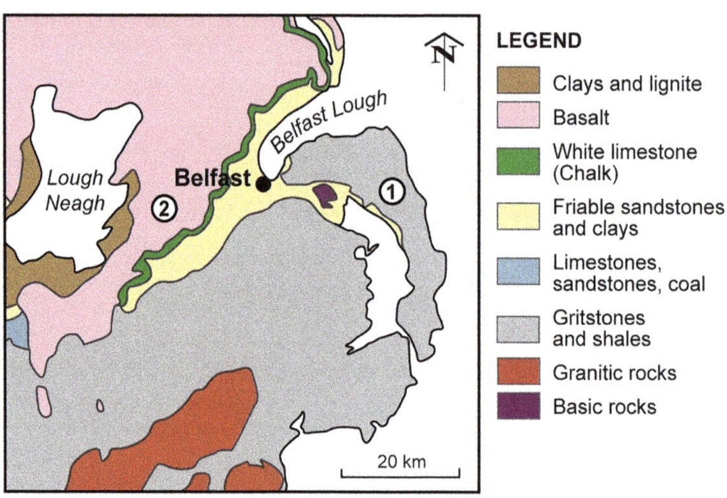

Fig. 10. Simplified geological map of the Belfast area, Northern Ireland showing major quarry locations worked by 853 Quarrying Company Royal Engineers: 1, Carrowdore; 2, White Mountain. Redrawn from part of a map accessed at www.northstonematerials.com/basic_geological_classification © Northstone (NI) Ltd. From Rose (2018*c*), courtesy of Extractive Industry Geology Conferences Ltd.

Fig. 11. 853 Quarrying Company Royal Engineers in Northern Ireland at White Mountain Quarry, Antrim. Lorries being loaded by hand rather than by an excavator. After Rose (2014*b*); from the photographic albums of the late Major J. Setchell, Royal Engineers, courtesy of Peter Setchell.

arrival on 5 May 1941 of 83 men from 681 General Construction Company Royal Engineers and later by transfers from infantry and other units.

Scotland

As described more fully by Rose (2018*c*), after initial training 855 deployed on 31 May 1941 to NW Scotland to Poolewe on Loch Ewe (Fig. 3). From February 1942 Loch Ewe was used as an assembly point, and later the main return point, for Arctic convoys supplying war materiel to the Soviet Union. From its date of formation, speedy deployment to Scotland and date of departure, it seems clear that 855 was raised to help provide an appropriate land-based infrastructure.

Three quarries were opened in June: (1) Riverside, to work 'close grained sandstone'; (2) Lochside, 'basalt with quartzite'; and (3) Point, 'heavy sandstone conglomerate'. This region formed part of the Hebridean terrane (Fig. 3), mostly early Precambrian (Archean) gneiss overlain by later Precambrian Torridonian sandstone, mapped as such by the Geological Survey of Scotland in the nineteenth century. No. 3 quarry was closed in September when Glen Quarry opened, near Gairloch to the SW, and No. 1 temporarily closed in October 1941..Quarrying continued at quarries 1 and 2 and at Glen Quarry through November 1941 to supply tarmac and broken stone for enhancement of the Gairloch to Achnasheen road. December was interrupted by military training, in preparation for assignment overseas, and all quarries were closed down on 31 December 1941. The unit left Poolewe for England on 1 January 1942 (WO 166/8340).

About 4250 tons had been generated in June 1941, increasing to 12 500 tons in July, material used as pitching, macadam and crushed stone

Fig. 12. 853 Quarrying Company Royal Engineers in Northern Ireland, at White Mountain Quarry, Antrim. A 'pneumatic drifter' making shot holes down from the top of the rock face. After Rose (2014*b*): from the photographic albums of the late Major J. Setchell, Royal Engineers, courtesy of Peter Setchell.

aggregate for the nearby Port A (Appendix A to WO 205/1190). Quarrying continued until the end of November 1941 and, although detailed output figures are not recorded, if the rate for July 1941 was maintained, a total of >60 000 tons would have been achieved (part of the minimum figure listed for 'UK miscellaneous' on Table 4).

Gibraltar

The establishment strength for Quarrying Companies by 1942 was four officers and *c.* 180 other ranks, and most of 855 Quarrying Company was deployed to work quarries in Yorkshire from January to May 1942. However, on 3 January 1942 orders were received to mobilize a detachment consisting of one officer plus 47 other ranks by 17 January 1942 for service overseas, evidently to form part of an Excavator Company Royal Engineers in Gibraltar at the western entrance to the Mediterranean Sea (Fig. 5) (as described by Rose 2017). On 5 June 1942 another detachment (the officer commanding the company, another officer and 71 other ranks) reported for posting overseas to join them. On 10 June 1942 the remnant of the company still within the UK was deployed to Wales, half to Sennybridge (Brecon) to work quarries in Old Red Sandstone, the other half to Castlemartin (Pembroke) to work a quarry in limestone (presumably Carboniferous limestone) to support Road Construction Companies Royal Engineers.

An airfield was under construction on the isthmus that linked the Gibraltar peninsula north to mainland Spain. This was important for military aircraft in transit between the UK, Malta and Egypt, and later for aircraft supporting the Allied invasion, first of North Africa (via French Morocco and Algeria in November 1942) and then of southern France (in August 1944). The geology of the famous Rock of Gibraltar had been studied by British pioneers of geology for *c.* 200 years by that time. Its principal features (a mass of Jurassic 'limestone' flanked by superficial scree breccias and sands of Quaternary age) were well established, although geologists working on the Rock in 1943 under military auspices were able to make significant refinements to earlier observations (Rose 2014*c*). Some of the rock fill needed to extend the runway westwards into the sea was provided by the spoil from tunnels then being excavated through the main ridge of the Rock (Rose 2018*b*), but most of it was quarried from scree breccia that flanked its sides (Fig. 13), initially by the Excavator Company (a unit almost unique within the Royal Engineers) and subsequently by 855 when assembled in sufficient strength to re-establish its identity as a separate unit.

Contemporary record describes the screes as a heterogeneous mixture of gravel, sand and limestone boulders, with a varying calcareous cement, the whole mass cut by fissures, both large and small. It proved impossible to judge from the surface what the rock would be like underneath. As a result of the mixed and fissured nature of the scree, much time was wasted clearing jammed drills and withdrawing casing. The bores often fell in between withdrawal and charging. Also, it proved impossible to judge the correct charge for a given burden. In most cases either the charge was dissipated in fissures, or the spoil was thrown all over the vicinity, causing a great deal of damage and much alarm. The difficulty of routine quarrying by blasting therefore led, successfully, to the use of water jets rather than explosives to bring down the scree, by adapting the pumps supplied originally to operate static flame throwers.

Quarrymen worked long hours until near the end of July 1943, but the pressure eased in August. The airfield was essentially complete. Extension to the length of 1550 yards (1417 m) initially planned had required *c.* 750 000 cubic yards (*c.* 573 400 m^3) of fill; extension to its final length of 1800 yards (1646 m) had increased this to 1.5×10^6 cubic yards (*c.* 1 147 000 m^3). Of this, *c.* 0.5×10^6 cubic yards had been generated as tunnel spoil (table 2 of Rose 2017) and the rest by quarrying. Using the conversion factor adopted in the War Diary of the Chief Engineer Gibraltar (WO 176/162), 1 cubic yard = *c.* 2 tons, the quarrying output is therefore estimated at *c.* 2×10^6 tons (Table 4). Work completed, 855 Quarrying

Fig. 13. The 400 m high north face of the Rock of Gibraltar, of Early Jurassic limestone and dolomite, viewed south from the current airport terminal in 2016. This shows (in the foreground) part of the airfield and (left-central in the background, rising above the modern buildings) the remaining part of a massive Quaternary scree breccia slope quarried during World War II to provide fill for the extension of the airfield westwards from the coast.

Company moved across the Mediterranean Sea to North Africa in August 1943.

Faroe Islands

The Faroe Islands, a Danish colony since 1380 and consisting of *c.* 22 islands in the North Atlantic Ocean halfway between Scotland and Iceland, were occupied by British troops on 12 April 1940 following the German occupation of Denmark (Dear & Foot 1995). Between 1942 and 1943, as the maritime Battle of the Atlantic reached its peak and Arctic convoys came under sustained attacks, Royal Engineers built the only airport on the archipelago and two Quarrying Companies Royal Engineers were deployed to aid this and associated construction work, as described more fully by Rose (2018*c*).

It had long been known that the islands were composed almost entirely of a thick sequence of basaltic lavas and tuffs. Geikie (1881) had published an account in English that described the main characteristics of the geology and island topography; the Danish geological map available pre-war was less enlightening (Fig. 14). A more detailed account of the geology was published long after the war had ended (Noe-Nygaard 1962), confirming that the basalts of the Faroes and Northern Ireland formed part of the same North Atlantic Igneous Province. King (2015) later summarized a more precise correlation.

125 Quarrying Company arrived at Sørvágur on Vága (Vaaga, Fig. 14) on 13 May 1942 and started working two quarries nearby (numbered 2 and 3 in its War Diary) on 18 May 1942 (Appendix A, WO 166/8081). 856 arrived soon afterwards, on 28 May 1942. By early June seven quarries were in operation to provide stone (WO 176/89): those numbered 1 and 2 (worked by 125) for construction at a flying boat base; 3 (by 856) for a sea wall; 4 (by 125) for an airfield runway; 5 (by 125) for a petrol dump; 6 (by 856) for a camp area; and 7 (by 856) for a bay-head jetty. Ultimately, only quarries 2, 6 and 7 were in use, as work came to an end in August 1943 (WO 166/12057).

125's priority task was the construction of an airfield, completed but for its finishing touches in

Fig. 14. Faroe Islands inset to the Geological Map of Denmark, scale 1:500 000, by J.P.J. Ravn, published in 1922 by the Generalstabens topografiske Afdeling, Kjøbenhavn (*Carte géologique du Danemark. Les formations préquaternaires*). The islands are shown uniformly coloured as basalt. From Rose (2018*c*).

July 1943. Work involved levelling ground to within 3 or 9 inches (75 or 225 mm) of finished head (248 000 square yards, *c.* 207 400 m^2); excavating rock to create a level surface (451 091 tons); and quarrying for fill (505 091 tons). Two runways were constructed, although No. 1 was abandoned before completion. No. 1 runway, with seven hard standings, taxiing tracks, plus hangar floor and apron, involved 68 000 square yards (*c.* 57 000 m^2) of levelling and 38 615 tons of excavation and filling. No. 2 runway involved 180 000 square yards (150 000 m^2) of levelling, 412 476 tons of excavation, 466 476 tons of filling and 1250 × 1 × 1 yards of drain construction. 125 also worked on the construction of parking areas and a slipway at a flying boat base; the excavation of pits for siting heavy anti-aircraft guns and coastal artillery; the excavation of footings for wireless masts and of pipe channels for a petrol store; and the supply of crushed stone for miscellaneous other tasks.

Output for the two companies produced a total of 902 917 tons of stone during their 15 months of operation (Table 4). 125 produced 596 499 tons between beginning work on 13 May 1942 and ending operations on 31 August 1943. 856 produced 306 418 tons before its earlier departure in July 1943, stone supplied principally for roadworks. Quarrying in total consumed *c.* 40 tons of explosive (mostly gelignite or ammonal), 100 miles (*c.* 160 km) of cord, 8.5 miles (*c.* 13.5 km) of safety fuse, 48 000 detonators and 7 tons of drill steel. Much of the work was achieved during the first 12 months, under 'unbelievably bad weather conditions', by working long hours. 'A splendid effort by all ranks.' On 2 September, 125 (four officers and 161 other ranks) sailed for Scotland from the Faroes' capital town Thorshavn on the island of Streymoy (Strømø, Fig. 14) to work on Scottish military ports.

England: the Central Ordnance Depot, Bicester

Four Quarrying Companies were deployed successively to quarry stone for the construction of a new Central Ordnance Depot at Bicester in Oxfordshire, as described by Rose (2014*b*). The depot was to supply materiel (primarily soft-skinned vehicles and small arms) for the Allied landings in Normandy as well as other overseas campaigns supported from the UK. The companies worked Mid-Jurassic limestone at the Kirtlington (Fig. 15) and Gibraltar (Inn) quarries near Woodstock, taken over by military authority from their civilian owners. Many similar quarries existed in this area at the time, close to the university city of Oxford, so the geology was well known (cf. Arkell 1947).

856 Quarrying Company was the first to arrive, but precisely when is not known. Its War Diaries prior to that beginning in September 1943 could not be found in the National Archives (Appendix A). However, it was formed in June 1941 at Barton Stacey (a training camp of the Royal Engineers near Winchester in Hampshire) and, after staffing and training, was assigned from July 1941 to quarrying and airfield construction tasks near Salisbury (Wiltshire), Bournemouth (then Hampshire, now Dorset) and Woodstock (Oxfordshire). The time of formation of the unit and the completion of its first assignments implies that 856 could not have arrived at Woodstock until late 1941 or early 1942. Certainly it left Woodstock in May 1942 for deployment to the Faroe Islands.

853 Quarrying Company succeeded 856 at Woodstock, arriving from Northern Ireland between 28 May and 3 June 1942 (Appendix A, WO 166/8338). Its monthly War Diaries, with their appendices, provide an exceptionally meticulous record of activities and quarry condition from the date of arrival. The company faced several immediate

Fig. 15. Kirtlington Quarry in 1923, with the informal lithostratigraphic units long recognized locally in the Middle Jurassic indicated at the left: Lower Cornbrash, Forest Marble and Great Oolite (=White Limestone) plus sub-units. From Rose (2014*b*); reproduced from a photograph by C.J. Bayzand, by permission of the Oxford University Museum of Natural History.

problems. 856, on departure, had taken all its tools and equipment with it, whereas 853 had been required to leave its equivalent materiel in Northern Ireland. 853 needed to be speedily re-equipped. Previous production had achieved 10 000 tons of stone per month, but although 853 was only tasked for 4000 tons per month, so few lorries were available to transport stone to Bicester that even this smaller tonnage could not be delivered. Moreover, both Kirtlington and Gibraltar quarries were reported as being in very poor working condition, Kirtlington especially carrying 'a large proportion of useless clay and overburden'. The problems were overcome and an instruction issued on 12 October 1942 that the company was to focus on quarrying work for 6.5 days per week (men being allowed Saturday afternoons off), discontinuing the practice of military training on Sunday mornings, indicated the 'extreme urgency of the job'. Captain Setchell, who had commanded 853 from its foundation in 1940, relinquished command on 11 February 1943 on posting (WO 166/12261). The new officer commanding maintained production until June 1943, but on 30 July 1943 the company moved on for extensive military training.

858 Quarrying Company arrived in succession, between 27 and 31 July 1943, and began work at Gibraltar Quarry on the latter date (WO 166/12264). 858 was then a new unit, the last of the Quarrying Companies to be formed (Table 1). It had been founded on 23 March 1943 and Woodstock marked its first deployment on completion of training. War Diaries completed for 858 are minimal in their content compared with those formerly compiled for 853 by Captain Setchell. They record that the monthly total outputs from Gibraltar Quarry consisted of 9447 cubic yards (7223 m^3) for August, 9851 for September, 8543 for October, 5366 for November, 6737 for December and 6572 for January 1944, 29 928 cubic yards (22 882 m^3) in total, but give no detail of the different qualities of stone produced or the reasons for the variation in output. From January 1944, officers commanding Quarrying Companies were ranked as majors rather than captains, but the then officer commanding was not promoted. Rather, he remained with 858 as second-in-command and a new officer commanding took up appointment on 31 January 1944: Acting Major J. Setchell, the former officer commanding of 853. Major Setchell introduced very firm control of activities and detailed records. Monthly total outputs achieved 8560 cubic yards for February, 8328 for March and 7337 for April, 24 225 cubic yards (*c.* 18 521 m^3) overall. However, from 15 April 1944, *c.* 25 men were withdrawn daily from quarrying activities to undergo military training and between 13 and 15 May 1944, 858 moved to Ravenfield Camp in Yorkshire to prepare for service in Normandy.

855 Quarrying Company had been founded on 22 April 1941 (Table 1). It had quarried initially at Poolewe in Scotland, but then mostly been sent to Gibraltar, moving on to North Africa and embarking for return to England at Algiers in December 1943. 855 disembarked at Liverpool on 4 January 1944, went on leave for most of that month and then began a period of military training, initially at Staindrop in County Durham (WO 171/1717). The unit eventually moved to Oxford on 4 May 1944; moved to a tented camp at Bicester on 9 May 1944; took over works at Gibraltar Quarry from 858 on 10 May 1944; and moved into accommodation at Woodstock on 15 May 1944. Work proceeded normally at Gibraltar Quarry, although *c.* 20% of

Table 5. *Stone produced at Gibraltar Quarry (per month, in cubic yards by grade), according to War Diary monthly appendices*

Month (1942–43)	Filling	2 inch/ 4 inch	Pitching	Total
June	78	138	117	333
July	515	254	575	1344
August	739	665	1062	2466
September	1115	1385	2245	4745
October	2729	1818	2994	5526*
November	2849	1450	2353	6352*
December	2095.5	2243.5	4642.5	8981.5
January	5592.5	2866.5	5811	14 270
February	1825	2990	5601	10 416
March		2618	6291	8909
April		1878	4058	5936
May		2285	4801	8718*
June	?	?	?	?
July		2953.5	6519	9472.5
Total	17 538	23 544.5	47 069.5	87 469*

*Inconsistent totals.

the personnel were employed in early June 1944 on military training until 14 June 1944. The quarry and stores were then handed over to 170 Tunnelling Company Royal Engineers and 855 moved on 15 June 1944 to join 858 at Ravenfield. Work by the Quarrying Companies for Central Ordnance Depot Bicester was finally at an end.

Overall output is estimated at *c.* 280 000 tons (Table 4). The War Diary for 856 Quarrying Company could not be traced in the National Archives, but that for 855, its successor in operation from the end of May 1944, refers to production during the previous month achieving 10 000 tons. Whether that was achieved also in the month or few months before that is not known, but it seems likely that 856 may have achieved a total output of at least 20 000 tons, perhaps 30 000 tons. 853 achieved a total output of >87 000 cubic yards from Gibraltar Quarry (Table 5), probably *c.* 96 000 cubic yards if output in June 1943 matched May or July, and >17 000 cubic yards from Kirtlington (Table 6): >100 000 cubic yards so *c.* 200 000 tons in total. 858 Quarrying Company achieved 29 928 tons by the end of 1943 and a further 24 225 tons in the first three months of 1944, so >54 000 tons in total and more (when not on military training) in April. 855 took over quarrying duties for May 1944, but ceased operation when mobilized for Normandy in June 1944, without recording details of output, which presumably did not exceed 10 000 tons.

Table 6. *Stone produced at Kirtlington Quarry (per month, in cubic yards by grade), according to War Diary monthly appendices*

Month (1942)	Filling	2 inch/ 4 inch	Pitching	Total
June	276	54	–	330
July	1628	235	797	2660
August	1228	905	936	3069
September	212	909	1344	2465
October	569	1336	2532	4437
November	647	962	2145	3754
December	45	119	369	533
Total	4605	4520	8123	17 248

NW Europe: the Quarry Group Royal Engineers

During 1944, five Quarrying Companies were united to form the Quarry Group Royal Engineers, to support 21st Army Group following the Allied landings in Normandy on D-Day, 6 June 1944 (Williams 1950; Rose & Pareyn 1995, 2003; Rose 2012*a*).

853 had moved from Woodstock in August 1943 to a Royal Engineers Training Camp for intensive training until early September. In early October 1943 it moved again to a camp at Otterburn, near Newcastle-on-Tyne, and there spent 2.5 days per week on general military and engineering training, the rest of the working time quarrying stone, primarily for the construction of gun spurs and ammunition dumps. Military training culminated in a six-week period away in February–March 1944, followed by more quarrying in late March through most of April 1944. Thereafter the company was in transit camps at Selkirk, Scotland, and Tonbridge, Kent, before embarking for Normandy on 17 June 1944.

The other four companies staged through Ravenfield Camp, near Rotherham, Yorkshire. 858 moved there from Woodstock on 10 May 1944, before staging southwards and finally embarking on 24 June 1944. 856 had been based in Scotland, quarrying for Military Ports 1 and 2, but by 16 May 1944 it too had completed a move to Ravenfield, before staging south in June and embarking on 10 July 1944. 125 had also been quarrying for the military ports in Scotland, until it too moved to Ravenfield, arriving on 1 June 1944. Despite a warning order on 29 June 1944 to prepare to move at 48 hours' notice, it did not move south until mid-July and embarked on 4 August 1944. 855 moved from Woodstock to Ravenfield on 15 June 1944, beginning its subsequent move south on 22 July 1944 and embarking on 31 July 1944.

All five of these companies had by this time gained long experience in their specialist military role and had benefited from periods of additional military training in the six months prior to embarkation. 853, scheduled to be the first Quarrying Company in action, was temporarily put under direct command of the British 2nd Army shortly after a Quarry Group 'formation' conference on 15 March 1944 and did not join the Group as such until early July. The other four companies were all routed through Ravenfield Camp, where their preparatory equipment and training was directed by the Group's commanding officer, Lieutenant-Colonel A.R.O. Williams, Royal Engineers, a man who had already distinguished himself in command of Tunnelling Companies Royal Engineers on Gibraltar (Rose 2018*b*).

Each company then consisted of four officers and 173 other ranks (52 quarrymen, 50 drivers and sundry tradesmen). With its headquarters, the Group thus had an established initial strength of just over 900 troops of all ranks. However, each company was to be supplemented in theatre by *c.* 100 unskilled men (from the Pioneer Corps, prisoners of war or local civilians). Based on operational experience earlier in the war, a standardized scale of equipment was at last agreed (Table 7).

As part of his own preparation, 'Bill' Williams was issued with a set of Technical Intelligence Service Local Resources map sheets, which later proved to be rather inaccurate with regard to quarry intelligence (although such intelligence had been gathered since at least mid-1943; Rose, this volume, in press). He was, however, put in touch with Major F.W. Shotton, Royal Engineers as Staff Officer (Geology) at the headquarters of the 21st Army Group (Rose *et al.* 2006; Rose & Clatworthy 2008*a*), from whom he received a set of French geological maps at scale of 1:80 000 reprinted by the Geographical Section General Staff, together with what he later acknowledged was much useful information regarding rock types in the beach-head area. Indeed, after Victory he concluded, 'Field reconnaissance yielded much useful information of a detailed nature but for an overall picture geological maps proved invaluable' (Williams 1950, p. 223).

Table 7. *Scale of major items of plant and equipment per company adopted by the Quarry Group Royal Engineers for service within Normandy and the subsequent NW Europe Campaign*

Excavators, ⅝ or ¾ cubic yard	
Base machines	3
Face shovel equipment	3
Dragline equipment	1
Angle dozer (7 ft blade)	1
Compressors, mobile	
200 ft^3 capacity	7
100 ft^3 capacity	1
Rock drills	
Drifters, 4 inch	3
Jackhammers, 50/60 lb	3
Jackhammers, 25/35 lb	8
Drill sharpener and furnace	1
Drill steels	
1¼ inch, round	3 tons
1 inch, hexagonal	6 tons
⅞ inch, hexagonal	6 tons
Crushing and grinding units, mobile	
10 tons per hour	5
25 tons per hour (Iowa)	1
Vehicles	
Dumpers, 3 cubic yard (Muirhill)	12
Tippers, 3 ton	20
Trucks, 15 cwt	2
Van, utility	1

After Williams (1950) and Anon. (1957).

Normandy

Components of 853 arrived at Courseulles (to the east of Arromanches, Fig. 16) between 18 and 22 June 1944; the first half of 856 arrived a few miles to the west of this, at Le Hamel, between 29 and 30 June 1944 and *c.* 40% of 858, plus 855 and 125, all arrived via Arromanches on 29 July, 31 July and 6–7 August 1944, respectively. The leading elements were thus in theatre 12 days after D-Day, 6 June 1944, during the build-up rather than the assault phase of the campaign, and the final three companies only after the break-out from the beach-head had begun on 25 July 1944. All arrived safely, except for the main part of 858, from which three men were killed and five injured when the ship carrying its heavy equipment was torpedoed *en route*. The ship was beached in England to prevent it sinking and it was many weeks before the men of 858 were re-united with their own heavy equipment.

Stone was required for road widening and repair and for the construction of new roads. On D-Day, >8000 vehicles landed in Normandy and by 50 days later there were >150 000 in an area little more than 30 km broad by 16 km deep, generating intensive road use. The weight of some of the vehicles and the passage of those equipped with caterpillar tracks proved destructive to road surfaces in a region that had previously experienced only relatively light traffic; road maintenance (as expected) became a serious problem. Much of the land within the developing beach-head area was put to military use, not only for airfields, but also for field hospitals, store depots, vehicle and gun parks, most of which required hard standings and so added to the military demand for aggregate.

The focus of the Quarry Group's activities lay within the British sector of operations. Initially, it supported the British 2nd Army. Later it supported the 21st Army Group as a whole as, from late July, this developed into an Anglo-Canadian rather than

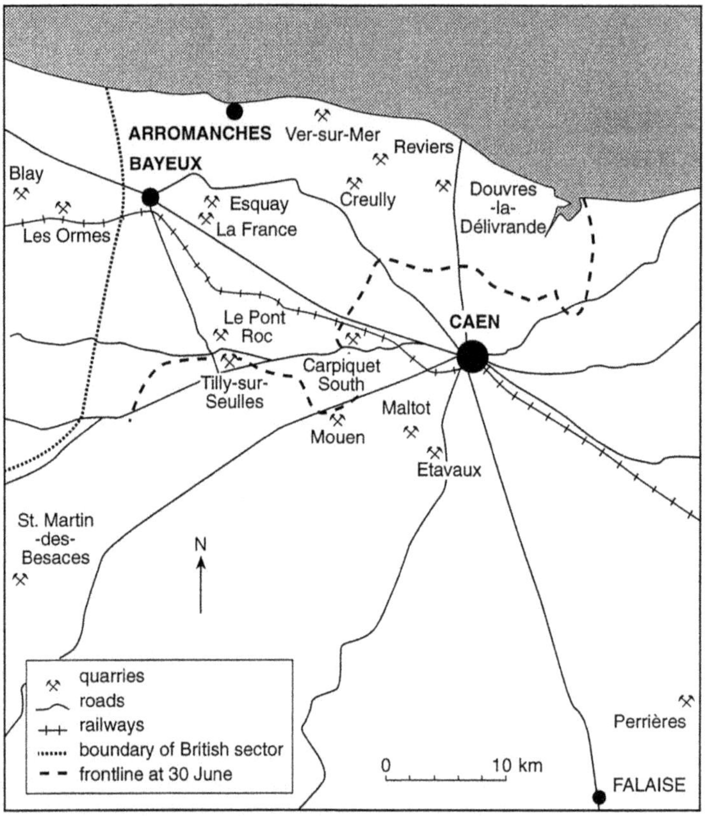

Fig. 16. Map showing positions of main roads, railways and British military quarries in Normandy during 1944. From Rose & Pareyn (1995), courtesy of Blackwell-Wiley Publications.

Anglo-American force, but there was close British–American cooperation as appropriate.

Initially, production was started in existing (but partly disused) quarries at Creully, Esquay, Douvres, Reviers, Carpiquet, Blay and Tilly-sur-Seulles (Fig. 16). Later, a new quarry was opened at Ver-sur-Mer. Apart from two quarries near Esquay, in Quaternary sands or gravels, all these quarries were within Jurassic limestone. They produced a rock that was weak and generally unsuitable for road repair, but within the restrictions imposed by the frontline and British–American sector boundary, there was no alternative. At Creully and comparable sites the rock quarried was a Mid-Jurassic limestone, worked previously for building or rough walling stone and similar to the Mid-Jurassic limestones worked in England at the Gibraltar and Kirtington quarries. These strata dipped gently to the NE, exposing Early Jurassic (Liassic) strata further inland, towards the SW margin of the Jurassic outcrop. Quarries were developed at Blay, Les Ormes and Tilly-sur-Seulles, working well-cemented limestone beds 0.10–0.35 m thick alternating with thinner marls, the limestones having previously been worked to burn for lime production.

Good roadstone became available only with the capture of Mouen, SW of Caen, towards the end of July. This provided access to strong, steeply dipping Cambrian quartzites flanking a NW–SE prolongation of the Precambrian Armorican basement.

'Bill' Williams was diligent in visiting sites during operation and in reconnaissance for new sites, sometimes accompanied by 'Fred' Shotton to provide geological guidance. Weak rock quarries were set aside as those yielding stronger rocks became accessible with the progress of battle. Creully, Douvres, Reviers, Ver-sur-Mer, Blay and Tilly were successively abandoned, and the quartzite or well-cemented Paleozoic sandstone quarries at Perrières, Etavaux, Maltot and St Martin-des-Besaces were brought to production. The choice of these particular quarries from the many that became potentially available with the expansion of the beach-head was dictated primarily by considerations of their proximity to the areas most needing stone and the rapidity with which production could be started.

The Perrières quarry was deemed the best, with good road and rail access. Production began at the end of August and continued until the end of the war in May 1945, supplying British and American forces with >70 000 tons of stone, nearly all crushed, from this one source. However, as the Allied armies advanced, the French quarries were replaced as sources of aggregate by quarries in Belgium and Germany. East of Caen, the bedrock in France is principally of Cretaceous sediments, mostly chalks, capped in places with a significant thickness of clay with flints, with some thick deposits of Quaternary gravel in the major river valleys. Otherwise there was no potential source of good aggregate other than the urban areas reduced to rubble by aerial or artillery bombardment. The Marquise group of quarries near Boulogne worked an inlier of older, stronger rocks, but although these were important to British troops in World War I (Rose 2018*a*), they were quickly bypassed in World War II.

Belgium and Germany

The Normandy phase of quarrying came to an end on 15 September 1944. Allied troops had broken out of the extended beach-head area and advanced >600 km eastwards across northern France between 25 July and 7 September 1944 to liberate Belgium. They consolidated their position in the Low Countries during the winter months and, between 16 September and 30 November 1944, the Quarry Group extended its activities to Belgium, at the same time maintaining some production in the Caen–Bayeux–Falaise region to meet demands for aggregate in the Rear Maintenance Area of the 21st Army Group. By early November 1944, 125 was quarrying at Etavaux; 853 at Mouen; 855 at Blay, St Martin and La France; 856 at Perrières, Carpiquet and Maltot; and 858 at Fontaine, Cocherel and Rouen (in urban rubble), east of the region shown on Figure 16. In mid-November, part of 855 was sent to Ostend in Belgium to crush Carboniferous limestone (found in a large dump near the docks) for tarmac manufacture; elements of the Quarry Group progressively took over responsibility for organizing the supply of all stone and tarmac from Belgian sources. The huge stocks of crushed rock found in Belgium, as well as the need to form other Royal Engineer units within the personnel ceiling imposed on the 21st Army Group, led to the disbandment of 858 in December 1944. The last Quarrying Company Royal Engineers to be formed, it was deemed appropriate that it was the first to be disbanded, despite protests from the Quarry Group against any diminution of its overall numbers.

From 1 December 1944 to 16 April 1945, the Quarry Group concentrated activities in Belgium, where it controlled the supply of stone to two armies (1st Canadian and 2nd British) and their Line of Communications. Most came from a dozen quarries in the long-famous 'porphyry' near Brussels: porphyritic quartz diorites of the Brabant Massif (Denaeyer & Mortelmans 1954), a basement structural high that stretches from western Germany across Belgium to southern England, formed principally during the Caledonian orogeny. These quarries, mostly within a sill at Lessines NW of the city and a plug at Quenast to the south, were mostly of impressive size and yielded an impressive output (Fig. 17). Quarries at Dongelberg to the east, and Opprebais, were operated from February 1945 onwards to meet local demands and relieve the rail lift position at Lessines, which was always difficult, but their output was small in comparison. Initially, stone was distributed from stock, but demand required that quarrying be implemented and civilian labour involved. By 1 April 1945, 125 was controlling stone production and/or issues at Lessines, Bierghes, Quenast, Soignies and Antwerp; 853 was quarrying at Cosyns; 855 (from its headquarters in Valkenswaard) was controlling activities such as stone crushing or tarmac production at Nijmegen, Weerpelt, Schaft, Hamont, Weert, Waterscheide and Ben-Ahin; and 856 was quarrying at Dongelberg and Opprebais (while still with one platoon operating at Perrières). Moreover, 4854 Flight Royal Air Force was also brought under command by this time, quarrying at Dongelberg and crushing dumped material at Le But.

From 17 April 1945 the Quarry Group began to move forward into Germany. Its headquarters had prepared for the move prior to the Allied crossing of the River Rhine in mid- to late March by making a study of all the available information regarding stone resources in northern Germany, by means of geological maps and publications and the help of the geologists then at the headquarters of the 21st Army Group (Majors F.W. Shotton and D.R.A. Ponsford: Rose *et al.* 2006; Rose & Clatworthy 2008*a*). Its task on arrival was the organization of stone supplies from local sources, initially for the British 2nd Army and later for the whole of the British forces east of the Rhine. By the time Victory in Europe was achieved, on 8 May 1945, only three quarries were in operation, in a line to the north of Osnabruck, but almost all of the Quarry Group had moved from Belgium to Germany before the end of May 1945. It was to continue operations until the end of August 1945, at times at a further 11 sites: quarries at Bentheim to the NW and Bissendorf to the south of Osnabruck; two at Steinbergen to the SE towards Hanover; five in basalt in the Harz Mountains near Göttingen to the south of Hanover (Bramburg, Barterode, Guntersen, Dransfeld and Meensen); a 'diabase' quarry near Langelsheim NE of this group; and a gravel washing plant at Rinteln near Steinbergen.

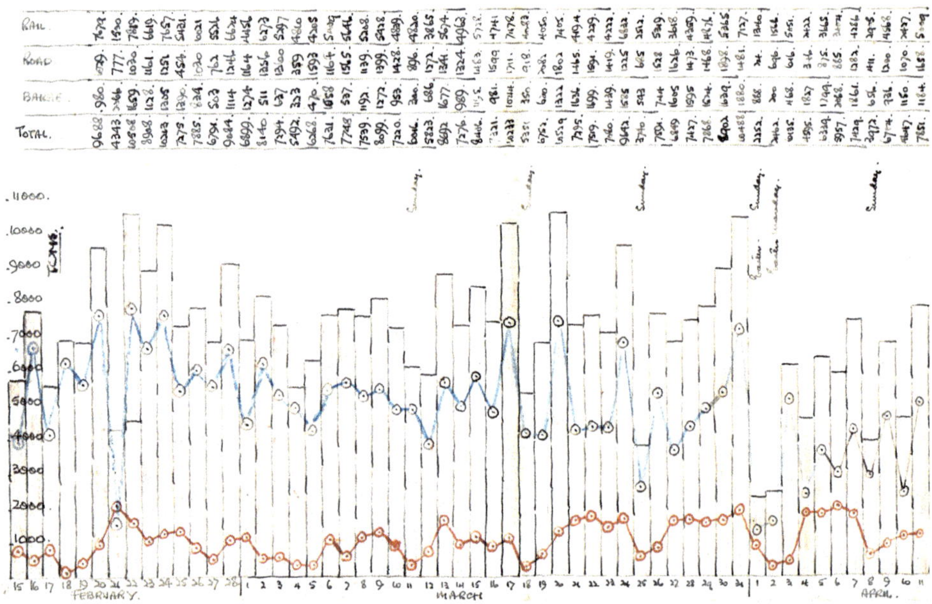

Fig. 17. Part of original hand-drawn sheet 2 (of four sheets) recording daily stone deliveries (in tons) from Belgian quarries (Lessines, Quenast, Bierghes, Ghent and Soignies) from 13 December 1944 to 31 July 1945: by rail, the largest amount, upper line, in blue; by barge, a lesser amount, lower line, in red; by road, too similar to barge amounts to be clearly shown separately; and total, by histogram. Figures for daily tonnages (too small to read at the scale of reproduction) are shown above the graph. From scale at left-hand side, it is clear that daily deliveries in this period could exceed 10 000 tons and that they did not normally fall below 3000 tons, even on Sundays (as indicated), except for Easter Sunday and Easter Monday (*c.* 2000 tons each). The totals decreased, especially on Sundays, after April 1945. From the personal records of the late Major J. Setchell, Royal Engineers, courtesy of Peter Setchell.

By then, three of the Quarrying Companies had been withdrawn to England with a view to service with SE Asia Command (in Burma, Malaya or Sumatra). 856 embarked for the UK on 20 May 1945, 125 entrained in Germany on 21 July 1945 and 855 on 25 July 1945. However, Victory over Japan on 15 August 1945 speedily removed the need for their further operational deployment. The Quarry Group briefly consisted of a headquarters plus 853 Quarrying Company thereafter. 'Bill' Williams was released from the Army on 23 August 1945 and succeeded in command by Joseph Setchell, who had founded 853 and later taken command of 858. The Group ceased to be responsible for any stone production outside Germany on 31 August 1945 and was disbanded as such on 30 September 1945. 853 remained in Germany to support the British Army of Occupation of the Rhine until adequate civilian support was in place. Then it too was disbanded on 8 March 1946.

Output

The Quarry Group had worked or controlled *c.* 40 quarries successively in France, Belgium and Germany during the 11-month NW Europe Campaign and delivered in total $>1.5 \times 10^6$ tons of stone, sand and tarmac (Fig. 18; Tables 4 & 8). Quarries in France had yielded 670 073 tons, Belgium 825 802 tons and Germany 6334 tons (Williams 1950; Rose 2012*a*). Excluding June 1944 (the opening phase of the campaign, when very little work was possible), output averaged 4789 tons per day.

Work done in the four months following the end of hostilities brought the total number of sites worked or controlled to *c.* 50 and total deliveries to about 2×10^6 tons before the military requirement came to an end and the Group was disbanded.

According to Williams (1950), the greatest difficulty the Quarry Group had to contend with throughout the campaign was transport:

> To meet the many and often unpredictable changes that occurred in the stone demand picture, the personnel, plant and equipment of the Group had to be moved, sometimes at short notice, over long distances. Although a Company could carry its personnel and much of its equipment in its own transport, it had not the tractors and low-loaders to move its plant
>
> (Williams 1950, p. 224).

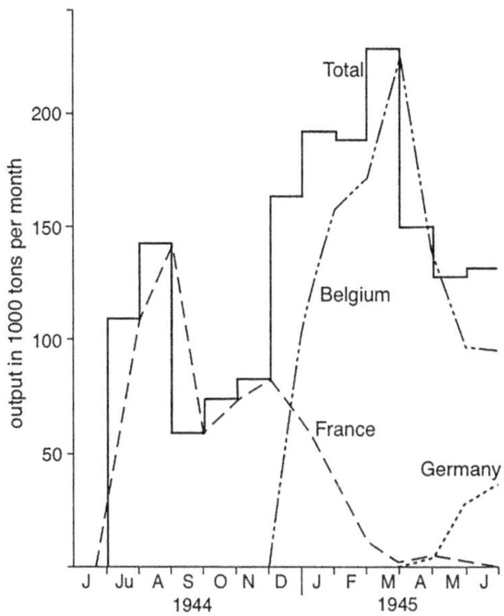

Fig. 18. Monthly quantities of stone produced or distributed by the Quarry Group Royal Engineers between 23 June 1944 and 30 June 1945. From Rose & Pareyn (1995), courtesy of Blackwell-Wiley Publications.

This required special transport, normally used for carrying tanks, for which there was always heavy demand.

Concluding remarks

There were significant differences between the roles of Quarrying Companies Royal Engineers in the two world wars, notably in terms of their regional deployment, their personnel and equipment, and the output they achieved. Military geologists, however, provided some guidance as staff officers at general headquarters within the major overseas theatres of operation in both wars, a role to be endorsed in an authoritative Royal Engineers text-book post-war.

Regional deployment

Quarrying Companies in World War I were mostly static, focused on the Marquise group of quarries near Boulogne in northern France to support the British Expeditionary Force on the Western Front (Rose 2018*a*). Those in World War II were more widespread and ultimately much more mobile in their use. Following their early assignment to bomb disposal, Quarrying Companies contributed to work associated with nine major projects or regions (Fig. 2; Table 4), as well as to miscellaneous

Table 8. *Output of broken stone by Quarry Group Royal Engineers from quarries successively in France, Belgium and Germany from late June 1944 to 7 May 1945*

Year	Month	Total (tons)	Tons per day
1944	June	12 905	
	July	107 270	3460
	August	145 498	4694
	September	68 433	2281
	October	85 886	2771
	November	136 927	4564
	December	156 307	5042
1945	January	198 263	6396
	February	182 356	6513
	March	229 306	7397
	April	134 866	4351
	May	44 191	6313
Average per month/day		150 000	4789

From Williams (1950) and Anon. (1957).

shorter term tasks within the UK. Overall they were used to support the construction of military facilities in the UK, the development of military outposts in the Faroe Islands and Gibraltar, and operations in North Africa, Italy and NW Europe.

- Quarrying as such began in 1941, primarily for port construction and associated road and railway works for two military ports in western Scotland, and to enhance defensive facilities, notably air-fields, in Northern Ireland.
- Three companies were deployed overseas in 1942, two to facilitate airfield construction on the Faroe Islands and one on Gibraltar, while another helped initiate the construction of the massive Central Ordnance Depot at Bicester in central England.
- Three companies were deployed overseas in 1943 for service in NW Africa (Algeria and/or Tunisia) and from there onwards in 1944 to support the Italian Campaign, continuing to support construction or reconstruction work in Italy and Austria, even after the end of hostilities.
- Five companies deployed to Normandy soon after D-Day in 1944, as components of the Quarry Group Royal Engineers, to support the 21st Army Group as it advanced across NW Europe by working quarries successively in France, Belgium and Germany.
- After Victory in Europe in May 1945, three companies from the Quarry Group were recalled to the UK for potential onward movement to the Far East, but the war against Japan ended before their operational deployment.

Personnel and equipment

Military quarrying in World War I was labour intensive: 13 Quarrying Companies Royal Engineers were raised in total, each of four officers and 264 other soldiers, creating a total workforce of >3000 men (mostly under the command of a lieutenant-colonel) assisted by a labour force of *c.* 4000 less skilled men (Rose 2018*a*). Less labour was involved in World War II. Only eight Quarrying Companies Royal Engineers were raised in total and, although their establishment varied slightly during the course of the war, by 1944 each consisted of four officers and 173 other ranks, creating a total simultaneous workforce of <1500 men (including a Quarry Group of five companies plus headquarters, and so 900 men, commanded by a lieutenant-colonel). Companies operational during the North African, Italian and NW Europe Campaigns were commonly supplemented by unskilled men (from the Pioneer Corps, prisoners of war or local civilians); *c.* 100 men per company in NW Europe. The total additional labour for all eight companies simultaneously thus probably averaged <1000 men. Overall, military quarrying during World War II seemingly involved little more than a third of the number of men similarly involved in World War I.

This reduction in personnel was offset by an increase in mechanical equipment. Whereas in World War I each Quarrying Company Royal Engineers left for service overseas equipped only with one wagon, two horses or mules to pull it and four bicycles to facilitate communication (Rose 2018*a*), it was realized from the beginning of World War II that Quarrying Companies would need to be equipped to a far higher standard (Table 3). However, the War Diaries indicate that there were variations in the equipment actually supplied. Thus the officer commanding 851 Quarrying Company reported that his unit came to Italy in March 1944 less satisfactorily equipped than 857 (Appendix A, WO 170/1781). Moreover, when the Quarry Group Royal Engineers was formed in March 1944 in preparation for service in Normandy, it was found that none of its five companies had been issued with a full scale of heavy plant and equipment (WO 205/1190); indeed, that the scales for a company's heavy plant and special equipment had still not been formally specified and that considerable tolerance had been allowed in the types and sizes of many items actually supplied. Nevertheless, one item in early and widespread use was the Ruston Bucyrus 19 face shovel (Rose 2018*b*, fig. 9). This was used on the Faroe Islands (WO 176/89), in Scotland (WO 166/12263), Oxfordshire (WO 166/12261), North Africa (WO 169/10779) and so presumably Italy, and NW Europe (WO 205/1190). Produced by an Anglo-American company from 1937, its use indicates that modern equipment had been adopted for the efficient movement of loose aggregate.

A scale devised for the Quarry Group's companies, based on operational experience to March 1944, represented the equipment standardized for the final year of the war (Table 7). Small items could be loaded onto each company's own vehicles (20 3-ton tippers and two 15-cwt trucks): pneumatic drilling equipment, drill steels up to 14 ft long and some of the heavier plants' lighter spares. For the heavier items (face shovels, drag lines, angle dozers, crushing units, large compressors, dumpers and long drill steels), the companies required help with transportation. Overall this equipment enabled the Group to meet its targets for stone required, with items being loaned from one company to another when necessary.

Output

In World War I, the Quarrying Companies concentrated in the Marquise area of northern France provided stone primarily to the British Expeditionary Force Directorate of Roads (Rose 2018*a*). Their output was mostly transported from the quarry face by wagons running on track and over long distances by rail. It reached a total of *c.* 10 000 tons per day, supplemented by 5000 tons per day from other sources, from March until at least November 1918. The overall total during this period was therefore >4.5 × 10^6 tons, used mainly as road stone and for railway ballast. Two companies began work in theatre during late 1916, 11 more during 1917, so, allowing for a build-up to the full production achieved in March 1918, it seems likely that the grand total output of Quarrying Companies in World War I exceeded 6 × 10^6 tons.

Output for World War II is more difficult to quantify overall. The companies were dispersed over a much wider region (Fig. 2) and provided stone for a correspondingly wider range of clients. Stone was typically transported from the quarry face to crusher and onwards by untracked vehicles rather than by rail (Figs 7 & 11) and the quarries were worked as close as possible to the client to minimize the demand for transport. There were also considerable differences between the companies in the data selected for inclusion in their War Diaries. With the exception of the Quarry Group in NW Europe, the companies lacked overall technical control and standardized records; paperwork was not always given high priority in wartime. None of the Quarrying Companies serving in the Mediterranean region listed output information in its War Diary.

Precise figures are available for the NW Europe Campaign, the Faroe Islands and Northern Ireland (Table 4). These total >2.5 × 10^6 tons by the end of the war and 3 × 10^6 tons by the disbandment of the Quarry Group Royal Engineers post-war. An amazing further 2×10^6 tons was quarried on Gibraltar, either by an element of 855 Quarrying Company when incorporated as a section within the Excavator Company Royal Engineers, or later by the company itself, when increased to about three-quarters of its established strength. However, the quarrying conditions were unusual; slopes of Quaternary scree breccia worked close to the client site, and by hydraulic means as well as blasting, with a generous assignment of transport vehicles. The output was thus exceptional. Wartime output from these four regions, by conventional and less conventional methods, therefore certainly achieved 4.5×10^6 tons, probably *c.* 4.8×10^6 tons when the estimated output required for construction of the Central Ordnance Depot at Bicester in Oxfordshire is added.

Quarrying for the two military ports in Scotland was intense by two companies for the first two years, less intense by three companies thereafter, and although the overall output is unknown, it seems likely to have reached at least 0.5×10^6 tons and so increased the estimated wartime Quarrying Company output to $>5 \times 10^6$ tons.

Unfortunately, estimates cannot be made for work in the North African and Italian Campaigns, other than to note that so many quarries were worked, and for so long, that the output from the three companies primarily involved is likely to have been considerable. One company in Italy is estimated to have achieved 300 000 tons in six months and, although that may have been atypical, analogy with the NW Europe Campaign suggests that perhaps 1×10^6 tons were generated in North Africa and perhaps $>1 \times 10^6$ tons in Italy. If so, the overall wartime output by Quarrying Companies on major tasks would have exceeded *c.* 7×10^6 tons and thus be greater than that in World War I.

In addition, five companies in total spent ten periods in the UK designated as 'miscellaneous' on Figure 2. Of these, the longest period (1 year 5 months) was that by 125 Quarrying Company from January 1941 training and then quarrying in its region of recruitment (Yorkshire) while waiting for mobilization for service overseas, a deployment repeatedly deferred as plans changed. In the eight months from mid-May to the end of December 1941, the company produced 55 033 tons by quarrying near Skipton and continued to produce stone there through the first four months of 1942, at weekly rates varying from *c.* 600 to 1100 tons, until assigned for service in the Faroe Islands. It achieved a final total reliably estimated at >70 000 tons. The next longest period was that by 855 (1 year 2 months), but this included its period of formation, followed by six months quarrying at Loch Ewe in Scotland (where it is estimated to have generated an output of *c.* 60 000 tons), and a short tour in England and Wales while being deployed, in stages, mostly to Gibraltar (while detachments continued to quarry in Wales). Two companies (856 and 858) spent short periods in England while they were raised and initiated into their quarrying role, prior to assignment to quarrying in Oxfordshire for construction work for the Central Ordnance Depot, Bicester. Five companies (125, 853, 855, 856 and 858) were assigned to training and quarrying tasks (six periods in total) in preparation for active service in Normandy: 853 for the longest period and 855 for two periods punctuated by quarrying in Oxfordshire. Miscellaneous tasks are therefore likely to have generated a further total output >150 000 tons (Table 4).

Post-war legacy

The UK extractive industry became increasingly mechanized in the 20 years post-war and it is at least arguable that this was significantly influenced by methods introduced during the war, both by the Quarrying Companies Royal Engineers and as a response to the wartime demands placed on the industry.

Quarrying Company experience guided the compilation of a textbook to provide the information required by Royal Engineers units for opening and operating service quarries and gravel pits in war. This concluded that:

> The quantity of stone required, in modern land warfare, for the construction, repair and maintenance of roads, airfields, and other engineering works, is likely to exceed that of any other item of engineer stores. Winning stone from the earth's surface and processing it to provide aggregate for constructional purposes are activities in which the military engineer is unlikely to gain much practical experience in peace-time. Furthermore, the civil engineer will find that under operational conditions, owing to the urgency of demands and probable limitations of equipment, the planning and development of Service quarries differ in many respects from commercial practice
>
> (Anon 1957, p. xiv).

The textbook identifies quarrying as a highly specialized branch of civil engineering. In civil life, quarries are typically operated as long-term investments. In military quarries, time rather than the cost of delivery is especially important and also sufficient quantity rather than the optimum quality of stone. Equipment and plant must have the mobility and adaptability to meet requirements that may change at short notice. Military quarries may have to operate in conditions unacceptable to civilian practice, e.g. where a shortage of plant makes it necessary to resort to working by hand, or where military needs so restrict the choice of sites that those with an excessive overburden must be used.

One particular lesson was apparently well learnt from wartime experience:

> Sound geological knowledge is required for the selection and development of a quarry. Information and advice should be sought from the geologist at Army headquarters, as regards both the reconnaissance and the quarry plan … The most promising areas are usually indicated in geological maps and records, but field reconnaissance by a trained geologist will often prevent great waste of effort …
>
> (Anon 1957, p. 8).

In World War I, two geologists had been based at the General Headquarters of the British Expeditionary Force in France (Rose 2018*a*). As World War II entered its final stages, two British geologists were based with the General Headquarters in Italy and a further two with the General Headquarters of the 21st Army Group in NW Europe. A need for military geologists to help guide military quarrying outside the UK had thus been firmly established, although the role was not to be mentioned in the official *History of the Corps of Royal Engineers* (Pakenham-Walsh 1958*a*, *b*).

Acknowledgements I am grateful to staff at the National Archives for access to documents in their care; to the copyright owners as acknowledged in figure captions for permission to reproduce images not yet in the public domain; to Mark Patton of the Geological Survey of Northern Ireland for confirming that the rock being worked in Figures 11 and 12 can be identified as basalt; and to Geoffrey Walton and an anonymous reviewer for their constructive comments on the manuscript as first submitted.

Appendix A

1939–45 War Diaries held in the War Office file series of the UK National Archives at Kew, London for Quarrying Companies Royal Engineers.

Unit no.	Date range of War Diaries	Command area	War Office file reference
125	September 1940–December 1941	Home Forces, UK	WO 166/3622
	January–May 1942	Home Forces, UK	WO 166/8081
	June 1942–July 1943	Faroe Islands	WO 176/89
	August–December 1943	Home Forces, UK	WO 166/12057
	January–December 1944	UK and AEF, France	WO 171/1551
	January–July 1945 (**to UK**)	AEF, Belgium/Germany	WO 171/5467
851	*August–December1940*	*Home Forces, UK*	
	January–December 1941	Home Forces, UK	WO 166/3927
	January–December 1942	Home Forces, UK	WO 166/8337
	January–June 1943	UK and Africa, Algeria	WO 175/703
	July–December 1943	NW Africa, Algeria	WO 169/10777
	January–December 1944	Africa and CMF, Italy	WO 170/1781
	January–December 1945	CMF, Italy	WO 170/5303
853	July1940–August 1941	Home Forces, UK	WO 166/3928
	September–December 1941	*Home Forces, UK*	
	January–December 1942	Home Forces, UK	WO 166/8338
	January–December 1943	Home Forces, UK	WO 166/12261
	January–December 1944	UK and AEF, France	WO 171/1716
	January–December 1945	AEF, Belgium/Germany	WO 171/5664
	January–March 1946 (**disbanded**)	BAOR, Germany	WO 171/9476
854	July 1940–December 1941	Home Forces, UK	WO 166/3929
	January–December 1942	Home Forces, UK	WO 166/8339
	January–June 1943	NW Africa, Algeria	WO 175/704
	July–December 1943	NW Africa, Algeria	WO 169/10778
	January–December 1944	CMF, Italy	WO 170/1782
	January–December 1945 (**disbanded**)	CMF, Italy	WO 170/5304
855	April–December 1941	Home Forces, UK	WO 166/3930
	January–July 1942	Home Forces, UK	WO 166/8340
	August 1942–August 1943	*Gibraltar*	
	August–December 1943	*NW Africa, Algeria*	
	January–December 1944	UK and AEF, France	WO 171/1717
	January–July 1945 (**to UK**)	AEF, Belgium/Germany	WO 171/5665
856	*June–December 1941*	*Home Forces, UK*	
	January–May 1942	*Home Forces, UK*	
	June 1942–August 1943	*Faroe Islands*	
	September–December 1943	Home Forces, UK	WO 166/12262

(Continued)

Unit no.	Date range of War Diaries	Command area	War Office file reference
	January–December 1944	AEF, France	WO 171/1718
	January–May 1945 (**to UK**)	AEF, Belgium/Germany	WO 171/5666
	June 1945	Home Forces, UK	WO 166/17011
857	October–December 1942	Home Forces, UK	WO 166/8341
	January–July 1943	Home Forces, UK	WO 166/12263
	July–December 1943	NW Africa, Algeria	WO 169/10779
	January–December 1944	Africa and CMF, Italy	WO 170/1783
	January–December 1945	CMF, Italy/Austria	WO 170/5305
	January 1946 (**disbanded**)	CMF, Austria	WO 170/8153
858	March–December 1943	Home Forces, UK	WO 166/12264
	January–December 1944 (**disbanded**)	UK and AEF, France	WO 171/1719

Diaries have not been found for dates shown in italics.
AEF, Allied Expeditionary Force; CMF, Central Mediterranean Force.

References

Anon 1957. *Military Engineering. Volume V. Part III. Quarries and Gravel Pits. WO Code No. 9422*. War Office, London.

Arkell, W.J. 1947. *The Geology of Oxford*. Clarendon Press, Oxford.

Dear, I.C.B. & Foot, M.R.D. (eds) 1995. *The Oxford Companion to the Second World War*. Oxford University Press, Oxford.

Denaeyer, M.-E. & Mortelmans, G. 1954. Les roches éruptives. *In*: Fourmarier, P. (ed.) *Prodrome d'une description géologique de la Belgique*. Société géologique de Belgique, Liège, 745–792.

Geikie, J. 1881. On the geology of the Faeröe Islands. *Transactions of the Geological Society of Edinburgh*, **30**, 218–269, https://doi.org/10.1017/S0080456800029033

Herries Davies, G.L. 1995. *North from the Hook: 150 Years of the Geological Survey of Ireland*. Geological Survey of Ireland, Dublin.

King, C. 2015. *A Revised Correlation of Tertiary Rocks in the British Isles and Adjacent Areas of NW Europe*. Gale, A.S. & Barry, T.L. (eds) Geological Society, London, Special Reports, **27**.

Mitchell, W.I. 2004. *The Geology of Northern Ireland: Our Natural Foundation*. Geological Survey of Northern Ireland, Belfast.

Nelson, C.M. & Rose, E.P.F. 2012. The US Geological Survey's Military Geology Unit in World War II: 'the army's pet prophets'. *Quarterly Journal of Engineering Geology and Hydrogeology*, **45**, 349–367, https://doi.org/10.1144/1470-9236/11-054

Noe-Nygaard, A. 1962. The geology of the Faroes. *Quarterly Journal of the Geological Society of London*, **118**, 375–383, https://doi.org/10.1144/gsjgs.118.1.0375

Pakenham-Walsh, R.P. 1958*a*. *History of the Corps of Royal Engineers. Vol. VIII. 1938–48. Campaigns in France and Belgium, 1939–40, Norway, Middle East, East Africa, Western Desert, North West Africa, and Activities in the UK*. Institution of Royal Engineers, Chatham.

Pakenham-Walsh, R.P. 1958*b*. *History of the Corps of Royal Engineers. Vol. IX. 1938–48. Campaigns in Sicily and Italy; the War against Japan; North-West Europe, 1944–45; Minor and Non-Operational Areas; Post-war, 1945–48*. Institution of Royal Engineers, Chatham.

Rose, E.P.F. 1996. Geology and the army in nineteenth century Britain: a scientific and educational symbiosis? *Proceedings of the Geologists' Association*, **107**, 129–141.

Rose, E.P.F. 1999. The military background of John W. Pringle, in 1826 founding superintendent of the Geological Survey of Ireland. *Irish Journal of Earth Sciences*, **17**, 61–70.

Rose, E.P.F. 2009. Military men: Napoleonic warfare and early members of the Geological Society. *In*: Lewis, C.L.E. & Knell, S.J. (eds) *The Making of the Geological Society of London*. Geological Society, London, Special Publications, **317**, 219–241, https://doi.org/10.1144/SP317.13

Rose, E.P.F. 2012*a*. Aggregates pave the way to victory: work of Royal Engineers geologists and Quarrying Companies during World War II, especially for the liberation of Normandy. *In*: Hunger, E. & Walton, G. (eds) *Proceedings of the 16th Extractive Industry Geology Conference*. EIG Conferences, Charlbury, 24–40.

Rose, E.P.F. 2012*b*. Groundwater as a military resource: development of Royal Engineers Boring Sections and British military hydrogeology in World War II. *In*: Rose, E.P.F. & Mather, J.D. (eds) *Military Aspects of Hydrogeology*. Geological Society, London, Special Publications, **362**, 105–138, https://doi.org/10.1144/SP362.7

Rose, E.P.F. 2014*a*. Boring and quarrying for military ports in Scotland: contributions to two major Royal Engineers construction projects of World War II. *Royal Engineers Journal*, **128**, 193–202.

Rose, E.P.F. 2014*b*. Royal Engineers quarrying from Woodstock during the Second World War. *In*: Jenkins, S.C. (ed.) *Bugle & Sabre VII: the Military History of Oxfordshire and Buckinghamshire*. Friends of the Oxfordshire & Buckinghamshire Light Infantry, Woodstock, 30–39.

Rose, E.P.F. 2014*c*. British pioneers of the geology of Gibraltar, part 3: E.B. Bailey and Royal Engineers 1943 to 1953. *Earth Sciences History*, **33**, 294–332.

Rose, E.P.F. 2016*a*. Foundation and roles of Quarrying Companies RE during World War II: a neglected footnote to history. *Royal Engineers Journal*, **130**, 119–126.

ROSE, E.P.F. 2016*b*. Oxford's Secret Sappers of the Second World War: terrain intelligence by military geologists. *In*: JENKINS, S.C. (ed.) *Bugle & Sabre IX: the Military History of Oxfordshire and Buckinghamshire*. Friends of the Oxfordshire & Buckinghamshire Light Infantry, Woodstock, 44–53.

ROSE, E.P.F. 2017. Quarrying for Gibraltar's airfield: an enduring Royal Engineer construction achievement of the Second World War. *Royal Engineers Journal*, **131**, 135–141.

ROSE, E.P.F. 2018*a*. Quarrying Companies Royal Engineers in World War I: a geologically constrained innovation to support British armies on the Western Front. *In*: ROSE, E.P.F., EHLEN, J. & LAWRENCE, U.L. (eds) *Military Aspects of Geology: Fortification, Excavation and Terrain Evaluation*. Geological Society, London, Special Publications, **473**. First published online March 6, 2018, https://doi.org/10.1144/SP473.1

ROSE, E.P.F. 2018*b*. Tunnelling Companies Royal Engineers in World War II: excavation of bomb-proof facilities in France, Gibraltar, Malta and the UK. *In*: ROSE, E.P.F., EHLEN, J. & LAWRENCE, U.L. (eds) *Military Aspects of Geology: Fortification, Excavation and Terrain Evaluation*. Geological Society, London, Special Publications, **473**. First published online March 6, 2018, https://doi.org/10.1144/SP473.3

ROSE, E.P.F. 2018*c*. The Battle of the Atlantic in World War II: contributions to victory by Quarrying Companies Royal Engineers and by geologist D.L. Linton, Professor of Geography at Birmingham 1958–1971. *In*: HUNGER, E., BROWN, T.J., SMITH, G. & ANDERSON, P. (eds) *Proceedings of the 19th Extractive Industry Geology Conference 2016 and Technical Meeting 2017*. EIG Conferences, Charlbury, in press.

ROSE, E.P.F. In press. Aerial photographic intelligence during World War II: contributions by some distinguished British geologists. *In*: ROSE, E.P.F., EHLEN, J. & LAWRENCE, U.L. (eds) *Military Aspects of Geology: Fortification, Excavation and Terrain Evaluation*. Geological Society, London, Special Publications, **473**.

ROSE, E.P.F. & CLATWORTHY, J.C. 2007*a*. The Sicilian and Italian campaigns of World War II: roles of British military geologists in Allied engineer 'Intelligence' and 'Works'. *Royal Engineers Journal*, **121**, 94–103.

ROSE, E.P.F. & CLATWORTHY, J.C. 2007*b*. Specialist maps of the Geological Section, Inter-Service Topographical Department: aids to British military planning during World War II. *The Cartographic Journal*, **44**, 13–43.

ROSE, E.P.F. & CLATWORTHY, J.C. 2008*a*. Fred Shotton: a 'hero' of military applications of geology during World War II. *Quarterly Journal of Engineering Geology and Hydrogeology*, **41**, 171–188, https://doi.org/10.1144/1470-9236/07-034

ROSE, E.P.F. & CLATWORTHY, J.C. 2008*b*. Terrain evaluation for Allied military operations in Europe and the Far East during World War II: 'secret' British reports and specialist maps generated by the Geological Section, Inter-Service Topographical Department. *Quarterly Journal of Engineering Geology and Hydrogeology*, **41**, 237–256, https://doi.org/10.1144/1470-9236/07-016

ROSE, E.P.F. & PAREYN, C. 1995. Geology and the liberation of Normandy, France, 1944. *Geology Today*, **11**, 58–63.

ROSE, E.P.F. & PAREYN, C. 2003. *Geology of the D-Day Landings in Normandy, 1944. Geologists' Association Guides*, **64**. Geologists' Association, London.

ROSE, E.P.F. & ROSENBAUM, M.S. 1993. British military geologists: through the Second World War to the end of the Cold War. *Proceedings of the Geologists' Association*, **104**, 95–108.

ROSE, E.P.F., CLATWORTHY, J.C. & NATHANAIL, C.P. 2006. Specialist maps prepared by British military geologists for the D-Day landings and operations in Normandy, 1944. *The Cartographic Journal*, **43**, 117–143.

TREWIN, N.H. (ed.) 2002. *The Geology of Scotland*. 4th edn. Geological Society, London.

WILLIAMS, A.R.O. 1950. The Royal Engineer quarrying companies in the North-west European campaign. *Quarry Managers' Journal*, **34**, 214–225.

WILSON, R.B. 1977. *A History of the Geological Survey in Scotland*. Institute of Geological Sciences, London.

Tunnelling Companies Royal Engineers in World War II: excavation of bomb-proof facilities in France, Gibraltar, Malta and the UK

EDWARD P. F. ROSE

Department of Earth Sciences, Royal Holloway, University of London, Egham, Surrey TW20 0EX, UK

0000-0003-4182-6426
ted.rose@earth.oxon.org

Abstract: 170 Tunnelling Company Royal Engineers left England in January 1940 to excavate bomb-proof military headquarters in northern France. Expansion into companies 170, 171, 172 and 173 was delayed when the British Expeditionary Force was defeated and evacuated, but completed in England in July to excavate accommodation underground for regional headquarters and artillery batteries – mostly in Cretaceous chalk. Companies 178, 179 and 180 formed in England in May–June 1940, and 178 and 180 soon deployed to Gibraltar, joined by 170 in 1941 and successively by 1st and 2nd Tunnelling Companies Royal Canadian Engineers. They excavated a major complex of tunnels and chambers within Jurassic dolomitic limestone by October 1943 to help fortify the rocky peninsula, work completed by a single company (172) thereafter. Companies 183, 184 and 185 joined 179 in 1941 to emplace 'Canadian pipe mines' to inhibit the invasion of Britain. 173 served in Malta 1941–43, and 171 during 1943–45, excavating >50 bomb-proof facilities in Oligo-Miocene limestones, plus works to enhance the civilian water supply. By late 1943, all companies except 172 (Gibraltar) and 171 (Malta) were based in Britain. At least five were converted for general engineering use and no longer needed geological assistance.

Tunnelling Companies were first raised within the Royal Engineers of the British Army during World War I, the Great War of 1914–18 (Anon 1922*a*; Grieve & Newman 1936; Cassels 2001). On initial foundation, each company consisted of five officers and 269 other ranks with appropriate experience, recruited primarily from civilian engineering firms and later from the British coal- and metal-mining industries. All were formed to serve within the British Expeditionary Force (BEF) deployed on the Western Front across Belgium and part of northern France. There they were complemented by three Tunnelling Companies from Canada, three from Australia and one from New Zealand, and by similar units in French forces, all opposed by German units with equivalent skills (Finlayson 2010; Jones 2010).

170–185 Tunnelling Companies Royal Engineers were formed in 1915 and those numbered 250–258 in 1915 or 1916, 25 companies in total. Initially, their primary purpose was mining and counter-mining at the front line. They excavated tunnels to be charged with explosives, which were then detonated to breach the enemy's fortified positions on the ground surface overhead or to destroy the enemy's mine tunnels. As the front line became held more by artillery fire power than infantry personnel, the purpose changed in 1917 to the construction of bomb-proof dug-outs, underground facilities to protect troops from artillery and aerial bombardment (Rose & Rosenbaum 2011). Excavation was mostly through relatively weak rocks: Upper Cretaceous chalk in France and Cenozoic sands and clays in Belgium (Anon 1922*b*). When the war ended, the BEF was progressively demobilized and all Tunnelling Companies Royal Engineers were disbanded.

For the UK, World War II began on 3 September 1939 and a new BEF began to move to France on 4 September 1939 (Dear & Foot 1995). This paper describes how 170 Tunnelling Company was quickly re-formed and expected to do much the same work in much the same place as in World War I. It was the first of only ten Tunnelling Companies Royal Engineers to be raised in total, using some of the World War I numbers. 170–173 formed between October 1939 and March 1940, 178–180 in May–June 1940 and 183–185 in 1941. From March 1940, two or more companies were commonly administered together under a unifying headquarters to create successively the 1st, 3rd and 4th Tunnelling Engineer Groups; plans to create 2nd Tunnelling Engineer Group were abandoned when the BEF was evacuated from France in 1940.

The first four companies served briefly in France. At various times thereafter, four companies constructed underground and other facilities within the UK; four served on Gibraltar, to extend a tunnel complex that is now a major feature of the famous Rock; two constructed underground facilities on the island fortress of Malta; and four were created as 'special tunnelling companies' to prepare underground

From: Rose, E. P. F., Ehlen, J. & Lawrence, U. L. (eds) 2019. *Military Aspects of Geology: Fortification, Excavation and Terrain Evaluation*. Geological Society, London, Special Publications, **473**, 201–232.
First published online March 6, 2018, https://doi.org/10.1144/SP473.3

landmines and other obstacles. The mines and obstacles were intended to deny the German invasion of Britain by sea or air, and road and rail movement post-invasion. In contrast with World War I, the ten companies sometimes worked in strong as well as weak rocks, and from first deployment their primary task was to provide protection from bombardment rather than to help breach fortified lines.

This paper provides an overall assessment of the role of Tunnelling Companies Royal Engineers in World War II. It is based on their monthly War Diaries (Appendix A), currently preserved in the War Office series of documents in the UK's National Archives at Kew, near London, and on documents held by the Royal Engineers Museum, Library and Archive (REMLA) at Chatham in Kent. Imperial rather than SI units of measurement are used for consistency with the historical sources, with metric equivalent where deemed necessary. Minor inconsistencies between documents have been ignored where these are evidently typographic or clerical errors.

Focused specifically on Tunnelling Companies Royal Engineers, there is no attempt to review the support given to the British Army by similar units raised in Commonwealth countries (although support from Canada, South Africa and India is known to have been especially significant), the construction of bomb-proof and other facilities underground by civilian agencies (although there are many examples of such work in the UK), or the tunnelling undertaken by Allied or enemy forces. The paper demonstrates only part of the range of underground excavation stimulated by World War II, but it is a part that is of particular military significance.

Origins of 1st Tunnelling Engineers in northern France, 1940

In World War II as in World War I, 170 Tunnelling Company Royal Engineers was the first to be raised, although in World War II it was named initially a Railway Tunnelling Company. It mobilized at Chatham (historically a major centre of Royal Engineers activities, *c.* 50 km east of London) from 19 October 1939, a month after the UK's declaration of war on Nazi Germany (Appendix A, WO 167/957; Table 1). Its war establishment, developed from that of Tunnelling Companies in World War I, consisted of 18 officers and 576 other ranks. Two officers, a major as officer commanding and a captain as second-in-command, plus 42 other ranks, formed the company headquarters. The remainder formed four sections, each consisting of four officers, a captain plus three subalterns (lieutenants or second lieutenants), and 135 other ranks. All were volunteers. Most of the other ranks transferred from other units of the armed forces, but were originally coal miners from northern England (Northumberland, Durham and Yorkshire), although some came from Wales and others from Scotland.

Memories of the exploits of the Tunnelling Companies of World War I had been kept alive for the 20 years post-war by a vigorous old comrades' association, so there was a pool of experience on which the armed forces could draw (Wilson 1992). For officers, the War Office in London duly approached the Institution of Mining and Metallurgy (the IMM, part of the Institute of Materials, Minerals and Mining since 2002), also in London, with a request for the names of veterans who might be available for service. According to the nominal roll for 31 January 1940 (annexed to War Diary WO 167/957 and to 170's file at REMLA, accession number 8293), the officers who formed the nucleus of the new 170 (Fig. 1) consisted of Major G. Maitland Edwards DSO; Captains W.S. Cowan (adjutant), D.G. McLea, D.M. Thomson MC, R.S.G. Stokes DSO OBE MC and A.S.W. Wood MC; with Second Lieutenants R.T. Brandt, StJ.H.M. Findlay, L.J. Foss, M.T. Keillar, J.H. Lander, R.G. Laver, G.A. Moorhead, J.E.A. Paterson, T. Pickering,

Table 1. *Foundation of Tunnelling Companies Royal Engineers 1939–45*

Company	Date	Place	Officer commanding (Major)
170	October 1939	Chatham, Kent	G.M. Edwards
171	April 1940	Northern France	A.C.B.M. Sim*
172	April 1940	Northern France	L.J. Foss
173	April 1940	Northern France	J.B. Simpson
178	May 1940	Chatham, Kent	G.A.[P.] Moorhead
179	May 1940	Chatham, Kent	T. Pickering
180	June 1940	Chatham, Kent	A.R.O. Williams
183	April 1941	Scarborough, Yorkshire	B. Llewellyn MC
184	July 1941	Scarborough, Yorkshire	S.V. Griffith
185	August 1941	Scarborough, Yorkshire	W.A. Smailes

*Major Sim commanded when 171 was refounded after evacuation from France.

Fig. 1. Officers of 170 Tunnelling Company Royal Engineers at Chatham in early January 1940, prior to departure for France. Identified by Phillip Robinson of the Durand Group (pers. com. 2017) as, from left to right: (inset rear) J.B. Simpson, N.B. Walker; (standing rear) G.A.P. Moorhead, T. Pickering, J.H. Lander, R.G. Laver, D.J. Rogers, L.J. Foss, StJ.H.M. Findlay, J.A.E. Paterson; (seated front) A.R.O. Williams, A.S.W. Wood, G.G. McLea, R.S.G. Stokes, G.M. Edwards, N.S. Cowan, D.M. Thomson, R.T. Brandt and T.M. Keillar. See Wilson (1992), but reproduced courtesy of Lieutenant Colonel Phillip Robinson.

D.G. Rogers, J.B. Simpson and A.R.O. Williams, plus Lieutenant N.B. Walker attached from the Royal Army Medical Corps.

Edwards, McLea, Thomson, Stokes and Wood had served as tunnellers in World War I (Phillip Robinson, pers. com. 2017). Brandt, Edwards, Moorhead, Paterson, Pickering, Rogers, Simpson, Stokes, Williams and Wood were members of the IMM (Wilson 1992). Four of these were men of particular ability, who became presidents of the IMM in the post-war years. Ralph Stokes (1882–1979) was a veteran who had hurried to the UK from South Africa as soon as he considered war to be inevitable. John Bell Simpson (1900–83) had flown with the Royal Air Force (RAF) in World War I before studying at the Camborne School of Mines in Cornwall, SW England, and enjoying a mining career in South America, India and West Africa before he too hurried back. Douglas John ('Jim') Rogers (1907–2001) was a non-veteran who had mined in Yugoslavia and South Africa. A.R.O. ('Bill') Williams (1904–89) was another non-veteran who had mined in Africa before employment in Canada, from where he hastily returned to volunteer for service. Rogers (according to Wilson 1992) and Williams (Rose 2012*a*) had certainly benefited from courses in geology during university education and other tunnelling officers had probably done so as well.

Tunnelling Companies in operation

It took time for the officers and men to be assembled and for their equipment to arrive (WO 167/957). 170 moved to France (via Southampton and Cherbourg) to join the growing BEF on 13 January 1940. For the first month it was billeted *c.* 35 km SE of Arras (Fig. 2) for regimental training. There it was soon apparent that the work would be very different from that which the tunnellers had initially been required to perform in World War I. German forces had already demonstrated, in Poland in 1939, a propensity for mobile Blitzkrieg rather than static trench warfare. There was to be little requirement for the mine warfare of 1915–18. However, air power now posed a major threat. Accordingly, the company spent most of its time in France, not in driving mines beneath enemy emplacements, but in excavating underground for military headquarters, operations rooms, casualty stations, and stores for ammunition and other supplies that needed protection from bombing attack. 'It was unexciting, laborious work on that first winter of the ['Phoney'] war when the main enemy was boredom – and the bitter cold' (Wilson 1992, p. 165).

Tunnelling as such began in early February, in Upper Cretaceous chalk. This was for underground battle headquarters *c.* 25 km north and ENE of Arras (Fig. 2) at Douai and Hulluch (Fig. 3) for the

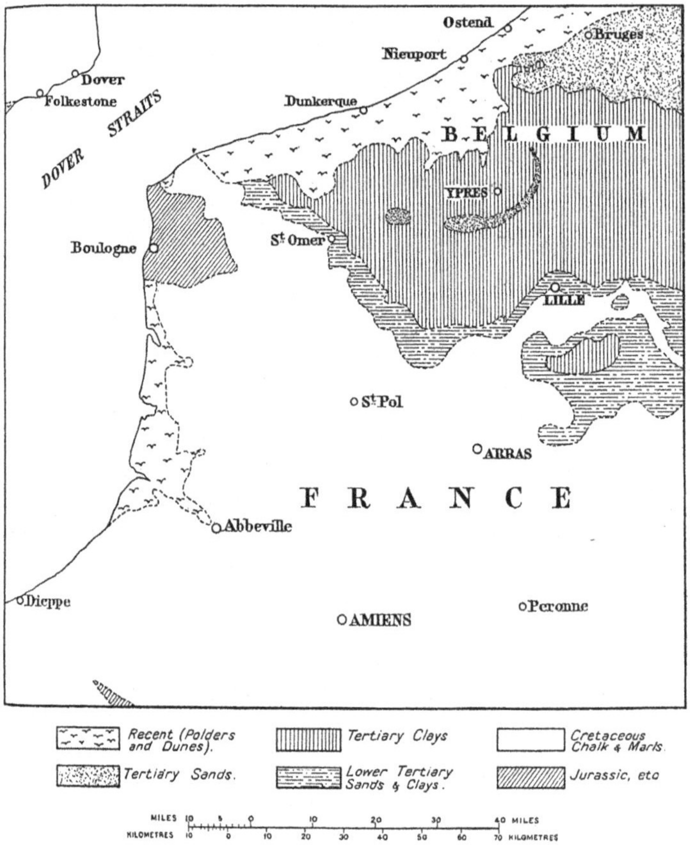

Fig. 2. Near-contemporary geological sketch map of Belgium and northern France, including the region contested by the British Expeditionary Force in World War I and similarly occupied by the British Expeditionary Force in World War II; see Rose (2012*b*).

Fig. 3. Piccadilly Circus, at Hulluch. This tunnel system in Upper Cretaceous chalk was initiated during World War I, but extended during World War II. Photographed during clearance by members of the Durand Group and reproduced by kind permission. © AnneSo Douchin/The Durand Group 2015.

British 1st and 2nd Corps (WO 167/957). Two tunnelling sections were assigned to each locality and the company headquarters moved closer to Arras for convenience. Early in March it was decided to excavate a shelter for the BEF's General Headquarters midway between Arras and Abbeville, under an old fortress at Doullens, and the number of personnel was progressively concentrated there.

It soon became apparent that the number of personnel needed to be increased and, on 26 March 1940, approval was given for 170 Tunnelling Company to expand to a lieutenant-colonel's command. It was to become 1st Tunnelling Engineers, a Royal Engineers Group with a headquarters and workshop sections plus four Tunnelling Companies (to be formed by expanding the existing four sections of 170).

During April 1940, while reinforcements were arriving to form the new companies and preparations were made for further work, No. 1 Section continued the work at Douai and Sections 2, 3 and 4 at Doullens. Indeed, the demand for tunnellers was such that the company's officer commanding, Major G.M. Edwards, was recalled to London at the end of April to discuss further expansion to four groups, making a total of 16 Tunnelling Companies. It was understood that 2nd Tunnelling Engineers would be required when the BEF expanded from one army to two. 3rd Tunnelling Engineers would be required immediately, for service in the Norwegian Campaign in response to the German invasion of Norway on 9 April 1940 (Dear & Foot 1995). Thomson (Fig. 1), by then a captain, was selected for promotion to lieutenant-colonel to command it, and Second Lieutenants Moorhead, Williams and Pickering (Fig. 1) were chosen to command the first three (of its potentially four) designated companies (WO 167/957). All were conse- [illegible]d back to England, while in France [illegible] Edwards was promoted to the rank of lieutenant-colonel to command 1st Tunnelling Engineers, then in the process of formation. In March 1940, No. 1 Section of 170 became 173 Tunnelling Company under the command of Acting Major J.B. Simpson (Fig. 1) (WO 167/958) and No. 4 Section became 172 (WO 166/3653). 170 and 171 were taking shape at about the same time, but the unit records of these changes had soon to be destroyed or abandoned.

A German attack westwards across the Low Countries and northern France began on 10 May 1940 (Dear & Foot 1995). Tunnelling work continued for a few days as normal (WO 167/957; Rogers 1986), but as German success became evident, on 17 May 1940 the tunnellers were ordered to prepare to move. On 19 May 1940 they began to fall back towards the port of Boulogne for evacuation to England. In the tumultuous days that followed, some tunnellers were ordered to use their explosives to destroy bridges in an attempt to halt the German advance. About nine officers and 350 other ranks were taken prisoner before embarkation, but the rest escaped via Dunkirk (Dunkerque on Fig. 2) or the port of Calais, *c.* 45 km westwards along the French coast, facing Dover.

Geological guidance

Geological features relevant to military tunnelling in northern France (Fig. 2) had been well documented in World War I. The most common problems in areas of Upper Cretaceous chalk derived from the constraints of seasonally fluctuating water tables (Anon 1922*b*). Records of any geological assistance received by 170 Tunnelling Company (later 1st Tunnelling Engineers) during its tunnelling through chalk in France in World War II were destroyed or lost during the retreat prior to evacuation. However, geological guidance for tunnelling activities had been available from BEF General Headquarters in much the same way in both wars. The first of the three geologists to serve at BEF General Headquarters in World War I, Lieutenant (later Captain) W.B.R. ('Bill') King, had been appointed ten days after the onset of World War II to serve (as a local major) in essentially the same role and place within the new BEF (Rose & Rosenbaum 1993; Rose 2012*b*).

1st Tunnelling Engineers in the UK, 1940–43

The survivors of 1st Tunnelling Engineers were joined by volunteer reinforcements from other arms in late June and early July 1940 to re-form as four companies, plus a group headquarters and a workshop section (Table 2) (WO 166/3535). They did so, after a few days at Newark in the East Midlands, at Chester (location 1 on Fig. 4), 140 km to the west, away from the southern area threatened by imminent invasion.

The group's headquarters and workshop section were based at Chester, with three of the four companies, until moved to the south of England on 30 October 1940 to keep pace with company movements. The priority tasks for all four companies became to excavate bomb-proof shelters for military headquarters and for troops manning batteries of heavy coastal artillery that threatened German shipping in the English Channel (WO 166/3652). An advance party had tried to find accommodation at Crowborough, the largest inland town in East Sussex, which was a location central to works in progress (WO 166/3535). However, this proved impracticable and the headquarters was therefore established at Brighton, a large coastal resort

Table 2. *1st Tunnelling Engineers: principal deployments 1940–43 in UK towns and Mediterranean countries*

Company	1940	1941	1942	1943
Headquarters	Chester Brighton	Brighton Tunbridge Wells	Tunbridge Wells *Egypt*	*Malta*
170	Chester Ringmer Dover	*Gibraltar*	*Gibraltar*	*Gibraltar*
171	Halton Aldershot	Aldershot Dover	Dover *Egypt*	*Malta*
172	Chester Dover	Dover Tunbridge Wells Canterbury	Canterbury Fareham Dover	Plymouth *Gibraltar*
173	Chester Dolgellau Newmarket	Tunbridge Wells Reigate *Malta*	*Malta*	*Malta*

Roman font denotes UK towns (see Fig. 4) and italic denotes Mediterranean countries.

c. 40 km to the SW (location 2 on Fig. 4). Lieutenant-Colonel L.C. Hill, DSO, MC became commanding officer on 5 August 1940, succeeding Lieutenant-Colonel Edwards when he was appointed Chief Engineer 2nd Corps. In April 1941, again to keep pace with company movements, the headquarters and workshops moved to Tunbridge Wells, a large town in western Kent, *c.* 60 km SE of London (location 3 on Fig. 4). They were to remain there until deployed to Egypt in May 1942 (WO 166/8042, WO 169/5389). By 1943, all four of the original companies were operational in the Mediterranean region.

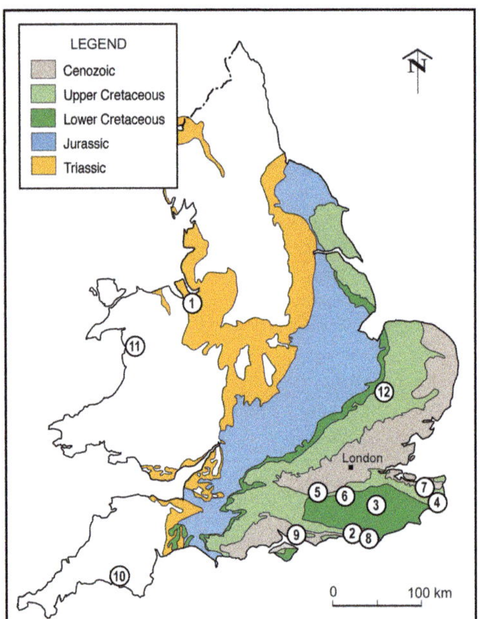

Fig. 4. Simplified post-Paleozoic geological map of England and Wales with areas of Paleozoic and earlier bedrock shown blank. The positions of the principal towns associated with the work of 1st Tunnelling Engineers 1940–1943 are shown: 1, Chester; 2, Brighton; 3, Tunbridge Wells; 4, Dover; 5, Aldershot; 6, Reigate; 7, Canterbury; 8, Newhaven; 9, Portsmouth; 10, Plymouth; 11, Dolgellau; 12, Newmarket.

170 Tunnelling Company

170 Tunnelling Company was employed on the construction of the Western Command Headquarters shelter at Chester from June until 20 September 1940, when it was moved south, first to work for 12th Corps at Ringmer in Sussex, 16 km ENE of Brighton, and next, on completion of that task, to join 172 working in Kent, near Dover (location 4 on Fig. 4) and also Deal, *c.* 11 km along the coast to the north (WO 166/3535). War Diaries for 170 for this period could not be found at the National Archives, but the War Diary for 172 (WO 166/3563) records that, on 30 December 1940, 172 took over tunnelling from 170 for a coastal artillery battery in St Margaret's Bay, north of Dover. Seemingly, 170 was focused on this project until departure for Gibraltar in early March 1941.

171 Tunnelling Company

171 was employed on work at Halton in Lancashire, a town *c.* 30 km north of Chester, from July until the end of September 1940 (WO 166/3535), when its War Diaries begin (WO 166/3652). Commanded by Major A.C.B.M. Sim, the company's four sections were employed on minor construction tasks

or in training. On 1 October 1940 it began to move to Aldershot (location 5 on Fig. 4), a garrison town in north Hampshire long regarded as the home of the British Army, in preparation for tunnelling a regional command shelter. Later that month, prior to excavation, a detachment from No. 1 Boring Section Royal Engineers (led by Captain G.R.S. Stow) drilled four exploratory boreholes to determine the ground conditions. There were also preliminary reconnaissance studies for the tunnelling of operational headquarters at three sites nearby (one on the Hog's Back, another at Hawley Hill and the third at Batt's Hall near Frensham). The site at Hawley Hill also required boreholes drilled by No. 1 Boring Section (see Rose 2012*b* for unit history). The Aldershot shelter (Fig. 5) was excavated through Cenozoic sands and clay between 26 October 1940 and 28 February 1941 and strengthening was completed by early April 1941.

As tunnelling at Aldershot drew to an end in February 1941, work began for another command shelter at a chalk quarry site near Reigate, *c.* 40 km to the east (location 6 on Fig. 4). However, this work was handed over to 173 Tunnelling Company in March and, after a few weeks of intensive military training, 171 moved to Dover during April 1941. Initially, it tunnelled shelters for gun positions nearby at St Margaret's Bay and at Wanstone Farm, and also at Capel, near Folkestone to the south, but as these tasks were completed, other sites were developed, notably at Dover Castle and at St Martin's Hill, near Canterbury (location 7 on Fig. 4). The company was still based at Dover for the first three months of 1942 (WO 166/8104), working principally on a medical dressing station being constructed beneath Dover Castle, but was soon subject to a period of intensive infantry training prior to mobilization for service overseas. Major W. Fowke succeeded Major Sim in command on 24 March 1941 and in April 1941 the company embarked for the long journey by sea around Africa to Egypt (WO 169/5284).

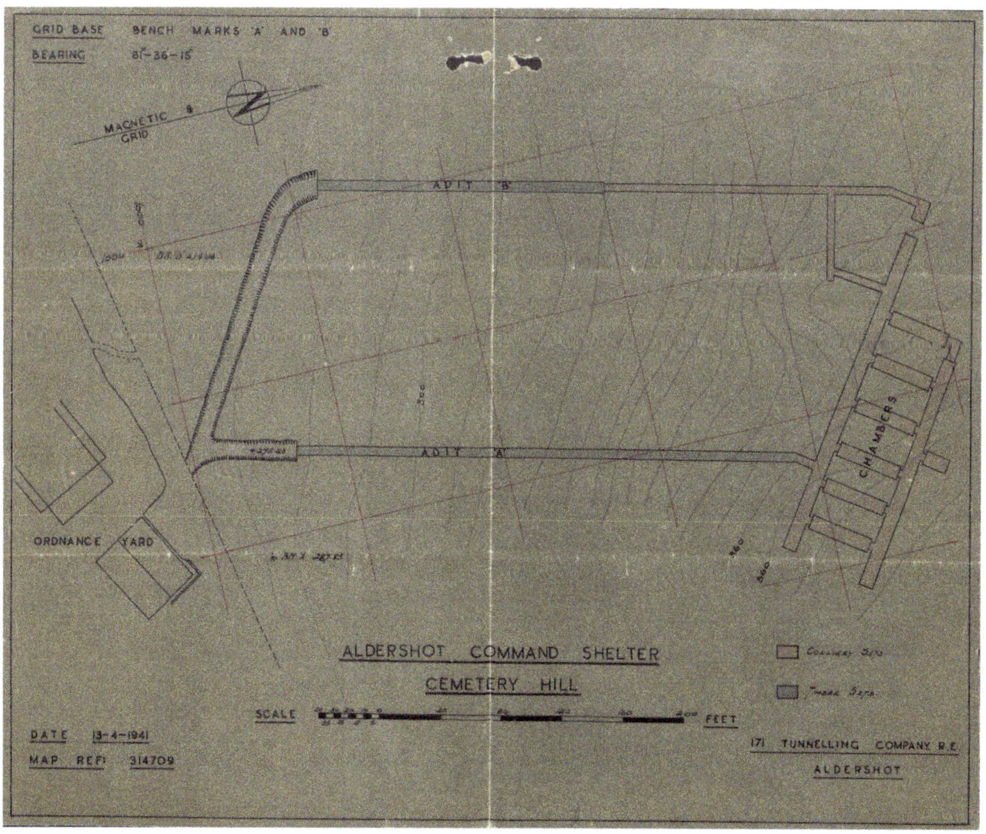

Fig. 5. Aldershot Command Shelter, completed by 171 Tunnelling Company Royal Engineers on 13 April 1941, one of many shelters excavated in England by Tunnelling Companies early in World War II. Reproduced by permission of The National Archives, Kew, from file WO 166/3652.

172 Tunnelling Company

Like 170, 172 was initially employed on construction of the Western Command Headquarters shelter at Chester, from mid-June 1940 until moving south to Dover on 16 November (WO 166/3535). Its War Diaries begin on 1 July 1940, opening with the record that the unit was formerly No. 4 Section of 170 Tunnelling Company (WO 166/3653). Major L.J. Foss (Fig. 1) was the officer commanding, having led the unit safely out of France.

At Dover (location 4 on Fig. 4), work quickly began on the construction of protective shelters for troops manning artillery batteries in Fan Bay and St Margaret's Bay to the north, an additional casemate at Dover Castle, and soon shelters also at nearby Lyddon Spout and Western Heights.

Work at Fan Bay was completed on 28 February 1941. It consisted of a system of tunnels designed to accommodate four officers and 185 other ranks. The tunnels were driven in Upper Cretaceous chalk as galleries 6.5 × 5.5 ft (2.0 × 1.7 m) in cross-section, opening into chambers of 9 × 8 ft (2.7 × 2.4 m) supported by steel arches. The whole system had a floor area of *c.* 3500 ft^2 (325 m^2). The task took *c.* 100 days and provided accommodation that included a medical centre. When operational, the site was visited approvingly by the Prime Minister, Winston Churchill. After the war it fell into disuse and its entrance was eventually blocked off with 30 tonnes of rubble. However, on 20 July 2015 (as reported in *The Independent* newspaper), the Fan Bay Deep Shelter complex opened regularly to visitors after a team of National Trust volunteers and advisers had re-excavated the site as a two-year conservation project.

On 12 March 1941 work began at Wanstone Farm on tunnelled shelters for troops operating a battery of 15-inch guns. Work at Western Heights – a tunnelled system in Upper Cretaceous chalk to accommodate four officers and 150 other ranks of 414 Coast Battery Royal Artillery, *c.* 3150 ft^2 (293 m^2) in floor area – was completed on 19 March 1941. Work in St Margaret's Bay was completed on 22 March 1941. This was another tunnelled system in chalk, to accommodate the four officers and 165 other ranks of 411 Coast Battery, with a floor area of *c.* 3480 ft^2 (323 m^2). Work at Lyddon Spout – tunnelled accommodation in chalk for another artillery battery, four officers and 170 other ranks, with a floor area of *c.* 3500 ft^2 (325 m^2) – was completed on 28 March 1941.

The company moved to Aldershot to begin a programme of military and field training on 21 April 1941 and then to Tunbridge Wells (location 3 on Fig. 4) on 19 May 1941 to take over work on an underground headquarters for 12th Corps at the nearby Broadwater Down. With its headquarters still based at Tunbridge Wells, 172 began tunnelling on 27 May 1941 for coastal facilities at Newhaven (location 8 on Fig. 4). Tunnelling at Broadwater Down through Lower Cretaceous 'medium hard sandstone' to create a shelter of 9000 ft^2 (836 m^2) in area was completed on 19 July 1941, but concreting was not completed until 14 September 1941. No. 3 Section moved to Reigate in August to tunnel an extension to the headquarters for South-Eastern Command. This consisted of 234 ft (71 m) of 9 × 8 ft (2.7 × 2.4 m) gallery in Upper Cretaceous chalk, supported by steel arches, and was completed on 29 September 1941. Between 8 and 12 October 1941 the Company moved to Canterbury to take over work from 171.

In February 1942, 172 moved from Canterbury to Fort Southwick, near Fareham and so *c.* 10 km NNW of Portsmouth in Hampshire (location 9 on Fig. 4) (WO 166/8105). There its prime task was to construct underground operational headquarters near the Portsmouth naval base. It was based in a hutted camp nearby from 23 May 1942. Work was progressively transferred to construct similar facilities near Dover in August and the company as such moved to Dover in September. The construction of two underground wireless transmitting stations was added towards the end of November 1942.

By the end of January 1943, all work at Dover was nearing completion and a new project at Plymouth in Devon (location 10 on Fig. 4) was in its early stages (WO 166/12075). Most of the company moved to Plymouth in February to excavate in strong rock (Devonian limestone overlain by shale) rather than in Cretaceous chalk 'by no means a straightforward task' according to the company's War Diary. The shallow depth of the water table required excavation within the limestone to provide adequate protective cover, but the limestone–shale boundary was found to be 'very irregular' and the limestone was penetrated by solution fissures. Moreover, the War Diary notes that 'Blasting in the tunnels disturbed the [Naval] Commander-in-Chief, Plymouth, and his staff in the Admiralty'. Despite diamond drilling to prove the ground, guided by expertise from 1st Tunnelling Company, Royal Canadian Engineers, work was intermittent and finally suspended in April 1943. The company then trained prior to mobilization for service overseas in Gibraltar.

173 Tunnelling Company

For 173, War Diaries begin on 9 July 1940, when the company was based at Chester and commanded by Major J.B. Simpson (Fig. 1) (WO 166/3564). Orders were received for the company to move to Dolgellau (location 11 on Fig. 4), a market town in NW Wales, to construct anti-tank defences in the region. Work took place on seven days each week because of the perceived urgency of the task. Anti-invasion defences were being established in Wales to counter the perceived threat of a German amphibious invasion from the Irish Republic, officially neutral, but seen by the UK government as possibly pro-German and therefore potentially hostile. A national defence plan included the rapid construction of a network of anti-tank obstacles to create inland stop lines and nodal defence points.

Company strength during the summer averaged seven officers and *c.* 200 men. The company moved SE to Newmarket (location 12 on Fig. 4) near Cambridge on 7 October 1940 (to work on tasks at Newmarket and at Euston in London), then to Tunbridge Wells on 31 January 1941 and later to Reigate in Sussex.

The Newmarket task was to construct an underground shelter in the shape of a grid 50 ft (15 m) below the land surface. It consisted of eight chambers, each 40 ft (12 m) long and with centres 30 ft (9 m) apart, between two access galleries. All chambers and galleries were lined by arcuate metal sheets and supported by steel colliery arches. The first gallery was driven from a vertical shaft at one end and a 30° incline at the other. Two access entrances were constructed as rises from the grid. All tunnelling was through Upper Cretaceous chalk, supposedly with a water level not higher than 60 ft (18 m) below ground level. Work was started on 9 October 1940, in daylight hours only. The vertical shaft (7 × 7 ft, 2.1 × 2.1 m) was put down to 50 ft, but water was struck at 48 ft (14.6 m). The incline (7 × 6 ft, 2.1 × 1.8 m) was sunk 80 ft (24 m). The pilot tunnel was started from both ends at 42 ft below the surface and 40 ft were completed. However, investigation was made at about this time to estimate how high the water table might rise in the spring and, consequently, the decision was made to make the shelter 17 ft (5.2 m) higher. Accordingly, the 40 ft of pilot tunnel, 25 ft of vertical shaft and 35 ft of incline were filled in and a new start was made 25 ft below the land surface. A new pilot tunnel of 298 ft (91 m) between the shaft and the incline was completed on 25 January 1941, when orders were received to stop all work because the cover of 25 ft had been calculated as insufficient. The pilot tunnel was made as safe as time allowed, to be used as an emergency shelter. Steps were placed in the incline and the shaft was covered except for ladder access. Over 13 000 ft^3 (368 m^3) of chalk had been excavated, at an average rate of 6.4 ft^3 (0.18 m^3) per person-shift.

The Tunbridge Wells task was again to construct an underground shelter at *c.* 50 ft (15 m) below the land surface, but in the shape of a grid of from five to nine chambers between two galleries, all 9 × 8 ft (2.7 × 2.4 m) in cross-section and supported by steel colliery arches and arcuate sheets. Access to the shelter was to be by three 35° inclines of 5 × 6.5 ft (1.5 × 2.0 m) cross-section, also supported by arches and sheets. Excavation was through a weak sandstone of Early Cretaceous age for the first 11 ft (3.4 m) and thereafter through stronger sandstone. Water was struck in all three inclines, requiring pumps to be used until the weather improved. An advance party started work on 4 February 1941, but was increased to half of the company three days later. It worked two shifts until joined by the other half of the company on 3 March 1941, after completion of the task at Euston, and three shifts then became the norm. When it was found necessary to concentrate the whole company on the Reigate task, on 17 May 1941, the Tunbridge Wells project was handed over to 172 Tunnelling Company. By then >50 000 ft^3 (1416 m^3) of material had been excavated, at an average rate of 5.0 ft^3 (0.14 m^3) per person-shift.

The Reigate task was yet again to construct an underground shelter with a cover of 50 ft (15 m), access being by three adits leading to a grid consisting of back and front galleries, nine chambers, a standby engine room and a gas filtration room. All chambers and galleries were to be supported by steel colliery arches and arcuate sheets. The site was accessed from a disused chalk quarry, which had been abandoned by the owners when deemed too dangerous to continue working. Excavation proved difficult because the chalk was closely jointed and could not easily be tunnelled without continuing support. Work had been started by 171 Tunnelling Company and was taken over by two sections of 173 from 7 March 1941, joined by the other two sections on 18 May 1941. Thus the full company was focused on the same project for the first time. The task was completed on 13 June 1941. More than 96 000 ft^3 (2718 m^3) of chalk had then been excavated, at an average rate of 9.0 ft^3 (0.25 m^3) per person-shift.

With all tasks completed, the company moved from Reigate to Fleet, near Aldershot, on 20 June 1941 and there received orders on 22 June 1941 to mobilize for a tropical climate. On 11 July 1941, 173 Tunnelling Company (seven officers and 223 other ranks) sailed for a destination unknown to them. Rescued by a destroyer from a troop ship torpedoed *en route* by a German E-boat, and repeatedly subject to air attack, the company reached its destination of Malta on 25 July 1941.

Geological guidance

By the end of World War I, all of England and Wales had been mapped by the Geological Survey of Great Britain at scales of one inch or six inches to one mile (1 : 63 360 or 1 : 10 560) (Bailey 1952). A programme of re-surveying areas in more detail was implemented post-war, although with priority for coalfield areas. By 1940, the UK was covered by published geological maps of the 'old' or 'new' series at a scale of 1 : 63 360 with sheets at 1 : 10 560 available for reference at Geological Survey offices. However, War Diaries indicate that the Tunnelling Companies needed some guidance by geologists, at least in the early months. This came both from geologists serving as officers in the Royal Engineers and from the civilian staff of the Geological Survey of Great Britain.

For 171, on 11 November 1940, 'Lieut. Young R.E. (Geologist) 141 O.C.T.U. [Officer Cadet Training Unit] examine[d] the beds exposed in the adits of the [Aldershot] Command Shelter' (WO 166/3652). A near-contemporary edition of *The Army List* records 11 junior Royal Engineers officers with the surname Young, but none known to be a geologist. The officer involved seems not to have contributed further to military geology. 141 OCTU was then based at Shorncliffe, on the coast near Folkestone in Kent, *c.* 150 km to the east of Aldershot (Edmonds 2002).

For Headquarters 1st Tunnelling Engineers, on 13 January 1941, the commanding officer (Lieutenant-Colonel Hill) and the adjutant (Captain Fowke) were accompanied on a visit to proposed shelter sites for 4th Corps at Paddockhurst and Westgate Place, *c.* 35 km north of its base at Brighton, by 'Major King OBE, MC' (WO 166/3535). King made further site visits before leaving for Tunbridge Wells on 16 January 1941. He had already contributed distinguished service as the pioneering British military geologist in France in both World War I and World War II (Rose & Rosenbaum 1993; Rose 2004). Following evacuation from France in 1940, he served as the Staff Officer (Geology) at Northern Command, whose area of responsibility effectively consisted of the NE quarter of England. From 1941 he held a similar appointment at Headquarters Home Forces, with responsibility for the UK as a whole. In October 1943 he was released from the Army, as a recently promoted lieutenant-colonel, to accept appointment as Woodwardian Professor of Geology at the University of Cambridge.

For 173, on 12 February 1941, an inspection of the groundwater situation at Tunbridge Wells was made by 'Dr Buchan (geologist)' (WO 166/3654). For Headquarters 1st Tunnelling Engineers, 'Dr. Buchan of the Geological Survey' visited the 12th Corps dug-out and the Canterbury dug-out with the commanding officer, when they arranged to allow 40 ft of cover at the Canterbury dug-out to keep its floor above the water table (WO 166/3535). Stevenson Buchan (1907–1996) was a senior hydrogeologist with the Geological Survey of Great Britain (from 1960 Assistant Director and 1968–1971 Deputy Director), whose duties included giving assistance to the armed forces in wartime (Gray 2004; Mather 2012). Most notably, he was involved in locating sites for boreholes to supply water to many military camps, airfields and wartime industries in the UK.

3rd Tunnelling Engineers in Gibraltar, 1940–45

Defeated by the German attack westwards in May 1940, the French government was forced to sign an armistice with Germany on 22 June 1940. As German victory over France became inevitable and victory over the UK seemed probable, Italy declared war on both countries on 10 June 1940, in the hope of making territorial gains from the French and British Mediterranean and African dependencies (Dear & Foot 1995). With German troops poised at the border between France and Spain, and the Italian fleet in the central Mediterranean not far away to the east, the British quickly strengthened the garrison defending their outpost of Gibraltar (Fig. 6), a 6 km^2 peninsula jutting south from Spain at the western entrance to the Mediterranean Sea. It had been progressively fortified as a base for the Royal Navy since ceded by Spain to Great Britain in 1713.

Tunnelling Companies in operation

The narrow peninsula is dominated by the Rock, a landmark famous to seafarers since antiquity. A mass of Early Jurassic dolomite and limestone that rises to a sharply ridged crest with peaks >400 m high, this had been subject to five phases of intensive tunnelling (Gale *c.* 1969; Rosenbaum & Rose 1991; Rose 2001):

(1) 1782–99, behind the steep north face of the Rock, which overlooks the low-lying sandy isthmus extending northwards into Spain, to provide vantage points from which cannon could fire;
(2) 1880–1915, when the army, Royal Navy and City Council all began underground excavation to provide ammunition magazines, communication tunnels or water reservoirs;
(3) 1933–38, when the construction of additional water reservoirs, plus air-raid shelters and underground hospitals, caused much tunnelling activity;
(4) 1939–45, when the greatly increased garrison required new accommodation space, and space for reserves of food, equipment and ammunition; and
(5) 1956–68, when the existing tunnels were linked together in a more efficient manner and additional storage chambers, reservoirs and access routes were hewn out of the bedrock.

Fig. 6. Aerial view of Gibraltar from the SE, prior to 1990 and therefore also prior to the subsequent urban development that has now masked some of the natural features. Two wave-cut southern plateaus (the Europa Flats and Windmill Hill Flats) form the foreground. Behind them, the main 2.5 km long main ridge of the Rock rises to peaks >400 m high. The slope west to the harbour is indicative of the steep westerly dip of the limestone bedrock. The steep scarp cliffs of the eastern Rock are fringed towards their base by slopes of Quaternary scree breccia and dune sand. A low sandy isthmus, dominated by the airfield, connects the peninsula northwards to the Spanish mainland and the town of La Linea, seen in the far distance. From Rose (2016), courtesy of John Wiley & Sons Ltd, the Geologists' Association and the Geological Society of London.

Phase 4 began in earnest in 1940. Reinforcements to the garrison to arrive that year included 178 Tunnelling Company Royal Engineers (Table 3). The foundation of 178 at Chatham had been authorized on 11 May 1940, with a core of officers recalled earlier from France for potential service in Norway (WO 176/223). However, Allied troops had been forced to evacuate from Norway at the beginning of June and on 7 June 1940 Norway's king and government departed for exile in the UK (Dear & Foot 1995). Up to strength on 29 May 1940, but ready for mobilization only on 15 June 1940, 178 was thus not needed for the Norwegian Campaign. It trained in England until embarking at Liverpool for Gibraltar on 20 August 1940 (WO 176/223).

178, consisting of five officers and 159 other ranks and led by Major G.A. Moorhead (Fig. 1), arrived at Gibraltar on 28 August 1940. Detachments from the company quickly began tunnelling in three areas of the northern Rock: behind the north face on

Table 3. *3rd Tunnelling Engineers: Company deployments to/from Gibraltar*

Company	1940	1941	1942	1943	1944
178	Arrived 28 August	Operational	Operational	Departed 22 April	
180	Arrived 3 October	Operational	Operational	Departed 6 July	
170		Arrived 10 March	Operational	Departed 9 November	
1 Canadian	Arrived 26 November	Departed 11 March			
2 Canadian		Arrived 10 March	Departed 14 December		
172				Arrived 26 May	Operational

30 August; at a chamber called Glen Rocky (to accommodate a distillery for fresh water) on 2 September; and at Willis's Gallery on 3 September 1940.

On 3 October 1940, 178 was joined by 180 Tunnelling Company, and by a headquarters company and workshop section, to found 3rd Tunnelling [Engineers] Group (Table 3) under the command of Lieutenant-Colonel D.M. Thomson (Fig. 1). Authority to form 180 had been issued on 4 June 1940 and Captain A.R.O. Williams (Fig. 1) was promoted to command it. (Authority to form 179 had been issued earlier, on 26 May 1940, with Acting Captain T. Pickering in command.) 180 grew to full strength in June 1940 and trained in July–August 1940 by tunnelling for a gun position at Langdon Battery, near Dover, Kent, before moving to Liverpool on 17 September 1940 for embarkation. It was at work tunnelling on Gibraltar from 8 October 1940, focused on projects at the south end of the Rock, while 178 concentrated on those to the north (WO 176/221). On 26 November 1940 a detachment of the 1st Tunnelling Company, Royal Canadian Engineers (RCE), arrived to begin work for an underground hospital.

Tunnelling work for the Group as a whole was pressed forward energetically, with breaks only for Christmas and New Year's Day. Excavation for the Glen Rocky water distillation plant was completed on 3 February 1941: 7000 tons of rock had been removed to form a chamber 108 × 32.5 × 27 ft (32.9 × 9.9 × 8.2 m) in size. By 6 February 1941 the cumulative total of diamond drilling had reached 1590 ft (485 m). On 18 February 1941, a fuel chamber for Glen Rocky was completed. Various other tasks were also completed or initiated during the month.

On 10 March 1941, 170 Tunnelling Company arrived. This consisted of seven officers and 217 other ranks, then or soon led by Major E.F.T. 'Jonah' Maunsell (Wilson 1946, p. 227; Edmonds 2002). 2nd Tunnelling Company RCE (Fig. 7) arrived on the same day to replace the diamond drilling detachment of 1st Tunnelling Company RCE by a full company. The detachment left for the UK the following day. The group was now complete with four companies (Table 3), all soon to be hard at work (Table 4).

Table 4. *Tunnel distance (in feet) and rock excavated (in tons) by Tunnelling Companies on Gibraltar in April 1941 as 3rd Tunnelling [Engineers] Group became fully operational*

Company	Week ending 3 April		Week ending 10 April	
	Footage	Tonnage	Footage	Tonnage
178	205	731	108	463
180	110	1209	103	1646
2nd RCE	63	1293	44	1133
170	64	590	68	467

Data from National Archives file WO 176/221.
RCE, Royal Canadian Engineers.

Fig. 7. 2nd Tunnelling Company Royal Canadian Engineers, photographed when serving as part of 3rd Tunnelling [Engineers] Group on Gibraltar during World War II. Reproduced by permission of The National Archives, Kew, from file WO 176/226.

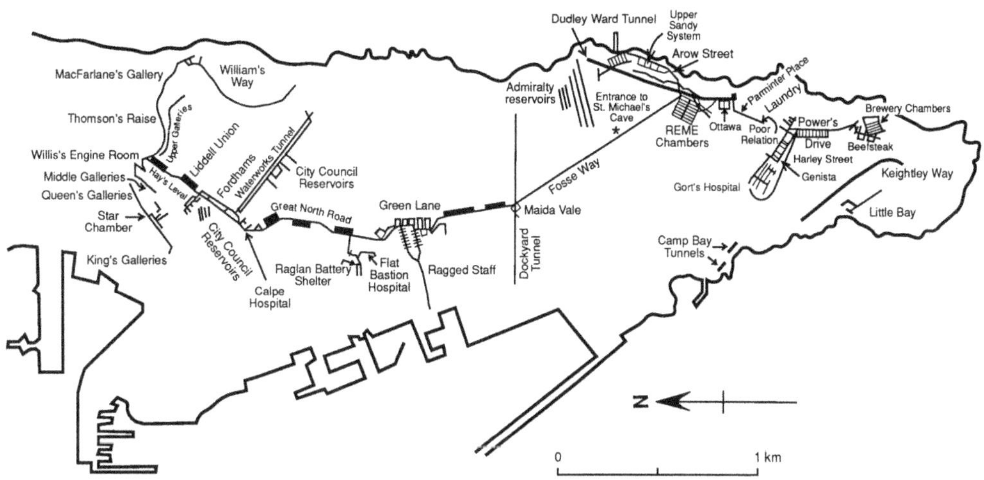

Fig. 8. Plan of the main tunnels on Gibraltar. The Upper Galleries behind the north face were constructed in 1782–99 and the Keightley Way and Dudley Ward tunnels in southern areas in 1956–68. The other tunnels shown all date from World War II. Reservoirs and minor tunnels cannot conveniently be illustrated at this scale, but see plans in Ramsey (1978) and Rosenbaum & Rose (1991). From Rosenbaum & Rose (1992), courtesy of John Wiley & Sons Ltd, the Geologists' Association and the Geological Society of London.

Weekly returns for these four units similar to those shown in Table 4 are preserved in WO 176/221 until it ends in October 1941. The official *History of the Corps of Royal Engineers* is thus in error to claim that reinforcements to the garrison during 1940 included 172 and 179 as well as 178 and

Fig. 9. The early stages of construction of the Royal Electrical and Mechanical Engineers workshops on Gibraltar during World War II (REME Chambers in Fig. 8). With a span of 16 m, a height to crown of arch of 10 m and length of 120 m, this unlined chamber was the largest excavated within the Gibraltar 'limestone'. From Rose (2001); © Imperial War Museum, photo RG128, and reproduced by permission.

Table 5. *Tons broken per company during excavation of tunnels/chambers on Gibraltar*

Company	1940		1941		1942		1943		Totals 1940–43	
	Total	Month	Total	Month	Total	Month	Total	Month	Total	Month
178	9 688	2 422	52 350	4 362	84 000	7 000	38 804	9 701	184 842	5 776
180	7 377	2 459	75 168	6 264	99 768	8 314	74 198	11 850	156 511	7 774
170			43 947	4 395	81 318	6 776	81 423	9 047	206 688	6 667
172							31 985	7 996	31 985	7 996
Canadian	3 190	1 595	58 410	4 867	92 239	8 385			153 839	6 149
Totals	20 255	6 475	229 875	19 888	357 325	30 475	226 410	38 594	833 065	38 362
Average		2 250		5 000		7 600		9 850		6 770

Data from National Archives file WO 176/162.
1 ton = 14 ft^3 (*c.* 0.5 cubic yard).

180 Tunnelling Companies Royal Engineers and a detachment of 1st Tunnelling Company RCE (Pakenham-Walsh 1958*a*, p. 324). Moreover, it is only partly correct that, in 1941, 170 Tunnelling Company Royal Engineers and 3rd Tunnelling Company RCE arrived because the Canadian company was in fact the 2nd rather than the 3rd Tunnelling Company RCE (Pakenham-Walsh 1958*a*).

In 1942, Lieutenant-Colonel Thomson was succeeded in command of the group by Lieutenant-Colonel A.R.O. Williams, who was promoted from command of 180 Tunnelling Company. Major D.H. 'Dai' Williams duly succeeded 'Bill' Williams in command of 180 Tunnelling Company (Edmonds 2002).

Although the National Archives do not preserve the 1942 and 1943 War Diaries, either for the Tunnelling Companies on Gibraltar or the 3rd Tunnelling Group, considerable detail may be gleaned from the War Diary of the headquarters of the Chief Engineer Gibraltar (WO 176/162). This indicates that the group was still operating at nearly full strength when the Gibraltar garrison as a whole reached its peak strength in May 1943. There were then 26 officers in post within an establishment for 28, and 756 other ranks within an establishment for 796.

The group was to construct a major complex of tunnels and chambers within the Rock (Fig. 8). A northern system was built first, largely developed behind the north face. A southern system was developed later, in the then military base area on the SE side of the peninsula facing the Mediterranean Sea. These two systems were later joined by a north–south access tunnel, the Great North Road, and its continuation known as the Fosse (or Foss) Way, a tunnel completed in 1944. (The Fosse Way was a major Roman road in England that linked Exeter in the SW to Lincoln in the NE. Major Lindsay Foss commanded 172 Tunnelling Company on Gibraltar until his death from a cerebral haemorrhage on 23 November 1943, shortly after he had sunk a German submarine while a 'guest gunner' on an RAF aircraft flying from the Rock; see Foss 1998.)

Tunnelling methods (Wilson 1946; Haycraft 1946; Cotton 1948, 1963; Ramsey 1978) were governed by the equipment available, which mainly consisted of hand-held jackhammers for the breaking of

Table 6. *Tunnel/chamber development on Gibraltar: feet advanced per company*

Company	1940		1941		1942		1943		Totals 1940–43	
	Total	Month	Total	Month	Total	Month	Total	Month	Total	Month
178	3132	696	5544	462	Incomplete		2240	547	10 916	520
180	693	231	5364	447	6828	569	5110	819	16 754	507
170			2390	239	4272	356	5950	661	12 620	407
172							2396	549	2090	522
Canadian	Records incomplete								5386	220
Totals	3825	927	13 298	1148	11 100	925	15 696	2626	47 756	2176
Average	11 912	309	4433	383	5550	462	3824	656	9553	435

Data from National Archives file WO 176/162.

ground and loaders for mucking out. Communication tunnels were cut to 7 × 7 ft (2.1 × 2.1 m) profile, with 13–18 holes of 4–5.5 inches (102–140 mm) depth per round of explosive. The speed of advance averaged *c.* 10 ft (3 m) per day, achieved by three shifts of four men each. Roads were cut to profiles of 12 × 12 ft (3.7 × 3.7 m) or 14 × 12 ft (4.4 × 3.7 m), with similar holes, and a similar speed of advance was achieved by three shifts of seven men each. Where roof-ripping could be used, this increased the speed of advance to 40 ft (12.2 m) per day per bench. Chambers (Fig. 9) were mined at an overall average rate of *c.* 300 cubic yards (229 m^3) per company section, the rate of mucking out being the main controlling factor. The tunnels and chambers were to remain generally unlined, but stable, for the rest of the war and long beyond (Rosenbaum *et al.* 1994).

The rate of excavation increased from an average of 2250 tons per company per month in 1940, to 5000 tons in 1941, 7600 tons in 1942 and 9850 tons in 1943 when the work passed its peak (Table 5). The speed at which the tunnels advanced increased correspondingly from 309 ft per month in 1940, to 383 ft per month in 1941, 462 ft per month in 1942 and 656 ft per month in 1943 (Table 6). During the years 1941 and 1942, plus the first nine months of 1943, the period for which most detailed records were kept, the consumption of explosives had in total reached 1 171 486 lbs (531 377 kg) of powder, 521 358 detonators and 3 698 879 ft (1 127 418 m) of fuse. Diamond drilling to facilitate excavation had achieved a cumulative total of 12 040 ft (3670 m), including holes at diameters of 1.5, 1.875, 2.5 and 3.625 inches (38, 48, 64 and 92 mm), consuming on average 0.252 diamonds per foot of hole.

Work completed after the outbreak of war to the end of September 1943 was to exceed that completed in all phases before the war (Table 7). By then, the excavation of tunnels and chambers had created a tunnel system measured as 67 680 ft (>20 km) in length and led to the removal of *c.* 794 820 cubic yards (607 683 m^3) of rock. From October 1943, it was scheduled that the remaining Tunnelling Company would complete excavation of a further 123 560 cubic yards (94 468 m^3) of rock, driving tunnels a further 6520 ft (nearly 2 km) in cumulative length (Table 8).

Geological guidance

The geology of Gibraltar had been studied since the eighteenth century (Rose 2013) and an unpublished geological map prepared by the Director of the Geological Survey of Great Britain in the late nineteenth century provided guidance for the initial tunnelling (Ramsay & Geikie 1878; Rose 2014*a*). It was

Table 7. *Summary of underground work on Gibraltar completed before the outbreak of war; after the outbreak to 30 September 1943; and in total to that date, by volume (cubic yards) and tunnel length (feet)*

Client	Pre-war				From outbreak to 30 September 1943				Total to 30 September 1943			
	Chambers	Tunnels	Total vol.	Tunnel length	Chambers	Tunnels	Total vol.	Tunnel length	Chambers	Tunnels	Total vol.	Tunnel length
The Admiralty	99 910	67000	166 910	12 600	7440	5600	13 040	1510	107 350	72 600	179 950	14 110
The Army	18 770	17 700	36 470	7470	310 800	106 000	416 800	36 000	329 570	123 700	453 270	43 470
City Council	97 570	14 300	111 870	6030	40 080	9650	49 730	4070	137 650	23 950	161 600	10 100
Total	216 250	99 000	315 250	26 100	358 320	121 250	479 570	41 580	574 570	220 250	794 820	67 680

Data from National Archives file WO 176/162.
Note: tunnel length is an accurate figure, but tunnel volume an estimate only because tunnels vary in cross-sectional area.

Table 8. *Programme of tasks for the Tunnelling Company Royal Engineers and City Council on Gibraltar to complete from 1 October 1943 onwards, following departure of 3rd Tunnelling Engineers Group, with estimated volume of rock to be removed and tunnel length to be created*

Location	Volume (cubic yards)	Length (ft)
Great North Road South	1790	
Great North Road South power station and cookhouse	1770	
Great North Road South Accommodation Chambers	21 770	
Ration store and cookhouse	4300	100
Bakery and flour store	1330	
Foss Way (connection Great North Road to Southern System)	20 700	3300
Norfolk House accommodation chambers	12 650	30
Royal Electrical & Mechanical Engineers (REME) workshop	33 240	1450
REME vehicle park (quarry)	11 000	
Powers drive accommodation	10 160	
Drive Northern System to Princes Lines	750	400
Drive Great North Road to Fordhams	820	450
Winze Great North Road to ragged staff magazine	600	330
Forbes Shaft	60	
Drive Ball Alley to North Gorge Tunnel	360	200
Rock accommodation	1030	60
MacPherson's Chambers	1230	200
Total	123 560	6520

Data from National Archives file WO 176/162.

evident that the Rock consisted of a thick sequence of massively bedded crystalline Jurassic 'limestone' (from 1943 known to be partly dolomite), cut by a fault zone that extended NW–SE from the south of the harbour. Strata in the main ridge of the Rock north of the fault dip to the west, whereas those in the southern plateaus to the south of the fault dip to the east. A 'shale' sequence (consisting of mudrocks and thinly bedded limestone) was believed to overlie the 'limestone' on the western coast.

Wartime tunnelling near the fault zone encountered bad ground and had to be abandoned. Contemporary quarrying at the northern and eastern margins of the Rock revealed more 'shales' and shales were also encountered in some low-level tunnels driven within the Rock, giving further problems. It was therefore realized in early 1943 by E.B. (later Sir Edward) Bailey, the then Director of the Geological Survey of Great Britain, in transit from England via Gibraltar to Malta, that a new geological map was required and that Sapper A.L. Greig (then stationed on the Rock as a driver) was well qualified to compile it (Bailey 1953; Rose 2014*b*). Greig was a graduate of Imperial College, University of London, who had taken courses in geology as part of his degree programme. His unpublished map (Fig. 10) and report completed later that year, which interpreted the 'shales' as a unit underlying, rather than overlying, the 'limestone', were to influence tunnelling on the Rock for the rest of the war, although to be superseded after it (Rose & Rosenbaum 1991; Rose 2016).

1st Tunnelling Engineers in Malta, 1941–45

173 Tunnelling Company, commanded by Major J.B. Simpson (Fig. 1), arrived in Malta on 25 July 1941 (WO 166/3654). The Maltese islands, of which Malta and Gozo are the largest, form a small archipelago in the central Mediterranean Sea, *c.* 95 km south of Sicily and 300 km east of the Tunisian coast of North Africa. The archipelago was then a British colony with a civilian population of 270 000, plus a garrison of British land, naval and air forces. By refusing to capitulate to intense aerial bombardment, the islands played a crucial part in the battle for the Mediterranean that raged from the summer of 1940, following Italy's entry into the war, until the Italian surrender in September 1943 (Dear & Foot 1995). Air attacks by Italian aircraft had begun on 11 June 1940 and were supported by Luftwaffe units based in Sicily from early 1941. During 1942 the intensity of the air attacks increased. There were 262 in January, 236 in February and, in March and April, twice the tonnage of bombs that London had suffered during the Blitz (a night-time bombing campaign from August 1940 to mid-May 1941) was dropped on Malta. From 1 January to 24 July 1942 there was only one 24-hour period in which no bomb fell. Blockaded also by mines and by naval forces, this siege of Malta brought great hardship to the population. On 15 April 1942, King George VI wrote to the Governor that 'To honour her brave people I award the George Cross [the supreme British award for civilian gallantry] to the Island

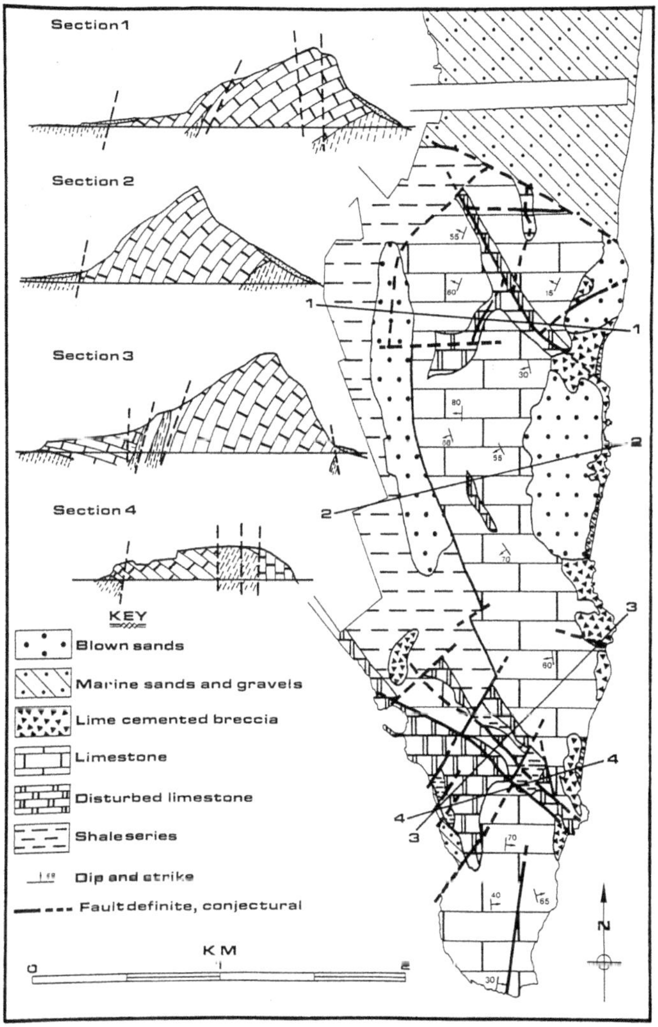

Fig. 10. Geological map of Gibraltar, together with four cross-sections, redrawn from an unpublished map and sections compiled by Sapper A.L. Greig in 1943. From Rose & Rosenbaum (1989), courtesy of the Institution of Royal Engineers.

Fortress of Malta to bear witness to a heroism and devotion that will long be famous in history'. Intensive air raids continued until November 1942, but diminished thereafter as Allied forces fought their way to victory in North Africa. The surrender of German/Italian forces in Tunisia in May 1943 effectively brought an end to the siege.

Malta was important because it had several airfields (Fig. 11) and the only British harbour between Gibraltar and Egypt. It was therefore crucial to British air and sea operations against the convoys supplying German and Italian forces fighting in the Western Desert of Egypt and Libya and subsequently in NW Africa.

The islands have a total land surface of *c.* 315 km^2 with a mid-Cenozoic bedrock succession of marine limestones and marls, ranging in age from Late Oligocene to Late Miocene and reaching a total exposed thickness of *c.* 500 m (Baldassini & Di Stefano 2017). Five formations had, by 1941, long been recognized in this sequence: Lower Coralline Limestone, succeeded by Globigerina Limestone, Blue Clay, Greensand and Upper Coralline Limestone (Fig. 12). The sequence, although disrupted by faulting, is essentially horizontal, so the pattern of outcrop is determined principally by topography. The youngest formations crop out in upland areas, which, on Malta itself, occupy the west of the island. The

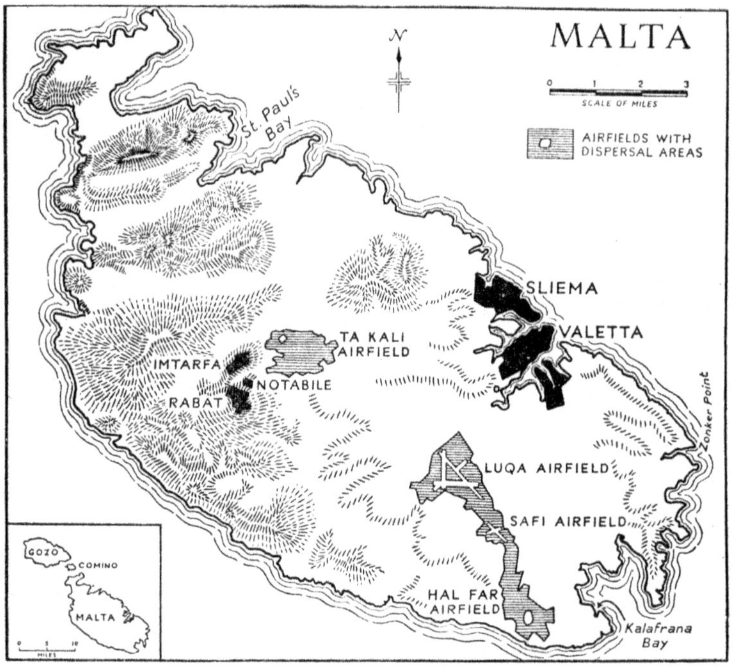

Fig. 11. Map of Malta during World War II showing locations of Royal Air Force airfields and major towns, and hachured relief. © Imperial War Museum, photo MH 4218, and reproduced by licence.

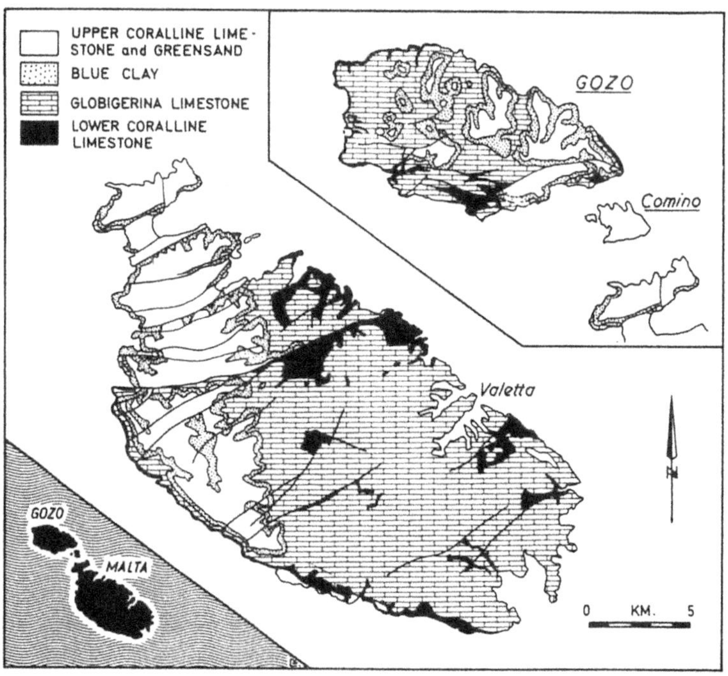

Fig. 12. Geological map of Malta and Gozo showing lithostratigraphic units as adopted for military use during World War II. From Rose (2012*b*), courtesy of the Geological Society of London, after Rose *et al.* (1992).

Grand Harbour with its naval dockyards, near the city of Valetta, and the RAF airfields were all sited on Globigerina Limestone, the rock of widest outcrop and the island's principal building stone, in which most tunnels were to be excavated.

Tunnelling Companies in operation

After unloading its equipment and settling into the new environment, 173's four sections were quickly at work, constructing underground facilities for a hospital, two army medical dressing stations and a Royal Army Service Corps depot, the first of 32 projects to be assigned by the end of 1942 (Table 9). A brief project report was completed for most of these, recording what was done and where. Each was typically accompanied by a plan (Fig. 13). These reports were subsequently filed as annexes to the July, August, September and October War Diaries of 1st Tunnelling Engineers for the year 1943 (WO 169/

Table 9. *Projects on Malta by 173 Tunnelling Company Royal Engineers*

Job	Project	Grid reference	Start	Finish
M/1	Imtarfa Hospital	363251	6 August 1941	1 November 1941
M/2	Musta Medical Dressing Station	398278	7 August 1941	19 September 1941
M/3	San Paul tat Tarja ADS	410290	7 August 1941	16 September 1941
M/4	Sa Maison (RASC)	468254	11 August 1941	18 August 1941
M/5	Luqa (Bren gun carriers)	455204	18 August 1941	6 October 1941
M/6	Poorhouse (Bren gun carriers)	457223	22 August 1941	27 December 1941
M/7	Central Infantry Brigade (battle HQ)	441258	17 September 1941	29 November 1941
M/8	St Andrews ADS (shelter)	448293	22 September 1941	29 November 1941
M/9	Command Supply Depot (store)	475262	29 October 1941	6 July 1942
M/10	Bomb disposal magazine	471257	7 October 1941	30 October 1941
M/11	St Patrick's Hospital	440299	5 November 1941	6 June 1942
M/12	Guardamangia (shelter addition)	462254	15 November 1941	29 November 1941
M/13	Cables and wireless (St George's)	458291	9 November 1941	6 December 1941
M/14	Gzira motor transport depot	460265	24 November 1941	25 July 1942
M/15	Gargur C (RAOC)	432285	2 December 1941	1 March 1942
M/16	South Infantry Brigade (battle HQ)	478223	5 December 1941	26 March 1942
M/17	223 (light anti-aircraft) Battery RA	475251	6 January 1942	28 February 1942
M/18	Ozpizic (RAOC depot)	471257	11 March 1942	29 August 1942
M/19	Bakery (RASC)	468255	11 March 1942	9 July 1942
M/20	War HQ – Lascaris (lower tunnel)	481257	28 March 1942	9 June 1942
M/21	Marina Pinto (RAOC)			5 September 1942
M/22	L'Inglin (REME)	482273	18 June 1942	14 November 1942
M/23	Luqa petrol	447206	12 June 1942	1 February 1943
M/24	Dingli (RAF)	349207	16 June 1942	23 July 1942
M/26	Battle HQ (Tal Verte)		17 August 1942	9 January 1943
M/27	Sa Maison II (RASC)	469254	2 September 1942	19 September 1942
M/28	Luqa personnel shelter	455204	7 October 1942	18 October 1943
M/29	Tal Balal	432284	3 October 1942	13 February 1943
M/30	St Patrick's II	440299	10 November 1942	26 February 1943
M/31	Imtarfa II	363251	26 November 1942	16 October 1943
M/32	Krendi		2 December 1942	
M/33	War HQ (RAF War Hill)	479257	18 January 1943	20 February 1943
M/34	Luqa (RAF bomb store)	451204	9 January 1943	20 February 1943
M/35	Hal Tar		8 February 1943	To 171 TC
M/36	War HQ	480257	16 February 1943	19 March 1943
M/37	Safi strip	468189	8 March 1943	27 March 1943
M/38	Attard Quarry (near Luqa)		2 March 1943	14 October 1943
M/39	Krendi strip	404184	2 March 1943	5 June 1943
M/40	Krendi Quarry		22 March 1943	
M/41	War HQ		29 March 1943	16 October 1943
M/43	Tal Balal		8 June 1943	2 July 1943
M/51	War HQ Calara Gardens	475253	1 April 1943	16 October 1943
M/52	Ta Kali strip	395252	6 April 1943	7 May 1943

Projects shown with grid references are known from completion reports/plans (in National Archives file WO 169/14601), others from War Diary entries only (in files WO 169/7435, 169/14603)

ADS, Army Dressing Station; HQ, headquarters; RA, Royal Artillery; RAF, Royal Air Force; RAOC, Royal Army Ordnance Corps; RASC, Royal Army Service Corps; REME, Royal Electrical & Mechanical Engineers; TC, Tunnelling Company.

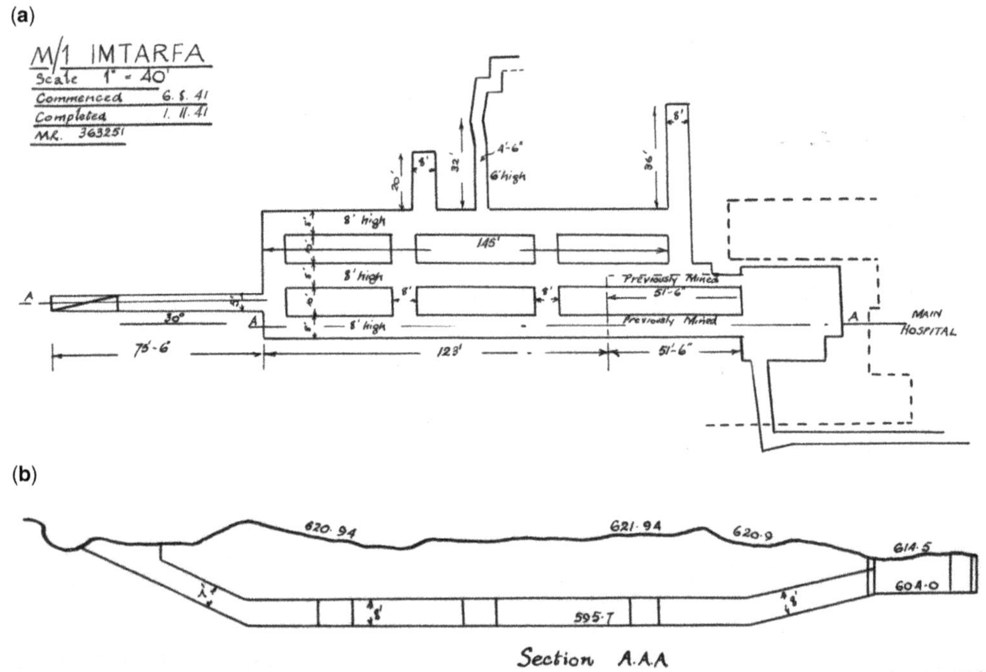

Fig. 13. Plan (a) and section (b) from completion report by 173 Tunnelling Company Royal Engineers of task M/1 Imtarfa Hospital (cf. Fig. 11; Table 9). Reproduced by permission of The National Archives, from file WO 169/14601.

14601). The first Malta tunnelling project (M/1) was one of the relatively few facilities to be excavated in Upper Coralline Limestone. It provided underground accommodation for beds and stretcher cases, access being by means of a ramp down from the main hospital. Three parallel 8 × 8 ft (2.4 × 2.4 m) wards with communicating passages were required, plus a small operating theatre and connection to an existing dug-out.

The company's War Diaries for 1941 and the first eight months of 1942 have not been found at the National Archives. The War Diary for the end of 1942 (WO 169/7435) contains only brief one-page reports for each of the four months and these primarily concern matters of administration. The Diary for 1943 (WO 169/14603) is also concise, but does note that Major Simpson was promoted to the rank of lieutenant-colonel in February to command 1st Tunnelling Engineers and that Major J.H. Lander (Fig. 1) succeeded him in command of 173. Excavation work by the Company gave way to the concrete lining of completed chambers during the year and to military training in October, the month in which the Diary comes to an end. From the records of 1st Tunnelling Engineers (WO 169/14601), it is known that 173 Tunnelling Company left the island by sea on 21 November 1943.

In 1943, Allied victories in North Africa and the neutralization and partial destruction of the Italian fleet largely ended the siege of Malta that had persisted since 1940 (Dear & Foot 1995). With reinforcements, the island began to prepare for its part in the Allied invasion of Sicily in July. In February 1943, Headquarters 1st Tunnelling Engineers arrived, together with 171 Tunnelling Company.

Headquarters 1st Tunnelling Engineers and 171 Tunnelling Company had left Scotland from the port of Greenock on 31 May 1942, their convoy taking the long, but relatively safe, route around Africa – via Freetown in Sierra Leone, Simonstown in South Africa and Diego Suarez in Madagascar – to reach Port Tewfik at the southern entrance to the Suez Canal in Egypt on 8 July 1942 (WO 169/5389, WO 169/5284). The strength of the headquarters consisted of three officers (Lieutenant-Colonel L.C. Hill, DSO, MC and two captains) and 42 other ranks. 171 consisted of seven officers (Major W. Fowke, one captain and five subalterns) and 212 other ranks. Quickly following their arrival, both units moved by train to Qassassin, a massive transit camp *c.* 50 km NE of Cairo, and to a camp nearby at Tahaq. There they settled in and were reunited with their equipment. In September 1942 the headquarters moved east to the Docks Construction Store Depot at El

Shatt on the Suez Canal, where the 'work required was far removed from tunnelling work, which is the normal function of this unit' (WO 169/5389). At this time, 171 moved nearby to Gebel Maryam, in the Suez Canal zone near the port of Ismailia, for construction and quarrying works rather than tunnelling. In January 1943 a new 'war establishment' came into force, changing the number of people authorized for tunnelling units Royal Engineers, and on 5 February 1943 both the headquarters and 171 left Egypt via Alexandria on the Mediterranean coast NW of Cairo, to arrive at Malta's Grand Harbour near Valetta on 10 February 1943 (WO 169/14601, WO 169/14602).

Three days later, Lieutenant-Colonel A.R.O. Williams (Fig. 1) arrived by air from Gibraltar to assess the tunnelling situation on Malta, with particular regard to stores and equipment. Also that month, Lieutenant-Colonel Hill was designated as Deputy Chief Engineer for Malta. Major Simpson, officer commanding the by now highly experienced 173 Tunnelling Company, was promoted to succeed him in command of 1st Tunnelling Engineers, whose headquarters now controlled both 171 and 173.

Considerable quantities of rock were to be excavated from then on (Table 10) to construct underground facilities for the army, the RAF and in the Royal Navy dockyard. There were surface excavations in addition to these underground works. The construction of an RAF airfield strip involved 10 510 cubic yards (8035 m^3) in April, 15 115 cubic yards in May and 2570 cubic yards in June 1943. Quarrying for relatively strong rock at Attard, near Luqa airfield, involved 12 986 cubic yards (9929 m^3) in April, 2185 cubic yards in May, 2717 cubic yards in June, 3219 cubic yards in July, 3366 cubic yards in August and 2682 cubic yards in September 1943.

Prior to 171's arrival, 173 had excavated rock primarily to create facilities for the RAF, to a lesser extent for the Royal Army Medical Corps and the Royal Army Ordnance Corps, plus war headquarters for all services, workshops for the Royal Electrical and Mechanical Engineers, shelters for the infantry and vehicle parks for the Royal Army Service Corps (Table 11). Work for the RAF continued to be a high priority after the arrival of 171, and work for the Royal Army Medical Corps and on war headquarters continued, but work for the Royal Navy dockyard replaced that of other former clients.

The records of work done by 171 Tunnelling Company are not in the form of completed project reports as previously for 173, but as diagrams showing progress to completion of various tasks attached to its monthly War Diaries (WO 169/14602, WO 169/16415, WO 169/18698, WO 169/22148) (Fig. 14). These tasks involved fewer excavations (cf. Table 12 with Table 9), but they were more extensive. The departure of 173 in November 1943 led to a reappraisal of the priority to be given to the few major tasks remaining.

On 18 April 1944 a conference convened by the Chief Engineer to the garrison agreed to a proposal that support be given to improve the Civil Government water supply (WO 169/18698). After careful health checks on the tunnellers involved, and stringent precautions with regard to hygiene, work on galleries for water abstraction began in the summer. A further conference, on 12 August 1944, involved not only representatives of Headquarters 1st Tunnelling Engineers and 171 Tunnelling Company, but also Captain Kneen and Lieutenant Morris

Table 10. *Cubic feet of rock excavated on Malta in 1943 by 1st Tunnelling Engineers units*

Month	For Army	For RAF	For dockyard	Total
February 1943	12 162	17 041		29 203
March 1943	53 225	41 636	21 493	116 354
April 1943	68 465	43 463	66 598	178 526
May 1943	66 881	42 036	98 794	207 711
June 1943	77 226	28 094	104 021	209 341
July 1943				276 351
August 1943	92 903	41 600	34 905	169 408
September 1943	71 448	4239	29 829	105 516
October 1943	Nil	10 269	45 384	55 653
November 1943	Nil	27 465	15 603	43 068
December 1943	Nil	38 488	26 872	65 360
Total	442 310+	294 331+	443 499+	1 456 491

1 cubic foot = 0.0283 m^3.
Data from National Archives file WO 169/14601.
Only a total figure is known for July.

Table 11. *Cubic feet of rock excavated by Tunnelling Companies Royal Engineers in Malta to end October 1943*

Client	173 1 August 1941– 13 February 1943	171 & 173 15 February 1943–1 May 1943	171 & 173 1 May 1943– 31 July 1943	171 & 173 1 August 1943– 31 October 1943	Total
RAMC	158 659	79 057	177 765	96 504	511 985
RASC	65 676				65 676
RAOC	144 590				144 590
REME	77 216				77 216
War HQ	79 907	59 489	113 741	88 907	342 134
Infantry	68 975				68 975
RAF	283 315	118 575	125 239	56 108	583 237
RN		88 721	335 869	116 561	541 151
Miscellaneous	6212				6212
Total	884 550	345 842	752 614	358 170	2 341 176

1 cubic foot = 0.0283 m^3.
Data from National Archives file WO 169/14601.
HQ, headquarters; RAF, Royal Air Force; RAMC, Royal Army Medical Corps; RAOC, Royal Army Ordnance Corps; RASC, Royal Army Service Corps; REME, Royal Electrical & Mechanical Engineers; RN, Royal Navy.

of No. 7 Boring Section Royal Engineers (Rose 2012*b*) and Captain Roberts of 42nd Geological Section South African Engineer Corps. Tunnelling

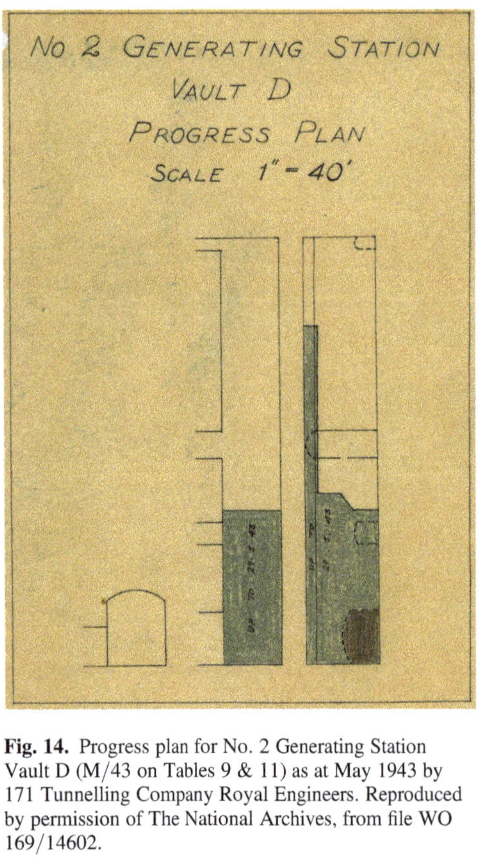

Fig. 14. Progress plan for No. 2 Generating Station Vault D (M/43 on Tables 9 & 11) as at May 1943 by 171 Tunnelling Company Royal Engineers. Reproduced by permission of The National Archives, from file WO 169/14602.

expertise was thus supplemented by expertise in the drilling of boreholes and in hydrogeology. Work in the dockyard came to an end that month and thereafter the tunnellers were focused on work at only two sites (Wied-is-Sueda and Ta Kali, both near Ta Kali airfield; Fig. 11), in addition to water supply. This continued until February 1945, when blasting operations began for a lido in St George's Bay (north of Sliema: Fig. 11), with 8350 ft^3 (236 m^3) of rock excavated. With that, the meticulous War Diaries of 171 come to an end, as the war in Europe too approached its end on 8 May 1945. Morris (1952, p. 34) was later to note that water supply projects 'involved far more ambitious shaft sinking than had hitherto been contemplated on Malta, and it was fortunate that, in Captain S.E. Pugh, R.E., of 171 Tunnelling Coy [Company], R.E., an officer was available with considerable experience of this type of work'. 171 moved from Malta north to Italy, and then Austria; its presence there was noted in June 1945 by No. 10 Boring Platoon (WO 170/5118).

Geological guidance

Pre-war understanding of Malta's geology (e.g. by Rizzo 1932) was sufficient to guide work by the Tunnelling Companies. The need to enhance the water supply, however, prompted six weeks of fieldwork by the Director of the Geological Survey of Great Britain (E.B. Bailey), who arrived from the UK via Gibraltar on 8 February 1943 (Morris 1952; Rose 2012*b*). In addition, there were visits in April, May and June 1943 by two geologists from General Headquarters of Middle East Command in Cairo, Egypt: the Geologist Staff Officer Captain F.W. Shotton

Table 12. *Tasks on Malta by 171 Tunnelling Company Royal Engineers according to its War Diaries*

Job	Task	Start	Finish
M/35	Hal Far aerodrome control room	25 February 1943	June 1943
M/42	Dockyard boiler shops	22 February 1943	?August 1943
M/43	Vault 'D' boiler shops	22 February 1943	By October 1944
M/49	Cospicua: services tunnel	By May 1943	By October 1944
M/53	Wied Dalam: cable tunnel	By July 1943	January 1944
M/54	Dockyard chamber: Manager Eng Dept	3 August 1943	By October 1944
M/55	Tal Balal/Ta Kali shaft	March 44	Not recorded
M/56	Wied-is-Sueda	By July 44	Not recorded
M/57	St. George's Bay lido	February 1945	Not recorded

National Archives files WO 169/14602, WO 169/16415, WO 169/18698, WO 169/22148.

and the officer commanding 42nd Geological Section of the South African Engineer Corps, Major G.L. Paver. From March until October 1944, one officer and 12 men were detached from 42nd Geological Section, based in Egypt, for service on Malta. This work enhanced the knowledge of Malta's geology, especially with regard to its hydrogeology and structural geology, although it was to be superseded in later years, most notably in terms of detailed stratigraphy, structure and palaeoenvironmental interpretation; see Baldassini & Di Stefano (2017) and references cited therein.

4th Tunnelling Engineers in the UK, 1941–42

4th Tunnelling Engineers was formed on 23 July 1941 at Scarborough in Yorkshire (Fig. 15) under the command of Lieutenant-Colonel D. Gilchrist MC (WO 166/3537). Its headquarters was given responsibility for the technical administration of 179 Special Tunnelling Company, then employed operationally with Northern, Scottish and Western Commands, 183 Special Tunnelling Company in the process of formation and training, and 184 and 185 Special Tunnelling Companies about to be formed and trained. It was proposed that one company would be allocated to each of the three northern commands (none to Southern or Eastern) (WO 166/3655). All companies were to be capable of undertaking straight tunnelling, flame barrage and field company work in addition to installing Canadian pipe obstacles. Orders were received in December that 179, 184 and 185 would join Western, Northern and Scottish Commands, respectively (WO 166/3537). 183 would remain at Scarborough until operational training was complete and then form part of the Home Forces General Headquarters reserve.

Flame barrage defence system

The flame barrage was an idea that originated with the Petroleum Warfare Department, an organization established in the UK in response to the invasion crisis of 1940 (Banks 1946). Based on experiments that began in August 1940, it was planned that, in the event of a German amphibious invasion, oil should be piped into the sea from sites on the coast and ignited. Ultimately, it was designed for use in locations where there were flat sandy beaches rather than in places, such as at Dover (Fig. 15), where there were steep cliffs that would bar easy inland access from an amphibious invasion (Cameron

Fig. 15. Map indicating areas in Britain protected during World War II by petroleum-fed flame defences, primarily the coastal flame barrage; the five regional command areas (a sixth, Aldershot Command, was renamed South-Eastern Command in 1941); and locations of principal towns associated with the work of 4th Tunnelling Engineers in 1941–42. Flame defences after Banks (1946), command areas after Dear & Foot (1995).

1999). Storage tanks were installed in the dunes near the sea and oil was pumped from them to headers along the shore. These headers were fitted with nozzles so that the oil, ignited by a pressure-actuated thermal charge or by wires heated by electricity, sprayed forward in flames. The nozzles were all positioned above the high-tide zone so that a wall of intensely hot flame could be produced whatever the condition of the sea.

On 24 February 1941, the Chiefs of Staff committee approved the installation of coastal facilities to create 50 miles (80 km) of flame barrage: 25 miles for SE Command, 15 miles for Eastern Command and 10 miles for Southern Command. However, the weapon could only be used when winds were favourable as it created a smokescreen potentially favourable to an attacker. Moreover, it was vulnerable to enemy bombing and shellfire. The resources required were considerable and there was a serious shortage of materials. The plans (Fig. 15) were therefore progressively cut back. Lengths of flame defence were ultimately completed in Kent at Deal, St Margaret's Bay and at Shakespeare Cliff near the Dover railway tunnel; in Sussex at Rye, where a remarkable system operated by remote control was installed across the marshes; and in Dorset at Studland Bay. In South Wales near Swansea, long stretches were put in hand at the time when the threat of invasion, potentially from the Republic of Ireland, was looming large. In Scotland, sections at Wick and Thurso (*c.* 30 km east and west, respectively, of the NE extremity of the Scottish mainland) were begun, but not brought to completion. At Porthcurno on the north Cornish coast, the small area shown on Figure 15 near the SW extremity of England where the important transatlantic cables came ashore, a gravity-fed section was put in as a security measure against raids. The flame barrage system was 'a category of defence provision for which nationally there is little surviving physical evidence' (Foot 2006, p. 69).

Canadian pipe mines

Canadian pipe mines, later known as McNaughton tube tank obstacles, were a type of landmine deployed in the UK when under threat of invasion in 1940–41. They consisted of pipes typically of 3 inch (76 mm) diameter and *c.* 55 ft (17 m) long packed with explosives, emplaced at a low angle at shallow depth (*c.* 8 ft (2.4 m)) and laid at intervals of 25 ft (7.6 m) (Fig. 16). When detonated, the resulting crater potentially formed an anti-tank obstacle or denied the use of a road or airfield to an invading enemy (WO 199/2661; Foot 2006). However, mine emplacement did not adversely affect Allied activities beforehand or reveal mine positions to enemy aerial reconnaissance until detonated. The

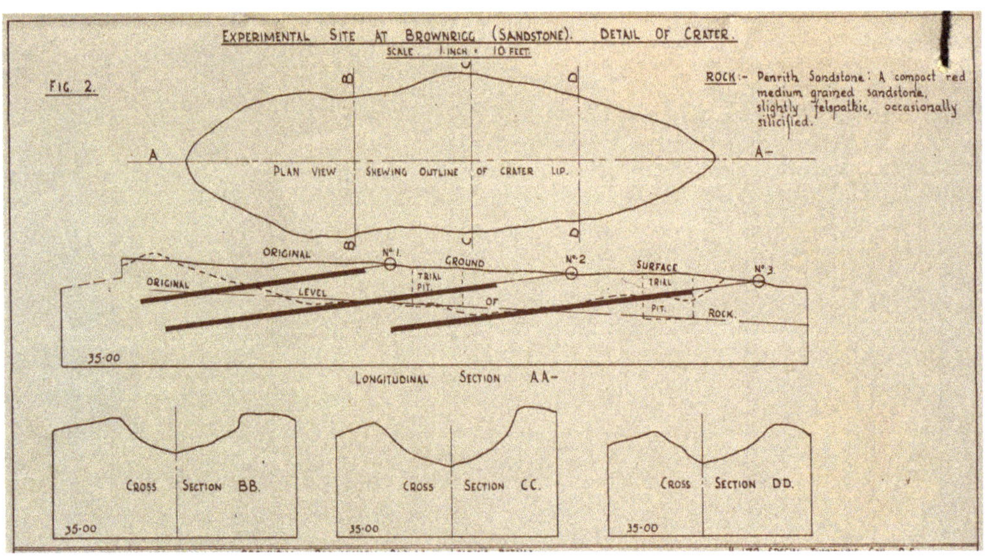

Fig. 16. Cross-section showing the emplacement of Canadian pipe mines (as typically) within low-angle boreholes, with plan and cross-sections of the crater formed after a trial detonation. Unusually for the War Diaries, this diagram also indicates the bedrock penetrated: part of the Bromsgrove Sandstone Formation, a unit (now obsolete) recognized at that time in the Lower Trias of the English Midlands (see Fig. 4). Reproduced by permission of The National Archives, from Special Tunnelling Technical Directive No. 15 Pipe Mines, annexed to the October 1942 War Diary in file WO 166/8043.

pipe could easily be pushed into position below the ground surface by a hydraulic jack fitted to a vehicle such as a tank transporter, or be emplaced within a shallowly inclined borehole.

Cameron (1999) has described how the idea for such a mine arose when the 1st Tunnelling Company of the RCE was given the task of preparing defences along England's south coast. Many of its men were from British Columbia and highly skilled in the use of diamond-tipped bits for drilling into metalliferous formations of strong rock. Such equipment and procedures could not be used in the weak rocks of southern England, but the drilling rigs were light and portable and easily moved on a truck or a road grader fitted with a flat steel platform. The gear ratios of the drills were therefore changed to allow for the quick penetration of soil and weak sedimentary rocks and the diamond bit was replaced by a simple steel collar into which teeth were cut. Using this modified equipment, the Canadian engineers prepared the approaches to bridges, narrow cuttings and high embankments for emergency demolition to create obstacles to enemy vehicular movement. The pipe mine was a further development of this procedure.

It was calculated (Special Tunnelling Technical Directive No. 15 in WO 166/8043) that aerodromes and landing grounds could be immobilized by the explosion of pipe mines emplaced to create a series of obstacles in the form of a grid with 100 yard sides. Demolition would be produced by pipe mines of 3 inch diameter, 60 ft in length. An aerodrome of average size (*c.* 1×10^6 square yards, requiring 375 holes) could be prepared by one of the four sections of a Special Tunnelling Company in 6–12 weeks, depending on the nature of the ground and other factors. Airfield construction had accelerated in World War II, achieving a historic peak of 856 in the UK (Blake 2000). Airfields were mostly concentrated in the southern and eastern counties (Fig. 17), potentially the area most vulnerable to seizure by German airborne forces because of their proximity to German bases in mainland Europe.

The pipe mine concept was promoted by the Canadian Corps Commander, Lieutenant-General Andrew McNaughton, after whom the mines came to be named. About 40 000 ft (12 km) of these mines were eventually installed in the UK, requiring 90 tons of explosives. They were expected to remain potent for several years, but, in some places, water led to deterioration of the explosives. These had to be removed and replaced. All the mines were supposedly deactivated at the end of the war, but a few were missed and deactivated subsequently. Foot (2006) has published plans of six existing defensive sites in England on which original pipe mine positions are indicated.

179 Tunnelling Company

The company formed on 22 May 1940 at Chatham in northern Kent, within 3rd Tunnelling Engineers (Group) commanded by Lieutenant-Colonel Thomson (Fig. 1) (WO 166/3655). Captain T. Pickering (also Fig. 1) was promoted to the rank of major to serve as 179's officer commanding. The company was up to strength in men, although not officers, by the end of the first week in June. Its first work was to put the Inner Lines at Chatham in a state of defence. Chatham, with maritime access to the Thames estuary and so westwards to London, had contained an important naval dockyard since the sixteenth century. The need to protect it by some kind of linear bastioned artillery fortifications had been appreciated since a raid by the Dutch navy in 1667. The Inner Lines formed an open space within the protective ramparts of the Great Lines created in the mid-eighteenth century. Subsequently, they were largely developed as a recreational area, but anti-tank obstacles were constructed when the Line system was massively refortified during World War II. 179's later tasks included additional defensive work and duties in the Chatham and nearby Rochester region.

The unit was re-formed in August as a 'Special Tunnelling Company' for training in 'pipe [mine] pushing', but continued with similar defensive duties. Lieutenant-Colonel Thomson, the headquarters of 3rd Tunnelling Engineers, and 180 Tunnelling Company left Chatham for Gibraltar supposedly *en route* to East Africa in September, and 179 began to receive intensive training in pipe pushing. A mixture of training and works continued until February 1941, when the company moved to Mickleham in Surrey, 45 km east of Aldershot (location 5 on Fig. 4), for training by 1st Tunnelling Company RCE. In May, the company moved again, to Scarborough in Yorkshire, with detachments in several nearby locations. In July 1941 Major Findlay (Fig. 1) succeeded Major Pickering in command, to be succeeded in turn by Major J.H. Bingley in November 1941.

Detachments from the company were to serve at locations in Scotland and Wales as well as northern England, but after assignment to Western Command in December 1941 (WO 166/3537), the company headquarters moved in 1942 from Atherstone in Warwickshire (in the Western Command region shown on Figure 15, but close to the triple boundary with the Northern and Southern Command regions in central England) to Cardiff in Wales (*c.* 55 km ESE of Swansea: Fig. 15) (WO 166/8017). Later it moved to the Mumbles area, 9 km SW of Swansea. However, the War Diaries provide details only of administration, not of the works undertaken. At the end of the year, on 25 December 1942, 179 was converted from a Tunnelling to a Field Company, the

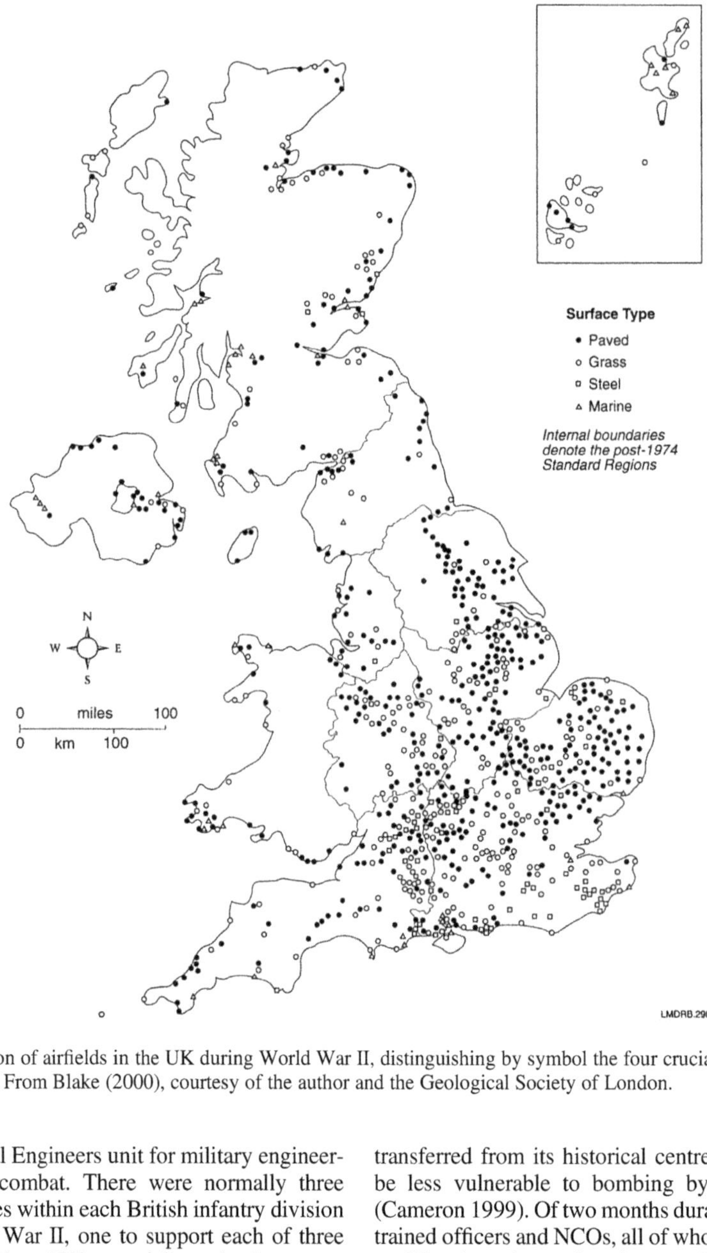

Fig. 17. Location of airfields in the UK during World War II, distinguishing by symbol the four crucial types of landing surface. From Blake (2000), courtesy of the author and the Geological Society of London.

standard Royal Engineers unit for military engineering work in combat. There were normally three field companies within each British infantry division during World War II, one to support each of three infantry brigades. 179's specialist role thus came to an end.

183 Tunnelling Company

183 Special Tunnelling Company formed at Scarborough on 15 April 1941, with Major B. Llewellyn MC in command (WO 166/3656). Key personnel were assigned from the Special Tunnelling Course at the School of Military Engineering. This was then based at Ripon (Fig. 15), also in Yorkshire, having transferred from its historical centre at Chatham to be less vulnerable to bombing by the Luftwaffe (Cameron 1999). Of two months duration, the course trained officers and NCOs, all of whom had previous qualifications in engineering, to Unit Instructors Standard (WO 166/3656). The unit as a whole trained in the following months, but was certainly operational by mid-November. No. 1 Section was then drilling at Eakring near Nottingham (Fig. 15) under the supervision of the D'Arcy [Petroleum] Exploration Company, a subsidiary of the Anglo-Iranian Oil Company, which became British Petroleum and later BP (Longhurst 1959). No. 2 Section was working under the supervision of 1st Tunnelling Company RCE at Hythe aerodrome in Kent,

c. 28 km NE of Rye (Fig. 15). No. 3 was drilling 6-inch holes *c*. 50 km east of London in Kent at Gravesend and Bradwell aerodromes. No. 4 Section was drilling holes at Dover. Similar work was to continue for the rest of the year. The oil well drilling at Eakring attracted a visit by the managing director of the D'Arcy Exploration Company, accompanied by a geologist and a geophysicist in December 1941 (WO 166/3537). The need to develop sources of petroleum within the UK was given high priority to reduce dependence on importation by ships prone to attack by the German Luftwaffe and navy, especially by submarines (Longhurst 1959).

During 1942, the company was still based at Scarborough, but undertook work at many localities (WO 166/8109). One section continued to be employed drilling oil wells at Eakring until June 1942. Other section work involved 'straight tunnelling', for auxiliary unit (Home Guard) shelters, or on attachment to 172 Tunnelling Company. However, a significant proportion of the work was 'special': preparing airfields for emergency demolition. These were primarily located on the Isle of Wight in the central south coast area of England (Fig. 15) and the adjacent mainland area of southern Hampshire. This part of Hampshire includes the New Forest, one of the largest tracts of unenclosed pasture land, heathland and forest in the otherwise heavily populated south of England and therefore used for the construction of several airfields (e.g. at Hurn, Beaulieu and Holmsley). Major C.H. Pitt, formerly the second-in-command, was promoted to succeed Major Llewellyn in command at the end of March 1942. In December 1942 the company moved south from Scarborough to Leigh-on-Sea, on the northern Thames estuary 50 km east of London, and on 1 March 1943 it converted to a Field Company Royal Engineers. Major Pitt relinquished command later that month and the company's 'special' role thus completely ended (WO 166/12083).

184 Tunnelling Company

184 Tunnelling Company began forming at Scarborough on 30 July 1941, with Major S.V. Griffith in command from 3 September 1941 (WO 166/3657). It remained at Scarborough until 24 February 1942, its sections deployed on a variety of tasks in the region, including the construction of auxiliary unit shelters and the preparation of obstacles (WO 166/8110). Thereafter it moved to Harton, near Newcastle (Fig. 15). Its officers are recorded as making various visits connected with the emplacement of pipe mines and attending a flame barrage demonstration in June 1942. Two sections were moved on the completion of work in Northumberland to aerodrome immobilization tasks for Southern Command in July 1942. On 29 November 1942 the company headquarters moved 60 km south to Gainford in County Durham (Fig. 15), and specialist work was replaced by training. On 8 January 1943 Major Griffith took part in a discussion on the inspection and maintenance of pipe mines in the Northern Command area; on 15 January 1943, 184 converted to a Field Company Royal Engineers and on 19 January 1943 Major O.J.R. Orr succeeded Major Griffith to command the company in its new role (WO 166/12084).

185 Tunnelling Company

185 Tunnelling Company began forming at Scarborough on 13 August 1941, with W. A. Smailes in command from 13 September 1941 (WO 166/3658). A programme of technical and military training began on 15 September 1941, the technical training (under the supervision of company officers) including pipe pushing and auger hole surveys. Training completed, the company moved to Falkirk in central Scotland (Fig. 15) on 21 December 1941, with sections soon detached for work nearby to the south and west in Lanarkshire and Dunbartonshire. No straight tunnelling work was to be undertaken in 1942 (WO 166/8111). 'Special' tunnelling obstacle work continued through the year, with flame barrage work between September and November, but there were insufficient tasks to employ the whole company. Accordingly, detachments were involved in a range of general construction tasks and one was deployed to the Orkney Islands north of the Scottish mainland on a water supply task from February to September 1942, drilling boreholes for groundwater abstraction. The com pany base moved *c*. 20 km NW to Stirling (Fig. 15) in February 1942, briefly south to Liberton (Edinburgh) in June 1942, and then back to Stirling. On 15 January 1943, 185 converted to a Field Company Royal Engineers and on 25 January 1943 Major R.E. Holden arrived to lead the company into its new role (WO 166/12085).

Concluding remarks

The year 1942 marked the peak of Tunnelling Company activities during World War II. Ten companies had been raised (Table 1), trained and operationally deployed to appropriate tasks within the UK, Gibraltar and Malta. They were grouped mostly within 1st, 3rd and 4th Tunnelling Engineers. Initial plans to raise 2nd Tunnelling Engineers had been abandoned when defeat frustrated plans to create a second army within the BEF in France.

Decline began when the Special Tunnelling Companies of 4th Tunnelling Engineers were progressively converted to Field Companies Royal Engineers. 179 was converted in December 1942, 184 and 185 in January 1943, and 183 in March

1943. Apart from 179, they were the last companies to be raised, primarily to emplace anti-invasion facilities (pipe mines and a flame barrage), and were converted when their mission was accomplished or deemed no longer necessary. The threat of a German invasion passed when German might became progressively focused on the Eastern Front against the Soviet Union. As conversion took place, 4th Tunnelling Engineers ceased to exist. This group of companies was used for drilling rather than tunnelling, but their initial training was largely based on techniques developed by Tunnelling Companies of the RCE.

Of the first-formed units that became 1st Tunnelling Engineers, during 1943 Tunnelling Companies 170 and 172 excavated underground in Gibraltar, and 171 and 173 in Malta (Table 2). However, Allied defeat of the German and Italian forces in North Africa in May 1943, the conquest of Sicily in July and August 1943 and the invasion of mainland Italy from 3 September 1943 forced the Italian government to surrender on 8 September 1943. Allied naval and air forces had thus become dominant in the Mediterranean region and the threat of invasion of both Gibraltar and Malta had passed. Tunnelling to provide underground facilities that were safe from naval or aerial bombardment was no longer required. Work planned or in progress was therefore completed by a single company in each case from late 1943: by 171 on Malta and 172 on Gibraltar. 170 returned from Gibraltar to the UK in November 1943, but not to any tunnelling work of significance. Its War Diaries for this period could not be found in the National Archives. The only known record of its activities is that, still officially as a Tunnelling Company, it took over from 855 Quarrying Company Royal Engineers in providing stone for construction work at the Central Ordnance Depot, Bicester (near Oxford) on 14 June 1944 when the Quarrying Company was mobilized for deployment to Normandy (WO 171/1717). Evidently, there was no need to use its tunnelling expertise as such, either in the UK or Normandy, at this time. 170 was assigned for service in SE Asia by February 1945, but as a Field rather than a Tunnelling Company (WO 172/7865). Presumably, 173 was also used for general engineering works on withdrawal from Malta, but the relevant War Diaries could not be found in the National Archives. It was assigned for service with Middle East Command by March 1945, but again as a Field rather than Tunnelling Company (WO 170/5194). 1st Tunnelling Engineers was disbanded on Malta, with some elements merging with 171.

Of the two other companies, formed initially to constitute 3rd Tunnelling Engineers, 178 Tunnelling Company was still in existence in May 1944. The War Diary of 125 Quarrying Company Royal Engineers (WO 171/1551) records that it handed over work being completed at Military Port No. 1 at Garelochhead in western Scotland to 178 on 31 May 1944 (Rose 2018). It was the company's familiarity with the use of explosives and the transportation of rock debris that made it suitable for the task, rather than a requirement for skills in tunnelling. 180 had by then (in December 1943) been converted to a Workshop and Park Company Royal Engineers, so no more than five of the ten Tunnelling Companies raised during the war were still in existence with their original designation. 3rd Tunnelling Engineers ceased to exist following withdrawal from Gibraltar. Edmonds (2002) notes that its tunnellers mostly returned to their former civilian roles mining for coal, because this was then a high national priority, although the remainder were used to form a company sent under Major Maunsell to SE Asia to quarry for roadstone in Burma.

Of the five Tunnelling Companies that survived into 1944, 171 continued until disbanded (in Austria) in December 1945. Like some other Royal Engineers units used for reconstruction tasks after the war ended, it had remained operational for several months after the end of hostilities. All but one of the other companies are presumed to have disbanded about this time. One company (173 according to Lauder (1963), but 172 according to the evidence presented here) remained on Gibraltar and was reduced to troop rather than company strength,

Fig. 18. A plaque on Gibraltar commemorating the end of tunnelling as a trade within the Royal Engineers in 1968.

becoming the Tunnelling Troop of 1st Fortress Squadron Royal Engineers (Lauder 1963). However, this was disbanded following the completion of its last tunnel, Molesend Way, in 1967 (Fig. 18) and all the remaining tunnellers left Gibraltar on posting to other units (Ramsey 1978). The history of the Tunnelling Companies Royal Engineers that began in 1915 was thus finally at an end. Apart from the tunnels on Gibraltar, Tunnelling Company achievements in World War II were to receive scant recognition in the official *History of the Corps of Royal Engineers* (Pakenham-Walsh 1958*a*, *b*).

Acknowledgements Grateful thanks are due to the National Archives for access to company War Diaries; to the copyright owners indicated in figure captions for permission to reproduce illustrations; to Wendy Cawthorne, Assistant Librarian of the Geological Society of London and also to the Royal Engineers Museum, Library and Archive, for bibliographic and biographical information; to Phillip Robinson for identifications shown in the caption to Figure 1; and to Thomas E. Eastler, Judy Ehlen and an anonymous reviewer for their constructive reviews of the paper.

Appendix A

1939–45 War Diaries for Tunnelling Companies and Groups Royal Engineers catalogued at The National Archives, Kew; and for Gibraltar only, Tunnelling Companies of the Royal Canadian Engineers.

Unit no.		Diary date range	Command area	File reference	Termination
170	1939	1 October–31 December	Home Forces	WO 167/957	
	1940	1 January–31 May	BEF, France	WO 167/957	
	1940	1 June–31 December	Home Forces	[WO 166/3535]	
	1941	1 January–28 February	Home Forces	[WO 166/3535]	
	1941	10 March–31 December	Gibraltar	[WO 176/221]	
	1942	1 January–31 December	Gibraltar	[WO 176/162]	
	1943	1 January–30 September	Gibraltar	[WO 176/162]	
	1943	10 November–31 December	Home Forces		
	1944	1 January–31 December	Home Forces	[WO 171/1717]	
	1945	1 February–31 December	SE Asia	WO 172/7865	as a Field Company
171	[1940	1 April–31 May	BEF, France	WO 167/957]	
	1940	1 July–31 August	Home Forces	[WO 166/3535]	
	1940	1 September–31 December	Home Forces	WO 166/3652	
	1941	1 January 31 December	Home Forces	WO 166/3652	
	1942	1 January–31 March	Home Forces	WO 166/8104	
	1942	1 April–31 May	Home Forces	WO 169/5284	
	1942	1 June–31 December	Middle East (Egypt)	WO 169/5284	
	1943	1 January–31 December	Malta	WO 169/14602	
	1944	1 January–31 March	Malta	WO 169/16415	
	1944	1 April–31 December	Malta	WO 169/18698	
	1945	1 January–28 February	Malta	WO 169/22148	
	1945	1 March–31 December	Italy/Austria	WO 170/5193	disbanded in Austria
172	[1940	1 April–31 May	BEF, France	WO 167/957]	
	1940	1 July–31 December	Home Forces	WO 166/3653	
	1941	1 January–31 December	Home Forces	WO 166/3653	
	1942	1 January–31 December	Home Forces	WO 166/8105	
	1943	1 January–30 April	Home Forces	WO 166/12075	
	1943	26 May–31 December	Gibraltar	WO 176/222 [WO 176/162]	
	1944	1 January–31 December	Gibraltar	[WO 176/162]	
	1945	1 January–31 December	Gibraltar	[WO 176/242] [WO 176/269]	converted
173	1940	1 March–31 May	BEF, France	WO 167/958	
	1940	1 July–31 December	Home Forces	WO 166/3654	
	1941	1 January–31 July	Home Forces	WO 166/3654 [WO 166/3535]	
	1941	1 August–31 December	Malta		
	1942	1 January–31 August	Malta		
	1942	1 September–31 December	Malta	WO 169/7435	
	1943	1 January–31 October	Malta	WO 169/14603	
	1943	1 November–31 December	Home Forces		

(*Continued*)

Unit no.		Diary date range	Command area	File reference	Termination
	1944	1 January–31 December	Home Forces		
	1945	1 March–31 December	Middle East	WO 170/5194	as a Field Company
178	1940	1 May–28 August	Home Forces	WO 176/223 [WO 176/221]	
	1940	28 August–31 December	Gibraltar	WO 176/223 [WO 176/221]	
	1941	1 January–31 March	Gibraltar	WO 176/223 [WO 176/221]	
	1941	1 April–31 December	Gibraltar	[WO 176/162]	
	1942	1 January–31 December	Gibraltar	[WO 176/162]	
	1943	1 January–22 April	Gibraltar	[WO 176/162]	
	1943	1 May–31 December	Home Forces		
	1944	1 January–31 December	Home Forces	[WO 171/1551]	
179	1940	1 May–31 December	Home Forces	WO 166/3655	
	1941	1 January–31 December	Home Forces	WO 166/3655	
	1942	1 January–31 December	Home Forces	WO 166/8107	converted to Field Company
180	1940	1 May–1 October	Home Forces	[WO 176/221]	
	1940	3 October–31 December	Gibraltar	[WO 176/221] [WO 176/162]	
	1941	1 January–31 October	Gibraltar	[WO 176/221] [WO 176/162]	
	1941	1 November–31 December	Gibraltar	[WO 176/162]	
	1942	1 January–31 December	Gibraltar	[WO 176/162]	
	1943	1 January–6 July	Gibraltar	[WO 176/162]	
	1943	1 August–30 November	Home Forces		
	1943	1 December–31 December	Home Forces	WO 166/12081	converted to Workshop & Park Company
183	1941	1 April–31 December	Home Forces	WO 166/3656	
	1942	1 January–31 December	Home Forces	WO 166/8109	
	1943	1 January–1 March	Home Forces	WO 166/12083	converted to Field Company
184	1941	1 August–31 December	Home Forces	WO 166/3657	
	1942	1 January–31 December	Home Forces	WO 166/8110	
	1943	1 January–15 January	Home Forces	WO 166/12084	converted to Field Company
185	1941	1 August–31 December	Home Forces	WO 166/3658	
	1942	1 January–31 December	Home Forces	WO 166/8111	
	1943	1 January–15 January	Home Forces	WO 166/12085	converted to Field Company
1 Tunnelling Engineers	1940	1 July–31 December	Home Forces	WO 166/3535	
	1941	1 January–31 December	Home Forces	WO 166/3535	
	1942	1 January–31 May	Home Forces	WO 166/8042	
	1942	1 June–31 December	M East (Egypt)	WO 169/5389	
	1943	1 January–31 December	Malta	WO 169/14601	
	1944	1 January–31 December	Malta	WO 169/18697	
	1945	1 January–31 May	Malta/Italy	WO 169/22147	disbanded
3 Tunnelling Group	1940	1 May–31 December	Gibraltar	WO 176/221	
	1941	1 January–31 October	Gibraltar	WO 176/221	
	1941	1 November–31 December	Gibraltar	[WO 176/162]	
	1942	1 January–31 December	Gibraltar	[WO 176/162]	
	1943	1 January–31 October	Gibraltar	[WO 176/162]	disbanded
4 Tunnelling Engineers	1941	1 July–31 December	Home Forces	WO 166/3537	
	1942	1 January–31 December	Home Forces	WO 166/8043	disbanded
1 Tunnelling Coy RCE Det	1940	1 November–31 December	Gibraltar	WO 176/225	
	1941	1 January–31 October	Gibraltar	WO 176/225	
2 Tunnelling Coy RCE	1941	1 April–31 December	Gibraltar	WO 176/226	
	1942	1 January–14 December	Gibraltar	[WO 176/162]	

Note: the series is incomplete, for reasons unknown. A file reference shown within squared brackets indicates a Diary that gives information of relevance, although it is not itself a Diary for that particular Company.
BEF, British Expeditionary Force.

References

ANON 1922*a*. *The Work of the Royal Engineers in the European War, 1914–19: Military Mining*. Institution of Royal Engineers, Chatham.

ANON 1922*b*. *The Work of the Royal Engineers in the European War, 1914–19: Geological Work on the Western Front*. Institution of Royal Engineers, Chatham.

BAILEY, E.B. 1952. *Geological Survey of Great Britain*. Thomas Murby, London.

BAILEY, E.B. 1953. Notes on Gibraltar and the Northern Rif. *Quarterly Journal of the Geological Society of London*, **108**, 157–175, https://doi.org/10.1144/GSL.JGS.1952.108.01-04.08

BALDASSINI, N. & DI STEFANO, A. 2017. Stratigraphic features of the Maltese archipelago: a synthesis. *Natural Hazards*, **86**(suppl. 2), 203–231.

BANKS, D. 1946. *Flame Over Britain*. Sampson Low, Marston, London.

BLAKE, R.N.E. 2000. Geological influences on the siting of military airfields in the United Kingdom. *In*: ROSE, E.P.F. & NATHANAIL, C.P. (eds) *Geology and Warfare: Examples of the Influence of Terrain and Geologists on Military Operations*. Geological Society, London, 275–312.

CAMERON, A.B. 1999. *Under Sand, Ice and Sea: the Memoirs of an Oil Man*. Trafford Publishing, Victoria, BC.

CASSELS, G.R. 2001. Memoirs of a tunnelling officer during the First World War. *Royal Engineers Journal*, **115**, 167–174.

COTTON, J.C. 1948. The tunnels in Gibraltar. *In*: ANON (ed.) *The Civil Engineer in War*. Vol. 3. Institute of Civil Engineers, London, 229–248.

COTTON, J.C. 1963. Military and other war-time tunnels. *In*: PÉQUIGNOT, C.A. (ed.) *Tunnels and Tunnelling*. Hutchinson, London, 459–483.

DEAR, I.C.B. & FOOT, M.R.D. (eds) 1995. *The Oxford Companion to the Second World War*. Oxford University Press, Oxford.

EDMONDS, P.E. 2002. Memoirs of a tunnelling officer during the Second Word War. *Royal Engineers Journal*, **116**, 141–147.

FINLAYSON, D. 2010. *Crumps and Camouflets: Australian Tunnelling Companies on the Western Front*. Big Sky Publishing, Newport.

FOOT, W. 2006. *Beaches, Fields, Streets and Hills ... the Anti-Invasion Landscapes of England, 1940*. Council for British Archaeology, York.

FOSS, L.J. 1998. A Sapper sinks a submarine: off Gibraltar in 1943. *Royal Engineers Journal*, **112**, 60–61.

GALE, M.C. 1969. History of Tunnelling at Gibraltar (1782–1968). Unpublished typescript. Royal Engineers Museum, Library and Archive, accession no. 8841.

GRAY, D.A. 2004. Buchan, Stevenson (1907–1996). *In*: MATTHEW, H.G.C. & HARRISON, B. (eds) *Oxford Dictionary of National Biography. Vol. 8. Brown–Burstow*. Oxford University Press, Oxford, 457–458.

GRIEVE, W.G. & NEWMAN, B. 1936. *Tunnellers: the Story of the Tunnelling Companies, Royal Engineers, During the World War*. Herbert Jenkins, London.

HAYCRAFT, T.W.R. 1946. The Gibraltar tunnels. *Royal Engineers Journal*, **60**, 310–320.

JONES, S. 2010. *Underground Warfare 1914–1918*. Pen & Sword Military, Barnsley.

LAUDER, J.G. 1963. Tunnelling in Gibraltar. *Royal Engineers Journal*, **77**, 339–369.

LONGHURST, H. 1959. *Adventure in Oil: the Story of British Petroleum*. Sidgwick & Jackson, London.

MATHER, J.D. 2012. War as a catalyst for change: groundwater studies in the Geological Survey of Great Britain before 1950 and the impact of two World Wars. *In*: ROSE, E.P.F. & MATHER, J.D. (eds) *Military Aspects of Hydrogeology*. Geological Society, London, Special Publications, **362**, 139–160, https://doi.org/10.1144/SP362.8

MORRIS, T.O. 1952. *The Water Supply Resources of Malta*. Government of Malta, Valetta.

PAKENHAM-WALSH, R.P. 1958*a*. *History of the Corps of Royal Engineers. Vol. VIII. 1938–1948. Campaigns in France and Belgium, 1939–40, Norway, Middle East, East Africa, Western Desert, North West Africa, and Activities in the U.K.* Institution of Royal Engineers, Chatham.

PAKENHAM-WALSH, R.P. 1958*b*. *History of the Corps of Royal Engineers. Vol. IX. 1938–1948. Campaigns in Sicily and Italy; the War against Japan; North-West Europe, 1944–45; Minor and Non-operational Areas; Post-war, 1945–48*. Institution of Royal Engineers, Chatham.

RAMSAY, A.C. & GEIKIE, J. 1878. On the geology of Gibraltar. *Quarterly Journal of the Geological Society of London*, **34**, 505–541, https://doi.org/10.1144/GSL.JGS.1878.034.01-04.37

RAMSEY, W.G. (ed.) 1978. *After the Battle, No. 21: Gibraltar*. Battle of Britain Prints, London.

RIZZO, C. 1932. *Report on the Geology of the Maltese Islands, including Chapters on Possible Ground-Water Tables and Prospecting for Mineral-Oil and Natural-Gas*. Government Printing Office, Malta.

ROGERS, J. 1986. *Tunnelling into Colditz: a Mining Engineer in Captivity*. Robert Hale, London.

ROSE, E.P.F. 2001. Military engineering on the Rock of Gibraltar and its geoenvironmental legacy. *In*: EHLEN, J. & HARMON, R.S. (eds) *The Environmental Legacy of Military Operations*. Reviews in Engineering Geology, **XIV**. Geological Society of America, Boulder, CO, 95–121.

ROSE, E.P.F. 2004. King, William Bernard Robinson (1889–1963). *In*: MATTHEW, H.G.C. & HARRISON, B. (eds) *Oxford Dictionary of National Biography. Vol. 31. Kebell–Knowlys*. Oxford University Press, Oxford, 691–692.

ROSE, E.P.F. 2012*a*. Aggregates pave the way to victory: work of Royal Engineers geologists and Quarrying Companies during World War II, especially for the liberation of Normandy. *In*: HUNGER, E. & WALTON, G. (eds) *Proceedings of the 16th Extractive Industry Geology Conference*. EIG Conferences, Charlbury, 24–40.

ROSE, E.P.F. 2012*b*. Groundwater as a military resource: development of Royal Engineers Boring Sections and British military hydrogeology in World War II. *In*: ROSE, E.P.F. & MATHER, J.D. (eds) *Military Aspects of Hydrogeology*. Geological Society, London, Special Publications, **362**, 105–138, https://doi.org/10.1144/SP362.7

ROSE, E.P.F. 2013. British pioneers of the geology of Gibraltar, Part 1: the artilleryman Thomas James (*ca*1720–1782), infantryman Ninian Imrie of Denmuir (*ca*1752–1820), and ex-militiaman James Smith of Jordanhill (1782–1867). *Earth Sciences History*, **32**, 178–212.

Rose, E.P.F. 2014*a*. British pioneers of the geology of Gibraltar, Part 2: cave archaeology and geological survey 1863 to 1878. *Earth Sciences History*, **33**, 26–58.

Rose, E.P.F. 2014*b*. British pioneers of the geology of Gibraltar, Part 3: E.B. Bailey and Royal Engineers 1943 to 1953. *Earth Sciences History*, **33**, 294–332.

Rose, E.P.F. 2016. The Rock of Gibraltar re-visited. *Geology Today*, **32**, 99–106.

Rose, E.P.F. 2018. Quarrying Companies Royal Engineers in World War II: contributions to military infrastructure within the United Kingdom and to Allied forces during the North African, Italian and North-West Europe campaigns. *In*: Rose, E.P.F., Ehlen, J. & Lawrence, U.L. (eds) *Aspects of Military Geology: Fortification, Excavation and Terrain Evaluation*. Geological Society, London, Special Publications, **473**. First published online March 6, 2018, https://doi.org/10.1144/SP473.2

Rose, E.P.F. & Rosenbaum, M.S. 1989. Royal Engineer geologists and the geology of Gibraltar Part I – tunnelling through the Rock. *Royal Engineers Journal*, **103**, 142–151.

Rose, E.P.F. & Rosenbaum, M.S. 1991. *A Field Guide to the Geology of Gibraltar*. Gibraltar Museum, Gibraltar.

Rose, E.P.F. & Rosenbaum, M.S. 1993. British military geologists: through the Second World War to the end of the Cold War. *Proceedings of the Geologists' Association*, **104**, 95–108.

Rose, E.P.F. & Rosenbaum, M.S. 2011. British geological maps that guided excavation of military dug-outs in Belgium during World War I. *Quarterly Journal of Engineering Geology and Hydrogeology*, **44**, 293–306, https://doi.org/10.1144/1470-9236/08-111

Rose, E.P.F., Pratt, S.K. & Bennett, S.M. 1992. Evidence for sea-level changes in the Globigerina Limestone Formation (Miocene) of the Maltese islands. *Paleontologia i Evolució*, **24–25**, 265–276.

Rosenbaum, M.S. & Rose, E.P.F. 1991. *The Tunnels of Gibraltar*. Gibraltar Museum, Gibraltar.

Rosenbaum, M.S. & Rose, E.P.F. 1992. Geology and military tunnels. *Geology Today*, **8**, 92–98.

Rosenbaum, M.S., Rose, E.P.F. & Wilkinson (née Buchanan), F.W. 1994. The influence of excavation technique on the integrity of unlined tunnel walls in Gibraltar. *In*: Oliveira, R., Rodrigues, L.F., Coelho, A.G. & Cunha, A.P. (eds) *Proceedings of the Seventh International Congress of the International Association of Engineering Geology*, Lisbon, 5–9 September 1994. Vol. 6. Balkema, Rotterdam, 4137–4144.

Wilson, A.J. 1992. *The Professionals: the Institution of Mining and Metallurgy 1892–1992*. Institution of Mining and Metallurgy, London.

Wilson, W.H. 1946. Tunnelling in Gibraltar during the 1939–1945 War. *Transactions of the Institution of Mining and Metallurgy*, **55**, 193–269.

Engineering geological considerations for the 'Old' Beacon Hill Railway Tunnel, Hong Kong Special Administrative Region

ALEXANDER D. MACKAY

HSS Integrated, Malaysia, 3rd Floor, Tower Block, Plaza Dwitasik, Jalan Permaisuri, Bandar Sri Permaisuri, 56000, Kuala Lumpur, Malaysia

Admackay1@gmail.com

Abstract: The 'Old' Beacon Hill Tunnel was constructed through granite from 1906 to 1910 as part of the Kowloon Canton Railway in the then British colony of Hong Kong. It was the first railway tunnel to be constructed in this region and, at 2.2 km long, longer than any Chinese railway tunnel. Prior to Japanese occupation in December 1941, the British detonated charges to collapse the tunnel to deny its use as an invasion route. Following Japanese defeat in 1945 and return to British government, the tunnel was re-excavated and the damaged sections were supported using steel arches based on Terzaghi's rock mass classification system, which uses geological descriptions to classify loading onto arched tunnel supports. The railway was subsequently realigned during the 1980s and placed in a new tunnel running alongside and west of the 'Old' Beacon Hill Tunnel, which was converted to accommodate a pipeline connecting gas supplies. Ground investigations implemented after return of the territory to Chinese sovereignty in July 1997 established that groundwater flow and tunnel stability were influenced by a dyke within the granite, joint set orientation, faulting and weathering as well as the wartime detonations.

The 'Old' Beacon Hill Tunnel, part of the Kowloon Canton Railway, was the first railway tunnel to be constructed in the former British colony of Hong Kong, now the Hong Kong Special Administrative Region of the People's Republic of China, on China's SE coast (Fig. 1). At the time of opening in 1910, at 2.2 km in length it was the longest tunnel in China and the fifth longest tunnel in the world outside Europe. Its route (Fig. 2) presently runs NNW–SSE beneath Lion Rock Country Park and the portals are located immediately adjacent to densely populated areas at Hin Keng Housing Estate to the north and City University to the south.

The Kowloon Canton Railway was expanded to form a network during the 1980s, which included upgrading the railway alignment through the 'New' Beacon Hill tunnel, parallel to the old tunnel alignment (Fig. 2). The present Kowloon Canton Railway, renamed the Mass Transit Railway, East Rail Line, runs for 41.5 km between east Tsim Sha Tsui and Sheung Shui passing through three underground sections (Fig. 1).

The 'New' and 'Old' Beacon Hill Tunnels pass beneath the densely vegetated Lion Rock Country Park, which rises to a maximum elevation of 450 m at Beacon Hill (Fig. 2). The topography above the tunnel crown rises to about 400 m, 375 m above the crown about midway along the alignment (Fig. 3).

History

Construction and usage of the 'Old' Beacon Hill Tunnel

The 'Old' Beacon Hill Tunnel was excavated using standard drill and blast techniques to form an arched profile (GEO 2013). The portals (e.g. Figs 4 & 5) were located in poorer quality ground with water inflows up to 45 l/min. (Eves 1908, 1911). The initial excavation comprised a 2.6 m tall by 5.5 m wide tunnel. This was followed by multiple-face excavation that increased the outer excavation periphery to 6.8 m tall by 5.5 m wide.

A permanent brick lining was installed upon completion of every 4.5 m full-face advancement. Three ventilation shafts were constructed: 91 m from the south portal to 28 m depth; close to the north portal to 15 m depth; and 396 m from the north portal to 81 m depth. The shafts were backfilled upon tunnel completion.

During the Second World War, the Kowloon Canton Railway was dismantled by the British to deny this as an access route to Japanese forces that invaded from the north early in December 1941. Charges were detonated within the tunnel, resulting in two main collapses within its southern portion, between chainages 1615 and 1635 (20 m length) and at chainage 1857 (Figs 2 & 3). After defeat

From: Rose, E. P. F., Ehlen, J. & Lawrence, U. L. (eds) 2019. *Military Aspects of Geology: Fortification, Excavation and Terrain Evaluation*. Geological Society, London, Special Publications, **473**, 233–239.
First published online October 9, 2018, https://doi.org/10.1144/SP473.12

Fig. 1. Map of part of Hong Kong Special Administrative Region, SE China, showing the 'Old' Beacon Hill Railway Tunnel (circled) and route of the present East Rail Line from East Tsim Sha Tsui north to the Sheung Shui Border Crossing. The distance across the map is 50 km. Adapted from Mackay *et al.* (2015, fig. 1).

and final surrender of the British garrison on 25 December 1941, the territory was occupied by Japanese troops who carried out temporary repairs to re-open the tunnel, including installing wooden props to support the roof. After the Second World War and reoccupation by the British, the tunnel was reinstated over an 18 month period. Work was completed by August 1947. The two collapsed sections, which became referred to as the 'Japanese Arches', were supported using steel arches embedded in concrete. An additional 0.8 km of tunnel was also reinstated, which included a crude, water-resistant membrane that was installed between the ribs and brick lining to divert water from the crown and walls into side drains to keep inflow away from the ballast and tracks.

During 1984 the Kowloon Canton Railway was diverted into twin tunnels running alongside the original alignment (Fig. 1). The 'Old' Beacon Hill Tunnel was converted to accommodate a gas pipeline supply (Mackay *et al.* 2015). As the demand for gas supply has increased, an additional gas pipeline installation was proposed during 2011 (Mott Macdonald 2011).

Support

After the Second World War, the tunnel support was assessed using the Terzaghi rock mass classification system (Terzaghi 1946). This was the first classification to relate geological descriptions to rock mass loading and the corresponding support requirements.

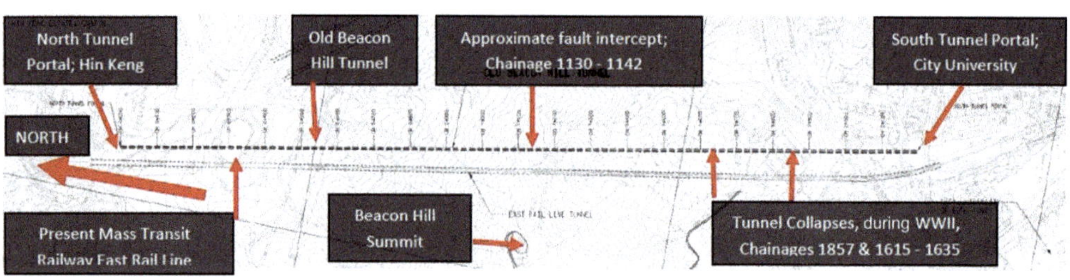

Fig. 2. The 'Old' Beacon Hill Tunnel alignment from north (left) to south (right), *c.* 2.2 km long and adjacent to the present East Coast Rail Line, showing chainages (numbered above, from north to south), positions of tunnel collapse and fault intercept (from Mackay *et al.* 2015, fig. 3).

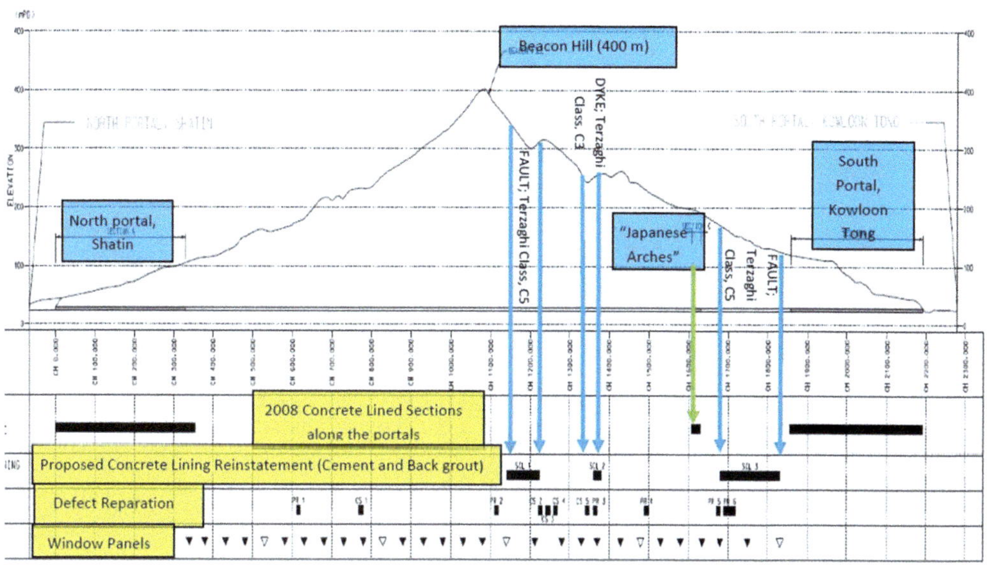

Fig. 3. 'Old' Beacon Hill Tunnel long section, showing concrete lined sections installed at the portals in 2008, areas for defect reparation, and window panels.

The tunnel support during this period comprised steel sets and brick lining installed into an arched profile excavation and the degree of support was estimated from loading from the surrounding ground allocated from seven Terzaghi rock mass categories, ranging from the best quality 'intact rock', Categories D1 and D2, through to 'blocky and seamy' ground, Categories C3–C5 and B6, to the poorest quality, 'swelling' ground, Category A7.

The 1984 tunnel diversion also used Terzaghi's rock mass classification system to provide an economical tunnel upgrading approach and comparison

Fig. 4. Current south portal entrance (2011) to the 'Old' Beacon Hill Tunnel (from Mackay *et al.* 2015, fig. 5).

Fig. 5. The south portal under construction in 1909 (from Mackay *et al.* 2015, fig. 7).

with the support used for the 'Old' Beacon Hill Tunnel. The geological descriptions used in this system were adapted to include specific weathering conditions and geological features encountered during the tunnel excavation and updated during the 1980s (GCO 1988). Terms such as 'partial weathering conditions' (GCO 1988; Whiteside 1988; GEO 2007; Fig. 6) were used instead of 'blocky' for Support Categories C3 to C5, 'faulting' and 'dyke intrusions' for 'stratified' for Support Category B6, and

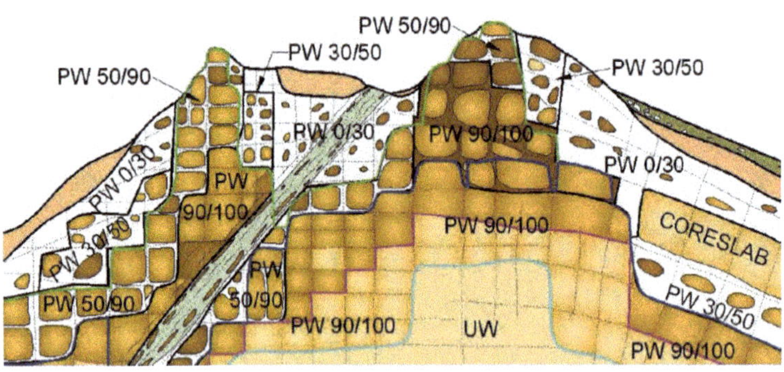

Fig. 6. Granitic weathering: PW, partially weathered; UW, unweathered. Adapted from Geotechnical Engineering Office (2007), with permission of the Head of the Geotechnical Engineering Office and the Director of the Civil Engineering and Development Department, the Government of the Hong Kong Special Administrative Region (SAR). The Government of the Hong Kong SAR does not accept responsibility for the accuracy, completeness or up-to-date nature of any reproduced versions of the material.

'extensive zones of completely decomposed granite' for the poorest Support Category A7.

Following the railway realignment, the 'Old' Beacon Hill Tunnel was upgraded by installing cast *in situ* concrete lining for *c.* 1.5 km. This included relining the 'Japanese Arches' and extending weep-holes from the brick lining through the cast *in situ* lining (Mott Macdonald 2011). Because the brick lining continued to deteriorate with increased water inflow, further upgrading works were carried out during 1998. Despite this, the water inflow increased up to 2700 l min^{-1} (GEO 2013), requiring concrete lining reinstatement for the remaining tunnel sections along the initial 0.35 km of tunnel inward from each portal during 2008 (Fig. 3) and drip-shed installations to divert water inflow into the side drains. Further site investigation works were carried out at this time to verify the tunnel's suitability to accommodate the additional pipeline installation (Mott Macdonald 2011). Recommendations from the site investigation included concrete lining installation along the entire tunnel length, and localized grout injection and concrete reinstatement within the central portion of the tunnel to arrest water inflow and infill voids behind the brick lining.

Geological conditions along the tunnel

The Hong Kong terrain is dominantly of igneous rocks, volcanic or granitic (Fig. 7). The majority of the 'Old' Beacon Hill Tunnel passes through the medium- and coarse-grained Kowloon Granite and a lesser part through the typically coarser-grained Shatin Granite, intruded by the Kowloon Granite to the south and by quartz monzonite to the north of the route. There is a geological interface between the Kowloon and Shatin Granites intercepting the tunnel alignment towards the north which has been intruded by a subvertical, porphyritic quartz monzonite dyke that trends ENE to WSW. Faults run sub-parallel and perpendicular to the dyke, NW–SE and NE–SW, dipping subvertically and intercepting the central portion of the tunnel alignment. Photo lineaments, linear surface depressions potentially representing faults (possibly the documented faults) identified by aerial photographic interpretation, intercept the central portion of the tunnel. The faults and dykes intercepting the tunnel, interpreted from Sewell *et al.* (2000), are shown in Figure 3.

These geological features were not recognized during the initial tunnel construction, and only became apparent from geological observations recorded during construction of the East Railway Line in the 1980s and increased understanding of the engineering geological conditions during the recent decades. The effects of each of the geological features intercepting the tunnel and the detonations during the Second World War and their impacts on the tunnel stability are detailed in the following sections, with inferred rock mass categories.

Quartz monzonite dyke

The dyke is composed of finer-grained constituents and therefore weathers to a finer-grained soil with a

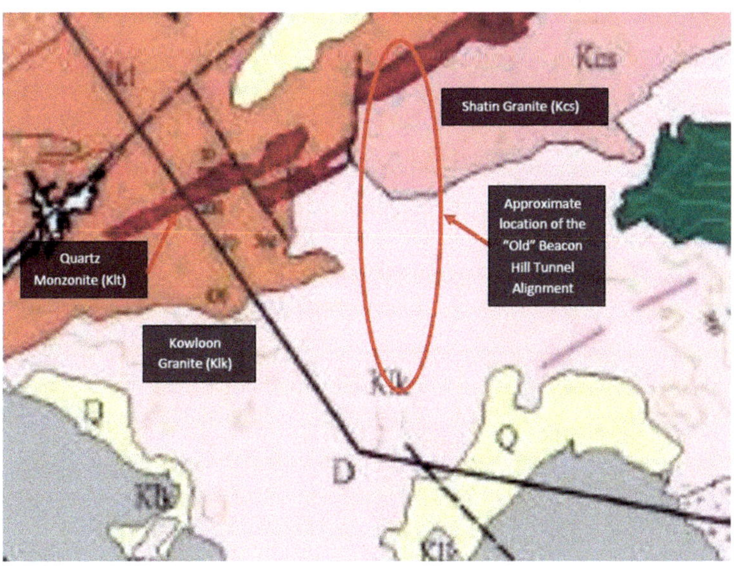

Fig. 7. Geology of the Beacon Hill area, along the 'Old' Beacon Hill Tunnel alignment. The distance across the map is 10 km and north is to the top of the map. Adapted from Mackay *et al.* (2015, fig. 9).

reduced permeability. The surface expression of the dyke is situated *c.* 200 m above the tunnel crown, in a depression at the foot of a 150 m vertical, steeply dipping hillside. Owing to the relatively decreased elevation and the presence of finer-grained constituents, the dyke forms an aquitard for groundwater flowing from higher to lower elevations, increasing the groundwater levels beneath the higher-level terrain. The dyke was allocated a C3 rock mass category.

Joint set orientation within coarse-grained granite

The joint sets generally have an orthogonal pattern consisting of three sets: two subvertical sets that are generally aligned parallel and perpendicular to the main fault orientation, and one subhorizontal joint set aligned sub-parallel to the ground surface. Joint orientation is therefore influenced by the proximity to the ground surface and major fault locations. Joint orientation influences the water flow, either increasing or decreasing the flow depending upon the permeability of the weathered materials along the joint traces and the connection with water run-off from the ground surface. The rock mass category improves away from the dyke and fault locations and was allocated D1 to D2, 'tight joints, slightly loosened'.

Faults

Faults are generally weaker and more fissile than the adjacent rock. Within the Kowloon Granite, joint frequency and persistence increase significantly along the fault traces with joint spacing decreasing in closer proximity to the fault trace, causing an increased susceptibility to weathering which, in turn, influences the hydrogeological regimes (Fletcher 2004). As the faults dip subvertically, this also allows connection to water run-off from the ground surface. Furthermore, the presence of the faults intercepting the tunnel allows localized increases in groundwater inflow. The approximate location of the fault in relation to the tunnel is shown in Figure 3. Owing to the 'seamy and blocky' characteristics of the faults, the rock mass was allocated to Category C5.

Partial weathering effects

Increased weathering intensity was observed along the joints during the initial 100–140 m of excavation from the portals for the East Railway Line. This is termed partial weathering (GCO 1988) and generally increased in intensity as the excavation approached the ground surface. Typically, corestones of moderately to slightly decomposed granite are present within a completely to highly decomposed granite soil matrix (Fig. 6).

Detonations during the Second World War

These detonations opened up discontinuities and, over time, allowed the finer-grained, looser constituents to be removed from within the joints by the increased water inflow. The geological features and impacts during the Second World War increased the water inflow from the 45 l min^{-1} recorded in the early 1900s tunnel construction to the current inflow of 2400 l min^{-1}.

Twenty-first century site investigations and findings

Site inspections and condition surveys carried out during 2000 and 2003, prior to the additional tunnel lining placement in 2008, revealed an increased water inflow during the wet season. This increased water inflow exacerbated deterioration, through loss of mortar from the brick lining joints and in the ground behind the outer brick lining. The voids thus created were subsequently filled with accumulated debris.

During 2010 and 2011 a ground investigation consisting of further condition surveys was carried out. Forty-eight cored boreholes were drilled to a maximum length of 3.5 m and 28 window panels (each 0.5×0.5 m) were opened through the tunnel side walls (Fig. 3). The data obtained were used to characterize the inner concrete and outer brick tunnel linings and to identify the voids, debris accumulation and *in situ* ground conditions. The concrete lining was unreinforced and averaged 0.25 m thick and the brick lining ranged from 0.34 to 0.57 m thick, averaging 0.5 m. Water-resistant geotextile installed in the late 1940s was identified locally between the concrete and brick linings. Moderately to slightly decomposed granite blocks had accumulated behind *c.* 25% of the lining and the void thickness ranged from 0.12 to 0.6 m.

The increase in void volume behind the tunnel lining generally corresponded with an increased water flow, increased groundwater pressure towards the central portion of the tunnel and an increase in the degree of weathering towards the portals. The voids also increased towards the 'Japanese Arches' where the early 1940s detonations had probably opened existing discontinuities surrounding the tunnel. Weathering intensity also increased in the vicinity of the quartz monzonite dyke along which completely to highly decomposed weathered bands up to 1.75 m thick are present.

The ground investigation and extrapolation of the Terzaghi rock mass classification from the East Rail Line tunnel were used to provide additional geological detail, to describe the tunnel lining conditions, and to verify the support requirements along the

'Old' Beacon Hill Tunnel. The poorer ground conditions, allocated C5 to C3 and B6, generally corresponded with the increased water inflow associated with the poorer geological conditions and the 'Japanese Arches'. The recommended reinstatement (Fig. 3) included back-grouting to infill voids behind the lining, mainly towards the 'Japanese Arches', and localized lining reinstatement.

Conclusions

The 'Old' Beacon Hill Tunnel was originally constructed over a century ago and, despite the militarily induced tunnel collapses during the Second World War, and following reinstatements during and since the war, is still in operation to facilitate a gas supply network beneath the Lion Rock Country Park. The work described here shows how the use of precise engineering geological descriptions related to rock mass classification support and associated groundwater inflow can guide recommendations for an economical method of tunnel support. The tunnel has been in almost continuous use for over a century and is therefore now approaching the end of the 120 year design life typically specified for tunnel construction projects.

Acknowledgements The author would like to thank Mr Tymon Mellor for his assistance in providing archived data and editorials.

Funding This research received no specific grant from any funding agency in the public, commercial or not-for-profit sectors.

References

Eves, G.W. 1908. *Progress report on the construction of the British section of the Hong Kong – Canton Railway to 31 December 1907.* Session Paper No. 4/1908, 31–39, Legislative Council, Hong Kong Special Administrative Region.

Eves, G.W. 1911. The Beacon Hill Tunnel, Kowloon–Canton Railway. *The Engineer*, **111**, 428–431.

Fletcher, C.J.N. 2004. *The Geology of Site Investigation Boreholes from Hong Kong: A Practical Guide for Geologists and Engineers.* Applied Geoscience Centre, Hong Kong Construction Association (Site Investigation Contractors Committee) and Association of Geotechnical and Environmental Specialists, Hong Kong.

Geotechnical Control Office (GCO) 1988. *Guide to Rock and Soil Descriptions (Geoguide 3).* Reprinted 1994. Geotechnical Control Office, Hong Kong Special Administrative Region.

Geotechnical Engineering Office (GEO) 2007. *Engineering Geological Practice in Hong Kong.* Publication 1/2007, General Engineering Office, Hong Kong Special Administrative Region.

Geotechnical Engineering Office (GEO) 2013. *Catalogue of Tunnels in Hong Kong.* Geotechnical Engineering Office, Hong Kong Special Administrative Region.

Mackay, A.D., Wightman, N.R. & Ishola, T. 2015. Engineering considerations for a railway tunnel conversion, Hong Kong. *In*: *Proceedings of International Conference & Exhibition on Tunnelling & Underground Space (ICETUS 2015)*, Kuala Selangor, Malaysia, The Institution of Engineers Malaysia, Selangor Darul Ehsan, 162–167.

Mott Macdonald 2011. *Project profile for installation of a 500 mm NB gas pipeline inside the existing disused tunnel (old Beacon Hill tunnel).* Report no. 243563121/F, The Hong Kong and China Gas Company Ltd., Hong Kong Special Administrative Region.

Sewell, R.T., Campbell, S.D.G., Fletcher, C.J.N., Lai, K.W. & Kirk, P.A. 2000. *The Pre-Quaternary Geology of Hong Kong.* Hong Kong Geological Survey, Geotechnical Engineering Office, Civil Engineering Development Department, Hong Kong Special Administrative Region.

Terzaghi, K. 1946. Rock defects and loads on tunnel supports. *In*: Proctor, R.V. & White, T.L. (eds) *Rock Tunnelling with Steel Supports*, **1**. Commercial Shearing and Stamping Company, Youngstown, OH, 17–99.

Whiteside, P.G.D. 1988. Structural control in the development of granitic corestones. *Geological Society of Hong Kong Newsletter*, **6**, 4–10.

Geological considerations of contemporary military tunnelling near Mosul, northern Iraq

MARK H. BULMER[1,2]

[1]*Roedown Research R²*, *3856 Wayson Road, Davidsonville, MD 21035, USA*

[2]*Department of Earth Sciences, University College London, Gower Street, Bloomsbury, London WC1E 6BT, UK*

0000-0003-1970-9485

roedown@gmail.com

Abstract: Members of the so-called Islamic State (IS) have used subterranea in Syria and Iraq as a core part of their control over territories and their military capability. Around Mosul they exploited existing natural sinkholes, caves and tunnels. The use of tunnels and the completion of new ones between 2014 and 2017 impacted the advance of Iraqi and Peshmerga forces in the battle for Mosul. IS created complex tunnel networks using hand tools, power tools and improvised tunnel-boring machines. To avoid detection, tunnels were started in buildings and spoil was hidden in rooms. The tunnels range from those large enough for one person to crouch in to those large enough for a vehicle. The most sophisticated tunnel complexes had dormitories, hospitals, armouries, kitchens, latrines, and command centres complete with CCTV, solar panels, computers, phones, electricity, respirators and ventilation facilitating underground occupation for long periods. IS tunnelled in almost all urban areas they controlled, demonstrating an enormous effort: moving from hand tools to tunnel-boring machines was a logical progression. The size of tunnel-boring machines and the amount of spoil generated defined the type of building they started in and the length of the tunnel they bored. Buildings with tunnels under them risk collapse.

The city of Mosul in northern Iraq, known in ancient times as the city of Nineveh, is geographically divided into east and west by the Tigris River (Fig. 1). In 2014, it was mostly inhabited by Sunni Muslims, and their widespread anger at the alleged corruption and mismanagement of the Shiite-dominated government in Iraq's capital Baghdad helped extremists to gain a foothold in the city. It was overrun by so-called Islamic State (IS) in a matter of days in the summer of 2014, and from Mosul's Grand Mosque IS leader Abu Bakr al-Baghdadi declared a 'caliphate' spanning regions of Iraq and Syria. Mosul had a pre-insurgency population of nearly two million and as an ancient city had many tunnels under the east and west sides that existed before IS. In the first weeks of October 2016, at the start of the government-led operation to recapture Mosul, Iraqi and Kurdish forces recaptured towns and villages east and south of the city. It was at this time that an extensive use of tunnels by IS became apparent and on a scale unexpected and largely unprepared for by the advancing forces, particularly armoured units. This raised the question as to how IS members were able to undertake so much digging between 2014 and 2017.

In this paper, the use of tunnelling by IS in northern Iraq and especially around Mosul is examined. One of the challenges to preparing it has been the difficulty in determining place names, especially when dealing with small towns that may have two or three names, in Arabic, Kurdish, Turkish or Syrian. Reports from Iraqi, Kurdish and foreign media often used culturally specific names. In addition, coverage was often specific to Iraqi forces, Kurdish Peshmerga, local militia, ethnic or religious groups. Standardization has been attempted here by cross-reference to multiple sources such as official geology and administrative maps for Iraq (Fouad & Sissakian 2011; Fouad 2012; Sissakian & Al-Khalidi 2012; Sissakian & Fouad 2015) and the Kurdistan Region of Iraq reference map as of February 2017 (REACH 2017).

Geology around Mosul

Mosul lies on part of the Arabian Plate and in a fold-thrust belt that is still active (Fig. 1). It is situated in the Low Folded Zone on the NE flank of the Albu Saif structure and near to its northern plunge (Sissakian *et al.* 2014). Figure 2 shows a portion of the most recent 1:1 million-scale geological map of northern Iraq, on which known IS tunnel complex locations and mass graves are shown. The sequence exposed around Mosul is Miocene in age and represented by the Injana Formation (600 m maximum

From: Rose, E. P. F., Ehlen, J. & Lawrence, U. L. (eds) 2019. *Military Aspects of Geology: Fortification, Excavation and Terrain Evaluation*. Geological Society, London, Special Publications, **473**, 241–265.
First published online June 22, 2018, https://doi.org/10.1144/SP473.11

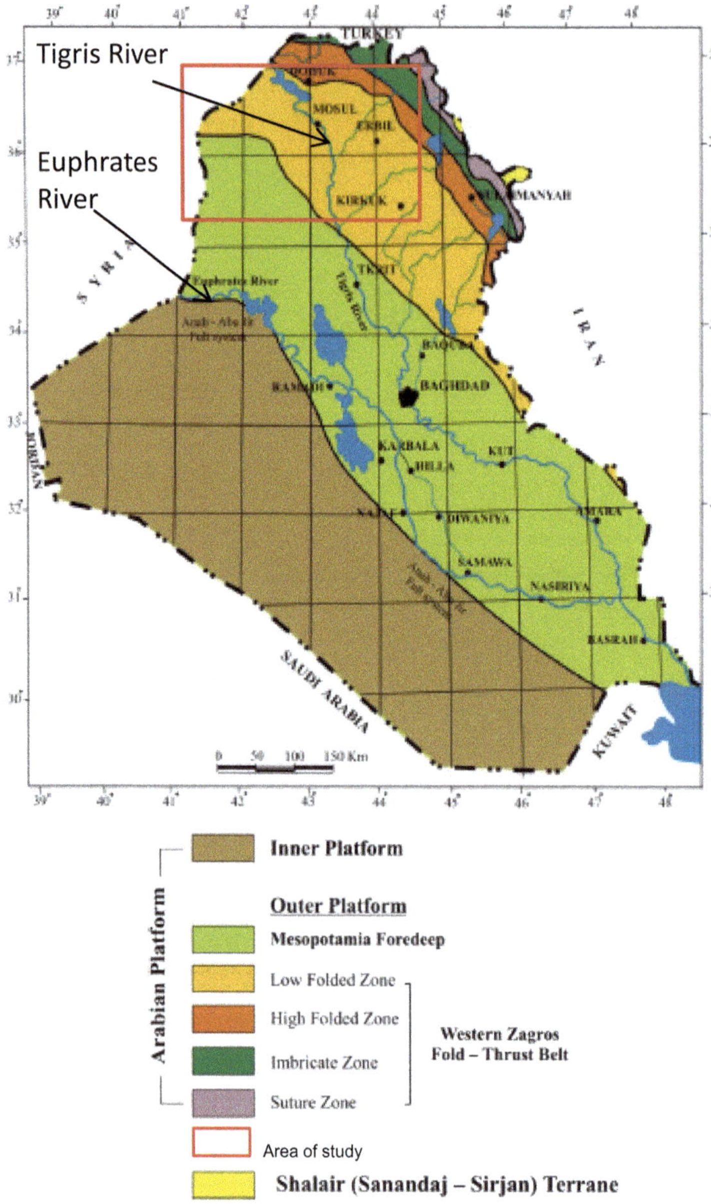

Fig. 1. Tectonic framework of Iraq, from Sissakian & Fouad (2015), after Fouad (2012), with the area of study and major rivers indicated.

thickness) and the underlying Fatha Formation (350 m maximum thickness) (Jassim *et al.* 1997; Al-Juboury 2009). The Injana Formation comprises fine clastic sediments (carbonate-rich sandstones, siltstones and claystones), its thickness varying locally (Al-Juboury 2009). The upper member of the Fatha Formation is a green and red claystone with gypsum, while the lower member (90 m in thickness) comprises carbonate (dolomite or dolomitic limestone with thin clay seams) under gypsum capped by a limestone marker bed. Anticlinal structures with Cretaceous and Paleogene rocks in their cores trend NW–SE, parallel to the Zagros Fold-Thrust Belt of Iran to the east, but arc to an east–west trend at Sinjar in the north. These provided high ground where people fled from conflict, as well as observation sites for combatants with views over large surrounding areas. The highest point of the Sinjar anticline is about 1462 m above sea-level. The location of karst fields associated with the anticlines identified in Jassim *et al.* (1997) can be related to radar images from the European satellite Sentinel-1A which are shown in orange (Fig. 3).

The geology and climate around Mosul are favourable to the formation of dolines, caves and sinkholes (Al-Dewachi 2005). Acidic rainfall dissolves the carbonate rock, and, as it moves through the zone of saturation, it gradually forms horizontal pathways and caves. In the active salt flat 'sabkha' settings, the amounts and distributions of calcite, gypsum and anhydrite change depending on groundwater chemistry and movements. Chemical and physical changes to these evaporites continue after burial. Anhydrite above the water table is replaced by gypsum through hydration (Schnabel 2016). As oxygen (O_2)-rich surface waters seep into deep anoxic karst systems, they bring oxygen that reacts with sulphide present in the system (H_2S) to form sulphuric acid (H_2SO_4). Sulphuric acid then reacts with calcium carbonate causing increasing solution within the limestone formation. The Fatha Formation is an excellent source for H_2S, which is emitted due to the presence of native sulphur (Jassim *et al.* 1999; Al-Ansari *et al.* 2015).

Deposits within the Tigris channels (old and modern) around Mosul are loosely cemented conglomerates, sands and marls. The lower terraces of the Tigris River around Mosul are composed of sandy, clayey silt and are up to 20 m thick (Wakeley *et al.* 2007). Caves in conglomeritic rock near Mosul University show massive, indurated conglomerate but in other localities it is well stratified with imbricated clasts and fining-upwards sequences (Al-Dewachi 2005). In addition to the presence of water, lithology is the main reason behind the formation of caves and horizontal pathways as the calcareous material is easily dissolved.

Within the city of Mosul this natural process was reinforced after installation of a water system in the 1980s. Water consumption rose to 200 litres/person/day and with no sewer system there was a dramatic rise in water percolation downwards, passing through permeable and fractured limestone into underlying gypsum, creating new dissolution cavities and entering the older karstic system. It was not uncommon for buildings over these cavities to fracture and collapse.

Military tunnelling

Evidence exists that tunnels have been used in warfare from ancient (crusades: e.g. Frumkin & Shimron 2006) to modern times (World War I and II, Vietnam, Korea, Afghanistan, Palestine–Israel, Lebanon–Israeli, and Syrian civil wars). IS has utilized pre-existing caves and tunnels and constructed new ones at an industrial scale. Research has identified tunnel complexes in over 60 towns and villages formerly controlled by IS in parts of Nineveh, Erbil and Kirkuk governorates (Figs 2 & 3). These are in addition to those in Anbar and Salah Al-Din governorates. IS used subterranea in Syria and Iraq as a core part of their control over territories and in their military capability. Caves and natural tunnels used by locals for shelter and storage quickly became known to IS after they took over any territory. Sinkholes and caves around Sinjar, Mosul and Tikrit were used by IS as execution sites and mass graves (Morris & Salim 2017). The Khasfa sinkhole 8 km south of Mosul was 400 m deep and IS filled it with an estimated 4000 bodies (Neuhof 2017). IS used the abandoned 2.5 km rail tunnel outside Mosul for a training site, setting up a shooting range, obstacle course, mosque and clinic within it (Reuters 2017). This tunnel south of Mosul was excavated in the early twentieth century for the Baghdad railway. Tunnelling later took place in the region using modern boring machines for underground oil and gas storage, as well as for road tunnels. In rural areas the ancient qanat (kārēz) technique of water supply from a deep conduit with a series of vertical access shafts (Safar 1947; Laessoe 1951; Garbrecht 1983) is still used for irrigation, providing cattle with water and for drinking water supply. Control of water has been significant in recruiting IS fighters locally and gaining legitimacy (King 2016). IS used the kārēz underground irrigation system to control water but also to fight in ways similar to the mujahedeen in Afghanistan (Grau 1996). Military aspects of hydrogeology are discussed in Mather & Rose (2012).

When IS crossed over the Syria–Iraq border and captured the town of Sinjar (Fig. 2) they used Kurdish Ezidi hostages to dig tunnels 10 m down in the local thinly bedded detrital limestone, chalky limestone, dolomitic limestone and marl (Aljubouri & Aldabbagh 1980). This provided them with protection from artillery and airstrikes. Over 70 tunnels have been identified in the complex below the town (Wyke 2016). Tunnelling to and under buildings is likely to have helped the maintenance of alignment by construction in short segments between successive intermediate shafts. Tunnels captured from IS in February 2016 are arch shaped with straight sides and hand-tool marks can be seen on the walls (Fig. 4). One has approximate tunnel dimensions wall to wall of 0.9 m and a floor to roof height of 1.8 m extending for 3 km. The lithology in the tunnel shows angular clasts and some indication of A-axis imbrication and bedding. Another tunnel, in Sinjar town, is 2 km long, has approximate dimensions of 0.6 m wall to wall and 1.2 m floor to roof. In places,

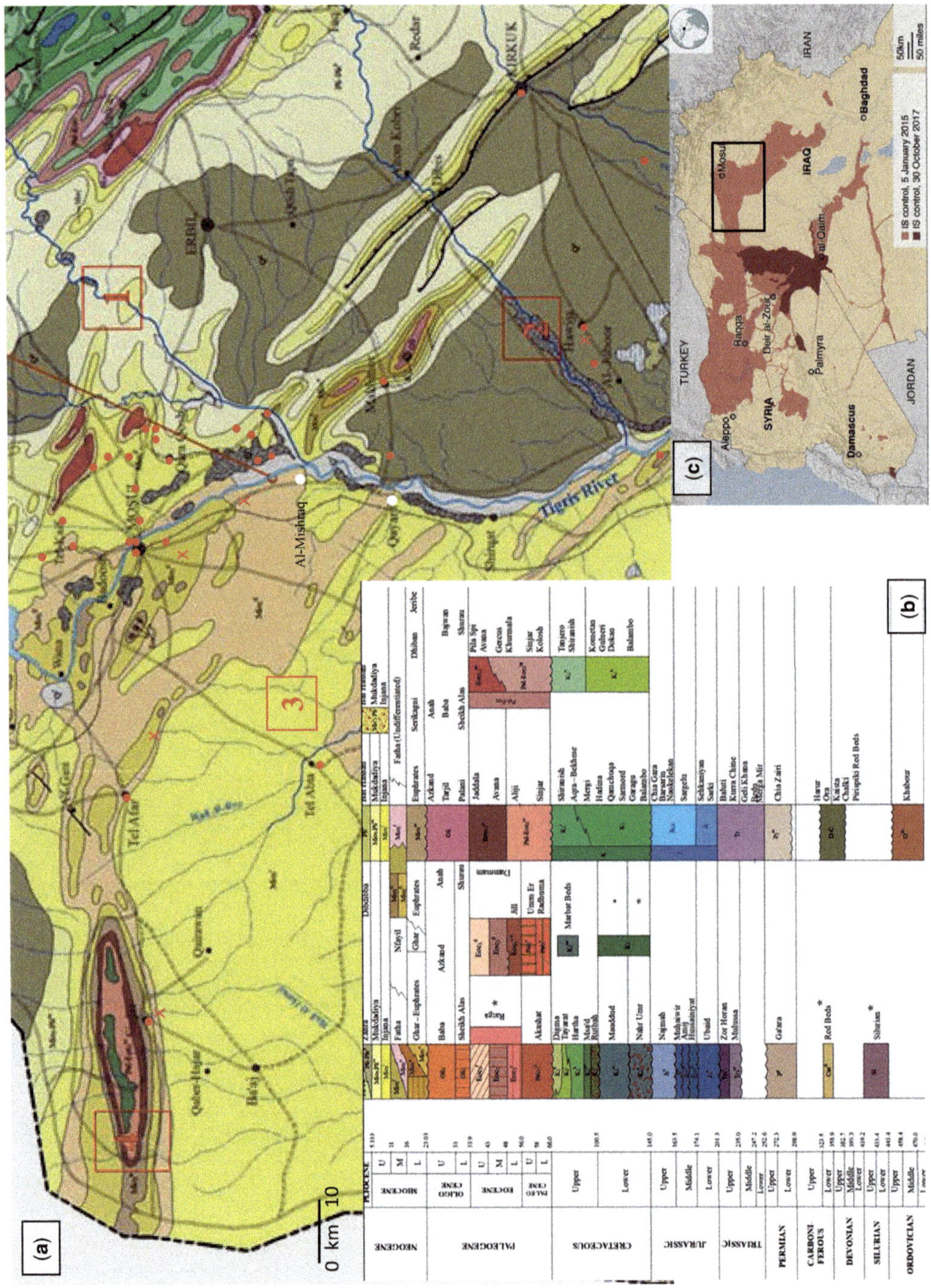
(a)
(b)
(c)
0 km 10
Al-Mishraq
Tigris River
MOSUL
ERBIL
KIRKUK
Tel Afar
TURKEY
SYRIA
IRAQ
IRAN
JORDAN
Aleppo
Raqqa
Deir al-Zour
Palmyra
Damascus
al Qaim
Baghdad
IS control, 5 January 2015
IS control, 30 October 2017
50km
50 miles

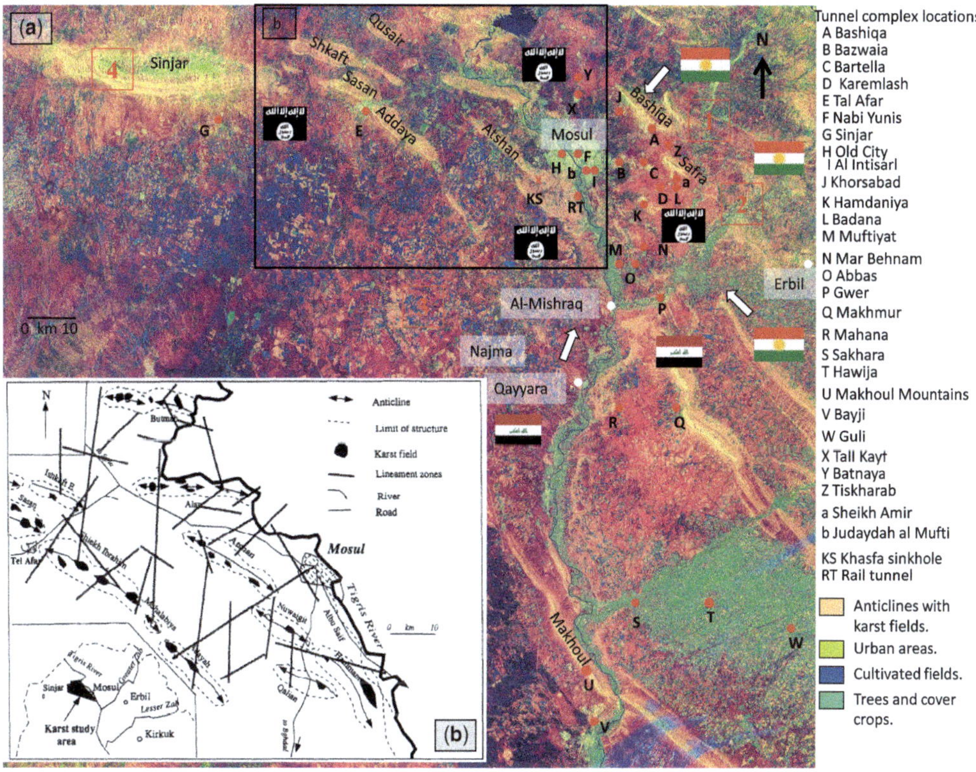

Fig. 3. Locations of anticlinal structures with karst fields. (**a**) Combined scans acquired on 3 and 15 October 2014 from Sentinel-1A's radar over Mosul. Colours come from combined coherence and intensity information. As in Figure 2, IS tunnel complex locations are shown by red dots, and mass graves by the red X. The directions of advance by anti-IS forces are shown by white arrows (from Chulov & Summers 2016). (The Sentinel-1A image is reproduced by permission of Copernicus data/ESA.) (**b**) The locations of anticlinal structures with karst fields are indicated and related to those identified by Jassim *et al.* (1997).

wooden roof supports were used and fans for ventilation were found along the lengths of the tunnels connected to an electric cable running along tunnels up to a generator located above ground.

East of Mosul to the Greater and Lesser Zab rivers (red boxes 1 and 2 in Figs 2 and 3), the Kurdish Peshmerga knew IS was digging because from Mount Bashinqa and Zartak (Safra) they had watched people coming out of the ground for 2 years. There are reports that IS paid some locals 4000 Iraqi dinar (about $4.65) per day to tunnel with other labour from hostages. The advance to Mosul by the anti-IS coalition started on 17 October 2016, moving along the key roads across the Lesser and then Greater Zab rivers on the eastern bank of the Tigris. On the western bank of the Tigris, the advance from the south moved up through Qayyara (Fig. 3). In the last week of October 2016, advancing Iraqi and Peshmerga forces had crossed the Greater Zab River and captured tunnel complexes in the villages of Sheikh Amir, Bartella and Bazwaia (Fig. 3). These were along the main road from Erbil to Mosul.

Documents found in a tunnel complex in Sheikh Amir reveal written IS operational practices for provisioning tunnels. Solar panels would be placed near an entrance to charge smartphones and other devices,

---

Fig. 2. (**a**) A portion of the 1:1 million scale geological map of northern Iraq (Sissakian & Fouad 2015) on which known IS tunnel complex locations (red dots) and mass graves (red X) are shown. Box 1 indicates the Greater Zab River; box 2 indicates the Lesser Zab River; box 3 indicates the offset of the Tharthar Valley; box 4 indicates the Sinjar anticline. (**b**) Legend to the 1:1 million scale geological map. (**c**) Territory under IS control in January 2015 and 30 October 2017, with area of (a) outlined (BBC, http://www.bbc.com/news/world-middle-east-27838034, 20 July 2017).

Fig. 4. **(a)**–**(c)** Examples of tunnels captured from IS in Sinjar February 2016. (The still frames for a and c were extracted from the video https://www.youtube.com/watch?v=acxldPgTI-8 2016; (b) is from Reuters/Ari Jalal 2015, https://www.reuters.com/news/picture/islamic-state-tunnels-idUSRTX1X0JN).

and a month's worth of food was to be kept in a storeroom. Men should not gather at the entrance or in the open, and entrances were to be concealed in houses.

Tunnels at Bartella are 10 m deep, excavated down through sand and clayey silt (Fig. 5). Below ground, tunnels have arched roofs similar to those in Sinjar. Drills, shovels and wheelbarrows were found in one section of the tunnel complex.

The entrance to the tunnel complex in Bazwaia is in a large building and the floor appears to have been cut using a concrete saw, enabling a 10 m-deep shaft to be dug and then a tunnel. The building was full of spoil, showing how IS tunnellers were able to conceal their activities. The spoil appears to be sandstones, siltstones and claystones from the Injana Formation. Below ground, some tunnels at Bazwaia are lined with wooden planks, showing knowledge and skill. Tunnels are wider than those under Bartella and have lighting, as well as fans, strung along them.

In the first week of November 2016, another large tunnel complex was captured in Karemlash (see Figs 3 & 6). One of the tunnels dug 10 m under St Barbara church was 9 km long. The tunnel characteristics at Karemlash are similar to those at Sinjar and Bartella but with evidence of tool marks that fit those made by drills (Fig. 6). The tunnels appear to be in the upper member of the Fatha Formation. In Bashiqa at the foot of the Bashiqa Mountains, 20 km east of Mosul, 10 tunnels were discovered in a complex 10 m underground. Some have rooms connected to CCTV on the surface, makeshift hospitals, rest areas, kitchens and weapons storage. Small generators in the house above the tunnel entrance powered lights and fans affixed to the walls. In early November 2016, Colonel Ibrahim of the Iraqi Army's 9th Division reported the existence of a network of tunnels extending over a total length of 72 km. At Judaydah al Mufti, on the eastern edge of Mosul, another long tunnel was found that was 9 km long.

By late 2016, captured tunnel complexes demonstrated that IS knew how to tunnel in the Injana and Fatha formations (Fig. 2). However, it was the capture of two different types of improvised tunnel-boring machines (ITBMs), one by Kurdish Peshmerga east of Mosul and the other by Iraqi forces, that demonstrated a significant and unexpected change in IS capacity and capability (Bulmer 2016*b*). These machines could potentially increase the amount of tunnelling and tunnel faster than could hostages or paid labour; using fewer tunnel workers would enable them to be used for other work such as moving spoil out of tunnels. ITBMs are now discussed in the following subsections.

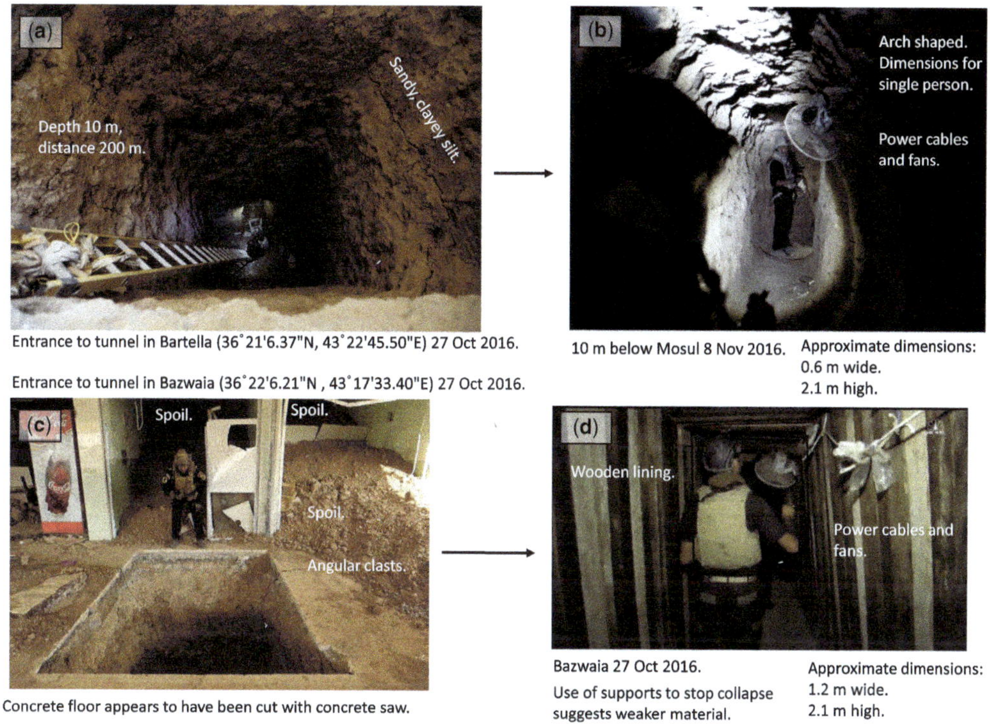

Fig. 5. Tunnels captured in the villages of Bartella and Bazwaia. (**a**) Entrance to a tunnel in Bartella (Reuters/Zohra Bensemra 2017, https://www.reuters.com/news/picture/portfolio-of-work-zohra-bensemra-idUSRTX3RNFV). (**b**) Tunnel below image (a) (Reuters/Zohra Bensemra 2016, https://www.reuters.com/news/picture/portfolio-of-work-zohra-bensemra-idUSRTX3RNFV). (**c**) Entrance to a tunnel in Bazwaia (Zohra Bensemra/Reuters 2016, https://www.reuters.com/news/picture/portfolio-of-work-zohra-bensemra-idUSRTX3RNFV). (**d**) Tunnel below image (c) (Associated Press/Ali Abdul Hassan 2016).

Improvised machines

Fabrication of ITBMs should not have been unexpected given the level of engineering and mechanical skill IS had demonstrated in Syria and Iraq in making weapons, uparmouring vehicles such as mining trucks, earth movers, tractors and pick-up trucks (Fig. 7), and the large number of vehicles, parts and sites of fabrication they had captured. There was an appreciable improvement in the skill and quality of mechanical work from 2014 to 2017 (Fig. 7). In the village of Bartella, 23 uparmoured vehicles laden with improvised explosive devices (vehicle-borne improvised explosive devices: VBIEDs) were driven at Pershmerga forces during the advance, and IS fighters appeared unexpectedly from the tunnel complex causing delay and casualties. This was to be the norm, demonstrating an industrial effort in making VBIEDs (Triebert 2017). Given the size and number of tunnel complexes that IS was constructing, the need to tunnel faster would have been apparent. Moving from hand tools to ITBMs can be seen as a logical progression.

Improvised small tunnel-boring machines

An improvised small tunnel-boring machine (ISTBM) was captured by Kurdish Peshmerga in the village of Tiskhrab on 24 October 2016 (Fig. 8). The ISTBM is composed of a circular cutting head estimated to be 0.3 m in diameter fixed to a drive shaft spun by an electric motor. The teeth in the cutting head appear to be set up for use in poorly consolidated material, and are thus well suited to the Injana and Fatha formations. The motor sits on four small wheels that in turn sit on tracks on a trolley that has four wheels. A support frame fixed to the trolley has a collar through which the drive shaft can spin and move. This allows the ISTBM to run along the tracks of the trolley. In this way, the angle of the cutting head and drive shaft is maintained. The trolley can be moved using a handle. The dimensions of the cutting

Fig. 6. Hand-dug or drilled tunnels in Karemlesh captured by Pershmerga forces in November 2016. Images are from *The Washington Post* 2016. (**a**) A tunnel entrance in a house (Nur Photo via Getty images 2016). (**b**) Narrow and low tunnel (Reuters/Zohra Bensemra 2016, https://www.reuters.com/news/picture/portfolio-of-work-zohra-bensemra-idUSRTX3RNFV). (**c**) Longest tunnel in Karemlash, with a higher ceiling than (b) (Reuters/Zohra Bensemra 2016, https://www.reuters.com/news/picture/portfolio-of-work-zohra-bensemra-idUSRTX3RNFV).

head of the ISTBM appear to fit with the width dimensions of tunnels in the villages of Karemlash and Bartella (Figs 5 & 6). It is not known how the ISTBM was pushed to keep the cutting head in contact with the face but this could have been one of the functions of the operator(s). Also unknown is how the operator(s) monitored tunnel alignment but vertical shafts could have been used in a way similar to the kārēz technique. It appears that the bore spoil was dry and was moved back to the head of the tunnel using human labour. Again, it has not been verified but the height of a tunnel such as those in Karemlash and Bartella (1.5–2 m) could have been made using a shaft dug to an initial depth (10 m) down which the ISTBM was lowered. The ISTBM would then have been used to make the horizontal bore. Once this had been completed, the ISTBM could be raised up another shaft or moved back to the entrance shaft and the floor dug down so another horizontal bore could be made. In this way a tunnel shoulder wide and tall enough to crouch through or to stand in could be made using two or three horizontal bores.

Improvised tracked tunnel-boring machines

An improvised tracked tunnel-boring machine (ITTBM) was captured in the first week of November 2016, in a tunnel outside Judaydah al Mufti (Fig. 3). There was a hidden ramp down to the tunnel that appears to be 10 m deep or more. The existence of the ramp shows that the ITTBM was not lowered down a shaft. The walls of the ramp down to the floor are light grey/brown and have the characteristics consistent with the Injana Formation (Fig. 9a). The ramp is next to buildings, and was camouflaged from overhead using roofing tin covered with soil, sand and rubble. The ITTBM was recovered from the bottom of the ramp at a tunnel face with the A-frame of an American-made M88A2 recovery vehicle from the 9th Armoured Division of the Iraqi Army (Fig. 9b). Photographs released by the media on 3 November 2016 enabled an assessment to be made of some of the key features on the ITTBM and their possible functions (Bulmer 2016*b*). Further media footage was released (Robinson 2016) and has been examined in an attempt to

Fig. 7. Improvised vehicles created by IS in Syria and Iraq. The level of sophistication, skill and quality in images (a)–(f) shows significant improvement over the period 2015–17. (**a**) A Belaz dump truck once used to build dams in Syria; used as a suicide vehicle-borne improvised explosive device (SVBIED) on the T4 airbase in August 2015 (Rogoway 2015). (**b**) Bulldozer with welded plates for protection. Location unknown (Dan Beaumarchais pers. comm. 2016). (**c**) Front loader with multiple steep plates welded to provide protection. Location unknown (Dan Beaumarchais pers. comm. 2016). (**d**) Front-end loader armoured with steel plates; used as a SVBIED in NW Mosul on 15 March 2017 (Rogoway 2017). (**e**) Pick-up truck surrounded with steel plates and angled to provide further protection; built as a VBIED and captured near Tal Afar, 6 November 2016 (Frantzman 2016). (**f**) Jeep Cherokee with add-on armour, captured in Mosul, 18 March 2017 (Captain Jerry Champion pers. comm. 2016).

understand the design of the ITTBM and how it was operated.

Cutting head. Images of the ITTBM show a conical cutting head (Fig. 10) with seven rows of six individually replaceable teeth similar in characteristics to commercially available 'Christmas tree' designs (Fig. 10b). Between each row of teeth, up where the diameter of the head is greatest, there is one more replaceable tooth. This makes a total of 49 replaceable teeth. Ten larger teeth appear to be fixed to the cutting head shroud. The nose of the cutting head has four large teeth, within which is a central auger running back through the ITTBM. Seven vertical pans, each positioned between a row of cutting teeth along the head, empty the spoil scraped off by the teeth into the central and exterior augers. The direction of rotation at the tunnel face was clockwise from the perspective of the operator. As for the ISTBM, the teeth in the cutting head appear to be set up for use in poorly consolidated material, and are thus well suited for tunnelling in the Injana and Fatha formations. Using the soldier seen in Figure 10 for scale, the height for the ITTBM is estimated from the cutting head diameter to be 2.1 m, and the length for the cutting head 1 m. This means that a single bore using the ITTBM would produce a tunnel tall enough for a person to stand in and wide enough for a car or pick-up truck to move in.

Tracks and chassis. The ITTBM is mounted on two tracks with metal treads (Fig. 11). There are drive sprockets on either side, at the front closest to the cutting head, for better traction and control. The design of the tracks looks very similar to the Hitachi EX120 hydraulic excavator deck or the Barbco BD100HP horizontal drilling machine (Fig. 11b), equipment in common civilian use. On either track, there are four running wheels on the bottom track and one on top. The grease point is evident and this is where the track was tensioned. At the back of the ITTBM is a blade that is behind the spoil drop from the central auger. It can therefore be used to move spoil but can also provide resistance to the ITTBM moving backwards when boring. Just forward of the midpoint of the track is a steel pin plate with four holes. These holes appear to be used in conjunction with a pivot pin at the back of the ITTBM that enables adjustment of the angle of the cutting head. Figure 11 shows that the second hole from the top in the pin plate is larger, possibly indicating worry of the pin in that position due to extensive vibration. This in turn may indicate the main configuration angle of the cutting head during boring. The top right of Figure 11 shows a rubber sheet with the word 'IVECO' on it and an arrow in the direction of cutting head rotation next to which are the numerals 100. On the opposite side of the ITTBM the same rubber sheet has another arrow in

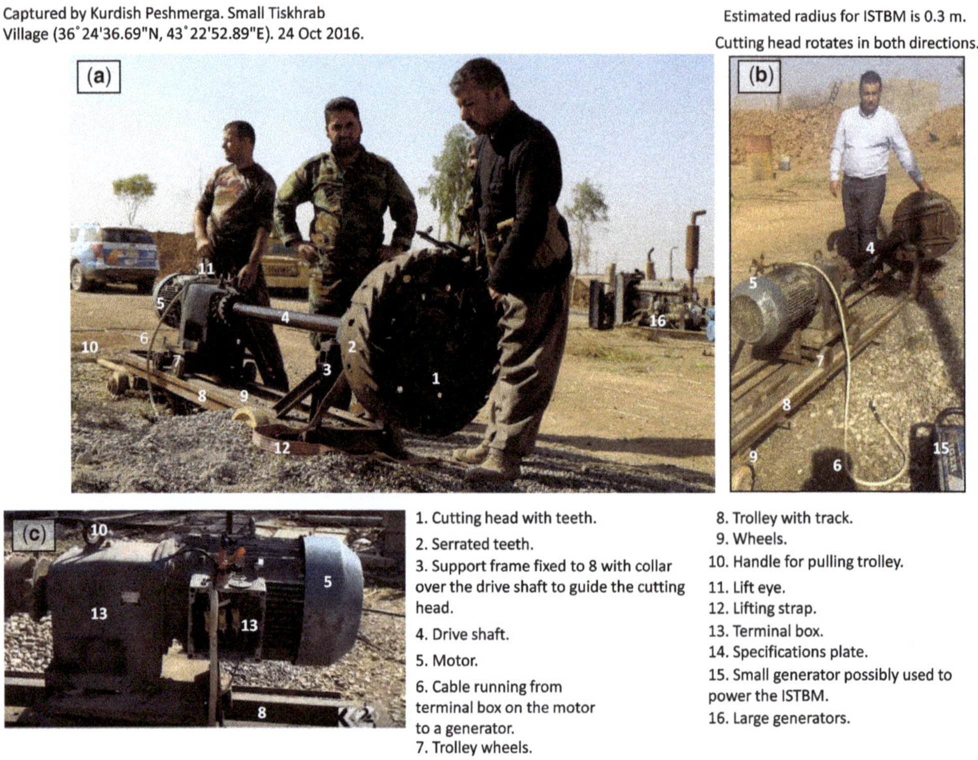

Fig. 8. View of the front, back and side of the improvised small tunnel-boring machine (ISTBM) captured by Kurdish Peshmerga in small Tiskhrab village (36°24′36.69″ N, 43°22′52.89″ E) on 24 October 2016. (**a**) View of the front of the ISTBM showing the cutting head; the soldiers provide the scale (Kurdistan24). (**b**) View of the back of the ISTBM from the motor and trolley system (Kurdistan24). (**c**) Side view of the electric motor and terminal box. The still frames were extracted from the video http://www.kurdistan24.net/en/video/f77c602e-b1bc-4c3f-8a08-8205c43f46c2

the direction of rotation and the number 70. It is not known whether these numbers relate to operational parameters or if the sheet is repurposed from commercial components.

Motors. One of the most striking aspects of the ITTBM when it was suspended from the M88A2 (Fig. 9) was the existence of six motors. Figure 12 shows the motor on the forward left of the ITTBM behind the cutting head. It is paired with an identical one on the forward right and separated by the central auger. Each of these two motors connects to a drive wheel that rotates with teeth on the back of the cutting head. Given the size and functions of these paired motors, they are likely to be three phase with four wires, three hot and one ground. Wiring from the terminal boxes of these paired electric motors runs back to a circuit panel at the rear right of the ITTBM. Situated behind the forward right electric motor, at the approximate midpoint of the ITTBM, is a smaller electric motor whose drive shaft in turn connects to a hydraulic filtration pump connected to a hydraulic reservoir. This appears likely to be a three-phase motor with four wires, three hot and one ground with wiring again running back to the circuit panel at the rear of the ITTBM. On the opposite side of the ITTBM to the midpoint electric pump is an electric motor that runs a fan for cooling the hydraulic fluid. Electric cable from the terminal box on the motor to the cooling fan runs back to the circuit panel at the rear right of the ITTBM. Hydraulic supply and relief lines run to and from the hydraulic reservoir, and also to levers connected to hydraulic control valves and drive linkages. These are likely to connect to the tracks and the cutting head. Two additional electrical motors run the two augers. Both are situated at the back of the ITTBM: one is on the end of the external augur on the left side, and the other, on the right side, runs the central auger. The back of the ITTBM (Fig. 13) has the seat for the operator on the left side in front of the control levers and to the right is the circuit panel. The back of the central auger is driven by a motor and chain that is covered by a protective

Fig. 9. (**a**) View of the back of an ITTBM underground at a tunnel face (the still frame was extracted from the video https://www.liveleak.com/view?i=1a7_1478292320 2016). (**b**) One of the first photographs published in the international media of the top side of the improvised tracked tunnel-boring machine (ITTBM) suspended from the M88A2 vehicle used to lift and drag it from a tunnel (Defence Ministry of Iraq).

panel. The circuit panel can be understood by relating the circuit breakers to the six electric motors. Triple circuit breakers are likely to connect to the motors. It is not known what the single pole circuit breakers ran.

Functional interpretation

All the different elements of the ITTBM identified in Figures 10–13 are brought together in Figure 14, which shows schematics for the plan view (Fig. 14a), right side view (Fig. 14b) and cross-section of the cutting head to machine frame connection (Fig. 14c). It appears that the ITTBM was steered and operated from the seat using the four levers. Steering would likely have been by moving left and right tracks forward and back, but how pitch, roll, and yaw were monitored is unknown. The connection of the drive wheels 3A and 3B from paired motors 7A and 7B to the out inner diameter of the cutting head provided the high torque needed for boring, but how the turn speed of the motors and the cutting head was controlled is unknown. The hydraulic system, motors and circuit panel would have required monitoring, but oil, temperatures or pressure level gauges are not evident. The design of the ITTBM suggests that the primary power source was the hydraulic power pack connected to a three-phase generator. This would have eliminated engine emissions in the tunnel. While there is a small indication of slurry, the evidence supports the inference that bore spoil conveyed by the continuous augers would have been dry when conveyed to the back of the ITTBM. It appears that the spoil was moved back to the entrance of the tunnel using human labour but there would have been room for a conveyor belt or rail system. This would have been relatively easy to construct, operate and power. At the head of the tunnel, spoil was likely to have been moved by hand into rooms, but a conveyor might have been set up.

Analysis of spoil generation and storage

Using the knowledge of the characteristics and dimensions of the ISTBM and ITTBM discussed above, it is possible to conduct first-order analysis

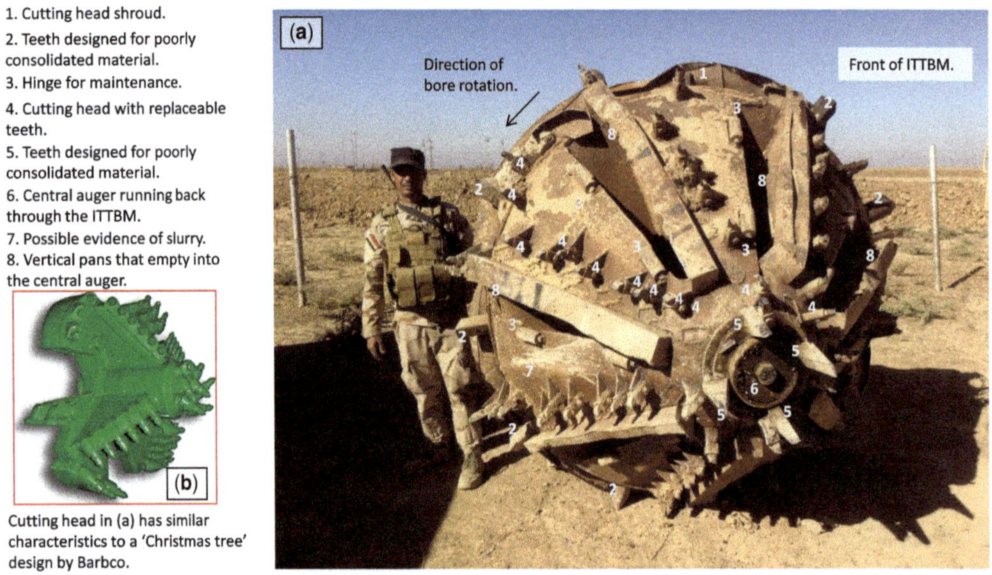

Fig. 10. (**a**) View of the cutting head of the ITTBM (Fig. 9b) seen on the surface after capture; the soldier provides scale (Defence Ministry of Iraq). (**b**) The conical shape with fixed and replaceable teeth has similar characteristics to a 'Christmas tree' designed by Barbco (http://www.barbco.com/cuttingheads/christmas-tree-rock/).

of spoil generation using:

$$\text{Volume (m}^3) = \pi r^2 L$$

where r is the radius of the cutter head and L is the length of the tunnel and for a single bore.

For the ISTBM the estimated radius r of the cutting head is 0.3 m and for the ITTBM it is 1 m. Individual tunnel lengths within complexes ranging from 25 m to 9 km have been identified in IS-controlled parts of Nineveh, Erbil and Kirkuk, Anbar and Salah Al-Din governorates. Using these figures, the amount of material excavated for a range of tunnel lengths using a single bore has been calculated (Table 1).

In the absence of knowledge of actual geotechnical values for spoil from tunnels around Mosul, coarse sand was used as a surrogate (1.6 t (metric tonnes) in 1 m^3) to calculate the tonnage removed. Bulking of the spoil was not included. Material removed by the ISTBM for a single 25 m bore is calculated to be 11.2 t. Since it takes two bores with the ISTBM to make a tunnel large enough to crouch in, the combined tonnage is 22.4. Over a tunnel of the same length, the ITTBM removed 125 t.

The first-order calculations of the volumes and weights of spoil generated by the ISTBM and ITTBM were next related to transport and storage. Using a truck with a capacity to carry 16 m^3, the number of trucks that it would take to move spoil for a range of tunnel lengths has been calculated. The ISTBM requires one truckload for construction of a 25 m-long tunnel large enough to crouch in; for the same bore length, the ITTBM requires eight truckloads. Table 1 shows that as the tunnel lengths became longer, the amount of spoil generated would have required transportation planning with all the related issues of how many trucks and drivers were in the Mosul area, where they came from, how far they had to travel to unload, road and bridge weights, fuel, spares, and maintenance. Given the evidence that numerous tunnels were being dug at the same time in IS-controlled areas around Mosul, the logistical requirement becomes significant even in a non-conflict environment. A future study should be able to examine evidence of truck haulage in remotely sensed images acquired from 2014 to 2017.

The calculations for trucking required for spoil haulage from ISTBM and ITTBM operations provide a practical rational for IS keeping excavated spoil as close to a tunnel as possible. This, as well as concealment from observation, can explain rooms within buildings being used to store spoil (Fig. 5c). The average house in the villages of Karemlash, Bartella and Bazwaia, with three bedrooms (each 2.4 m wide, 3.6 m long and 2.4 m high) and one living room (4.8 m wide, 6 m long and 3 m high), was used to calculate the volumes of spoil that could be stored within them (Table 1). Using these room dimensions, the spoil generated by the ISTBM from a bore 25 m long would fill 0.3 of a

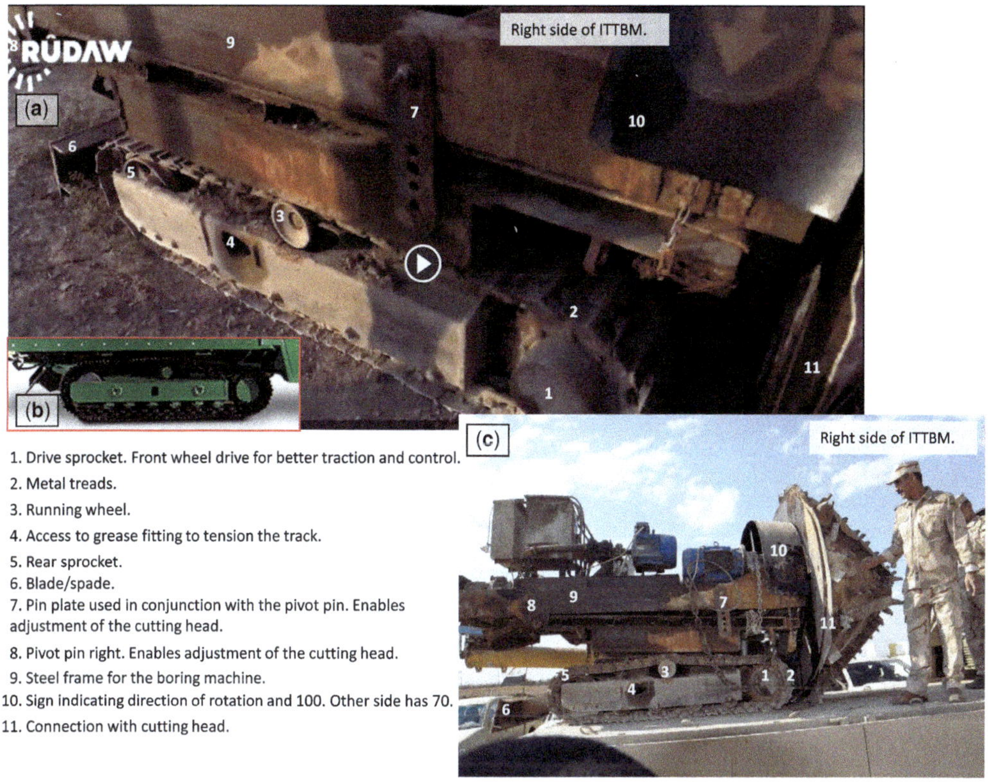

Fig. 11. (**a**) View of the right side of the track of the ITTBM (Fig. 9b) seen on the surface after capture (the still frame was extracted from a video by the Kurdish media network RUDAW (http://www.rudaw.net/english/middleeast/iraq/081120164 2016). (**b**) The design of the ITTBM tracks looks very similar to the Barbco BD100HP (http://www.barbco.com/directionaldrills/bd100/). (**c**) Side view of the ITTBM on a flat-bed truck showing the same side as in (a) (Hugo Kaman pers. comm. 2016).

bedroom or 0.1 of a living room and combined would fill 0.05 of a house. The spoil generated by the ITTBM from a 25 m-long tunnel would fill four bedrooms or one living room and combined would fill one house. Using the volume space of 148.5 m^3 for a house and, again, assuming the excavated spoil to be similar in weight to coarse sand, a house will be able to hold 238 t. These numbers show that the advantage of ITBMs in digging faster than with hand tools resulted in a logistical burden around the need to move or store spoil. It seems logical that IS would have stored spoil in houses rather than using trucks. The implication of this is that many houses and buildings around tunnel complexes, not destroyed in the fighting, would be filled with spoil.

Tunnelling rate

Having conducted a first-order analysis of spoil generation over defined distances of tunnels for ISTBMs and ITTBMs, and how much space is required for spoil storage, it is possible to make first-order estimates of the rate of tunnelling for the ITTBM. In order to calculate tunnelling rates for TBMs in a given geology, it is necessary to determine penetration (mm) of advance per rotation of the cutter head. To obtain net boring speed (m/h), penetration is multiplied by the speed of rotation of the cutter head (revolutions per minute: rpm) multiplied by 60 (seconds) and divided by 1000 (mm). The calculation from net boring speed to rate of advance must take account of obstructions, faults and interruptions to tunnelling (Schneider *et al.* 2010). The hope was that the paired motors (Fig. 14, components 7A and 7B) connected to the cutting head could provide details on horsepower, voltage and speed, enabling determination of cutting head rpm and how this is controlled (Hatch Mott MacDonald 2001). However, this was not possible, so, in lieu, data from tunnels created by commercial Robbins SBU-A Boring machines, with a similar diameter cutting head, were used to calculate a

1. Electric motor on forward left to drive cutter head.
2. Back of cutting heads. Note the teeth that engage with the drive wheel from the paired motors.
3. Electric motor on forward right to drive cutter head.
4. Central auger shaft connected to the cutter head. Casing is in the open position.
5. Specification plate (missing).
6. Electric motor for hydraulic pump.
7. Drive shaft from motor to the hydraulic pump.
8. Hydraulic fluid reservoir. Natural head to reduce cavitation.
9. Hydraulic supply and relief lines.
10. Cooling fan for hydraulic oil.
11. Electric motor for cooling fan for hydraulic oil.
12. Levers connected to hydraulic control valves and drive linkages. Likely connected to tracks and cutter head.
13. External auger screw conveyor.
14. Pivot pin left. Enables adjustment of the cutting head.
15. Pivot pin left. Enables adjustment of the cutting head.
16. Hydraulic actuator for the spade/blade
17. Eyebolts for lifting forward right electric motor.
18. Eyebolts for lifting forward left electric motor.

Fig. 12. (**a**) View of the left side of the ITTBM (Fig. 9b) looking forward to the cutting head. (**b**) View of the right side of the ITTBM looking rearwards (the still frames were extracted from a RUDAW video obtained of the ITBM on the surface after capture and transport to an Iraqi base (http://www.rudaw.net/english/middleeast/iraq/081120164 2016).

possible range of tunnel daily, weekly and yearly advance rates (Anon 2017). These data are derived from excavation in hard rock (6 m/day), boulders and cobbles (12–18 m/day), clay, alluvium and dry soil (42–45 m/day). Operations of the ITBMs were assumed for 2 × 10 h shifts for 5 days each week (Fig. 15), leaving 4 h/day for maintenance. The type of support system installed immediately behind the cutter would also have a significant influence on the rate of advance. The tunnel at Bazwaia shows that IS tunnel teams were skilled in using wooden and steel arches. One IS tunnel in Raqqa shows they used steel mesh, and one tunnel 20 m deep has prefabricated concrete and steel forms. The latter has dimensions (height, width and shape) very similar to those produced by the ITTBM found outside Judaydah al Mufti. Based on the examination of the ISTBM and ITTBM, it seems a team would have consisted of one member driving the TBM, probably three installing ground support and three running the spoil movement. This is consistent with the size of operator teams recommended for Robbins machines (Anon 2015).

When tunnelling, penetration depends upon machine, rock mass and their interaction. Like the ISTBM, it is apparent that the ITTBM is a robust design, and both are well suited to interact with the Injana and Fatha formations. Given the nature of these formations, advance rates of 24 m/day for the ITTBM seem realistic but ranges from 6 to 45 m/day are shown in Figure 15a, b. An ITTBM with a 2 m diameter cutting head advancing at 6 m/day (hard rock) can tunnel 1560 m in 1 year, compared to a tunnel 6240 m long at an advance rate of 24 m/day in variable geology or 11 700 m long in soil at an advance rate of 45 m/day over the same period. To move the respective 7845 t, 31 378 t and 58 834 t of excavated ITTBM spoil off-site would have taken 306, 1226 and 2298 truckloads (Fig. 15c). To store this volume of material would have required 33, 132 and 248 houses. At the time of the Iraqi and Peshmerga advance to recapture Mosul in October 2016, IS had been in control of the city and surrounding villages for 2 years. So over this period, advance rates of 6, 24 and 45 m/day would have enabled tunnel lengths of 3120,

1. Levers connected to hydraulic control valves. Likely connected gear box to tracks and cutter head.
2. Hydraulic hose bend pipe fittings. Likely connected gear box to tracks and cutter head.
3. Cooling fan powered by electric motor for cooling hydraulic oil.
4. Central auger shaft connected to the cutter head.
5. Electrical panel.
6. Electrical wires from the 3 blue electric motors. Likely connected to 3 phase generator.
7. Triple circuit breaker.
8. Single pole circuit breaker.
9. Circuit breaker for main starter. Likely 3 phase.
10. Electric motor to drive chain for central auger.
11. Protective panel over the drive chain and motor for the central auger.

Fig. 13. (**a**) View of the back of the ITTBM (Fig. 9b) looking forward showing the control levers and circuit panel. (**b**) View of the back of the same ITTBM just prior to it being extracted by the M88A2 vehicle. Note the difference in circuit breaker numbers between (b) and (a). It appears that circuit breakers came off during extraction and transit or were removed once on the surface. The still frame was extracted from a RUDAW video obtained of the ITTBM on the surface after capture and transport to an Iraqi base (source as for Fig. 12).

12 480 and 23 400 m to be completed. To move the respective 15 689 t, 62 757 t and 117 668 t of excavated spoil off-site would have required 613, 2452 and 4597 truckloads. To store this would have required 66, 264 and 496 houses, respectively. At least two, and possibly three, ITTBMs were reportedly found by Iraqi or Peshmerga forces, so it is possible that others exist. In aggregate over 1 year, three tunnels bored using the ITTBM at 6 m/day would have produced a total tonnage of 23 534, requiring 919 truckloads or 99 houses. For comparison, in 1 year, three tunnels bored using the ITTBM at 24 m/day would have produced a total tonnage of 94 135, requiring 3677 truckloads or 396 houses. Three tunnels bored at 45 m/day would have produced a total tonnage of 176 503, requiring 6895 truckloads or 743 houses.

While the numbers calculated here use advance rates taken from Robbins SBU-As applied to the ITTBM, they provide an insight into just what the use of one, or three, ITTBMs could mean to increasing the capacity and capability of IS to tunnel. Circles in Figure 15a show that an ITTBM with an advance of 24 m/day starting in Karemlash could connect with tunnel complexes at Bartella, Hamdaniya, Badana and Sheikh Amir within 1 year, and with Bazwaia in just over 1 year at a rate of 45 m/week. In 2 years, at a rate of between 42 and 45 m/week, connection could be made with Bashiqa and Al Intisarl. One of the implications of this analysis is that these machines would have needed a robust supply chain of spares, oils and lubricants in sufficient volumes that would have to come to, or near to, the tunnel site on the road network.

Searching for tunnels

While it is possible to identify locations of ongoing tunnelling due to noise, dust, vibration and subsidence, there are no media reports that any were found in this way. Radar systems and synchronized electromagnetic wave gradiometers have been used for searches of cross-border tunnels in the Korean DMZ (Demilitarized Zone), the Gaza Strip and the US southern border (Stolarczyk *et al.* 2005), but

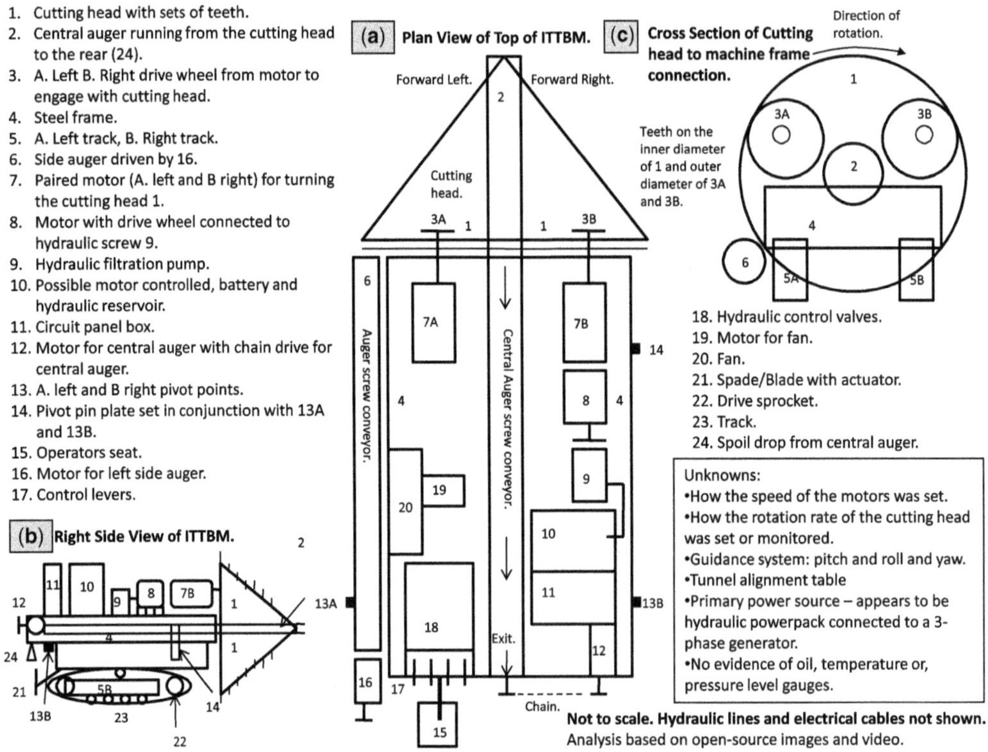

Fig. 14. (**a**) Plan view of the top of the ITTBM similar to Figure 9. (**b**) View of the right side similar to Figures 11 and 12. (**c**) Cross-section of the cutting head to machine frame connection. The schematic for ITTBM is not to scale and is compiled from analysis of open-source images and video. Hydraulic lines and electrical cables are not shown. From Bulmer (2016*b*).

tunnels around and in the urban setting of Mosul were largely found by advancing soldiers. The first-order analyses of the ITTBMs shows a way to search for tunnels using knowledge of their design combined with geology suited to their design. The size of boring machine and amount of spoil generated defined the type of buildings they started in, their possible alignment and the length of the tunnel. Figure 16a shows an area selected on the eastern edge of east Mosul with a variety of building types. The dimensions of five different building types were measured (Fig. 16b) to determine the amount of spoil that could be stored in them. These buildings were then examined as the start and ends of tunnels bored by either the ISTBM or the ITTBM to determine whether there was enough space to store/hide the spoil generated from the length of the tunnel bore (Fig. 16c). The analysis shows that the spoil excavated from two bores of the ISTBM can be fitted into all five of the buildings proposed, making it a suitable machine to use. For the ITTBM, two tunnels (A–B and A–C) can be completed starting at building A. However, three (B–C, B–D and D–C) can only be completed if the start of the tunnel is changed to go from C to B and from C to D. It is not possible to hide the spoil from a tunnel from B to D or D to B, making the ISTBM – and not the ITTBM – the machine to use if this tunnel is required. Each of these buildings for possible tunnel locations is close to a road, so getting trucks to them would have been feasible. No evidence of repeat truck haulage was noted by those examining remotely sensed imaging acquired from 2014 to 2016. This indicates that IS stored spoil in structures that proved to be very effective in providing camouflage.

Tunnelling programme

The very large number of tunnel complexes that IS constructed between 2014 and 2017 would have necessitated connection of the logistics associated with the tunnelling programme to the wider economic and military activity in the 'caliphate' and planning by IS 'Ministries'. Additionally, running

Table 1. *First-order analysis of spoil generation from improvised small tunnel-boring machine (ISTBM) and improvised tracked tunnel-boring machine (ITTBM) operations with truckloads and storage requirements for the excavated spoil*

	Radius of cutting head (m)	Length of tunnel (m)	Spoil from tunnel (m^3)	Spoil tonnes*	Number of 16 m^3 trucks	Living room 86.4 m^3	Bedroom 20.7 m^3	House 148.5 m^3
ISTBM	0.3	25	7	11.2	0.7	0.1	0.34	0.05
	0.3	50	14	23	1	0.2	1	0.1
	0.3	100	28	45	3	0.3	1	0.2
	0.3	200	57	90	6	1	3	0.4
	0.3	300	85	136	8	1	4	0.6
	0.3	400	113	181	11	1	5	0.8
	0.3	500	141	226	14	2	7	1.0
	0.3	1000	283	452	28	3	14	2
	0.3	2000	566	905	57	7	27	4
	0.3	3000	849	1358	85	10	41	6
	0.3	4000	1131	1810	113	13	55	8
	0.3	5000	1414	2263	141	16	68	10
	0.3	6000	1696	2714	170	20	82	11
	0.3	7000	1979	3166	198	23	96	13
	0.3	8000	2261	3618	226	26	109	15
	0.3	9000	2544	4070	254	29	123	17
ITTBM	1	25	78	125	8	1	4	0.5
	1	50	157	251	16	2	8	1.1
	1	100	314	502	31	4	15	2
	1	200	629	1006	39	7	30	4
	1	500	1571	2514	98	18	76	11
	1	1000	3143	5028	196	36	152	21
	1	2000	6286	10 057	393	73	304	42
	1	3000	9429	15 086	589	109	455	63
	1	4000	12 571	20 114	786	146	607	85
	1	5000	15 714	25 143	982	182	759	106
	1	7000	22 000	35 200	1375	255	1063	148
	1	8000	25 133	40 213	1571	291	1214	169
	1	9000	28 274	45 238	1767	327	1366	190
	1	10 000	31 429	50 286	1964	364	1518	212
	1	15 000	47 143	75 428	2946	546	2277	317
	1	20 000	62 857	100 571	3929	728	3037	423

*Tonnes calculated for dry coarse sand. Bulking of the spoil is not included.

all these tunnel teams at the same time would have required knowledgeable, skilled and competent project managers to oversee them. The level of training for tunnel workers would be related to the tunnel construction methods, using hand tools to using power tools and ITBMs. Crews working in a tunnel would have probably endured very hazardous conditions of heat, dust, noise, poor visibility and machine breakage, with constant risks of tunnel collapse. These conditions would have necessitated discipline and teamwork. This in turn suggests that the tunnel face operators were skilled (IS or paid), and that unskilled workers (either forced or paid) were most probably used to move spoil. The highest skilled would have been the operators of the ITBMs. If these machines were operated in shifts, then two teams of trained operators were required per ITBM. This suggests IS having a tunnelling training cadre that was able to teach tunnelling, learn, adapt and disseminate lessons. This is supported by the very similar tunnel designs and adaptations observed in IS-occupied areas of Iraq and Syria (Fig. 17).

This existence of an IS training cadre run by a senior leadership points to a coordinated and well-resourced tunnelling programme. It seems likely that recognition of the need to tunnel faster and for TBMs to meet that demand would have been staffed by the tunnelling cadre. IS appears to have attracted foreign geological and tunnel engineer expertise to their cause, similar to attracting skilled oil sector specialists (Bulmer 2018). The 'emir of tunnelling' has been named as Hatem Salman, a Jordanian. In Syria, rebel groups like the Islamic Front, Al-Nusra and IS dug tunnels and used explosives to

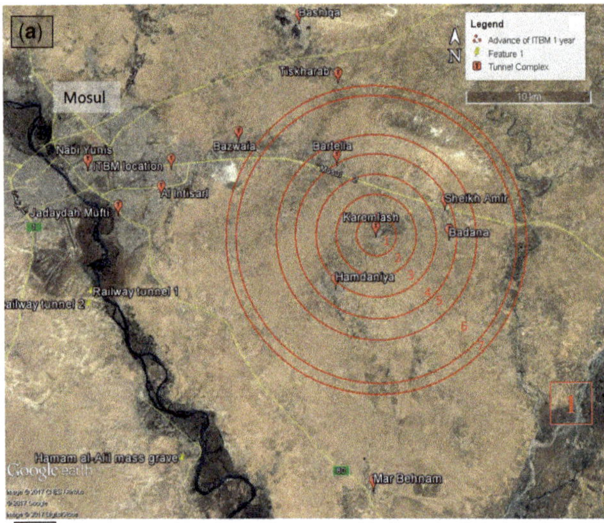

(b)

Circle	rate m/day	rate m/week	1 year m	2 year m
1	**6**	**30**	**1560**	**3120**
2	12	60	3120	6240
3	18	90	4680	9360
4	24	120	6240	12480
5	30	150	7800	15600
6	42.2	211	10972	21944
7	**45**	**225**	**11700**	**23400**

(c)

Duration	Circle	Cutting head r m	Tunnel Length m	Spoil m^3	Tonnes of spoil	16 m^3 truck	House 148.5 m^3	3 tunnels (tonnes)	16 m^3 truck	House 148.5 m^3
1 year	1	1.0	1,560	4,903	7,845	306	33	23,534	919	99
	2	1.0	3,120	9,806	15,689	613	66	47,067	1,839	198
	3	1.0	4,680	14,709	23,534	919	99	70,601	2,758	297
	4	1.0	6,240	19,611	31,378	1,226	132	94,135	3,677	396
	5	1.0	7,800	24,514	39,223	1,532	165	117,669	4,596	495
	6	1.0	10,972	34,876	55,802	2,180	235	167,405	6,539	705
	7	1.0	11,700	36,771	58,834	2,298	248	176,503	6,895	743

Fig. 15. (**a**) Radius (red circles) of seven potential tunnel lengths bored at different rates over a 1 year period by the ITTBM (Fig. 9b) starting at the village of Karemlash. Box 1 indicates the Greater Zab River. (**b**) The seven red circle numbers are related to different rates of boring advance for the ITTBM based on rates of a Robbins SBU boring machine. (**c**) Spoil generation calculated for ITTBM operations at the seven different advance rates along with truck loads and house storage requirements for the excavated spoil. Modified from Bulmer (2016*b*).

attack fixed military positions of the Syrian armed forces and allied militias. These efforts were countered using tunnelling expertise gained by Hezbollah who had experience from the Lebanon–Israel war in building a sophisticated underground network of tunnels with North Korean assistance (Lappin 2016). These had close resemblance to North Korea's own network of tunnels. It therefore seems likely that tunnelling experience, designs practices and, possibly, operators came into IS-controlled Iraq from conflicts in Syria, Gaza, the Lebanon (Roshkin 2015; Cohen *et al.* 2017) and, possibly, Turkey. They were then adapted and disseminated around Mosul.

Fighting from the tunnels

The tunnels enabled IS to avoid detection from air and space surveillance, to protect their personnel and materiel from artillery and air munitions plus chemical fallout, and to effectively attack, delay and defend against ground forces. The distribution of tunnel complexes shows they covered the main routes of advance towards Mosul and that they occurred all the way from the Lesser Zab (Fig. 3). Covered trenches were found running from the villages of Bartella and Bazwaia to the outskirts of Mosul city. Tunnels designed for denial of routes were shallow and often filled with IEDs. Networks of tunnels dug under urban areas enabled neighbourhoods and critical infrastructure to be accessed (UNEP 2017; UN Habitat 2017*a*, *b*). Deeper tunnels were used for mobility, storage, dormitories, hospitals, factories, command and control, and escape. Tunnel complexes were built for IS to live underground for long periods. The most sophisticated, with commander's private room, quarters, kitchens, water storage, armoury, hospital, mosque, computers and recreation room, were serviced by transformers, converters and diesel engines. These are very similar to Viet Cong tunnel complexes, and enabled IS fighters to come up on to the surface behind, in flanking positions, or amongst advancing forces, achieving

(b) Possible buildings for a tunnel.

Building	Area m²	Width m	Length m	Height* m	m³	Tonnes^
A	495.3	21.4	25.3	6	3248.5	5197.6
B	128.9	9.4	12.3	4.5	520.2	832.32
C	1285.8	22.9	62.5	9.1	13024.3	20838.88
D	107.5	9.5	12.4	4.5	530.1	848.16

(c) Possible start and end buildings for a tunnel.

Building	Distance m	ISTBM m³	ITTBM m³	Tonnes^	Comment for ITTBM
A to B	124	35	389.7	623.52	
A to C	313.3	88.6	984.6	1575.36	
B to C	287.8	81.4	904.5	1447.2	Go C to B
B to D	245.6	69.4	771.8	1234.88	
D to C	408	115.4	1282	2051.2	Go C to D

Fig. 16. An example of how knowledge of ISTBM (Fig. 8) and ITTBM (Fig. 9b) spoil generation, road access for haulage, and storage needs can be used to search for tunnels. (**a**) A region in east Mosul with four different building types of measured floor area. Note that all sites have good road access. (**b**) The dimensions of the four different building types, and calculation of the volume and weight of spoil that could be stored within them. (**c**) Possible start and end buildings for five tunnels to be bored by either the ISTBM or the ITTBM with corresponding volume and weight calculated for the tunnel spoil, and prediction of routes for which it would not be possible to hide spoil in nearby buildings. Modified from Bulmer (2016*a*).

great military effect. Combining the evidence for open and covered trenches, shallow and deep tunnels, and sophisticated complexes points to IS having conceived of, and built (some part of), a coordinated tunnel network at a range of depths. This greatly enhanced its ability to fight against Peshmerga and Iraqi forces backed up by a US-led coalition. Using small arms and IEDs, they were able to delay and, at times, destroy superior armoured and mechanized forces advancing on Mosul and were very effective in urban areas. Deeper tunnels bored by ITTBMs were large enough to move VBIEDs around the battlespace undetected until they appeared on the surface. This, along with hidden garages, may explain the often sudden appearance of VBIEDs and their ability to kill and maim ground forces whether armoured, mounted or dismounted infantry.

Tunnelling at depths of 10 m and deeper shows that IS understood how to protect from indirect-fire strikes and airstrikes (Rohani 1975; Kaufmann *et al.* 2011). To reach a tunnel, munitions needed to penetrate the roof and floor of buildings and then through 10 m plus of overhead (burster and cushion layer) (Fig. 18). Locating tunnels in buildings below 10 m of overburden and tunnel reinforcement enabled survivability from 150 mm rounds (standard artillery), 240 mm rounds (howitzer), 300 mm rounds (multi launch rocket systems) and, possibly, 400 mm (16-inch) rounds from howitzers. In addition, the size and design of the tunnel networks enabled fighters to move from an area under bombardment. Given this, penetrator air munitions presented the greater threat to tunnels but these needed to be guided to the target to be effective and minimize collateral damage. To deter airstrikes and thwart ground forces advancing on Mosul, IS set fire to oil wells in the nearby Qayyara and Najma oil fields (Fig. 3) that lasted from May 2016 until March 2017 (Bulmer 2018). Examination of

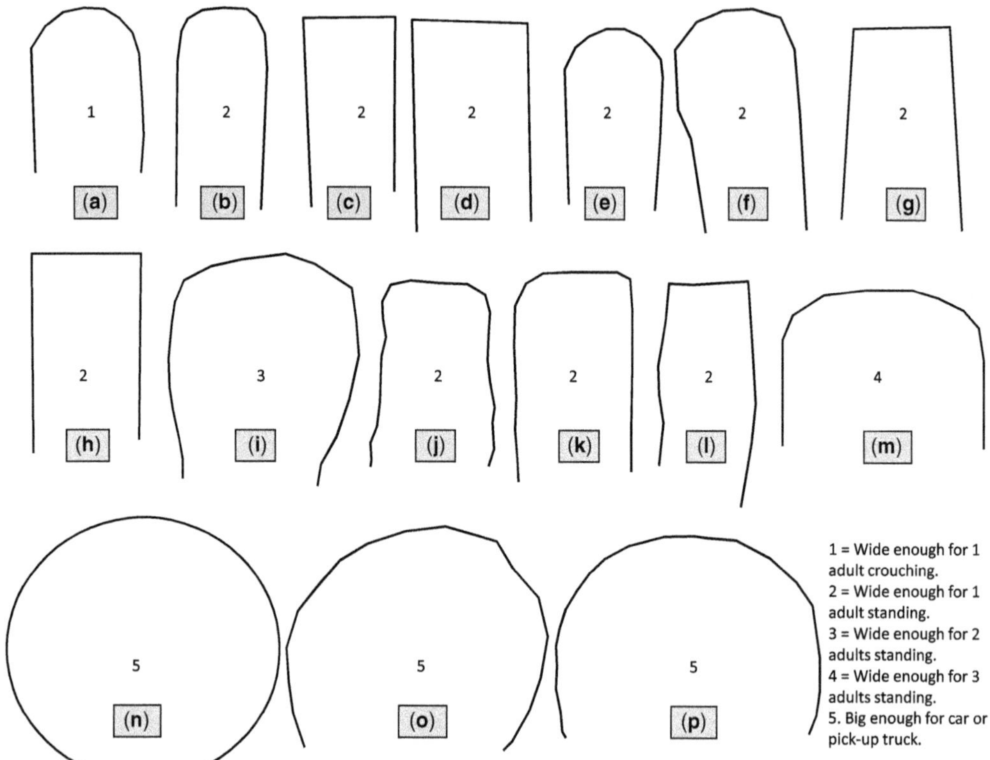

Fig. 17. Shapes of a selection of IS tunnels in Iraq and Syria, with the location and date of capture by Iraqi government or allied forces, compared with the cutting head area for an improvised tracked tunnel-boring machine (ITTBM). (**a**) Sinjar 3 February 2016, from Figure 4b. (**b**) Bartella 27 October 2016, from Figure 5b. (**c**) Bartella (with wooden supports) 27 October 2016, adjacent to Figure 5b. (**d**) Bazwaia (with wooden supports) 27 October 2016, from Figure 5d. (**e**) Karemlash November 2016, from Figure 6c. (**f**) Shrine of Nabi Yunis, Mosul, 8 March 2017, from AFP/Getty images. (**g**) Khordabad (with wooden supports) 23 October 2016, from a still taken from http://en.alalam.ir/news/1877495/video–isis%E2%80%99s-secret-tunnels-hidden-beneath-mosul. (**h**) Mufti (with wooden and steel supports) 4 June 2016, from *The Telegraph*. (**i**) Hamdaniya 24 October 2016, from *The Arab Weekly*. (**j**) Hawija 11 October 2017, from *Iraqi News*. (**k**) Kabisa 24 March 2016, from *Iraqi News*. (**l**) Badana 18 October 2017, from *The Boston Globe*. (**m**) Eastern Raqqa city, Syria (lined with prefabricated concrete) 5 September 2017, from the YPG Press Office. (**n**) ITTBM cutting head outline, 3 November 2016, from Figure 10. (**o**) Outside Judaydah al Mufti November 2016, from Figure 9a (the still frame was extracted from the video https://www.liveleak.com/view?i=1a7_1478292320 2016). (**p**) Mosul Old City 10 October 2017, from *The Baghdad Post*.

daily MODIS satellite images over the region shows that winds to the east or NE spread the plumes to Kirkuk, Erbil or Mosul. In addition, fires were started at wells between Bayji and Kirkuk in late September and October. On 20 October, fires were ignited at the Al-Mishraq sulphur plant, creating plumes of sulphur dioxide and hydrogen sulphide to impede Peshmerga and Iraqi forces advancing to the tunnel complexes in the villages of Sheikh Amir, Badana, Karemlash, Mar Behnam, Abbas and Muftiyat (Fig. 2). Fortunately, winds blew the plumes southwards towards Qayyara but respirators found in tunnels show that IS was prepared to operate in the conditions they had created.

Fighting against tunnels

After artillery bombardment and airstrikes, advancing Iraqi, Peshmerga and militia forces attempted to deal with threats from tunnel fighters by throwing grenades and burning tyres through the tunnel entrance(s). However, given the complexity of the tunnels, their extent and the extremist ideology of IS, this was not enough. For example, in the village of Mufti, Peshmerga forces from two different directions linked up in the village centre. While they were static an IS suicide bomber wearing an explosive belt emerged from a small tunnel, shot dead a brigadier and wounded his bodyguard in the leg. The bomber

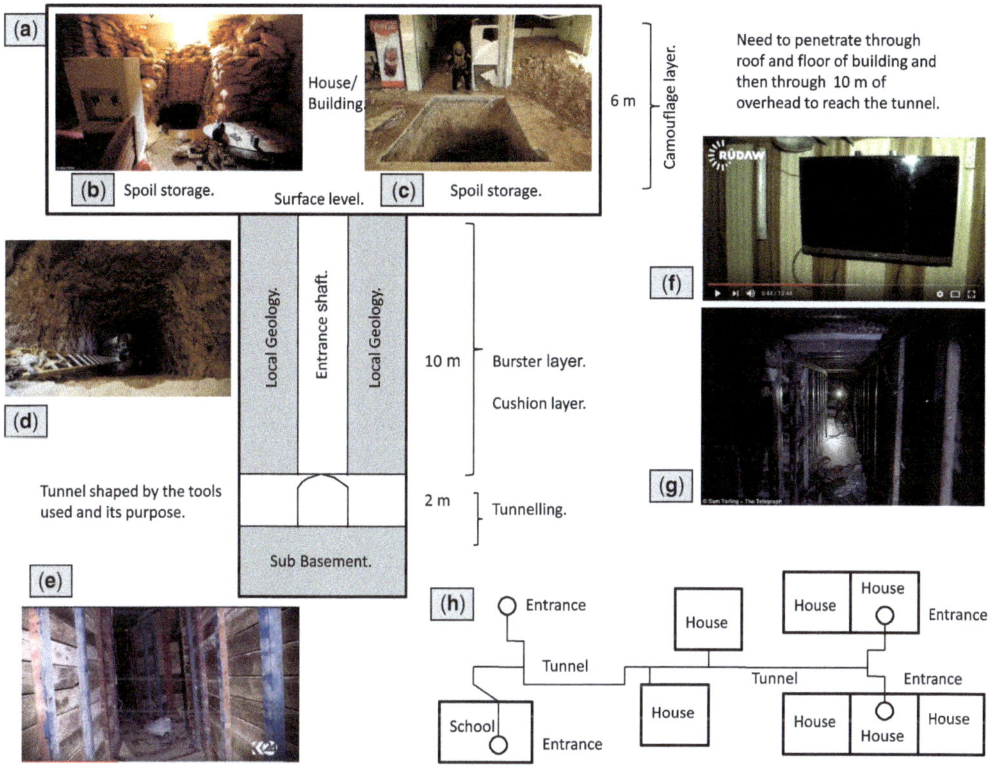

Fig. 18. (**a**) Schematic showing the profile of a house at surface level with an entrance shaft down to a tunnel level. (**b**) Sand bags used to reinforce the entrance; from Reuters. (**c**) Spoil stored in the building, from Figure 5c. (**d**) Ladder down a vertical shaft, from Figure 5a. (**e**) Wood and steel supports (https://www.youtube.com/watch?v=Q8pZPiTanhM 2016). (**f**) TV, lights and fan in a commander's room, from the video http://www.rudaw.net/english/middleeast/iraq/081120164 2016. (**g**) Dormitories along a section of tunnel (Sam Tarling/*The Telegraph*). (**h**) Plan view showing buildings with tunnel entrances and then the network of tunnels (cf. Fig. 16). 45° changes to the tunnel alignment provide blast protection.

detonated his explosives belt as he was shot by other Kurdish soldiers. Shallow tunnels designed for denial of routes filled with IEDs often collapsed under the weight of a main battle tank or armoured troop carriers. At all times there was a constant threat to advancing forces from VBIEDs that would appear suddenly travelling at speed. To counter this, advancing forces acquired earth-moving equipment to build berms any time they stopped. Peshmerga engineers often cleared and then defused IEDs in tunnels they captured but there were not enough of them to meet the scale of the need and to maintain the rate of advance. Bulldozers were used to close off tunnels but not to make them safe. Colonel Dorrian, Operations Inherent Resolve Spokesman for the US Department of Defense, reported in early November 2016 that in the first month of the advance on Mosul 70 tunnels had been 'destroyed'.

The Iraqi 9th Armoured Division advancing on the left of the Peshmerga was able to take ground with its superior surface mobility, firepower and protection but was not configured to clear and secure ground. This was exploited by IS with fighters who emerged from tunnels and successfully attacked armoured vehicles confined in streets. An example of this occurred in early December 2016, when the 9th Division advanced towards the al-Salam hospital in Mosul and commanders, meeting only light resistance, decided to use their mobility to push forward without clearing the ground. They seized the hospital but several hours later it was surrounded by hundreds of IS fighters with heavy weapons and rocket-propelled grenades. They had concealed themselves in houses and moved using tunnels. The IS attack was also coordinated with 15 VBIEDs, and IS advanced into the hospital killing or wounding nearly all the reportedly 100 Iraqi soldiers and capturing armoured vehicles. US-led airstrikes to support the trapped Iraqi force were not possible since there was no separation between opposing forces.

A ground force had to fight its way to the hospital. Iraqi commanders frequently reported that even after a street was cleared, IS fighters would appear out of a house on that street hours later having used the network of tunnels.

Post-conflict

Now that IS forces have been defeated in Mosul and surrounding areas it will take months and possibly years to determine the true extent of tunnels and use of ITBMs. The existence of tunnels is an ongoing and unknown risk to security and stability that must be addressed during reconstruction (Cordesman 2017). What to do with captured tunnels in urban settings is challenging to military commanders and civilian officials at all levels. It is further complicated by utility lines and deep foundations, as well as considerations of rebuilding and redevelopment (UN Habitat 2017*a*, *b*; World Bank 2017). There is also evidence of tunnelling under the Mosque of Jonah (recaptured from IS in January 2017) and under Mosul's museum (Campbell 2017). In these latter two cases artefacts were destroyed in the name of IS ideology but also sold to raise money. In areas of tunnel complexes, the risk of building collapse from loss of support and from the weight of spoil stored in them will remain a long-term problem. A comprehensive investigation will need to be undertaken. The first-order calculations of spoil tonnage in Table 1 and Figure 15 are for one and three tunnels. These numbers can be multiplied by the multiple tunnels in each IS complex and the resulting tonnage then added to 10 million tonnes of surface debris from damaged buildings in Mosul calculated in August 2017 (UNEP 2017); and with that for Aleppo (14.9 million tonnes) and Homs (5.3 million tonnes) calculated in January 2017 (World Bank 2017). Removal of tunnel spoil from buildings will require large amounts of trucking in addition to that needed for the disposal of building debris. However, like building debris, the tunnel spoil can be screened for reuse in Mosul and surrounding villages as engineering fill. Spoil can also be put back underground but will be a substantial and complicated undertaking (Dixon & Keto 2009). It must be anticipated that there are tunnels that have yet to be found and may be breached at some future time during a construction project. This is a real concern to civil engineering projects going forward given that IS stored IEDs, munitions and chemical agents in tunnels and tunnel complexes.

This account shows that desk-based studies can be undertaken to identify likely IS tunnel sites using knowledge of ITBMs combined with local geology. Site investigations should then look for buildings with rooms filled with spoil and other physical evidence associated with their operation such as containers for oil and lubricants, spare parts, tools, generators, and cables. These sites should be monitored for toxic and chemical agents before any descent is made into subterranean spaces. Locals were involved in tunnelling activities and they can be used to assist in finding tunnels. Those involved would have shown physical signs of being underground with their skin penetrated with dust. They may be recognizable today due to tunnelling-related injuries they sustained and in the future by respiratory diseases associated with high exposure to dust inhalation (e.g. Ulvestad *et al.* 2000), as well as psychological trauma.

Conclusions

IS ran a very large tunnelling programme across the 'caliphate' using subterranea in Syria and Iraq as a core part of their control over territories and their military capability. Caves and natural tunnels used by locals for shelter and storage quickly became known to IS after they took over any territory. Sinkholes and caves around Sinjar, Mosul and Tikrit were used by IS as execution sites and mass graves. More than 60 tunnel complexes have been identified in towns and villages formerly controlled by IS in parts of Nineveh, Erbil and Kirkuk governorates. These were constructed between 2014 and 2017, and demonstrate that IS possessed geological, tunnelling and engineering knowledge. The scale of the programme would have necessitated connection to the wider economic activity in the 'caliphate' and planning by IS 'Ministries'. Running these tunnel teams at the same time required knowledgeable, skilled and competent project managers to complete them. This suggests that IS had a tunnel training cadre, and moving from hand tools to ITBMs can be seen as a logical progression. From making weapons and uparmouring vehicles, IS have demonstrated engineering and mechanical skill and facilities necessary to fabricate an ITBM. The ISTBM and ITTBM found around Mosul are effective designs able to tunnel in the surrounding geology. While improvised, they show design maturity similar to the progression observed in IS VBIEDs. That more than one ISTBM and ITTBM were captured should not be unexpected given the scale of the tunnelling need. Using its substantial revenues, IS could purchase parts and expertise from around the world (CAR 2016) either on the open or black market.

Tunnels enabled IS to avoid detection from air and space surveillance, to protect from artillery and air munitions plus chemical fallout, and to effectively attack, delay and defend against ground forces. The locations of tunnel complexes covered the main routes of advance on Mosul all the way from the

Lesser Zab. Tunnel complexes enabled IS to live underground for long periods and for IS fighters to come to the surface behind, in flanking positions, or amongst advancing forces, achieving great military effect. Using small arms and IEDs, they delayed and, in places, destroyed superior armoured and mechanized forces advancing on Mosul. Open and covered trenches, shallow and deep tunnels, and sophisticated complexes combined to form a coordinated tunnel network at a range of depths. Deeper tunnels bored by ITTBMs were large enough for VBIEDs to move around the battlespace undetected until they appeared on the surface. Neither Peshmerga nor Iraqi forces were equipped or trained to find, secure, deny or destroy tunnels, other than in a token or rudimentary manner. This despite having watched IS digging for 2 years. This critical deficiency also reflects a lack of recognition of the scale of need to fight underground by those foreign militaries assisting Iraqi and Peshmerga forces. This was occurring at a time when these militaries were identifying as a potentially significant future problem the challenges in detecting tunnelling and fighting in subterranean environments (Cohen *et al.* 2017). The existence of tunnels large enough for cars presented a whole new and unexpected level of military threat, need and response. It must be recognized that some IS tunnel complexes were probably never detected during the fighting and IS were able to escape. The existence of tunnels is an ongoing and unknown risk to security and stability that must be addressed during reconstruction. In the same way that tunnelling expertise came into IS-controlled Iraq from conflicts in Syria, Gaza and the Lebanon (Cohen *et al.* 2017), it has probably now moved elsewhere. It should be anticipated that IS have other TBMs, and that these designs and accompanying expertise will appear in current and future battlespaces. In addition, TBMs are now increasingly available commercially and there is a growing market in used machines.

Acknowledgements The author thanks the reviewers for their comments. I would like to thank Paul Nathanail, Drew Craig, Daniel Scott and Christopher Schuman for their discussions regarding tunnelling and surveying.

Funding This work was supported by Roedown Research R2 and 170 (Infrastructure Support) Engineer Group, Royal Engineers.

References

Al-Ansari, N., Issa, I., Sissakian, V. & Knutsson, S. 2015. Mystery of Mosul Dam the most dangerous dam in the world: karstification and sinkholes. *Journal of Earth Sciences and Geotechnical Engineering*, **5**, 33–45.

Al-Dewachi, B.M. 2005. Origin of caves in Quaternary fluvial deposits at the left side of Mosul City-Iraq. *Iraqi Journal of Earth Science*, **5**, 58–67.

Aljubouri, Z.A. & Aldabbagh, S.M. 1980. Sinjarite, a new mineral from Iraq. *Mineralogical Magazine*, **43**, 643–645.

Al-Juboury, A.I.A. 2009. The Upper Miocene Injana (Upper Fars) Formation in Iraq: insight on provenance history. *Arab Journal of Geoscience*, **2**, 337–364, https://doi.org/10.1007/s12517-009-0045-1

Anon 2015. Compact tunnel boring machines are major players in North America. *Trenchless Technology Magazine*, 2 December 2015, https://trenchlesstechnology.com/compact-tunnel-boring-machines-major-players-north-america/ [last accessed 30 January 2018].

Anon 2017. *Project Categories: Small Boring Unit.* Robbins Company, Solon, OH, USA, http://www.therobbinscompany.com/project-category/small-boring-unit/ [last accessed 30 January 2018].

Bulmer, M.H. 2016*a*. *Geological Considerations of Tunnelling around Mosul on the Tigris River.* Unpublished Report 170 (Infrastructure Support) Engineer Group, Royal Engineers, Chillwell, Nottingham, UK.

Bulmer, M.H. 2016*b*. *Understanding the Improvised Tracked Boring Machine (ITBM) Captured Near Mosul on the Tigris River.* Unpublished Report 170 (Infrastructure Support) Engineer Group, Royal Engineers, Chillwell, Nottingham, UK.

Bulmer, M.H. 2018. Military use of environmental degradation by Islamic State, southern Nineveh, Iraq. *Scientia Militaria: South African Journal of Military Studies*, **46**, in press.

Campbell, S. 2017. Tunnels to Mosul's secret palace revealed: Hidden passages to 2600-year-old shrine are found buried under the ruins of an 'idolatrous' mosque blown up by ISIS as images also emerge of wrecked museum. *MailOnline*, 11 May 2017, http://www.dailymail.co.uk/news/article-4304450/Inside-museum-destroyed-looted-ISIS-jihadis.html#ixzz4ujUrehIK [last accessed 30 January 2018].

CAR 2016. Tracing the supply of components used in Islamic State IEDs. Evidence from a 20-month investigation in Iraq and Syria. *Conflict Armament Research*, February 2016, http://www.conflictarm.com/reports/tracing-the-supply-of-components-used-in-islamic-state-ieds

Chulov, M. & Summers, H. 2016. Mosul offensive: forces launch mass attack on Iraqi city in bid to oust Isis. *The Guardian*, 17 October 2016, https://www.theguardian.com/world/2016/oct/17/iraqi-forces-begin-assault-on-isis-stronghold-mosul [last accessed 30 January 2018].

Cohen, R.S., Johnson, D.E., Thaler, D.E., Allen, B., Bartels, E.M., Cahill, J. & Efron, S. 2017. *From Cast Lead to Protective Edge.* RAND Corporation, Santa Monica, CA, USA.

Cordesman, A.H. 2017. *After ISIS: Creating Strategic Stability in Iraq.* Center for Strategic and International Studies, Washington, DC, USA.

Dixon, D.A. & Keto, P. 2009. *Backfilling Techniques and Materials in Underground Excavations: Potential Alternative Backfill Materials in Use in Posiva's Spent Fuel Repository Concept.* Working Report 2008-56. Posiva, Eurajoki, Finland.

Fouad, S.F.A. 2012. *Tectonic Map of Iraq. Scale 1:1 000 000*. 3rd edn. GEOSURV, Baghdad, https://iasj.net/iasj?func=fulltext&aId=99665

Fouad, S.F.A. & Sissakian, V.K. 2011. Tectonic and structural evolution of the Mesopotamia Plain. *Iraqi Bulletin of Geology and Mining*, Special Issue No. 4, 33–46.

Frantzman, S.J. 2016. Tunnel vision: From Gaza to Mosul. *The Jerusalem Post*, 6 November 2016, http://www.jpost.com/Middle-East/Tunnel-vision-From-Gaza-to-Mosul-471784 [last accessed 30 January 2018].

Frumkin, A. & Shimron, A. 2006. Tunnel engineering in the Iron Age: geoarcheology of the Siloam Tunnel, Jerusalem. *Journal of Archeological Science*, **33**, 227–237.

Garbrecht, G. 1983. Ancient waterworks – lessons from history. *Impact of Science on Society, UNESCO*, **1983**, 5–16.

Grau, L. 1996. *The Bear Went Over the Mountain. Soviet Combat Tactics in Afghanistan*. National Defense University Press, Washington, DC, USA.

Hatch Mott MacDonald 2001. *Tunnel Boring Machines: Tunnel Assumptions and Cost Estimating Output*. Appendix D. Tunnel Estimating Database. Version 0.02.2. Hatch Mott McDonald, Iselin, NJ, USA.

Jassim, S.Z., Jabril, A.S. & Numan, N.M.S. 1997. Gypsum karstification in the Middle Miocene Fatha Formation, Mosul area, northern Iraq. *Geomorphology*, **18**, 137–149.

Jassim, S.Z., Raiswell, R. & Bottrell, S.H. 1999. Genesis of the Middle Miocene stratabound sulphur deposits of northern Iraq. *Journal of the Geological Society, London*, **156**, 25–39, https://doi.org/10.1144/gsjgs.156.1.0025

Kaufmann, J.E., Kaufmann, H.W., Jankovic-Potocnik, A. & Lang, P. 2011. *The Maginot Line. History and Guide*. Pen & Sword Books, Barnsley, UK.

King, M.D. 2016. The weaponization of water in Syria and Iraq. *The Washington Quarterly*, **38**, 153–169.

Laessoe, J. 1951. The irrigation system at Ulhu, 8th century BC. *Journal of Cuneiform Studies*, **5**, 21–32.

Lappin, Y. 2016. Security and defence: The North Korean connection. *The Jerusalem Post*, 12 March 2016, http://www.jpost.com/Arab-Israeli-Conflict/SECURITY-AND-DEFENSE-The-North-Korean-connection-447557 [last accessed 30 January 2018].

Mather, J.D. & Rose, E.P.F. 2012. Military aspects of hydrogeology: an introduction and overview. *In*: Rose, E.P.F. & Mather, J.D. (eds) *Military Aspects of Hydrogeology*. Geological Society, London, Special Publications, **362**, 1–18, https://doi.org/10.1144/SP362.1

Morris, L. & Salim, M. 2017. ISIS dumped bodies in a desert sinkhole. It may be years before we know the full scale of the killings. *The Washington Post*, 2 March 2017, https://www.washingtonpost.com/world/middle_east/isis-dumped-bodies-in-a-desert-sinkhole-it-may-be-years-before-we-know-the-full-scale-of-the-killing/2017/03/02/0a4c0fd4-fd2b-11e6-9b78-824ccab94435_story.html?utm_term=.eaeb8c4191ff [last accessed 30 January 2018].

Neuhof, F. 2017. Horror of Mosul where sinkhole became mass grave for 4000 of Isil's victims. *The Telegraph*, 27 February 2017, http://www.telegraph.co.uk/news/2017/02/25/horror-mosul-sinkhole-became-biggest-mass-grave-iraq/ [last accessed 30 January 2018].

REACH 2017. *Iraq – Kurdistan Region of Iraq*. Reference Map As of February 2017. REACH Resource Centre, Geneva, Switzerland, http://www.reachresourcecentre.info/system/files/resource-documents/reach_irq_map_iraq_kri_reference_19feb2017_0.pdf [last accessed 30 January 2018].

Reuters 2017. Heaven is jihad in the path of God': Amid harsh fight for Mosul, Iraqi forces uncover hidden ISIS training camp. *Business Insider UK*, 6 March 2017, http://www.businessinsider.com/r-mosul-train-tunnel-reveals-assault-course-for-elite-islamic-state-fighters-2017-3 [last accessed 30 January 2018].

Robinson, J. 2016. The Mad Max-style tunnelling machines ISIS made out of old farm equipment to bore a maze of rat-runs under Mosul. *MailOnline*, 4 November 2016, http://www.dailymail.co.uk/news/article-3904774/The-Mad-Max-style-tunnelling-machines-ISIS-old-farm-equipment-bore-maze-rat-runs-Mosul.html#ixzz4ujbyGyy7 [last accessed 30 January 2018].

Rogoway, T. 2015. ISIS used this huge crudely modified truck to attack a key Syrian air base. *Foxtrot Alpha*, 24 August 2015, https://foxtrotalpha.jalopnik.com/isis-used-this-huge-crudely-modified-truck-to-attack-a-1726246835 [last accessed 30 January 2018].

Rogoway, T. 2017. This ISIS armored loader just broke through Iraqi army lines and blew up. *The Drive*, 15 March 2017, http://www.thedrive.com/the-war-zone/8358/this-isis-armored-bulldozer-just-broke-through-iraqi-army-lines-and-blew-up [last accessed 30 January 2018].

Rohani, B. 1975. *Analysis of Projectile Penetration into Concrete and Rock Targets*. Soils and Pavement Laboratory, US Army Engineer Waterways Experiment Station. Miscellaneous Paper S-75-25. Prepared for Defence Nuclear Agency, Washington, DC.

Roshkin, J. 2015. Geoscience influences on the development of clandestine smuggling tunnels between Sinai (Egypt) and Rafah (Gaza Strip) and terror-guerilla tunnels between Gaza Strip and Israel. Paper presented at the 11th International Conference of Military Geosciences, 15–19 June 2015, Annapolis, MD, USA.

Safar, F. 1947. Sennacherib's project for supplying Erbil with water. *Shumer*, **3**, 23–25.

Schnabel 2016. *Mosul Dam Stability Assessment Nineveh Province, Iraq*. Unpublished report 11001001.25, dated 19 February 2016. Schnabel Engineering.

Schneider, E., Spiegl, M., Turtscher, M. & Leitner, W. 2010. Penetration, cutter wear and rate of advance for TBM operations in hard rock. *Tunnel*, **7**, 43–52.

Sissakian, V.K. & Al-Khalidi, Z.B. 2012. The lithological map of Iraq. *Iraqi Bulletin of Geology and Mining*, **8**, 1–13.

Sissakian, V.K. & Fouad, S.F.A. 2015. Geological Map of Iraq, Scale 1: 1 000 000. 4th Edition, 2012. *Iraqi Bulletin of Geology and Mining*, **11**, 9–16, https://www.iasj.net/iasj?func=fulltext&aId=99666

Sissakian, V.K., Al-Ansari, N. & Knutsson, S. 2014. Origin of some transversal linear features of NE–SW trend in Iraq, and their geological characters. *Natural Science*, **6**, 996–1011, https://doi.org/10.4236/ns.2014.612091

Stolarczyk, L., Troublefield, R. & Battis, J. 2005. Detection of underground tunnels with a synchronized electromagnetic wave gradiometer. *In*: Carapezza, E.M. (ed.) *Sensors and C3I Technologies for Homeland Security and Homeland Defence IV. Proceedings of SPIE*, **5778**, 994–1001.

Triebert, C. 2017. Mapping Mosul's SVIED attacks. *Bellingcat*, 9 January 2017, https://www.bellingcat.com/news/mena/2017/01/09/mapping-mosuls-vbied-attacks/ [last accessed 30 January 2018].

Ulvestad, B., Bakke, B., Melbostad, E., Fuglerud, P., Kongerud, J. & Lund, M.B. 2000. Increased risk of obstructive pulmonary disease in tunnel workers. *Thorax*, **55**, 277–282.

UNEP 2017. *Environmental Issues in Areas Retaken from ISIL, Mosul, Iraq. Rapid Scoping Mission July–August 2017.* Technical Note. United Nations Environment Programme, https://postconflict.unep.ch/publications/Iraq/Iraq%20Technical%20Note_September2017.pdf [last accessed 9 January 2018].

UN Habitat 2017*a*. *Damaged Sensitive Sites and their Potential Environmental Hazards.* 15 May 2017, http://unhabitatiraq.net/mosulportal/wp-content/uploads/2017/06/170515_Environmental-Hazards.pdf [last accessed 9 January 2018].

UN Habitat 2017*b*. *Roads Infrastructure Assessment.* 16 June 2017, https://reliefweb.int/map/iraq/iraq-mosul-road-infrastructure-assessment-16-june-2017 [last accessed 9 January 2018].

Wakeley, L.D., Kelley, J.R., Talbot, C.A., Pearson, M.L. & Broadfoot, S.W. 2007. *Geologic Conceptual Model of Mosul Dam.* ERDC TR-07-6. US Army Engineer Research and Development Center, Vicksburg, MS, USA.

World Bank 2017. *The Toll of War. The Economic and Social Consequences of the Conflict in Syria.* The World Bank Group, Washington DC.

Wyke, T. 2016. ISIS's rat-holes: Jihadis create network of secret escape routes to evade coalition airstrikes in Iraq. Ishtar TV, 4 March 2016, http://www.ishtartv.com/en/viewarticle,37861.html [last accessed 30 January 2018].

Geological influence of the Great Red River Raft on the Red River Campaign of the American Civil War

DANNY W. HARRELSON*, NALINI TORRES, AMBER TILLOTSON & MANSOUR ZAKIKHANI

US Army Engineer Research and Development Center, 3309 Halls Ferry Road, Vicksburg, MS 39180, USA

M.Z., 0000-0001-8532-153X

**Correspondence: danny.w.harrelson@usace.army.mil*

Abstract: About 2000 years ago, the Mississippi River avulsed westwards and captured the Red River of the South, uprooting vegetation and causing bank instability, resulting in a series of log jams known as the Great Red River Raft. The Great Raft presented a major navigation challenge during the settlement of the Red River Valley from the seventeenth century until its removal in the 1890s. The Great Raft extended *c.* 100–300 km upriver from Natchitoches, Louisiana to the Louisiana–Arkansas boundary. In the nineteenth century, it contributed to a major military failure by the Union Army and Navy during the 1864 Red River Campaign of the American Civil War. While planning the Red River Campaign, Union commanders made tactical errors by overlooking unique geological aspects of the Red River, including its flashy nature, high sediment load, anastomosing channels and rapids. The Confederates used their knowledge of the Great Raft and the Red River's morphology to attack the Union's river fleet while in a position of restricted manoeuverability. The Confederates used the topographically higher defensive positions of natural levee deposits along the Red River to keep the Union's land and naval forces from supporting each other during the Campaign.

The American Civil War was an internal conflict fought within the USA between 1861 and 1865. The 17 Union states of the North faced 11 secessionist states in the South, collectively known as the Confederate States of America. The primary issue dividing the two blocks was States' Rights: did the individual states have the right to make their own decisions on certain issues, including slavery, or did the Federal Constitution reign supreme? The first battle occurred on 12 April 1861, when Confederate artillery fired on the Union garrison at Fort Sumter, South Carolina, and the last battle took place on 9 April 1865 at Appomattox Court House, Virginia, immediately after which Confederate General Robert E. Lee surrendered to Union General Ulysses S. Grant. This four-year conflict, which resulted in an estimated 500 000 casualties, confirmed the supremacy of the Union and the Federal Constitution and ended slavery, the basis of the Southern agrarian economy, as an institution within the USA.

Military operations took place in three theatres: the Eastern Theatre, which consisted of the states along the Atlantic seaboard; the Western Theatre, which extended from the Appalachian Mountains to the Mississippi River; and the Trans-Mississippi Theatre, which consisted of the states west of the Mississippi River. The Red River Campaign was fought in the Trans-Mississippi Theatre late in the war, between 10 March and 22 May 1864. The Union Armies were commanded by Major General Nathaniel P. Banks and the Confederate Armies by Lieutenant General Richard Taylor.

The Great Raft

The first sighting of the Red River and the Great Raft by Europeans most likely took place when Spanish explorer Hernando De Soto's second expedition wandered through the valley in 1541 (Torres & Harrelson 2012). The expedition was poorly documented and ultimately ended in disaster with De Soto's death from a fever (probably malaria) at the mouth of the Red River in 1542.

The Red River originates in the Texas Panhandle region and flows SE along the Texas–Arkansas border, then through Louisiana to unite with the Atchafalaya and Mississippi rivers. It is characterized by a series of unique sedimentary and geomorphological environments, which resulted in major socio-economic impacts in this region from the seventeenth into the twentieth century. It also has one of the largest sediment concentrations of all major rivers in the world and currently requires a strict maintenance dredging programme to keep the 2.7 m deep channel open for navigation (Torres & Harrelson 2012).

The Great Raft was not a stationary log jam within the river; as pieces of the raft deteriorated or

From: Rose, E. P. F., Ehlen, J. & Lawrence, U. L. (eds) 2019. *Military Aspects of Geology: Fortification, Excavation and Terrain Evaluation*. Geological Society, London, Special Publications, **473**, 267–273.
First published online March 6, 2018, https://doi.org/10.1144/SP473.5

broke up and floated downstream (Fig. 1), new logs and debris were added at its upper end (Joiner 2006). This ecosystem of entangled logs, vegetation and sediments remained in place for almost two millennia, altering the flow regime of the Red River and causing a complete change in its geomorphic character from a single channel to a series of anastomosing channels. It is believed that the initial formation of the Great Raft was triggered by catastrophic flooding as the Red River was going through some major geomorphic threshold, such as a major avulsion. The main contributors to the development of the Great Raft are believed to be the shifting geomorphic conditions in conjunction with extensive precipitation, river bank rotational slips and slab failure, rapid lateral migration, copious, rapidly growing riparian vegetation, exceeding a geomorphic threshold, a flashy hydrograph and a very heavy sediment load. The raft modified the flow and behaviour of the river and produced some of the most challenging obstacles to river navigation (Foster *et al.* 1987).

Since 1542, the lower Red River has been the focus of interest because of its perceived value as a navigational route, as it was then the only practical route to northern Texas and to an enormous area of Indian territories, then a part of Mexico. The river was the logical route for westwards expansion and commerce, but unlike the Ohio and the Mississippi rivers, the Red River's unique geological and hydrological properties would prove to be major challenges to the numerous attempts to remove the raft and make the river suitable for navigation (Wright 1930).

Captain Henry Miller Shreve, appointed Superintendent of Western River Improvement in 1829, was hired by the US Army Corps of Engineers to remove the Great Raft to improve navigation on the Red River. Captain Shreve was a steamboat entrepreneur who had successfully invested in the new steam-power technology by developing the snag boat, a steam-powered boat used for raft removal. He had already used this technology to clear navigational paths in the Ohio and Mississippi rivers in 1827. Captain Shreve arrived at the toe of the Great Raft in April 1833 with four snag boats and a force of 159 men. His group began clearing a navigational path through 115 km of the Great Raft and, finally, by the spring of 1838, a path had been cleared; however, the remnants of the Great Raft along the river banks were not cleared and the Great Raft immediately began to reform, producing

Fig. 1. View of the Great Red River Raft, from Red River Raft (1985).

a 0.7 km long section by August 1838. The Great Raft continued to reform and, by 1841, it had grown to 32 km in length and, by 1855, to 48 km in length. Discussions about diverting the Red River through lateral channels instead of removing the Great Raft itself were considered again (Report of Red River Survey 1855), but the construction of these lateral channels was interrupted in 1857 due to the approaching American Civil War.

Red River geology

The Red River is *c.* 2188 km long and drains an area of *c.* 145 000 km^2 (Fig. 2). It flows through Paleozoic, Mesozoic and Cenozoic age sedimentary rocks. Large amounts of red sediment eroded from the Permian rocks give the river a reddish colour, particularly prominent during periods of high flow, which led to its name (Autin & Pearson 1993). Native Americans and Europeans referred to the river by different names, including Napleste (Native American), Rio Rojo (Spanish) and Riviere Rouge (French), depending on which region they visited. There are three geological constrictions caused by strong Oligocene sandstone in the lower part of the river near Texarkana (Texas), Grand Ecore (near Natchitoches, Louisiana) and at Colfax (Louisiana). Pleistocene fluvial deposits form terraces and upland deposits flanking the river (Autin & Pearson 1993).

Red River Campaign

At the time of the Red River Campaign in April 1864, the outcome of the Civil War appeared to be largely decided. The agricultural South had fought tenaciously against the industrial North, but the zeal and military prowess of the Confederates was not enough to prevail against the vast resources of the North (Davis 1890). The Red River Campaign, conducted from 10 March to 22 May 1864, was the Union's largest combined army and navy campaign west of the Mississippi River (Fig. 3). The Campaign was conducted mainly along the Red River in Louisiana between 30 000 Union troops under the command of Major General Nathaniel Banks and between 6000 and 15 000 Confederate troops commanded by Lieutenant General Richard Taylor. The land forces of Nathaniel Banks were to be supported by Rear Admiral David Dixon Porter's fleet of *c.* 90 boats.

Planning for the Red River Campaign began after the fall of Vicksburg, Mississippi, the 'Gibraltar of the Confederacy', on 4 July 1863 and the defeat of Confederate General Robert E. Lee's second

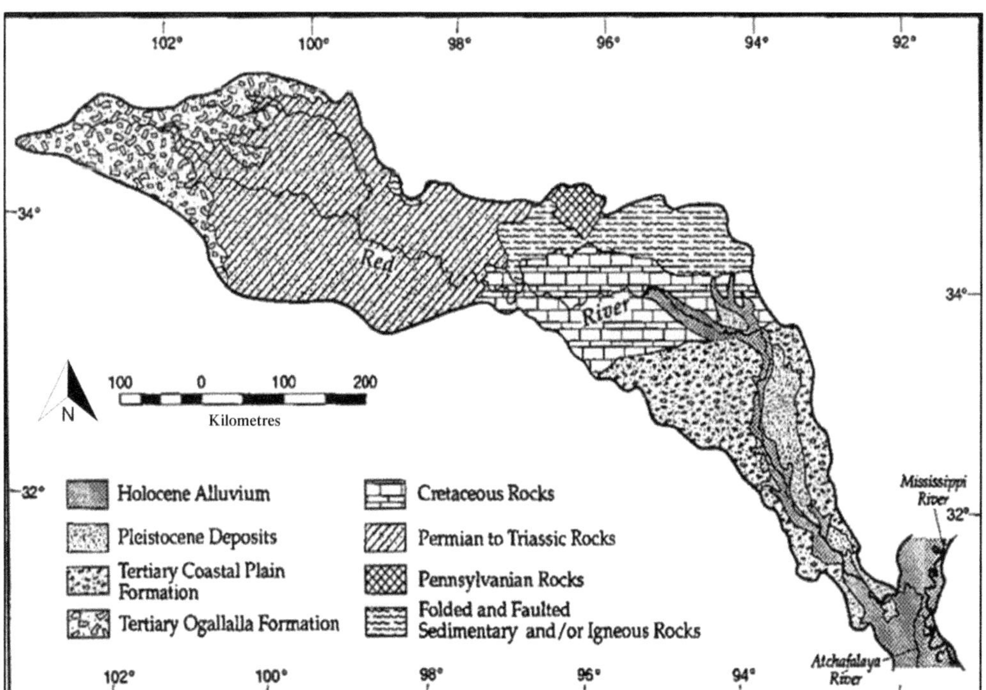

Fig. 2. Geology of the Red River Valley, modified from Autin & Pearson (1993).

Fig. 3. Red River Campaign Map, modified from Red River Campaign Map (1997).

invasion of northern territory at Gettysburg, Pennsylvania, on 3 July 1863 (Foote 1974). These two major defeats resulted in the first division of the Confederacy. The Union's plan for the Red River Campaign was finalized in January 1864, with the stated objectives of dividing the Confederacy a second time by moving up the Red River and capturing the Confederate Army headquarters for the Trans-Mississippi located at Shreveport, Louisiana and also seizing major stockpiles of cotton located in the area. Shreveport was the nexus of a small military–industrial complex in eastern Texas and southern Arkansas, which included armouries, foundries and a naval shipyard. It was also the gateway for a potential invasion of Texas, a tactic that President Abraham Lincoln was considering should the war drag on into late 1865.

Early in 1864, several Union generals were vying for the command of the Red River Campaign, largely as a ploy to seek political office. One Union general, Nathaniel P. Banks, was commander of the Department of the Gulf based in New Orleans. Banks was a contender for the Republican nomination for the presidency in 1864. He was a native of Massachusetts, had served as Speaker of the US House of Representatives and was an ally and confidant of several wealthy New England textile mill owners. A strictly political general with no formal military training, Banks wanted a battlefield victory to cement support for his run for the presidency. He outranked every Union general west of the Appalachians. Ulysses S. Grant (Lieutenant General and Commanding General of the Union Army) knew him well, having faced problems with Banks' lack of co-operation during the Vicksburg Campaign. President Lincoln and Henry Halleck, the Union Army's Chief of Staff, both wanted a campaign up the Red River and Banks decided to use it as his best opportunity for military success. Union General William T. Sherman wanted to lead the campaign, but Grant could not spare him from the upcoming Atlanta Campaign. The naval commander, Rear Admiral David Dixon Porter, had committed to the Campaign under the assumption that Sherman would lead it. Sherman loaned Porter 10 000 of his best veterans to protect the fleet from the Confederates and frankly told Porter that the men were there in case '... Banks abandoned the fleet ...', which Banks had implied he would do if his land force got into trouble (Brooksher 1998, p. 128).

By early March 1864, Porter had assembled a formidable battle fleet consisting of City Class iron-clads with 2.5 inch (6.35 cm) of casemate armour plating, lightly armoured gunboats (tin-clads), a timber-clad (wood as armour plating), a high speed ram, three river monitors (with revolving turrets) and several support vessels, including steamboats (Brooksher 1998). Porter's fleet consisted of 90 boats carrying a total of 210 heavy guns. This river fleet was to be supported by Major General Frederick Steele, who was to aid in the attack on Shreveport from the north while Banks would bring 20 000 troops from the south. Banks' contingent would enter the Red River Valley, eliminate any Confederate forces or fortifications there and then move up-river to Alexandria, Louisiana (Fig. 3). From there, Banks' land forces would move NW towards Shreveport using local river roads located on natural levee deposits, which more or less followed the course of the Red River. Once the naval and land forces were underway, Steele and his Union force of 10 000 men would leave Little Rock, Arkansas, and approach Shreveport from the north and east.

Banks' plan was complex and relied on coordination between three separate forces, two land armies and a naval force, which were spread over 650 km of unfamiliar and hostile terrain. Success would hinge on each force mutually supporting the other and completing their objectives following a stringent timeline. To make matters worse, neither President Lincoln nor his chief of staff appointed an overall commander for the Campaign; Banks, Steele and Porter all held separate commands and were, to a large degree, not answerable to each other. As for the naval component of the campaign, Porter's choice of vessels was based on a concentration of overwhelming firepower, which worked for the Vicksburg Campaign, but did not take into account the flashy nature of the Red River or the series of anastomosing channels that had developed in the decades following the first attempt to remove the Great Raft. In addition, Confederate forces used the series of interconnecting anastomosing channels as 'strategic shooting lanes', luring the Union gunboats into the narrow and shallow channels where they were then attacked. Porter's heaviest ships, the iron-clads, drew too deep a draft (2.1 m) to easily manoeuvre in the sandbar- and snag-filled Red River. Worse, Banks' political backers desperately needed cotton to run their textile mills, so boatloads of cotton speculators accompanied him on this military operation. Their presence antagonized Sherman's troops, who thought they were now part of a 'cotton raid' and not a proper military expedition.

As with numerous commanders from both the Union and Confederate sides, Porter and Banks did not like each other. Banks did not trust Porter because the Admiral was a close friend of both Grant and Sherman, and he was jealous of the powerful trio. Porter did not trust Banks because he thought the political general was a disaster waiting to happen, being both incompetent and unreliable; however, Banks' greatest problem was that he vastly underestimated the resolve of the Confederates and believed they were a defeated foe that '... would not fight him before or after his arrival in Shreveport' (Joiner 2003, p. 163).

The Campaign had an unfavourable beginning on 12 March 1864 when the USS *Eastport*, Porter's largest iron-clad, became stuck on a sandbar at the mouth of the river, thereby delaying the advance of the fleet. Once the USS *Eastport* was freed, the fleet moved up-river *c.* 4 km to the village of Simmesport, Louisiana, which Porter and Union General A.J. Smith used as a supply base. By 31 March 1864, Banks had reached Natchitoches, Louisiana, only 100 km south of Shreveport (Fig. 3). The Arkansas phase of the Red River Campaign, led by Major General Fredrick Steele, began around 23 March 1864. Called the Camden Expedition, the effort called for Steele's forces to march south to Shreveport where they would link up with Banks' amphibious expedition. Steele would garrison Shreveport while Banks continued up the Red River into northeastern Texas; however, Steele's route to Shreveport was through an area full of hostile Confederates, who immediately began to attack his forces. In addition, provisions in the area for Steele's troops were scarce and soon his forces were on half-rations.

Confederate General E. Kirby Smith focused his attention on fortifying the bluffs overlooking the Red River; he constructed Fort DeRussy, forts at Grand Ecore and forts at and surrounding Shreveport. Below Shreveport, Kirby Smith's soldiers blew up a dam, which forced the water from the river to move to a nearby channel, greatly lowering the water level. A steamboat was also used to block part of the Red River just below the dam. Because of Kirby Smith's tactics, Steele's Union troops were forced to retreat back to Little Rock, Arkansas, and abandon the attempts to link up with Banks at Shreveport. Steele's force of 10 000 suffered more than 2500 causalities and were so badly mauled by Kirby Smith's Confederates during their retreat back to Little Rock that the Camden Expedition is considered to be the largest Union defeat in the Western Theatre.

On 6 April 1864, Banks ordered his force west using a single road that followed the natural levee deposits along the banks of the Red River. This force, commanded by Brigadier General Albert L. Lee, stretched >30 km and was led by his cavalry division of *c.* 4000 troops. Half of Lee's men were classified as mounted infantry and did not know how to care for their horses, much less how to ride them with confidence. Lee called them his 'amateur equestrians'. Following these troops were the division's 300 supply wagons, artillery units, an infantry division, 700 additional wagons and the bulk of the Thirteenth and Nineteenth Corps. At the rear of this long procession were A.J. Smith's veterans (loaned from Sherman's army), relegated to eat the dust of the entire 30 km long column. The forces first marched west to Los Adaes and then turned north on the Shreveport–Natchitoches stagecoach road. As the lead elements of the Union forces were encamping at Pleasant Hill, Louisiana, the tail of the column had not yet left Natchitoches, Louisiana (Fig. 3).

Lee advanced from about 9.30 a.m. on the morning of 8 April 1864, crossing a ridge and descending into a small stream bottom, where he then traversed bands of trees and fields until he saw a long, sloping ridge occupied by Confederate cavalry near Mansfield (Fig. 3). As Lee prepared his attack, the Confederates fled. He then cautiously approached the top of the ridge and, once there, saw ahead of him the Confederate Army arrayed *c.* 1 km on both sides of the road and extending at a right angle to his right. Surveying his position, he discovered the Confederate cavalry and immediately requested help from Banks, then several kilometres behind his position. Typically, Banks was slow to give aid, so Lee dismounted his division and formed them into an 'L' to match the Confederate position. With Lee's wagons still on the road, it was difficult for reinforcements to reach him and, when help finally arrived, it was only one corps, the small Thirteenth Corps, so Lee was forced to deploy his two artillery units, Nims' Battery and the Chicago Mercantile Battery as well. About this time, Banks came forward to see what the fuss was about, still believing the Confederates would only offer token resistance.

Confederate Lieutenant General Taylor had chosen his place of battle well, mustering *c.* 10 000 men on the field. Hoping vainly that Banks would react in a rash manner and attack, Taylor waited almost six hours, until about 4 p.m., when finally his left flank began an en echelon charge. Whether this attack was by Taylor's direct order or a misinterpreted command, we will never know. The Louisiana units on his left attacked Lee's position and, in short order, overran it. The Louisiana officers had decided to ride on horseback so their men could see them. This proved to be a brave, but fatal, gesture, as 11 of the 14 officers were lost within the first 20 minutes of the attack, including Brigadier General Alfred Mouton, their beloved leader. As the Union right flank crumbled, Taylor's Texas troops advanced in line and, in short order, destroyed the Union left flank. Taylor's men next ripped through a secondary Union line at Sabine Crossroads *c.* 1 km behind their front line. Darkness ended the fighting as Taylor's men pushed the Union soldiers behind Chapman's Bayou. The Union line stiffened on the crest of a natural levee deposit above a river channel produced by the Great Raft, thereby ending the Battle of Mansfield with a resounding Confederate victory.

Banks' retreat

The next day, the Union Army began its retreat. Taylor tried valiantly, but without success, to trap them again, this time in the vicinity of Pleasant Hill, Louisiana (Fig. 3). At this point, one of the strangest battles of the Civil War was fought when 2500 men of Taylor's cavalry engaged a river monitor, timber-clad and transport belonging to Porter's fleet. After a short, but fierce, river battle, the Confederate cavalry was finally repulsed and Banks, short of supplies and receiving divided opinions from his staff, ordered a full retreat down-river to Natchitoches and Grand Ecore, Louisiana. There, Banks received confidential orders from Grant to move the army down-river to New Orleans and thereby avoid issues with the falling water level of the Red River. Banks obeyed and began moving his troops to Alexandria, Louisiana. At the same time, Porter, fearing his entire fleet would be stranded by the falling river, moved it downstream to Alexandria to regroup before the final retreat to New Orleans. Once there, Porter's worst fears were realized as the falling river trapped the entire Union fleet of *c.* 90 vessels above a series of waterfalls and rapids, because the boats needed 2.1 m of water to float (Mills 1978).

The rapids are hypothesized as having a major impact on the Union's retreat down the Red River. At this time, Joseph Bailey, a former lumberjack from Wisconsin and Lieutenant Colonel and Chief Engineer of the Nineteenth Corps, came forward and described a dam he could build to raise the water level sufficiently to save the fleet. Porter and Banks were sceptical because they were well aware of the flashy nature of the Red River. Scepticism aside, they agreed and allowed him to try what became to be known as Bailey's Dam. It was actually a series of small dam-like structures commonly used by lumberjacks to aid in the transportation of logs.

The structures were a series of four 'wing dams' across the river, which was 230 m wide at this point (Mills 1978). The wing dams raised the water level by *c.* 1 m, allowing all but one Union boat, the USS *Eastport*, to pass safely through the rapids. On 18 May 1864, Taylor made a final effort to trap the retreating Union force below Alexandria at Yellow Bayou, but again Bailey came to the rescue, building a 550 m wide pontoon bridge of transports, which allowed the Union army to safely cross the Atchafalaya River and escape once again.

Conclusions

While formulating the tactics of the campaign, Union commanders made several tactical errors by overlooking a number of unique geological and hydrological aspects of the Red River system, including its flashy nature and high sediment load and a series of rapids located slightly north of Alexandria, Louisiana. The rapids are hypothesized as having a major impact on the development and persistence of the Red River Raft by serving as a geological knick-point (Gastaldo & Degges 2007). In addition, the raft altered the flow of the river from a single channel to a series of interconnecting anastomosing channels, which Confederate forces used as strategic shooting lanes, by luring the Union gunboats into the narrow and shallow channels where they were then attacked. The Union's army and navy forces' ability to manoeuvre were severely restricted by the Great Raft, resulting in poor co-ordination and communication between these forces that were meant to mutually support one other.

The Confederates' advantageous knowledge of the Red River's behaviour, the location of the Great Raft and its environment, and the lack of clear Union objectives were the foundation of the Campaign's ultimate failure. The behaviour of the Campaign's leaders raised such concerns in Washington that a congressional investigation was called and General William T. Sherman best described the failure of the Red River Campaign as '... one damn blunder from beginning to end' (Joiner 2003, p. 169).

References

AUTIN, W.J. & PEARSON, C.E. (leaders) 1993. Quaternary geology and geoarchaeology of the Lower Red River Valley: a field trip. 11th Annual Field Conference, Alexandria, LA, 26–28 March 1993. Friends of the Pleistocene, South Central Cell, Alexandria, LA.

BROOKSHER, W.R. 1998. *War Along the Bayous: the 1864 Red River Campaign in Louisiana*. Brassey's Military Books, London.

DAVIS, J. 1890. *A Short History of the Confederate States of America*. Part III. Belford, New York.

FOOTE, S. 1974. *The Civil War: a Narrative, Red River to Appomattox*. Random House, New York.

FOSTER, J.E., O'DELL, C.R. & GLOVER, J.E. 1987. *Channel Development in the Lower Reach of the Red River, Hydraulic Mode Investigation*. Technical Report **HL-87-9**. US Army Corps of Engineers, Waterways Experiment Station, Vicksburg, MS.

GASTALDO, R.A. & DEGGES, C.W. 2007. Sedimentology and paleontology of a Carboniferous log jam. *International Journal of Coal Geology*, **69**, 103–118.

JOINER, G.D. 2003. *One Damn Blunder from Beginning to End: the Red River Campaign of 1864*. American Crises Series. SR Books, Wilmington, DE.

JOINER, G.D. 2006. *Through the Howling Wilderness: the 1864 Red River Campaign and Union Failure in the West*. University of Tennessee Press, Knoxville, TN.

MILLS, G.B. 1978. *Of Men and Rivers: the Story of the Vicksburg District, U.S. Army Engineer District*. US Army Corps of Engineers, Vicksburg, MS.

RED RIVER CAMPAIGN MAP 1997. *Museum of Historic Natchitoches*. Natchitoches, LA, www.civilwarhome.com

RED RIVER RAFT 1985. *Archives & Special Collections*. Noel Memorial Library, Louisiana State University-Shreveport, Shreveport, LA.

REPORT OF RED RIVER SURVEY 1855. January 18, February 17, 1855. 33rd United States Congress, Washington, DC. 2nd session, Senate Executive Document, iii, No. 62, 1–6.

TORRES, N. & HARRELSON, D.W. 2012. The Great Red River Raft and its sedimentological implications. *In*: MCLAURIN, B.T., ELLIOTT, A.C. & TORRES, N. (eds) *Reconstructing the Human–Landscape Interactions*. Springer Briefs in Earth System Sciences, **1**. Springer, New York, 35–55.

WRIGHT, M.H. 1930. *Early Navigation and Commerce along the Arkansas and Red Rivers in Oklahoma*. Chronicles of Oklahoma, **VIII**, No. 1, http://digital.library.okstate.edu/chronicles/v008/v008p065.html

Aerial photographic intelligence during World War II: contributions by some distinguished British geologists

EDWARD P. F. ROSE

Department of Earth Sciences, Royal Holloway, University of London, Egham, Surrey TW20 0EX, UK

0000-0003-4182-6426

ted.rose@earth.oxon.org

Abstract: During the war of 1939–45, intelligence was gleaned from aerial photographs by a newly founded organization that developed into the Allied Central Interpretation Unit. This was based primarily at Danesfield House (known as Royal Air Force Medmenham) some 50 km west of London, in Buckinghamshire. At least six British geoscientists (and at least one American, L. J. Simon) were amongst its pioneering photographic interpreters, all recruited from civilian life: palaeobotanist H. Hamshaw Thomas; geologists L. R. Wager, N. L. Falcon, P. E. Kent and P. Allen; and a geologist who became distinguished as a geographer, D. L. Linton. Of these six, all except Linton were to become Fellows of the Royal Society (FRS): the highest British academic accolade for a scientist. Work at Medmenham, although important for the war effort, required interpreters familiar with aerial photographs rather than geology as such – but geology did assist the search for storage sites for 'V' weapons, terrain interpretation for the 1944 Allied landings in Normandy, and in guiding plans to bomb German industrial complexes hidden underground.

During the 1939–45 World War, many British geologists continued with their pre-war civilian employment. Most of those employed by the Geological Survey of Great Britain did so, although they were given new priorities to assist the war effort, including increased assessment of natural resources and site investigations for military facilities, e.g. new airfields, army camps and depots for stores and munitions (Bailey 1952; Mather 2012). Geologists in industry, especially those concerned with fossil fuels or metal mining, typically helped to maintain or expand production in order to meet wartime needs, many being thus in a 'reserved occupation' from which combatants could not be drawn (cf. Falcon & Dunham 1987; Farrington 2009). Geologists who found ways to volunteer for service with the armed forces seldom found an outlet for their professional skills, so most served as any other volunteer (cf. Rose & Hughes 1993).

Only a few served in uniform as 'field force' geologists: solitary geologist officers within the Engineer staff at General Headquarters of the British Expeditionary Force in France (1939–40), Middle East Forces in Egypt (1941–43), Central Mediterranean Force in Italy (1943–45), and 21st Army Group in NW Europe (1943–45) (although an additional geologist officer was appointed to serve in Italy and another in NW Europe, as those campaigns developed (Rose & Rosenbaum 1993; Rose 2004, 2011; Rose & Clatworthy 2008*a*)). Other geologists were able to make indirect use of their professional skills by service within the ten Boring Sections created within the Royal Engineers (the 'RE'), primarily to access potable groundwater for British military use (Rose 2012). Engineers with geological training or association were amongst officers assigned to the eight Quarrying Companies RE, and to the ten Tunnelling Companies RE, also newly created during the war, but were typically not geologists as such (Rose 2018*a*, *b*).

Additionally, in 1943 as British forces moved increasingly from defence to counter-attack, geologists were used to found two units to compile terrain intelligence for military use. The Strategic Branch of the Geological Survey of India, based in the Survey's headquarters at Calcutta, grew from its foundation in June to comprise two officers at the rank of major (or equivalent) and six captains, plus a staff captain as administrator and appropriate clerical, drawing office and ancillary staff (Rose 2005). At Oxford in England, a Geological Section within the Inter-Service Topographical Department grew from four men appointed as Royal Engineers captains in November to a maximum effective strength of 12 geoscientists by the end of hostilities in 1945 (Rose & Clatworthy 2007, 2008*b*).

A hitherto largely neglected story is that of British geologists who served from earlier in the war in a different terrain intelligence role: those who adapted their civilian skills to help meet an urgent need for military aerial photographic interpretation. This account describes how L. R. Wager, D. L. Linton,

From: Rose, E. P. F., Ehlen, J. & Lawrence, U. L. (eds) 2019. *Military Aspects of Geology: Fortification, Excavation and Terrain Evaluation*. Geological Society, London, Special Publications, **473**, 275–296.
First published online July 10, 2018, https://doi.org/10.1144/SP473.13

H. H. Thomas, N. L. Falcon and P. E. Kent, who all were or became Fellows of the Geological Society of London (FGS), came to join and to rise to significantly influential positions within the Central Interpretation Unit (CIU) of the Royal Air Force (RAF), an organization that became the Allied Central Interpretation Unit (ACIU) in May 1944. By the end of the war the ACIU comprised *c.* 326 RAF and Women's Auxiliary Air Force (WAAF) officer interpreters and 1000 other ranks, 16 officers from the Royal Navy, 76 from the British Army, and at least 125 officer interpreters and 79 other ranks from the US Army Air Force (AIR 15/635); other records, e.g. Williams (2014), put the overall total even higher. It pioneered development of aerial photographic interpretation for the armed forces both of the UK and the USA (Stanley 1981). Like the 'field force' geologists, the geologists used for aerial photographic interpretation were few in number but included scientists of exceptional ability: one who had already been elected a Fellow of the Royal Society (FRS), the highest UK academic accolade for a scientist, and four who were later to be elected as Fellows.

Evolution of the Central Interpretation Unit (CIU)

Although units for aerial photographic interpretation were eventually developed by the Allies within all the main operational theatres of the war, by far the largest and most significant in the UK was the CIU, based at the RAF station at Medmenham in Buckinghamshire (Fig. 1) (AIR 15/635; AIR 40/1169; also Hinsley 1979–90; Powys-Lybbe 1983; Babington Smith 2004; Downing 2011).

When the UK declared war against Germany on 3 September 1939, the RAF had some capability for long-range photographic reconnaissance, using primarily Bristol Blenheim aircraft, but the Blenheim was soon found to be inadequate for this role. Trouble was also experienced with existing camera systems. The task of improving aerial photography from the UK was therefore assigned to an Australian freelance civilian photographer, Frederick Sidney Cotton (1894–1969), a colourful character who had been used for clandestine missions to photograph German naval activity and troop concentrations as war loomed (Watson 2004). He was quickly commissioned into the RAF Volunteer Reserve (RAFVR) as a squadron leader, honorary wing commander. Based at RAF Heston in Middlesex, from 15 September 1939, he created the 'Heston Flight' equipped with suitably adapted Lockheed, Beechcraft, Blenheim and later Spitfire aircraft. The 'Flight' was re-named No. 2 Camouflage Unit on 1 November 1939 (with a 'Special Survey Flight' detached for service with the British Expeditionary Force in France). It became the Photographic Development Unit (PDU) on 17 January 1940, and the Photographic Reconnaissance Unit (PRU) on 8

Fig. 1. Aerial view of RAF Medmenham in 1945, showing Danesfield House with adjacent complexes of Nissen and other huts temporarily constructed within its grounds by the Air Ministry Directorate of Works. Crown copyright; reproduced courtesy of the Medmenham Collection.

July 1940, when Cotton was succeeded in command by a regular serving officer. At this time, the unit was divided into four flights: 'B Flight' at St Eval in Cornwall, in the SW of England; 'A Flight' at Wick, in the NE of Scotland; with two Flights remaining at Heston until 27 December 1940, when the PRU moved its base to RAF Benson in south Oxfordshire, in central England. The PRU was renamed No. 1 PRU on 16 November 1940 (when No. 2 PRU was created in Egypt), but 1 PRU was disbanded on 19 October 1942 and its parts formed into 540, 541, 542, 543 and 544 (Photographic Reconnaissance) Squadrons RAF (Table 1). At the peak of activity, there were 13 squadrons (the five RAF squadrons, three Canadian, and five from the US Army Air Force) whose sole function was to obtain aerial photographs. Every bomber and coastal patrol aircraft also carried a camera.

Initially, the RAF was poorly equipped to deal with the films obtained. Between the wars, aerial photographs were normally passed to the army for interpretation reports or for use in map-making. The UK entered World War II with only eight trained photographic interpreters in post: one RAF officer in an intelligence cell at the Air Ministry; six at Bomber Command; and one army officer teaching the interpretation short course at the RAF School of Photography, at Farnborough in NE Hampshire (although there were in addition a number of both RAF and army intelligence officers who had passed the course (Powys-Lybbe 1983)). At the start of the war, to process and interpret films, and to make detailed drawings from aerial photographs by means of specialist equipment (notably the Wild stereographic plotter), Cotton used the Aircraft Operating Company (AOC): an aerial survey firm based on the western outskirts of London at Wembley in Middlesex. This unofficial arrangement was achieved with the help of a friend, the company's managing director.

However, the Air Ministry gave the company a formal contract to operate as part of the PDU from 1 April 1940, and later that month moved its own photographic interpretation cell into the AOC's premises at Wembley, thereafter known as Paduoc House (an anagram of PDU and AOC) (Powys-Lybbe 1983). The cell merged with the company in May and on 12 June 1940 the Wembley organization was given its own identity as the Photographic Development Unit – Interpretation and Intelligence (the 'PDUI').

On 18 June 1940 control of the PDU and PDUI passed from the Director of Intelligence, Air Ministry, to Headquarters Coastal Command and on 11 July 1940 a further re-naming of the two units took place. When PDU was re-named the Photographic Reconnaissance Unit (PRU), PDUI became the Photographic Interpretation Unit (PIU).

The PIU had an early establishment for 19 RAF and 18 WAAF officers, mostly for interpretation, plus 69 RAF and eight WAAF other ranks, mostly for processing and printing (Powys-Lybbe 1983). During 1940 the accommodation at Wembley was rapidly outgrown and its work was frequently disrupted by the need to shelter from German air force (Luftwaffe) bombing raids aimed at London. In April 1941, the PIU therefore moved *c.* 50 km westwards, to Danesfield House at Medmenham in southern Buckinghamshire (Fig. 1), and closer to the base for No. 1 PRU, which had moved to RAF Benson in Oxfordshire at the end of 1940. The PIU was here renamed the Central Interpretation Unit (CIU). Later that year, the Bomber Command Damage Assessment Section was absorbed, and amalgamation was completed when the Night Photographic Interpretation Section of No. 3 PRU, from RAF Oakington in Cambridgeshire (formed on 16 November 1940 to provide Bomber Command with its own interpretation unit), was integrated with the CIU in February 1942.

To meet an increasing demand for aerial photographic interpreters and enhanced skills, a training school was established in 1941 at RAF Nuneham Park in Oxfordshire, as a satellite station *c.* 35 km to the NW of Medmenham (Powys-Lybbe 1983). Nuneham House, an eighteenth-century villa, and the park around it, were requisitioned for the purpose. As at Medmenham, Nissen huts and other

Table 1. *Development during World War II of principal RAF units associated with aerial photography and aerial photographic interpretation*

Role/Date	Sept. 1939	Early 1940	Mid-1940	April 1941	Oct. 1942	May 1944
Photography	Heston Flight	PDU	PRU		PR squadrons	
Interpretation	AOC	PADUOC	PDUI/PIU	CIU		ACIU

ACIU, Allied Central Interpretation Unit; AOC, Aircraft Operating Company; CIU, Central Interpretation Unit; PADUOC, Paduoc House (an anagram of PDU and AOC); PDU, Photographic Development Unit; PDUI, Photographic Development Unit Interpretation and Intelligence; PIU, Photographic Interpretation Unit; PR, Photographic Reconnaissance; PRU, Photographic Reconnaissance Unit.
Aerial photography from a single flight via the PDU and PRU ultimately to specialist PR squadrons. Aerial photographic interpretation from the AOC via a specialist cell to the PDUI, the PIU and the CIU, ultimately to the ACIU.

temporary buildings were erected adjacent to the mansion to provide the additional space required.

During 1942 and 1943 the CIU gradually expanded and was concerned in the planning stages of practically every operation of the war, and in every aspect of intelligence. Officers could be drawn from the army and the Royal Navy but were primarily from the RAF. American personnel formed an increasing part of the organization. Accordingly, on 1 May 1944, the title was changed to the Allied Central Interpretation Unit (ACIU). The title reverted to CIU when the Americans returned to the United States in August 1945.

The three-phase structure of interpretation

A three-phase structure of interpretation (Fig. 2) was quickly developed, and then operated throughout the war.

1st Phase – initial 'immediate' debrief of the pilot and interpretation of his photographic images, a dialogue between the flight crew and the photographic interpreter (the 'PI'). 1st Phase interpreters were based at the airfields from which reconnaissance flights operated, in 1941 (Fig. 2) at RAF Benson (Oxfordshire, near Medmenham), RAF Wick (northern Scotland) and RAF St Eval (Cornwall).

2nd Phase – 24 hours after the flight landing, photographs were passed to a general PI and placed into the overall context of the enemy 'activity' that the pilot had been tasked to photograph. 2nd Phase interpreters were based at Medmenham, to provide more detailed assessment within 48 hours. They were divided into three 'shifts' or 'watches' (Fig. 2), so as to provide 24-hour cover.

3rd Phase – the most important phase, when specialist PIs examined photographs taken at different times, to identify changes in situations on the ground. 3rd Phase interpreters were also based at Medmenham and, grouped into specialist sections, produced 'detailed' interpretation based on long-term comparative study of suites of aerial photographs. These sections were progressively developed, although most of them (Fig. 2) were

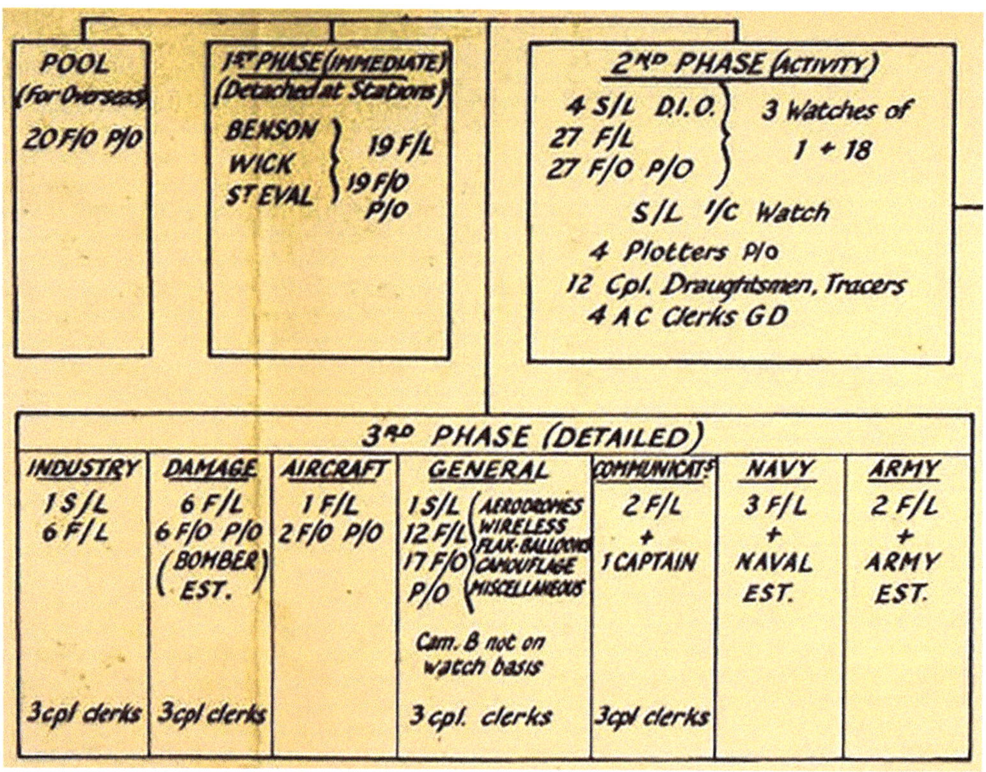

Fig. 2. Part of an organizational chart for the Central Interpretation Unit (CIU) at RAF Medmenham, as at 17 August 1941, showing how the Photographic Interpreters were grouped into 1st, 2nd and 3rd Phase sections, and a 'Pool' for overseas use. S/L, Squadron Leader; F/L, Flight Lieutenant; F/O, Flight (WAAF) or Flying (RAF) Officer; P/O, Pilot Officer; Cpl, Corporal. Reproduced by permission of The National Archives from file AIR 40/1169.

initiated by January 1941 prior to the move to Medmenham (Powys-Lybbe 1983). In final form they dealt with matters related to the navy, army, industry (including oil refineries), camouflage, communications (including railway activity), wireless (including radar installations), bomb damage, the aircraft industry, night photography, decoys, shipping (including submarine (U-boat) construction), and underground factories.

By 1945 Medmenham's daily intake of imagery averaged 25 000 negatives and 60 000 prints. By the end of the war in Europe (8 May 1945) its print library, which documented and stored aerial photographic coverage for operational areas worldwide, held five million prints from which 40 000 reports had been produced. Over 20 000 of those reports are currently held by the UK's National Archives, at Kew (Table 2).

Geologists were amongst the first interpreters to be recruited, since members of the profession were known to make use of aerial photographs to assist regional interpretation of geological structures and to facilitate ground surveys in areas lacking detailed topographical maps. At least six British geoscientists of established or potential distinction can be identified from Medmenham and other records. In near-successive dates of appointment these comprised L. R. Wager, D. L. Linton, H. H. Thomas, P. E. Kent, N. L. Falcon and P. Allen. One closely associated American geoscientist can also be identified, L. J. Simon.

Lawrence Rickard Wager (1904–65)

As described more fully by Rose (in press), Lawrence Wager (Fig. 3), commonly known as 'Bill', began work as an aerial photographic interpreter in June 1940. He was then 36 years of age, with an already impressive academic career. Born in Batley, Yorkshire, on 5 February 1904, in 1923 he had been awarded an Exhibition to Pembroke College in the University of Cambridge to read for the Natural Sciences Tripos (Deer 1967; Hargreaves 1991; Vincent 1994, 2004). He obtained first class honours in Part 1

Table 2. *The National Archives' index of air photo interpretation reports, categorized according to subject and the immediacy of requirement and by a one- or two-letter prefix*

Prefix	Subject	File reference	Number of reports
A	Navy	AIR 34/90-92	416
AR	Navy – immediate	AIR 34/98-100	256
AS	Navy – special	AIR 34/94-97	203
B	Army	AIR 34/102-114	1227
BR	Army – immediate	AIR 34/115-116	584
BS	Army – special	AIR 34/117-138	1089
D	Industry	AIR 34/199-207	425
D(C)	Oil refineries, etc.	AIR 34/700	145
DR	Industry – immediate	AIR 34/185-195	2099
DS	Industry – special	AIR 34/196-198	147
E	Camouflage	AIR 34/209	57
F	Communications	AIR 34/210-215	1015
FD	Railway activity	AIR 34/216	26
G	Wireless (Radar, etc.)	AIR 34/223-229	3088
GR	Wireless (Radar, etc.) – immediate	AIR 34/230	263
GS	Wireless (Radar, etc.) – special	AIR 34/221	106
K	Bomb damage	AIR 34/245-266, 542-543	3769
KS	Bomb damage – special	AIR 34/267-270	1633
L	Aircraft industry	AIR 34/272-274	299
LP	Aircraft production	AIR 34/275	152
LR	Aircraft industry – immediate	AIR 34/276-280	853
LS	Aircraft industry – special	AIR 34/281-282	386
N	Night photography	AIR 34/283-288	609
NS	Night photography – special	AIR 34/286, 747-748	661
Q	Decoy – special	AIR 34/289	132
S	Shipping	AIR 34/139-140	117
SA	Circuses, roadsteads & ramrods	AIR 34/208, 361-399	3193
SS	U-boat construction	AIR 34/101	20
U	Underground factories	AIR 34/271	68

Reports issued by the Central Interpretation Unit, subsequently developed as the Allied Central Interpretation Unit, based at RAF Medmenham.

Fig. 3. Squadron Leader L. R. Wager in 1941, wearing the white ribbon of the Polar Medal awarded to him in 1933. From Hargreaves (1991); reproduced by kind permission of his daughter, Jane Hargreaves.

of his degree course and graduated in 1926 with a first in Part 2 (in geology) for which he was awarded a Goldsmiths' Company Research Studentship. His consequent three years of research at Cambridge were supervised by a fellow Yorkshireman 'who became a lifelong friend' (Hargreaves 1991, p. 21): W. B. R. King, a pioneer of British military geology in World War I who was to gain distinction as the first and most senior British military 'field force' geologist of World War II (Rose 2004, 2011).

In October 1929 Wager was appointed a lecturer in petrology and mineralogy at the University of Reading. However, after only one term, he was invited to join a 1930–31 year-long expedition to Greenland, a Danish colony and the world's largest island, lying within the North Atlantic and largely within the Arctic Circle. He therefore completed only two more terms of teaching before leave of absence. In Greenland, he discovered and initiated geological surveys of igneous intrusions in its eastern coastal mountain region, studies to be reinforced by subsequent expeditions in 1932 and 1935–36. In 1933 he participated in another expedition, to climb Mount Everest. Leave of absence accounted in total for five out of his ten years at Reading.

As a university lecturer, Wager was in a 'reserved occupation'. However, in 1940 he was persuaded by Michael Spender, who had been working at Wembley as a photogrammetrist in the Air Survey Division of the AOC, to offer his services to the nascent PDU. Wager and Spender had both been members of the expedition to Greenland in 1932 and had become firm friends with a similar thirst for exploration in the years that followed.

Wager began work as a civilian (Hargreaves 1991) until his commission in the RAFVR became effective. This was duly published in the *London Gazette*, one of the official journals of record of the UK government, with seniority dated from 12 August 1940 (*London Gazette*, 27 September 1940). The *London Gazette* later records that he was promoted to flying officer on 12 August 1941 (for it was then normal for such promotions to be made after 12 months as a pilot officer) and that he actually reached the 'war substantive' rank of flight lieutenant on 1 September 1942. However, he achieved 'honorary', 'local' or 'acting' rank promotions long before this. During the war, substantive ranks were fully paid and held by career officers whereas 'war substantive' ranks were similarly fully paid but held only for the duration of the war and 'honorary' ranks were promotions granted as an honour, without pay appropriate to the rank. 'Local' ranks were temporary, restricted to service in a specific geographical region. Officers of 'acting' rank were entitled to the appropriate pay and allowances for that rank but might revert to their previous rank and pay on change in duty. 'Temporary' ranks could be assigned for specific tasks or missions, and so for service within specific time periods rather than within specific geographical regions. Normally, only substantive, war substantive and some temporary promotions were published in the *London Gazette*.

Wager joined about a week before the Wembley organization became the PDUI (on 12 June 1940), and was serving when it became the PIU (in July). From the known priorities of unit work through the summer (Babington Smith 2004), it can be inferred that he would have been involved in monitoring the build-up of enemy preparations for the invasion of Britain (Operation Sealion, scheduled for late September; Rose & Willig 2004), and the movements of the German Fleet. A letter home to his wife as he began to learn his new role advised 'My work is a good deal bottle-washing [i.e. elementary routine] but that is inevitable. For the most part I look over 500 photographs of a run … we discuss and refer to Spender as one of the experienced people' (quoted by Hargreaves 1991, p. 76). Also in June, he wrote to her that he had 'been dealing with oil tanks, bombed and unbombed, and trying to arrange the data we have' (Hargreaves 1991, pp. 77–78).

Wager was quickly employed as a 2nd Phase photographic interpreter, and soon appointed to lead one of the three 'shifts' that provided 24-hour cover. In April 1941, the PIU moved to RAF Medmenham and became the CIU. Wager's shift work continued. A colleague later remembered working with him that summer:

> His job was to allocate the sorties when they arrived from [RAF] Benson, St. Eval and Wick to interpreters on his shift. He would have the plots of the sorties and he knew who would best interpret subjects. Any difficulties or queries would be brought to him and he would check their interpretation with each interpreter, who would have dossiers, maps, and charts and previous covers from the library, in order to comment on changing progress. Finally, he would collate the parts and write his report on each whole sortie, or accept reports with or without any alterations
>
> (Quoted by Hargreaves 1991, p. 79).

Later, Squadron Leader Wager became 'responsible for the whole 24 hours of 2nd Phase' (Hargreaves 1991, p. 81).

In 1942, Wager learnt that a small photographic interpretation unit was being planned for deployment to Murmansk, on the Arctic coast of Soviet Russia, to keep watch on the battleship '*Tirpitz*' then hiding in Norwegian fjords. '*Tirpitz*' was rated so formidable a threat to the Allied convoys transporting supplies past Norway to the Russians that Medmenham generated a daily report on her situation. The base could not be permanent because the Arctic winters were too long, too cold and too dark. It was therefore planned to send RAF detachments for a few months only in three successive years. Wager volunteered to lead the first detachment and he duly left England on 14 August 1942. He arrived back in the UK at the end of October and returned to his duties at Medmenham.

During 1943, Wager clearly felt that he had done enough for the war effort and that there were by then enough trained interpreters to cope without him. He successfully applied for the chair of geology at Durham University and in September accordingly requested release from the RAFVR. This was granted and his service at RAF Medmenham came to an end, on 31 December 1943. He moved to Durham in January 1944, where his professorial career flourished and in 1946 he was elected FRS. Early in 1950 he was appointed to an even more prestigious chair in geology, as head of the Department of Geology and Mineralogy in the University of Oxford. He breathed new life into an old department and pioneered further advances in geosciences (Vincent 1994), but on 20 November 1965 he died suddenly from a heart attack, still in appointment.

David Leslie Linton (1906–71)

David Linton (Fig. 4) began work as an aerial photographic interpreter slightly later than 'Bill' Wager, in October 1940, and at the slightly younger age of 34, but also with a developing academic career.

Linton, born in London on 12 July 1906, was a geographer who had begun life as a geologist: he left Haberdashers' Askes Hatcham School for King's College in the University of London with a scholarship to read for a general honours degree in

Fig. 4. Squadron Leader D. L. Linton (right), with his colleague Villiers David, editing Medmenham's Monthly Interpretation Review. Crown copyright; reproduced courtesy of the Medmenham Collection.

chemistry, physics and geology (Goudie 2004). However, awarded first class honours in 1926, he then proceeded to a special honours degree in geography, graduating in 1927, again with first class honours. He remained at King's as a demonstrator in geology and geography until 1929, when he was appointed to the teaching staff of the University of Edinburgh. He collaborated in the 1930s with Sidney William Wooldridge (1900–63), a geologist graduate of King's subsequently renowned as a geomorphologist and geographer, who was a pioneer in the study of the geomorphology of SE England (Baigent 2004). In a classic account (Wooldridge & Linton 1939) of the structure, surface and drainage of SE England, they developed an influential model for the theory of landscape evolution.

Babington Smith (2004, p. 58) noted that in 1940 'valuable additions to the team at Wembley were Alfred Stephenson, a polar explorer, who later directed all interpretation training [at Nuneham Park]; and David Linton, an expert geographer'. Precisely how and why Linton was recruited to the unit at Wembley is not known. Certainly he joined in October, several months after Wager, seemingly on first appointment to a commission. The *London Gazette* of 19 November 1940 gives the date of his appointment as a pilot officer 'on probation' in the RAFVR as 22 October 1940. He was duly promoted to the rank of war substantive flying officer one year later, i.e. with effect from 22 October 1941 (*London Gazette* of 24 March 1942), to temporary flight lieutenant on 22 March 1943 (*London Gazette* of 8 January 1943), and to war substantive flight lieutenant (with seniority from 22 March 1943; *London Gazette* of 19 October 1945). He finally relinquished his commission 'under the provisions of the Navy, Army and Air Force Reserve Act, 1954' as a flight lieutenant 'retaining the rank of Squadron Leader' (*London Gazette* of 20 July 1954).

Babington Smith (2004, pp. 98–99) recorded that 'David Linton … arrived at St Eval [in Cornwall] on New Year's Day 1941, after only two months as an interpreter', consistent with first appointment to Wembley at the end of October. He was one of only 'A few interpreters [who] set off from Wembley' on assignment to the RAF station at St Eval, a strategic airbase for RAF Coastal Command. Its primary role was to provide anti-submarine and anti-shipping patrols off the SW coast of England. Additionally, aircraft from the base were used for meteorological flights, convoy patrols, air-sea rescue missions and protection of the airbase from the Luftwaffe, and it also operated aerial photographic reconnaissance missions. PDU's 'B Flight' had been based there from mid-1940. Whilst at St Eval, Linton's 1st Phase photographic interpretation unit found that the German heavy cruiser '*Admiral Hipper*', which had broken out from the Baltic Sea into the Atlantic Ocean to operate against Allied merchant shipping in December 1940, had arrived at the port of Brest in German-occupied France: a significant discovery at the time.

Later in 1941, but still concerned with shipping, Linton was sent north to Scotland, to the RAF Coastal Command station at Wick, whose aircraft mounted patrols over both the adjacent North Sea and part of the Atlantic Ocean. Although PDU's 'A Flight' had been based there from mid-1940, the station included a new Photographic Reconnaissance Unit of Spitfires fitted with long-range fuel tanks and cameras instead of their usual armament.

> Linton was chosen to start up the First Phase section … At Wick David Linton built up a most efficient section … Linton's pre-war experience as a geographer had given him a fund of knowledge about Scandinavia and also about meteorology and he could brief the pilots with real authority
>
> (Babington Smith 2004, p. 99).

On 21 May, two of Wick's Spitfires made a reconnaissance patrol over the Norwegian fjords in search of the '*Bismarck*' (sister ship of the '*Tirpitz*' and so one of the two largest German battleships ever built) and its escort, the heavy cruiser '*Prinz Eugen*'. One sighted and photographed two large warships in Kors Fjord, near Bergen. Whilst the pilot was being debriefed, the wet prints of the photographs he had taken were brought in. Linton was able to confirm identification of the two ships, and inferred that they were about to move (Babington Smith 2004, p. 100) – the first step in the Royal Navy's much celebrated hunt for and ultimate sinking of the '*Bismarck*'.

Linton's initial service was therefore in 1st Phase interpretation and focused on tasks associated with the Atlantic (Rose 2018*c*). However, in September 1942 he was posted (via the Coastal Command station at RAF Leuchars) to Medmenham. Promoted to the rank of squadron leader, he was given command within 3rd Phase interpretation of its Intelligence Section, formed by the amalgamation of the previous Ground Intelligence and Target Intelligence sections 'which had grown organically since their creation at Wembley' (Williams 2014, p. 129). As an outsider, it was hoped that he could provide the necessary unifying direction for the section, which was responsible for coordinating intelligence from ground sources as well as photographic interpretation reports. A 'C.I.U. Organisation' chart for June 1943 (in file AIR 40/1169) shows that his 'Intelligence and Targets Section' was responsible for

> Recording of all information received from all sources. Preparation of target material for dispatching to all R.A.F. Stations prior to air attack … The section prepared detailed Target Folders for every enemy objective selected by the Air Ministry, which had standing committees deciding on the attack priority for

enemy-controlled oil, industry and transportation facilities as well as other targets

(Williams 2014, p. 129).

From 1943 Linton's section was one of those at Medmenham that worked on identifying launching and storage sites for 'V' (Vergeltungswaffen = 'retaliation' or 'revenge') weapons in northern Europe, particularly northern France. The V-1 'flying bomb' was a cruise missile powered by a pulse-jet engine first tested in December 1942 that was ordered into mass production in 1943, and finally launched against England from 13 June 1944. The V-2 'rocket', which was developed in parallel, was a ballistic missile powered by a liquid fuel, first successfully fired in October 1942. It was ordered into production in May 1944, and first arrived in England on 8 September 1944. V-1 launch sites (e.g. Fig. 5) were surface constructions that could be identified from the ramps and associated buildings soon recognized as characteristic. The earliest sites to be constructed had massive ramps that were slow to construct, relatively easy to identify, and relatively easy to put out of action by bombing. Later versions had steel ramps that were quicker to assemble and easier to camouflage, so became more difficult bombing targets. The 'flying bombs' themselves came to be stored underground to preserve them from bombing, until assembled for flight, and it became a priority to identify and attack these storage sites. Linton himself became particularly successful in this when he realized that German engineers were making use of tunnels in Upper Cretaceous chalk excavated so that the stronger 'Chalk Rock' horizon within the sequence formed a roof. He borrowed a set of French geological maps and narrowed his aerial photographic search to the appropriate chalk outcrop and was soon locating many sites.

Fig. 5. Aerial photograph of one of the later type of V-1 'flying bomb' launch sites, hidden in the village at Vignacourt in the Somme department of northern France. The steel launch ramp is hidden in trees (top right), and a servicing building has been hidden under a pre-existing barn. Crown copyright; reproduced courtesy of the Medmenham Collection.

Following the end of the war in September 1945, Linton was released from his duties and appointed Professor of Geography in the University of Sheffield. He moved to a similar position at the then more prestigious University of Birmingham in 1958, thereby following in the footsteps of the distinguished 'field force' geologist F. W. Shotton: appointed Professor of Geology at Sheffield in 1945, and at Birmingham in 1949 (Rose & Clatworthy 2008*a*). Post-war, he became distinguished as a geographer with wide interests who achieved a significant number of national positions and academic honours (Garnett 1971, 1972; Steers 1971; Waters 1971; Gold *et al.* 1983; Goudie 2004). He was elected FGS on 10 January 1968, barely three years before he died, of cancer, on 11 April 1971.

Hugh Hamshaw Thomas (1885–1962)

Hamshaw Thomas (Fig. 6) also arrived at Wembley in October 1940, but at 55 years of age and with considerably greater experience in the interpretation of aerial photographs taken for military purposes than either Wager or Linton. The skill had been founded in World War I. The British Army had developed a system to make topographical maps based on mosaics constructed from aerial photographs, and 'Ham' Thomas (as he became known) had applied this skill in Egypt from 1917 to 1918.

Fig. 6. Wing Commander H. H. Thomas. Crown copyright; reproduced courtesy of the Medmenham Collection.

Thomas had been born on 29 May 1885 in Wrexham, in northern Wales (Butler 1963; Harris 1963, 2004). Admitted as an undergraduate to Downing College in the University of Cambridge in 1904, he read botany, physics and chemistry for two years (for Part 1 of the degree course), gaining first class honours, and then read history for Part 2, to fit himself for entry to a career in the Civil Service. However, he did not like the post offered, so remained at Cambridge earning a living by non-stipendiary teaching whilst studying fossil plants at the Sedgwick Museum, until in 1909 he was appointed Curator of the Botany School Museum; a post which he held until 1923. He was elected FGS on 18 December 1912 (Wendy Cawthorne, pers. comm. 2011) although he had no formal training in geology.

Thomas joined the Officers' Training Corps at Cambridge at the beginning of World War I, was subsequently commissioned in the Royal Field Artillery, and served with the British Expeditionary Force in France in 1915. However, on the eve of the Battle of the Somme he was sent to Egypt, where he was given charge of artillery supplies for a sector of the [Suez] Canal Defence Zone. As a lieutenant (later captain), he was then seconded to the Royal Flying Corps, as officer-in-charge of aerial photography, 'a side of military intelligence then in its infancy' (Harris 1963, p. 288). He was responsible for the compilation of maps at a scale of 1:40 000 drawn from aerial photographs for the Battle of Gaza in Palestine and, despite lack of specialized equipment, he subsequently mapped the whole of southern Palestine (Powys-Lybbe 1983). At first, almost single-handed, he was responsible for the mapping which enabled the British General Allenby to direct his campaign against the Ottoman Turks with great accuracy (Babington Smith 1962). Twice mentioned in despatches, as hostilities drew to an end he was appointed to the Order of the Nile, 4th Class (*London Gazette*, 9 November 1918), an award made by Egypt for exceptional services to the nation, and later MBE (a Member of the Order of the British Empire) (*London Gazette*, 3 June 1919).

After the end of hostilities, Thomas was sent to India to report on the possibility of making a survey using aerial photography. Subsequently, he returned to Cambridge and assisted in the direction of research at its Aeronautical Department, helping also to found the Cambridge University Air Squadron. He was quick to deliver a lecture to the Geological Society of London 'On some features in the topography and geological history of Palestine', illustrated by 'aeroplane photographs taken during the War' in which 'The Lecturer observed that a perfectly new method of illustrating and investigating some branches of physical geology is afforded by aeroplane photography' (Thomas 1920, p. ii). This illustrated how not only physical features, but also general geological structure of a region could be determined from aerial photographs, and how such photographs and 'the maps made from them, may throw much new light on the questions of climatic change and of topographical changes due to faulting in Palestine' (Thomas 1920, pp. ii–iii). He thus pioneered a major new technique in geology. In 1920 he delivered the Cantor Lectures to the Royal Society of Arts on the theme 'Aircraft photography in war and peace'.

Thomas was appointed to a university lectureship at Cambridge in 1923, and his academic career as a palaeobotanist flourished. He was elected FRS and a Fellow of Downing College in 1934. However, on the outbreak of war in 1939 he volunteered immediately for the RAFVR and tried to return to aerial photographic interpretation. He was commissioned 'for the duration of hostilities as [a] Pilot Officer on probation' on 11 October 1939 (*London Gazette*, 27 October 1939) soon after the start of the war, but spent nearly a year as a station intelligence officer at RAF Bomber Command Helswell in Lincolnshire before he requested transfer to photographic interpretation duties. Sent initially to the Air Ministry (to revise the RAF manual on aerial photographic interpretation), he arrived at the PIU Wembley in October 1940, to interpret photographs of enemy industry (Williams 2014).

Thomas was initially confirmed in his appointment and promoted to the rank of flying officer after the usual year (on 11 October 1940; *London Gazette*, 12 November 1940). However, his earlier military service was later taken into account: the *London Gazette* of 23 December 1941 announced that he was to be granted the rank of flying officer on probation (war substantive) with effect from 17 January 1940 and confirmed in this appointment as from 11 October 1940, notifications in the *London Gazette* of 12 November 1940 and 25 November 1941 being cancelled. He was promoted to temporary flight lieutenant on 1 March 1942 (*London Gazette*, 27 March 1942); mentioned in despatches as an acting squadron leader in the *London Gazette* of 11 June 1942; promoted war substantive flight lieutenant on 1 July 1942 (*London Gazette*, 11 September 1942); and promoted war substantive squadron leader on 1 September 1943 (*London Gazette*, 3 December 1943). He resigned his commission on 22 December 1943, but retained the rank of wing commander (*London Gazette*, 4 January 1944): evidence of his earlier promotion to this local or acting rank.

When the PIU moved from Wembley to Medmenham to become the CIU, Thomas was put in charge of all 3rd Phase interpretation. By December 1942, according to an organization chart (in AIR 40/1169), the CIU had grown and become re-structured to comprise an administration unit plus three squadrons. Thomas was then in command of 'C Squadron' in the rank of wing commander, subordinate only to the group captain who commanded CIU as a whole. Thomas thus directed the work of 14 sections including B (Army) Section led by Captain Falcon (see below) and the Intelligence Section led by Squadron Leader Linton. Thomas was still in command of 'C Squadron' in June 1943, when a new organizational chart indicated that the squadron then comprised 11 sections of which two, 'Military' and 'Intelligence and Targets', were led by geologists, Major Falcon and Squadron Leader Linton respectively. His overall responsibilities and long-term influence were therefore considerable.

However, they were to grow greater. On 19 April 1943, following credible reports that the Germans were attempting to develop rockets with which to bombard England, the Air Ministry directed the CIU to investigate the German secret weapons' programme. Thomas was appointed to lead the investigation (Williams 2014). The initial focus for aerial photography was to be a facility on Peenemünde (Fig. 7), an island in the Baltic Sea close to the coast of NE Germany.

'Operation Crossbow', the search for the 'V' weapons, has been described at length by Williams (2014), and features as a now classic story in many other historical accounts of the use of aerial photography in World War II (e.g. Stanley 1981; Powys-Lybbe 1983; Babington Smith 2004; Downing 2011). The V-1 'flying bomb' was first tested at Peenemünde (in December 1942). The V-2 'rocket' was first successfully fired also at Peenemünde (in October 1942). Staff in several sections of Thomas's 'C Squadron' contributed to studies that later confirmed Peenemünde as a key site for subsequently testing both weapons and, to slow development, it was bombed during the night of 17–18 August 1943 by a raid which, with diversions, involved 596 aircraft. Between 1 May 1943 and 31 March 1944, 40% of all Allied photographic reconnaissance sorties were directed against the V-weapons' programme, and a total of 1.25 million photographs were taken (Dear & Foot 1995). Thomas's staff (including David Linton, as described above) located all of the first 96 launching positions being constructed for the V-1, and an Allied bombing campaign to prevent them becoming operational duly began on 5 December 1943.

Thomas himself, however, was feeling the strain and was increasingly unwell towards the end of 1943. He resigned from the RAFVR on health grounds on 22 December 1943 and returned to academic life in Cambridge. In his later life he gave

Fig. 7. Aerial photograph of Test Stand Seven at the German experimental site on the Baltic island of Peenemünde, prior to Allied bombing during the night of 17–18 August. Two V-2 rockets are lying horizontally near the bottom entrance to the oval site. Crown copyright; reproduced courtesy of the Medmenham Collection.

much thought and time to the history of science, helping to found the British Society for the History of Science in 1947 and serving as its president. In 1958, the Darwin-Wallace Centenary year, he was judged to be among the 20 biologists in the world who had made the most outstanding contribution to knowledge of evolution, and was presented with a commemorative medal. He died at the age of 77, after a life in which he helped found and/or lead an astonishing number of academic organizations (Harris 1963).

Norman Leslie Falcon (1904–96)

Norman Falcon (Fig. 8) arrived at Wembley in September 1940, some three months after Wager but also at 36 years of age. An army rather than RAF officer, he was, like Linton, to pioneer development of a section of 3rd Phase interpretation within the squadron commanded from 1941 by 'Ham' Thomas.

Born on 29 May 1904 at Braunton in North Devon, Falcon had gone from Exeter School to Trinity College in the University of Cambridge in 1923, obtained first class honours in Part 1 of his degree course, and graduated with first class honours in Part 2 (in geology) in 1927. 'At Cambridge his mentors [included] W. B. R. King and ... His first expeditions abroad were climbing expeditions with L. R. Wager' (Martin & Lapworth 1998, p. 162).

After graduation, in late 1927, Falcon joined the Anglo-Persian Oil Company (renamed the Anglo-Iranian Oil Company in 1935, now BP), and went to work in Iran. He returned to the UK in 1937 and was employed in the team set up to explore the possibilities of locating onshore oil in the UK. Later, he was asked again to work on Iranian geology, at a

Fig. 8. N. L. Falcon, a photograph dated May 1960, from the Geological Society of London archives. © BP plc, archive reference ARC190350, and reproduced by kind permission.

time when the use of aerial photography was being developed extensively as a new and practical tool by the oil industry. He was seconded in 1939 to Shell Petroleum's centre at The Hague in the Netherlands, for training and to develop the aerial mapping of Iran and assist in planning a joint Anglo-Iranian/Shell venture in southern India – having himself learnt to fly in 1938. In May 1940 he managed to board one of the last ships to leave for the UK as the 'Phoney War' ended and the German Army advanced through the Netherlands.

On his return to England, Falcon joined the Local Defence Volunteers and was commissioned in the General List. He then transferred to the Intelligence Corps, being appointed to a Regular Army Emergency Commission as a second lieutenant (equivalent to an RAF pilot officer) on 14 August 1940 (*London Gazette*, 11 September 1940). He attended Intelligence Course No. 1 at Matlock in Derbyshire and was sent to the Army Interpretation Course at Farnborough in Hampshire before joining the new Army Photographic Interpretation Section (APIS) at Wembley. He arrived as a lieutenant and was the first member of the Intelligence Corps to be employed as a photographic interpreter with APIS, which was at that time a unit of 22 people (Williams 2014).

With the move to RAF Medmenham in 1941 and the establishment of the CIU, Falcon worked in 'B' Section of 3rd Phase interpretation – APIS by another name. According to Babington Smith (2004, p. 178), 'By the beginning of 1943 there was a large and thriving section of Army interpreters at Medmenham'. CIU organizational charts (AIR 40/1169) show Falcon as head of 'B' Section within 'Ham' Thomas's 'C Squadron' in the rank of captain by December 1942, and in the same post but as a major by June 1943. 'B' Section's role was then defined on a chart as the 'Interpretation of all military detail, primarily the enemy's fixed defences, equipment, etc. Selection of suitable sites for landing operations'. The contemporary *Army List* records that Falcon was promoted war substantive captain and temporary major on 31 March 1943, and war substantive major and temporary lieutenant-colonel on 31 August 1944; a rank equivalent to that achieved by 'Ham' Thomas before him.

Amongst the many tasks accomplished by 'B' Section, two were particularly important:

(1) The search for 'V' weapon launch and storage sites in 1943 and 1944. A cryptic War Office memorandum dated 13 February 1943, sent to Falcon for the eyes of 'B' Section only, was the first indication to Medmenham that the Germans were planning to bombard England with 'secret' weapons (Williams 2014). He became one of only six key people (another being 'Ham' Thomas) who began investigating aerial photographic evidence for the 'V' weapons' programme: contributing to a wider study initially designated 'Operation Bodyline', later 'Operation Crossbow'. The importance of the work was such that this small team gave rise to a new sub-section 'B2', created overnight on 8–9 November 1943 (Williams 2014, p. 169). B2 grew steadily in size throughout the remainder of the year and in 1944, at the height of the investigation at Medmenham, comprised two army officers, six RAF officers, 11 WAAF officers, two American officers and five other ranks from the army – 26 people in total. Moreover, it drew in assistance from other specialist interpreters from across the Medmenham spectrum as required: both David Linton and 'Ham' Thomas were amongst those involved in the search for 'V' weapons, as noted above. The interpreters developed great expertise in identifying launch and storage sites for 'V' weapons which were subsequently bombed. This helped

to delay launch of the first V-1 flying bombs against England until 13 June 1944, and although by the end of March 1945 over 10 000 bombs had been launched towards the UK from fixed sites in NW Europe, through countermeasures and mechanical faults only 3531 reached their target. The V-2 rocket developed in parallel first arrived in England on 8 September 1944. By the end of March 1945, a total of 1054 rockets had fallen on England, an average of about five per day. There were no protective countermeasures possible once the weapon was launched, and mobile launching pads made detection from aerial photographs in time to implement bombing difficult, but storage facilities for the rockets were detected and targeted to disrupt the launch programme.

(2) Terrain studies to facilitate the Allied landings in Normandy on D-Day. In 1944 Falcon commanded 'B' Section as a whole during preparations for the Allied invasion of Normandy that began on D-Day, 6 June 1944. The Section built up a very detailed picture of enemy defences (Fig. 9), enabling exact models of all the beaches and landing areas to be produced. He was subsequently awarded the US Army's Bronze Star in recognition of his work. Notably, one of the officers under his command, Geoffrey Stone, although a former instructor at the Army School of Military Intelligence (at Matlock in Derbyshire) rather than a geologist, was given a specifically geological assignment: one of his first tasks on posting to Medmenham 'was to list all the quarries in northern France and to analyse what kind of stone each one produced' (Downing 2011, p. 247). It was realized that an advancing Allied army would require stone for repair or enhancement of its supply routes and foundations for airfields and supply depots.

Maximum output from Medmenham took place in the final year of the war, 1944–45. During this time requests for interpretation reports from a wide range of 'customers' (including over 50 departments of the Admiralty, War Office, and Air Ministry) were filtered by the Joint Photographic Reconnaissance Committee, a sub-committee of the Joint Intelligence Committee answerable to the British Chiefs of Staff, before being passed to the Technical Control Office (TCO) at Medmenham. The TCO was led by an RAF wing commander, and otherwise comprised the Senior Photographic Officer (an RAF squadron leader), plus one representative each of the Royal Navy, the US Army Air Force, and the British Army: Lieutenant-Colonel Norman Falcon (Powys-Lybbe 1983). The TCO was responsible for all the activities at Medmenham and would allocate priorities.

Following the end of the war, Falcon returned to the Anglo-Iranian Oil Company (Martin 1998, 2004). Based in London, he developed an exploration programme for the UK. The Iranian oil industry was nationalized in 1951, and the company was therefore re-named British Petroleum in 1954 ('BP' in 2000). He was appointed its chief geologist in 1955, and in 1960 elected FRS. A founder member of the Medmenham Club (whose members comprise serving and former personnel associated with aerial photography and its interpretation) and one of the first four vice-presidents appointed at its inauguration meeting in 1946, Falcon was also the senior founding member of the Geologists' Pool created in the Territorial Army in 1949, accepting the lower rank of major, in the Royal Engineers. He served in the Pool for six years, until his appointment as chief geologist at British Petroleum made further service impracticable (Rose & Hughes 1993). Fiercely loyal to the company and nation throughout life, he died on 31 May 1996.

Percy Edward Kent (1913–86)

Percy Kent (Fig. 10), always known as 'Peter', arrived at Medmenham in about October 1941, a year later than Wager, Linton, Thomas and Falcon. At 28 years of age, he was also younger than the other four, but already an experienced petroleum geologist and destined for a career of great distinction.

Born on 18 March 1913 in West Bridgford, near Nottingham in the East Midlands of England, Kent had left West Bridgford Grammar School with a scholarship to University College Nottingham in 1931. He obtained a BSc degree with first class honours in geology in 1934 (Falcon & Dunham 1987) and then began postgraduate research, initially on Jurassic sedimentary rocks of Nottinghamshire and Lincolnshire but soon also on East African geology. Although he joined the Anglo-Iranian Oil Company in 1936 to work on the petroleum prospects of the East Midlands and southern England, his PhD on the fossiliferous Miocene beds of Kavirondo, Kenya (thesis submitted in 1939, degree awarded in 1941) was based on his African studies.

In a 'reserved occupation', Peter Kent continued to work for Anglo-Iranian in Nottinghamshire after the outbreak of war but was released in 1941 when the company cut back its geological surveying. He then joined the RAFVR, being 'interviewed after volunteering … by Sq. Ldr. Dr H. Hamshaw Thomas' (Falcon & Dunham 1987, p. 349). He was commissioned in the Administrative and Special Duties Branch as an acting pilot officer on probation on 6

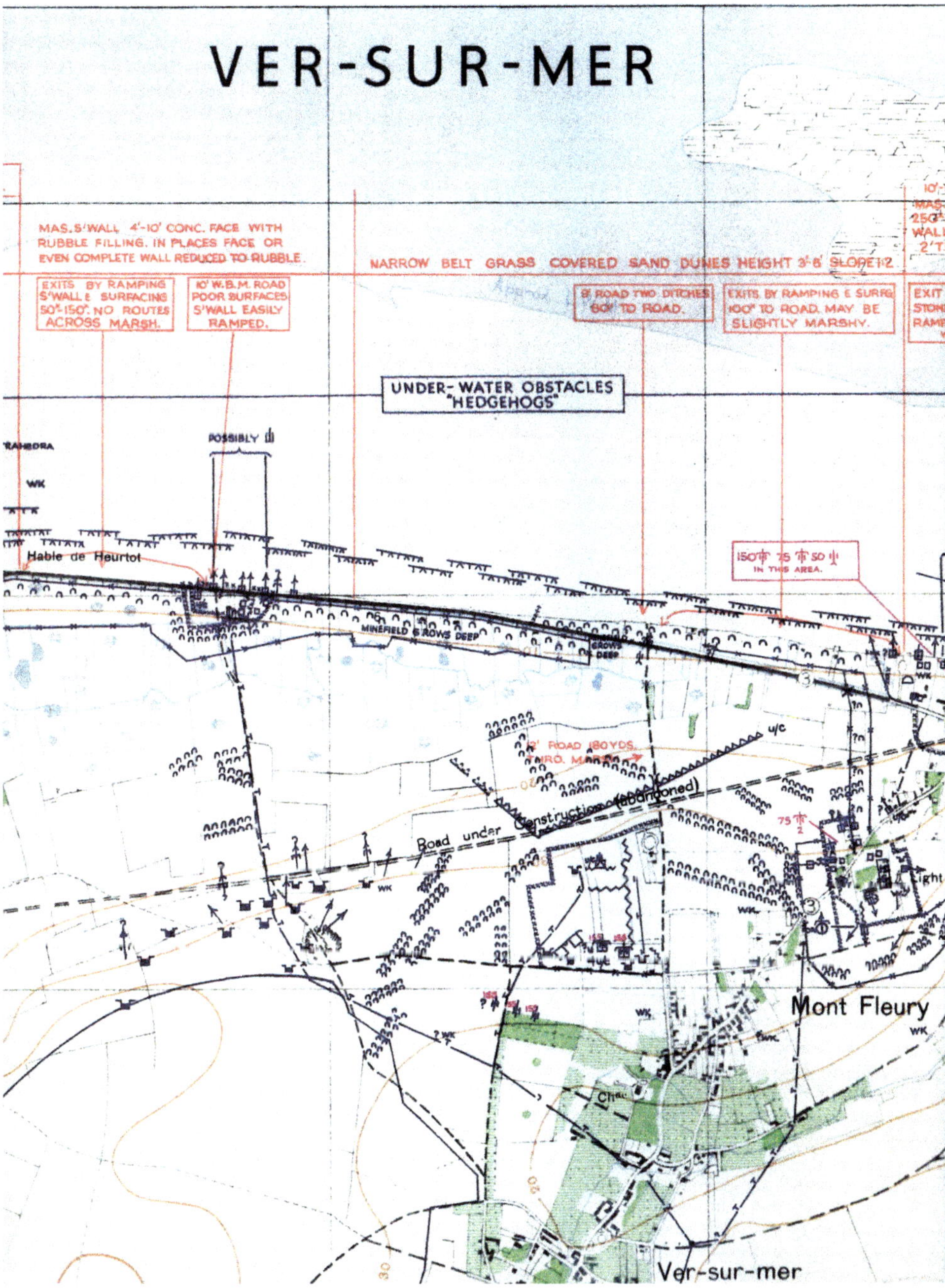

Fig. 9. Extract from one of a series of topographical maps of France at 1:12 500 scale with 'top secret' overprint in red and purple showing German 'defences' as at May 1944 (e.g. 'under-water obstacles "hedgehogs"'), used to guide Allied troops landing on D-Day, 6 June 1944. Part of the Normandy coast near Ver-sur-Mer; grid lines at 1 km intervals. Crown copyright; reproduced courtesy of the Medmenham Collection.

October 1941 (*London Gazette*, 11 November 1941); confirmed as a pilot officer on 6 December (*London Gazette*, 9 December 1941); and promoted war substantive flying officer on probation on 1 October 1942 (*London Gazette*, 27 November 1942). He was listed as an acting flight lieutenant

Fig. 10. Sir Peter Kent in 1973; reproduced by permission of the Geological Society of London.

when 'mentioned in despatches' (*London Gazette*, 1 January 1945), and when he finally relinquished his commission under the provisions of the Navy, Army and Air Force Reserves Act, he was granted permission as a flying officer to retain the rank of flight lieutenant (*London Gazette*, 29 April 1958).

Kent was assigned to the Industry Section at the CIU, established to gain information regarding the output of enemy war material. The discipline had been pioneered by 'Ham' Thomas at Wembley, and the section developed within Thomas's 'C' Squadron at Medmenham. Babington Smith (2004, p. 171) records that 'Early in May 1942 … Flight Lieutenant Peter Kent, a young oil geologist … had just been put in charge of interpreting the oil plants'. His task was to locate and measure oil storage tanks on the surface and to identify storage installations – concrete reservoirs of considerable size – that were being constructed underground.

Sixteen synthetic oil plants were constructed across Germany. The first to be identified was at Leuna, near Merseburg, west of Leipzig, in the summer of 1941, and the largest was at Brüx (now the city of Most in the Czech Republic, close to the German border) in 1942: production there could have reached 750 000 tons per annum. The patents for hydrogenation were internationally owned by the German IG Farben pharmaceutical and chemical industry conglomerate with the British ICI (Imperial Chemical Industries) and US Standard Oil companies. The photographic interpreters duly obtained information from ICI regarding German operations and methods. Also, Germanic insistence on order resulted in precise layouts for each operational unit, making it an easier task to pinpoint particular sections of a plant. Babington Smith (2004, p. 172) recorded that:

> Soon the volume of work on the oil industry had increased so much that Kent had six interpreters helping him, including two Americans. And it was the special experience acquired by Kent and his team that made possible the purposeful planning of the offensive against oil which began in May 1944 … Kent and his assistants could estimate how long repairs were likely to take after each raid.

One of the Americans was the geologist Louis Simon, who 'spent the war years [1943–45] in the Petrol Oil and Lubricants (P.O.L.) section working with two British oil geologists (one from Shell, and one from British Petroleum). His field was buried oil storage' (Carroll 1997, p. 1336). Louis Joseph Simon (1912–96), born in Indiana, had graduated from high school at the New Mexico Military Institute in Roswell, New Mexico, before studying geology as an undergraduate at Pomona College in Claremont, California, and from 1935 as a graduate student at Stanford University (Carroll 1997). When offered employment in the micropalaeontology laboratory of the Texas [Oil] Company (Texaco), however, he left academic life for a lifetime career as a micropalaeontologist in the petroleum industry. This was soon to be interrupted by World War II. Simon had presumably joined the Reserve Officer Training Corps programme while at Pomona since he was commissioned on graduation as a lieutenant in the US Army. He was therefore called up for active service after the USA entered the war in December 1941, following the Japanese attack on Pearl Harbor. Initially posted for service with anti-aircraft artillery, he was subsequently sent to the Air Force Intelligence School at Harrisburg in Pennsylvania: a school that was opened in April 1942, with benefit from experience at Medmenham. On graduating, Simon was posted to Medmenham itself. He began by attending a six-week language course to learn some of the niceties of 'British' English, and presumably arrived in the Industry Section in 1943, with its other US officers (Powys-Lybbe 1983).

Simon (1957) recorded that four rather than three geologists, three British and one American, all former oil company employees, worked in the POL unit of the Industry Section at Medmenham. Simon himself was evidently the American, and although he provides no clue to the names of two of the British geologists, the other (the British Petroleum geologist noted by Carroll 1997) must have been Peter Kent because he (Simon 1988, p. 493) later recorded:

> Flight Lieutenant P. E. Kent was head of P.O.L. (petrol-oil-lubricants) at ACIU. Peter was personally responsible for the interpretation of all photos of bombings of synthetic oil plants. This work included a weekly meeting in London to pick the bombing targets for the following week.

In addition to POL work, geologists associated with the Industry Section at Medmenham are known to have made use of geology as such on at least two important occasions:

(1) From the summer of 1943, when it was feared that the Germans might be developing atomic weapons (Powys-Lybbe 1983). If so, facilities for vast new water and power supplies would have to be built. Suspicion fell on 14 new factories, all identical in shape, that were identified in an area to the south of Stuttgart in SW Germany. However, each site was located next to mining operations. All featured a grid laid out on the ground, a chimney and an oil storage tank.

> The discovery that all the factories were located on the same contour, and adjacent to mining operations, prompted the idea that there might be a geological rationale behind their location. [Medmenham's Technical Control Officer, Wing Commander Douglas] Kendall sent one of the D [Industry] Section geologists to the Geological Museum in South Kensington – now part of the Natural History Museum – to consult their records
>
> (Williams 2014, p. 116)

Geological maps and literature indicated that the factories were sited adjacent to a thin bed of low-grade oil shale. With this clue, and the knowledge that by this time the Germans were acutely short of oil supplies, it became apparent that they were faced with exploiting this low-grade and expensive source of fuel. The factories thus did not represent an 'atomic menace'. A three-month study concluded that there was no evidence from the aerial photographs available at Medmenham that the Germans were working on an atomic bomb at that time.

(2) From 1944 especially, when the Germans increasingly located key factories underground to protect them from the increasingly destructive power of Allied bombing.

> That something as complex as the workings of a factory deep underground could be revealed by the interpretation of aerial photographs with the co-operation of the geologists at [Medmenham] may seem impossible to the uninitiated
>
> (Powys-Lybbe 1983, p. 143).

The Medmenham reports prove otherwise. A well-documented example is of the factory built partly underground in southern Germany at Kahla/Grosseutersdorf (Fig. 11), in mountainous country thickly covered by pine forests *c.* 100 km SW of Leipzig. Secret Interpretation Report No. U.55 dated 17 April 1945 described that a runway over 1000 m long had been constructed on top of a long, high ridge known as the Walpersberg (elevation 320 m), indicating that some sort of aircraft factory might lie beneath. Aircraft nearby were identified as the Messerschmitt Me262: fighters powered by a twin-jet engine whose performance was superior to that of the British Spitfire and a formidable threat as the war neared its end. Three tunnel entrances were identified on the north side of the ridge, at least ten more on the south side – evidence that facilities were being developed underground. The report described geological conditions in detail, giving the stratigraphical sequence of the Triassic ('Middle Bunter') sands/sandstones in the ridge and deducing the bed most suitable for tunnelling, inferring also that the ridge could be opened from both sides and the entrances be connected. Begun too late in the war to have posed a serious problem to the Allies, the facility was found postwar to have been a conversion of former 'sand mines': an existing tunnel system had been extended to 30 km in length by the use of over 12 000 forced labourers from Italy and Eastern Europe and some 3000 skilled workmen, but the factory produced only 20–30 completed aircraft before the Allied 'Victory in Europe', on 8 May 1945 (https://de.wikipedia.org/wiki/REIMAHG accessed 13 April 2018).

After service in Medmenham's Industry Section, Peter Kent was posted to the Mediterranean Allied Photographic Reconnaissance Wing and on return to Medmenham, to its 2nd Phase section. He was 'Mentioned in Despatches' for his work to the end of 1944 (*London Gazette*, 1 January 1945), and after the Allied victory in Europe he worked in the USA at the Pentagon (the headquarters of the US Department of Defense) on potential Japanese targets, work for which he was honoured by appointment as a Legionnaire of the US Legion of Merit (*London Gazette*, 18 June 1946). He went to Hiroshima and Nagasaki as a member of the US investigation team to study the effects of the atomic bombs, but this was an experience 'about which he rarely talked' (Falcon & Dunham 1987, p. 349).

Fig. 11. Present-day relic of REIMAHG (a facility named as an abbreviation of Reichs Marschall Hermann Goering), a factory constructed partly underground in the Walpersberg near Kahla/Grosseutersdorf, Germany, in 1944 to manufacture Messerschmitt Me262 jet fighter aircraft. © Alexander Hoernigk/CC BY-SA 4.0 (via Wikimedia Commons), https://commons.wikimedia.org/wiki/File:REIMAHG_Werk_Walpersberg_2014-04-28-11-57-56.jpg.

Demobilized in 1946, Kent returned to work at the Anglo-Iranian Oil Company, which in 1954 became British Petroleum (later BP). In 1966 he became the company's chief geologist and, in that year also, he was elected FRS. He retired from the company as 'assistant general manager (exploration)' in 1973, and was honoured by the Queen, dubbed a Knight Bachelor. He served as chairman of the UK Natural Environment Research Council from 1973 to 1977, and then returned to industry as a consultant and company director. Author of 145 publications, he received honorary degrees from eight UK universities for his considerable contributions to science and to the nation (Falcon & Dunham 1987; Martin 1987; Larminie 2004). Despite his many other commitments, he became and remained a member of the Medmenham Club that maintained links between alumni of the Central Interpretation Unit of RAF Medmenham until his death in 1986.

Percival Allen (1917–2008)

Percival Allen (Fig. 12), known to all as 'Perce', who became Professor of Geology at the University of Reading in 1952 and was elected FRS in 1973, had a less prominent and less well-known role at Medmenham.

Fig. 12. Percival Allen, a post-war photograph reproduced by kind permission of his eldest son, J.A. Allen.

Born on 15 March 1917 in the village of Brede, Sussex, in the Wealden district of SE England, Allen went from Rye Grammar School to Reading University in 1936. He began studies in geology and zoology, but under the influence of geologists including 'Bill' Wager, 'zoology never really had a chance' (Allen 1982).

Allen graduated in 1939, with first class honours, but remained at Reading to undertake research for a PhD degree on the Cretaceous sediments of the Weald, the focus of studies throughout his life. However, in 1940 he volunteered for service with the 'Photographic Interpretation Unit of the RAF' and 'seems to have worked round the clock, when on night shifts he used the day-time to do his field work and when on days he would do his research at night' (Allen 1982). In 1943 he achieved his PhD, the first to be awarded by Reading in geology, 'largely written on the train between his RAF station and the Weald' (Allen 1982; Anon 2008). His 'RAF station' must have been Medmenham, some 15 km NE of Reading, and it would be surprising if his appointment had not been influenced in some way by 'Bill' Wager, one of his teachers at Reading and a pioneer of photographic interpretation at Medmenham.

Allen served in the RAF 'for two years as an interpreter of aerial photographs' (Batten 1995, p. 601) but was released in 1942 to join the staff of the Geology Department at Reading as a Demonstrator (Batten 1995). He became an Assistant Lecturer in 1945 but in 1946 transferred to the University of Cambridge, appointed initially as a Demonstrator and later as a University Lecturer in Geology, in a department led, since 1943, by W. B. R. King. He returned to Reading to take up the Chair of Geology in 1952, still only in his mid-30s, remaining in post and serving with distinction for the next three decades, notably after his foundation of the prestigious Sedimentology Research Laboratory in 1962 (Batten 1995, 1996; Anon 2008). He died on 3 April 2008, survived by three sons and a daughter, who had been told little of his 'secret' work in wartime.

No record has been traced that Allen was ever commissioned as an officer. Two sons (John and Henry Allen, pers. comm. 2011) have no recollection of his rank, but John a vague feeling that he might have been a corporal. E. A. Vincent, a contemporary as an undergraduate with 'Perce' at Reading who also developed an academic career in geology, recalled that 'Perce' became ill in his efforts to work both day and night and was duly invalided out of the RAF after only a short time of service (Peter Worsley, pers. comm. 2012).

Concluding remarks

Although the Allied Central Interpretation Unit came to include a workforce of over 1700 people, and known geologists formed a very small proportion of that number, they held several positions of

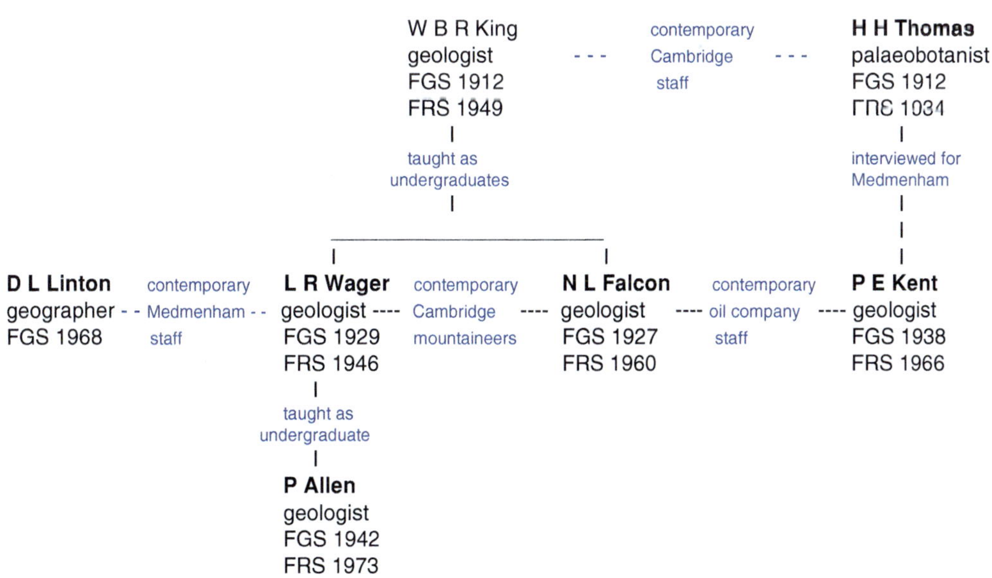

Fig. 13. Connections between the six geoscientists whose war service is described in this paper and 'Bill' King, the leading British military geologist of World Wars I and II, with year of election as a Fellow of the Geological Society and, where appropriate, as a Fellow of the Royal Society.

particular influence during the unit's development and operation. 'Bill' Wager, one of the earliest members of its 2nd Phase section, was soon leading one of the section's three 'shifts' and commanded the entire section for much of 1942 and 1943. David Linton helped to pioneer 1st Phase interpretation at two of the three RAF airfields first used for reconnaissance flights, and then commanded one of the major 3rd Phase sections, 'Intelligence and Targets', for the rest of the war. 'Ham' Thomas founded the important 3rd Phase 'Industry' section, and then commanded all 3rd Phase sections until the end of 1943, notably helping to identify sites associated with the German 'V' weapons programme. Norman Falcon helped to pioneer development of the Army Photographic Interpretation Service (='B' Section of 3rd Phase) and commanded this until the end of the war, thus making a major contribution both to guidance of the bombing campaign that degraded German ability to launch 'V' weapons and to terrain studies that facilitated the Allied landings in Normandy in 1944. Peter Kent led the Petrol/Oil/Lubricants sub-section of the 3rd Phase 'Industry' section that guided bombing to degrade the fuel resources essential to the German war effort. These geologists in leadership positions are known to have been assisted by at least a few other geologists serving in less prominent roles, either as officers, for example Louis Simon, or other ranks, such as 'Perce' Allen.

It is striking that all of the British men described here demonstrated their ability from an early age. Wager, Linton, Falcon, Kent and Allen had all graduated with first class honours degrees, Wager and Falcon with 'double firsts' (as graduates of the University of Cambridge, whose examination scheme made such achievements possible). Thomas, member of an earlier generation and also with a 'first' from Cambridge, showed his ability in the examination for entry to the Civil Service – coming top of the candidate list in botany.

Three of the six (Thomas, Wager and Falcon) had graduated from the University of Cambridge, one (Allen) from the then relatively new University of Reading where he was strongly influenced by two Cambridge graduates, 'Bill' Wager, and the Professor of Botany 'Tom' Harris, a former student of 'Ham' Thomas and so close a friend that he was to write Thomas's definitive obituaries (Harris 1963, 2004). Post-war Allen himself was briefly appointed to the teaching staff at Cambridge, under 'Bill' King. Linton and Kent as graduates of the University of London (Linton as an internal student via King's College and Kent as external student via the then University College Nottingham) were thus exceptional to the core group, which was dominantly Cambridge in background (Fig. 13).

At Cambridge, Wager and Falcon had been undergraduate students of 'Bill' King, the pioneer of British military geology in World War I and its leading proponent in World War II. They had learnt some of their basic geological mapping skills from him, and perhaps been influenced by his patriotism, but were linked by their undergraduate enthusiasm for mountaineering and, later, that for exploration of poorly mapped terrain and the consequent use of aerial photographs for terrain analysis rather than by any direct military guidance from King during the war.

The oldest and most experienced member of this group, 'Ham' Thomas, achieved the rank of wing commander RAFVR, and Falcon the equivalent army rank of lieutenant-colonel. The younger RAF officers Wager and Linton both became squadron leaders RAFVR. Their senior rank enabled all of them to have a formative influence on the work at Medmenham and so the war effort. Kent was only a flight lieutenant RAFVR, but the honours he received are testimony to his exceptional service.

General Werner Freiherr von Fritsch, Commander-in-Chief of the German Army 1935–38, had warned in 1938 that 'the military organisation with the best aerial reconnaissance will win the next war' (Williams 2014, p. 10). The Allies did develop a very proficient organization for aerial reconnaissance and photographic interpretation and did indeed win the war. Geologists at Medmenham contributed to that victory largely on the basis of their personal leadership qualities rather than their expertise as geoscientists, but geology did assist the search for storage sites for 'V' weapons, terrain interpretation for the 1944 Allied landings in Normandy, and in guiding plans to bomb German industrial complexes hidden underground. Indeed, the American geologist Louis Simon (1957, p. 1567) was to conclude from his service at Medmenham that 'geology is especially useful in the field of Intelligence. Experience suggests that the British appreciate this fact more than we did'.

Acknowledgements Thanks are due to librarians at The National Archives and to the Trustees of the Medmenham Collection held in the Military Intelligence Museum and at RAF Wyton (www.medmenhamcollection.org) (per Christine Halsall and Wing Commander (Retired) Mike Mockford) for access to unpublished documents cited in this paper; Wendy Cawthorne, Assistant Librarian at the Geological Society of London, for help in locating biographical data in the Society's archives and elsewhere; John and Henry Allen, sons of 'Perce' Allen, for their efforts to track down details of their father's wartime service; Paul Stewart, Jim Griffiths and Ursula Lawrence for helpfully reviewing the paper; and to copyright owners, as acknowledged in figure captions, for permission to reproduce illustrations.

The author was an undergraduate student of 'Bill' Wager at Oxford from 1960 to 1963, served with Sir Peter Kent as a member of the International Trust for Zoological

Nomenclature in the early 1980s, and served from 1969 to 1987 in the Engineer Specialist Pool (Territorial & Army Volunteer Reserve) that succeeded the Geologists' Pool pioneered by Norman Falcon in 1949–55, so knew all three men slightly albeit none well.

Funding This research received no specific grant from any funding agency in the public, commercial, or not-for-profit sectors

References

AIR 15/635. *History – P.R.U. and 106 Group*. Unpublished report, The National Archives, Kew.

AIR 40/1169. *CIU RAF Medmenham: organisation, control and function Jan 1941 – Aug 1943*. Unpublished report, The National Archives, Kew.

ALLEN, J.R.L. 1982. *Portrait of P. Allen*. Unpublished document issued on 2 October 1982 to mark the opening of the Allen Laboratory at the University of Reading. Seen at the Geological Society of London Library.

ANON 2008. Professor Percival Allen. [Obituary] *The Telegraph*, Monday 26 May.

BABINGTON SMITH, C. 1962. Service Aviation. Air Force, Naval and Army Flying News. Hugh Hamshaw Thomas. *FLIGHT International*, 26 July 1962, 142.

BABINGTON SMITH, C. 2004. *Evidence in Camera. The Story of Photographic Intelligence in the Second World War*. Sutton, Stroud [Originally published by Chatto & Windus, London, 1957].

BAIGENT, E. 2004. Wooldridge, Sidney William (1900–1963). *In*: MATTHEW, H.G.C. & HARRISON, B. (eds) *Oxford Dictionary of National Biography. Vol. 60. Woolmark–Zuylestein*. Oxford University Press, Oxford, 249–251 [with portrait].

BAILEY, E.B. 1952. *Geological Survey of Great Britain*. Murby, London.

BATTEN, D.J. 1995. A tribute to Professor P. Allen, F.R.S. Part 1. Editorial. *Cretaceous Research*, **16**, 599–601 [with portrait].

BATTEN, D.J. 1996. A tribute to Professor P. Allen, F.R.S. Part 2. Editorial. *Cretaceous Research*, **17**, 1–4 [with portrait and list of publications].

BUTLER, F.H.C. 1963. Hugh Hamshaw Thomas, M.B.E., Sc.D., F.R.S., F.L.S., F.G.S. Born 29 May 1885. Died 30 June 1962. *British Journal for the History of Science*, **1**, 280–283.

CARROLL, N.P. 1997. [Memorial] Louis Joseph Simon (1912–1996). *American Association of Petroleum Geologists Bulletin*, **81**, 1336–1337.

DEAR, I.C.B. & FOOT, M.R.D. (eds) 1995. *The Oxford Companion to the Second World War*. Oxford University Press, Oxford.

DEER, W.A. 1967. Lawrence Rickard Wager. *Biographical Memoirs of Fellows of the Royal Society*, **13**, 359–385 [with portrait].

DOWNING, T. 2011. *Spies in the Sky. The Secret Battle for Aerial Intelligence During World War II*. Little, Brown, London.

FALCON, N.L. & DUNHAM, K. 1987. Percy Edward Kent. *Biographical Memoirs of Fellows of the Royal Society*, **33**, 343–373 [with portrait].

FARRINGTON, P. 2009. Geological Section ISTD in World War II. *Royal Engineers Journal*, **123**, 82.

GARNETT, A. 1971. David Leslie Linton, 1906–1971. *Geography*, **56**, 341–343.

GARNETT, A. 1972. David Leslie Linton. *Transactions of the Institute of British Geographers*, **55**, 171–178 [including bibliography].

GOLD, J.R., HAIGH, M.J. & WARWICK, G.T. 1983. David Leslie Linton 1906–1971. *Geographers Bibliographical Studies*, **7**, 75–83 [including bibliography].

GOUDIE, A.S. 2004. Linton, David Leslie (1906–1971). *In*: MATTHEW, H.G.C. & HARRISON, B. (eds) *Oxford Dictionary of National Biography. Vol. 33. Leared–Lister*. Oxford University Press, Oxford, 940.

HARGREAVES, J. 1991. *L. R. Wager, a Life: 1904–1965*. Published privately.

HARRIS, T.M. 1963. Hugh Hamshaw Thomas 1885–1962. *Biographical Memoirs of Fellows of the Royal Society*, **9**, 286–226 [with portrait].

HARRIS, T.R. 2004. Thomas, Hugh Hamshaw (1885–1962). *In*: MATTHEW, H.G.C. & HARRISON, B. (eds) *Oxford Dictionary of National Biography. Vol. 54. Taylour–Tonneys*. Oxford University Press, Oxford, 336–337 [with portrait].

HINSLEY, F.H. 1979–90. *British Intelligence in the Second World War*. HMSO, London.

LARMINIE, G. 2004. Kent, Sir Percy Edward (1913–1986). *In*: MATTHEW, H.G.C. & HARRISON, B. (eds) *Oxford Dictionary of National Biography. Vol. 31. Kebell–Knowlys*. Oxford University Press, Oxford, 323–324.

M[ARTIN], A.J. 1987. Sir Peter Kent. *Annual Report of the Geological Society for 1986*, 25–27.

MARTIN, A.J. 1998. Norman Leslie Falcon. *Annual Report of the Geological Society for 1997*, 19–20.

MARTIN, A.J. 2004. Falcon, Norman Leslie (1904–1996). *In*: MATTHEW, H.G.C. & HARRISON, B. (eds) *Oxford Dictionary of National Biography. Vol. 18. Elu–Fancourt*. Oxford University Press, Oxford, 963–964.

MARTIN, A.J. & LAPWORTH, P.B. 1998. Norman Leslie Falcon. *Biographical Memoirs of Fellows of the Royal Society*, **44**, 161–174 [with portrait].

MATHER, J.D. 2012. War as a catalyst for change: groundwater studies in the Geological Survey of Great Britain before 1950 and the impact of two World Wars. *In*: ROSE, E.P.F. & MATHER, J.D. (eds) *Military Aspects of Hydrogeology*. Geological Society, London, Special Publications, **362**, 139–160, https://doi.org/10.1144/SP362.8

POWYS-LYBBE, U. 1983. *The Eye of Intelligence*. Kimber, London.

ROSE, E.P.F. 2004. King, William Bernard Robinson (1889–1963). *In*: MATTHEW, H.G.C. & HARRISON, B. (eds) *Oxford Dictionary of National Biography. Vol. 31. Kebell–Knowlys*. Oxford University Press, Oxford, 691–692.

ROSE, E.P.F. 2005. British military geology in India: its beginning and ending. *Royal Engineers Journal*, **119**, 46–53.

ROSE, E.P.F. 2011. Credit due to the few: British field force geologists of World War II. *In*: HÄUSLER, H. & MANG, R. (eds) *International Handbook Military Geography Vol. 2. Proceedings of the 8th International Conference on Military Geosciences, Vienna, Austria, 15–19 June*

2009. Truppendienst, Ministry of Defence, Vienna, 429–442.

Rose, E.P.F. 2012. Groundwater as a military resource: development of Royal Engineers Boring Sections and British military hydrogeology in World War II. *In*: Rose, E.P.F. & Mather, J.D. (eds) *Military Aspects of Hydrogeology*. Geological Society, London, Special Publications, **362**, 105–138, https://doi.org/10.1144/SP362.7

Rose, E.P.F. 2018*a*. Quarrying Companies Royal Engineers in World War II: contributions to military infrastructure within the UK and to Allied forces during the North African, Italian and NW Europe campaigns. *In*: Rose, E.P.F., Ehlen, J. & Lawrence, U.L. (eds) *Military Aspects of Geology: Fortification, Excavation and Terrain Evaluation*. Geological Society, London, Special Publications, **473**. First published online March 6, 2018, https://doi.org/10.1144/SP473.2

Rose, E.P.F. 2018*b*. Tunnelling Companies Royal Engineers in World War II: excavation of bomb-proof facilities in France, Gibraltar, Malta and the UK. *In*: Rose, E.P.F., Ehlen, J. & Lawrence, U.L. (eds) *Military Aspects of Geology: Fortification, Excavation and Terrain Evaluation*. Geological Society, London, Special Publications, **473**. First published online March 6, 2018, https://doi.org/10.1144/SP473.3

Rose, E.P.F. 2018*c*. The Battle of the Atlantic in World War II: contributions to victory by Quarrying Companies Royal Engineers and by geologist D. L. Linton, Professor of Geography at Birmingham 1958–1971. *In*: Hunger, E., Brown, T.J., Smith, G. & Anderson, P. (eds) *Proceedings of the 19th Extractive Industry Geology Conference 2016 and Technical Meeting 2017*. EIG Conferences Ltd, Charlbury, 37–50.

Rose, E.P.F. in press. Lawrence Rickard Wager (1904–1965): a distinguished geologist who helped to pioneer aerial photographic interpretation for Allied forces in World War II. *Earth Sciences History*.

Rose, E.P.F. & Clatworthy, J.C. 2007. Specialist maps of the Geological Section, Inter-Service Topographical Department: aids to British military planning during World War II. *The Cartographic Journal*, **44**, 13–43.

Rose, E.P.F. & Clatworthy, J.C. 2008*a*. Fred Shotton: a 'hero' of military applications of geology during World War II. *Quarterly Journal of Engineering Geology and Hydrogeology*, **41**, 171–188, https://doi.org/10.1144/1470-9236/07-034

Rose, E.P.F. & Clatworthy, J.C. 2008*b*. Terrain evaluation for Allied military operations in Europe and the Far East during World War II: 'Secret' British reports and specialist maps generated by the Geological Section, Inter-Service Topographical Department. *Quarterly Journal of Engineering Geology and Hydrogeology*, **41**, 237–256, https://doi.org/10.1144/1470-9236/07-016

Rose, E.P.F. & Hughes, N.F. 1993. Sapper Geology: Part 2. Geologist pools in the reserve army. *Royal Engineers Journal*, **107**, 173–181.

Rose, E.P.F. & Rosenbaum, M.S. 1993. British military geologists: through the Second World War to the end of the Cold War. *Proceedings of the Geologists' Association*, **104**, 95–108.

Rose, E.P.F. & Willig, D. 2004. Specialist maps prepared by German military geologists for Operation Sealion: the invasion of England scheduled for September 1940. *The Cartographic Journal*, **41**, 13–35.

Simon, L.J. 1957. Additional notes of the use of geology in the European theater of operations during World War II. *Geological Society of America Bulletin*, **68**, 1567.

Simon, L.J. 1988. (Memorial) Peter Kent (1913–1986). *American Association of Petroleum Geologists Bulletin*, **72**, 493.

Stanley, R.M. 1981. *World War II Photo Intelligence*. Sidgwick & Jackson, London.

Steers, J.A. 1971. Professor David Leslie Linton. *Geographical Magazine*, **43**, 658–659 [with portrait].

Thomas, H.H. 1920. [Abstract] On some features in the topography and geological history of Palestine. *Proceedings of the Geological Society of London*, **76**, i–iii, http://jgs.lyellcollection.org/content/jgsleg/76/1-4/i.full.pdf

Vincent, E.A. 1994. *Geology and Mineralogy at Oxford 1860–1986: History and Reminiscence*. E. A. Vincent, Department of Earth Sciences, University of Oxford, Oxford.

Vincent, E.A. 2004. Wager, Lawrence Rickard [Bill] (1904–1965). *In*: Matthew, H.G.C. & Harrison, B. (eds) *Oxford Dictionary of National Biography. Vol. 56. Usk–Wallich*. Oxford University Press, Oxford, 687–688.

Waters, R.S. 1971. [Obituary] David L. Linton. *Geographical Journal*, **137**, 432–433.

Watson, J. 2004. *Sidney Cotton: The Last Plane Out of Berlin*. Hodder Headline Australia, Sydney, Australia.

Williams, A. 2014. *Operation Crossbow: The Untold Story of the Search for Hitler's Secret Weapons*. Preface Publishing, London.

Wooldridge, S.W. & Linton, D.L. 1939. Structure, surface and drainage in south-east England. *Transactions of the Institute of British Geographers*, **10**, 1–124 [Reprinted 1955: George Philip, London].

One hundred years of cross-country mobility prediction in Germany

FLORIAN MALM

Bundeswehr Geoinformation Centre, Frauenbergerstrasse 250, 53879 Euskirchen, Germany

FlorianMalm@bundeswehr.org

Abstract: The potential for cross-country mobility (CCM) by vehicles has increasingly become an integral part of tactical planning for military operations and depends on a multitude of parameters. A comprehensive trafficability forecast requires the analysis of terrain, including slope, land cover and soil properties, and weather, as well as specific vehicle parameters. Terrain and weather interactions were recognized as important for the first time during World War I, when all types of vehicle would frequently bog down in muddy ground. The lessons learned influenced the orientation of future German military geology. During World War II, German military geologists and engineers prepared their first CCM maps. However, these often lacked a consistent format and standardized analysis. During the Cold War, the US Army carried out extensive research to create empirical mobility models based on physical properties and trafficability tests. These results were incorporated into the production of CCM maps. Present day computer-based data processing and analysis have opened up the possibility of enhancements to trafficability forecasting and thus make a vital contribution to providing the armed forces with advice on exercises and operations.

The success of military operations depends not only on combat power, leadership and morale, but also, to varying degrees, on environmental factors such as the weather, terrain and local soil conditions. Being able to deploy forces and elements quickly and efficiently across a given terrain, on and off the beaten track, has provided a decisive advantage influencing the outcome of many wars and battles in the course of history.

The local soil properties determine the mobility of combat forces as well as the successful and timely deployment of supplies and logistics (Halsall 2002). In the pre-mechanized age, inadequate trafficability mainly affected the supply trains accompanying the armies. Horse-, mule- or ox-drawn wagons (Fig. 1) were especially vulnerable to poor soil conditions (Halsall 2002). In addition, unfavourable interactions between weather and soil could slow the speed of infantry and cavalry, such as at the Battle of Agincourt, where the attack speed of the French cavalry was slowed down and horses slipped and fell on the rain-soaked soil (Barker 2005).

As part of tactical considerations, trafficability has gained priority since motorized combat forces were deployed during World War I (Fig. 2). This was the first time it became apparent that the physical properties of the soil resulting from the geology, soil cover and weather had a decisive influence on cross-country mobility (CCM). Over time, the effects of these properties and their interaction have been analysed and made available for use by military planners in a number of different ways – first, on simple maps, often drawn by hand, then by detailed plans and empirical trafficability models and subsequently by dynamic, computer-based digital data processing in geoinformation systems (GIS).

Trafficability prediction

The potential of CCM by both wheeled and tracked vehicles depends on a multitude of static and dynamic parameters (Zieger *et al.* 2015). A comprehensive trafficability forecast requires the examination and analysis of terrain features and weather as well as specific vehicle parameters. Important terrain features are slope, land use and physical soil properties (Eichrodt 2003). The soil properties, in turn, are substantially influenced by the weather conditions. To make a final trafficability assessment, these environmental parameters must be evaluated in relation to the technical parameters of the chosen vehicle, including its weight, tyre or track dimensions, vehicle cone index (VCI) and vehicle clearance.

Trafficability maps are used to separate the soil into the categories 'Go' (trafficable), 'No Go' (not trafficable) and 'Slow Go' or 'Not Sure' (moderately trafficable). Depending on the analytical method used for classifying trafficability conditions, the results may vary. Over time, the spectrum of methods has ranged from an author's subjective evaluation, especially concerning old maps such as those produced in World War II, to current, computer-based analyses of the dynamic interactions between soil and weather.

From: Rose, E. P. F., Ehlen, J. & Lawrence, U. L. (eds) 2019. *Military Aspects of Geology: Fortification, Excavation and Terrain Evaluation*. Geological Society, London, Special Publications, **473**, 297–306.
First published online March 6, 2018, https://doi.org/10.1144/SP473.7

Fig. 1. A mule team bogged down in thigh-deep mud near Potijze Farm in the Ypres Sector in Belgium during World War I. *Source*: unknown Australian Official Photographer 1917 (Australian War Memorial n.d.).

Fig. 2. A team of horses helps an Austro-Hungarian automobile bogged down in the unpaved roads of Russian Poland in 1915. From Baer (1917).

Static parameters

Static parameters are those properties undergoing minimal or no change over the period under consideration. Slope, land cover, soil type and surface roughness are especially important with respect to the assessment of the trafficability of geological surfaces.

Slope. Slope has a strong influence on the trafficability of geological surfaces. All vehicles have a specific tilt or climb angle. The tilt angle describes the maximum angle at which a vehicle can still move perpendicular to the slope without overturning. The climb angle is the maximum angle at which a vehicle can move parallel to the slope. When comparing slope with the tilt and climb angle of the vehicle, the terrain can be classified as:

- trafficable (tilt and climb angle > slope)
- untrafficable (tilt and climb angle < slope)
- moderately trafficable (tilt angle < slope < climb angle).

Combined with the direction of movement, soil type and soil condition, slope also affects mobility in general (e.g. spinning wheels, sliding).

Land cover. Land cover can have both direct and indirect influences on the local trafficability. Whereas open individual paved areas (e.g. car parks, streets) are generally classified as trafficable, built-up areas other than roads are mostly regarded as untrafficable. This is strongly dependent on the scale of the analysis. On a large scale it might be reasonable to take single buildings into account, whereas it might make more sense to mark a whole built-up area as untrafficable on a small scale (with roads being the exception).

Different types of vegetation cover (e.g. meadow, field or forest) can also have an indirect influence on the temporary properties of a location, such as soil moisture, and generally need to be considered in the analyses. Wooded areas are a special case because they have an indirect influence on the soil moisture budget, affecting the temporary soil conditions and, depending on the tree population and density, may be regarded as trafficable or not trafficable for some heavy vehicles. For instance, most modern main battle tanks are only capable of traversing forests when the tree spacing is >3 m and the tree diameter is <30 cm.

Soil type. The physical nature of the soil can be identified on the basis of the soil type. The most important aspects are the grain size distribution and the amount of organic components in the total volume of the soil. There are numerous ways to classify soils, with some based on very different scientific principles. In this context, the Unified Soil Classification System (ASTM 2011) is of particular interest because this kind of soil classification serves as the basis for the determination of trafficability using the most common cone index/vehicle cone index (CI/VCI) method. Soil classes in the Unified Soil Classification System are determined mainly by grain size, organic content, liquid limit and soil plasticity (Headquarters Department of the Army 1999). The system divides the soil into three major soil groups: coarse-grained, fine-grained and highly organic soils (ASTM 2011).

Soil classification, however, is not the only aspect required to assess trafficability. Soil moisture is one of the most important factors in evaluating whether a soil is trafficable or not. The soil moisture content depends not only on the soil type, but also on the precipitation and is therefore a highly time-dependent parameter. Another important factor is the penetration depth (commonly measured using the CI), particularly with respect to wheel or track slip.

Surface roughness. Soils with a high number of large rocks (>10 cm) or that are covered with small-sized dunes (Fig. 3), for example, have a high surface roughness. Whereas the terrain as a whole might be flat, numerous small obstacles can impede or even prevent the movement of vehicles. In general, surface roughness is difficult to determine because, in most cases, the conditions on site can only be evaluated by making a local inspection.

Dynamic parameters

Dynamic parameters are those properties undergoing actual or potential changes over the period under consideration. A decisive factor influencing the trafficability forecast of a geological surface is the weather because this has a direct influence on the soil moisture content.

Precipitation. Precipitation, combined with the local vegetation and the type of bedrock, directly influences the soil moisture content. Precipitation is the main factor that adds water to the moisture content of the soil, apart from flooding, artificial watering and ground water related soil moistening. The soil moisture content depends on both the current precipitation and weather history. If the weather conditions have been humid during the previous days, even a small rainfall event can saturate the soil, whereas the same amount of precipitation during a dry period may have little impact on the general soil moisture content.

Transpiration and evaporation. The transpiration of vegetation and the direct evaporation of water both reduce the water content of the soil, depending on the type and density of vegetation and the type of

Fig. 3. Loess dunes in Kunduz, Afghanistan in 2003. Although the terrain as a whole is flat and a superficial trafficability analysis would probably result in a positive evaluation, the large number of small dunes would prevent vehicles of most types from crossing this terrain. The limited spatial extent of these features often makes such areas difficult to identify and evaluate. Photograph from the collection of D. Willig.

surface. Forests, for example, extract more moisture from the soil than grassland. Other weather-related parameters – such as the air temperature, air humidity and solar radiation – also have a direct impact on the degree of soil moisture through their influence on the evaporation and transpiration of local vegetation. Weather conditions need to be determined as exactly as possible and must be included in the analysis to produce a sound trafficability forecast.

Ground water recharge. Soils lose moisture by the gravitational flow of water out of the soil layer into deeper layers underground. The velocity of the moisture loss is mainly influenced by the permeability of the soil, which is based on its texture and grain size distribution.

Trafficability prediction during World War I

The history of trafficability analysis began when the first tanks were deployed in combat by the British Army in September 1916 during World War I, and later also by the French and German armies. During the course of the war, military planners began to realize the significance of static and dynamic parameters and their influence on soil conditions. In the turmoil of trench warfare, all kinds of vehicles, as well as personnel and equipment, often got stuck in the muddy and churned-up ground. Many of the first tanks also got bogged down in the so-called No-Man's Land and were thus unable to achieve their full desired effect (Fig. 4). As trench lines stabilized, movement along the front lines almost ceased for German forces, clearly shifting the focus on geological activities to the vicinity of the trenches, with respect to excavation and slope stabilization. The primary focus of geoscientific advice, however, was on supplying the front lines with water, draining the trench systems and with soil conditions with regard to how easy it was to dig and work in the terrain. Soil trafficability was a topic of lesser importance in the work of the military geologists of the warring countries. This is reflected by the fact that discussions of trafficability are almost completely absent in the most common German textbooks written in the years between the two world wars (e.g. Mordziol 1938; von Bülow *et al.* 1938) and that there is no known production of real German CCM maps during World War I.

Trafficability prediction during World War II

Many of the lessons learned over the course of World War I were used to develop a re-orientation in military geology in Germany in the following generations. It can be noted, however, that assessing soil trafficability only played a subordinate part in military geological science at that time (e.g. Mordziol 1938; von Bülow *et al.* 1938). This can be ascribed to the experience of German military geologists in the previous war. Another reason for this point of view may have been the fact that the tanks of the earlier period were not advanced enough to turn trench warfare into mobile combat because they were slow,

Fig. 4. British tanks bogged in mud at 'Clapham Junction' in the Ypres Salient, Belgium, in World War I. From Institution of Royal Engineers (1922).

mostly not faster than walking speed, and therefore an easy target for artillery and infantry. They also often suffered technical problems and had an overall high failure rate, despite the fact that they were generally able to cross trenches and break up hardened front lines under good conditions.

During the more mobile warfare of World War II (Blitzkrieg), which was mainly based on deploying tanks and other motorized elements, trafficability forecasts for the area of operations suddenly became an important factor in strategic, operational and tactical planning, as reviewed by Häusler (2006). For this purpose, military geologists and engineers of both allied forces and the German Wehrmacht created the first CCM maps. German maps, however, were not standardized and were often subjective, relying on the knowledge and experience of the person preparing the map (e.g. Fig. 5).

The CCM assessment during World War II by the Wehrmacht was mostly based on soil and geological maps supplemented by aerial photographic interpretation and was ideally completed by targeted reconnaissance patrols. Libya in North Africa provides a well-documented example (Häusler 2003). Here, as was common, military geologists assigned physical properties to the types of soil specified on soil maps based on personal experience (cf. Figs 5 & 6). Some of the maps were almost exclusively based on route reconnaissance. During a lecture on 9 March 1962, Dr Leo J.M. Kuckelkorn, the geologist who led Wehrgeologenstelle 12 (Military Geology Centre/Team 12) in Rommel's German Africa Corps from 1941 until 1942, explained how his trafficability maps were created:

> It was of decisive importance for the Command to know where the desert was trafficable, under which circumstances, and at which times. As soon as I was free to act, we began reconnoitering the desert. A net of exploratory surveys was created, with a larger mesh than we would have liked, but time was short. To produce an areal representation, we had to interpolate between the traversed lines, not mathematically of course, but following the geological structure of the terrain
>
> (translated from Kuckelkorn 1962, p. 8).

Concerning the classification of trafficability maps and the general difficulties in creating and interpreting them, Kuckelkorn wrote in a letter dated 9 April 1942:

> Even the terrain itself has its quirks. Unfortunately, it does not simply consist of the three categories: well, moderately, and not trafficable, but there are many influences that intensify, diminish, or create otherwise blurred categories. In this case, the classification of an area often only depends on the vehicle, the driver or the personal taste of the geologist. Over the course of many years, we might be able to create an ideal special trafficability map on a good 25 000 map. But since we only had three months and two men with vehicles that kept breaking down to accomplish what 30 years of colonial policy had not even begun to manage, the only option was a general map indicating only an average value of the respective strip of terrain. Most of the time, it is not a problem to shift the boundaries of the different trafficability areas by several kilometres at one's own

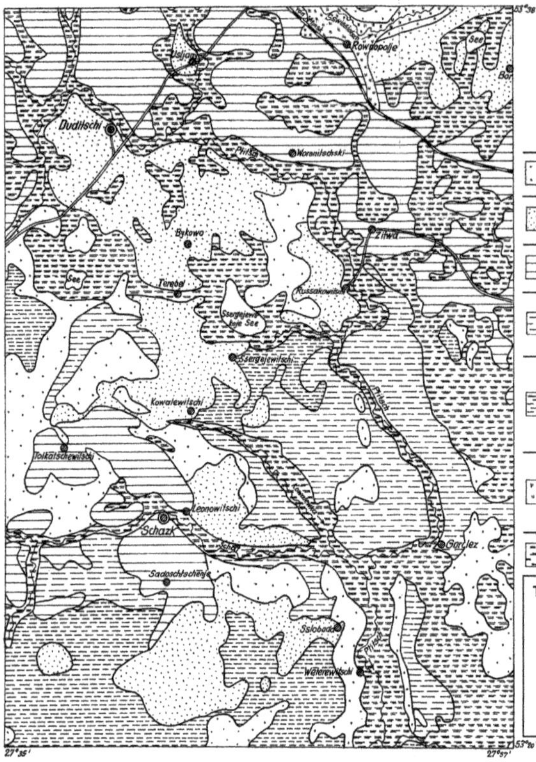

Cross-country mobility map*

	Soil	Tanks		wheeled vehicles	
		normal conditions	muddy conditions	normal conditions	muddy conditions
	Sand	++		-	+
	loamy Sand			+	o
	sandy Loam (Loess)	++	o	++	--- or --
	Sand, Groundwater level < 1 m			+[1]	--[2] or ---
	Sand, single parts with turf, shallow Groundwater	+[3]	-[3]	-- or ---	---
	Sand and clayey sand of the floodplains	+	-[4]	- or ---	---
	Moor	--[5]	--	---[5]	---

Trafficability conditions **Scale 1 : 126 000**

++	well trafficable	* interpreted from soil type map
+	trafficable	[1] partly difficult
o	slightly difficult	[2] persistent
-	difficult	[3] with untrafficable parts
--	very difficult	[4] or mostly untrafficable
---	untrafficable	[5] after strong, snowfree frost partly trafficable

Fig. 5. A World War II German CCM map prepared from a 1 : 126 000 scale soils map of territory in present day Belarus. German CCM maps created during World War II were rarely standardized and were heavily influenced by the working methods applied by the respective geologist compilers. Translated and modified from Stremme (1944).

> discretion. It may happen that a vehicle gets stuck in a sand pit in the middle of a 'well trafficable' area [Fig. 7], or you can find a few kilometres of flat race tracks in the middle of the rugged mountain wilderness. That is why I always ask planners not to regard these maps as special maps, but only as general maps with average values. This also means that the boundaries between an 'escarpment that is trafficable in many places' and one that is 'only trafficable on dirt roads' are fluid …
>
> (translated from Häusler 2003, p. 94).

It can be assumed that many maps were created under extreme time pressure. Considering these circumstances, it is striking that the quality and usability of many German World War II maps was relatively high. However, it was not until the years following World War II that a scientific, systematic approach was developed.

Trafficability prediction from the Cold War until the present day

Extensive basic scientific research and *in situ* trafficability tests (Fig. 8) were first conducted during the Cold War by the US military to create trafficability models based on physical properties.

These models were based on the CI/VCI approach to forecasting trafficability developed by the US Army Corps of Engineers Waterways Experiment Station, which had been applied since the end of World War II (Wong 2008). The CI (penetration resistance in pounds per square inch, psi) of the soil is compared with the VCI (specific value calculated from vehicle data, psi). If the value for the soil is larger than that for the vehicle, the soil is regarded as trafficable (Wong 2008). Although the VCI represents a fixed vehicle-specific value, the CI varies depending on the condition of the soil. Moist soil, for example, has a lower CI than the same soil under dry conditions. Depending on the type of soil, the CI values may vary substantially under different soil conditions. The soil moisture is therefore an indispensable parameter to be considered in all trafficability assessments.

The models were shared among North Atlantic Treaty Organization (NATO) countries. The German Army, the Bundeswehr, also started to use the CI/VCI approach. The results were included in the

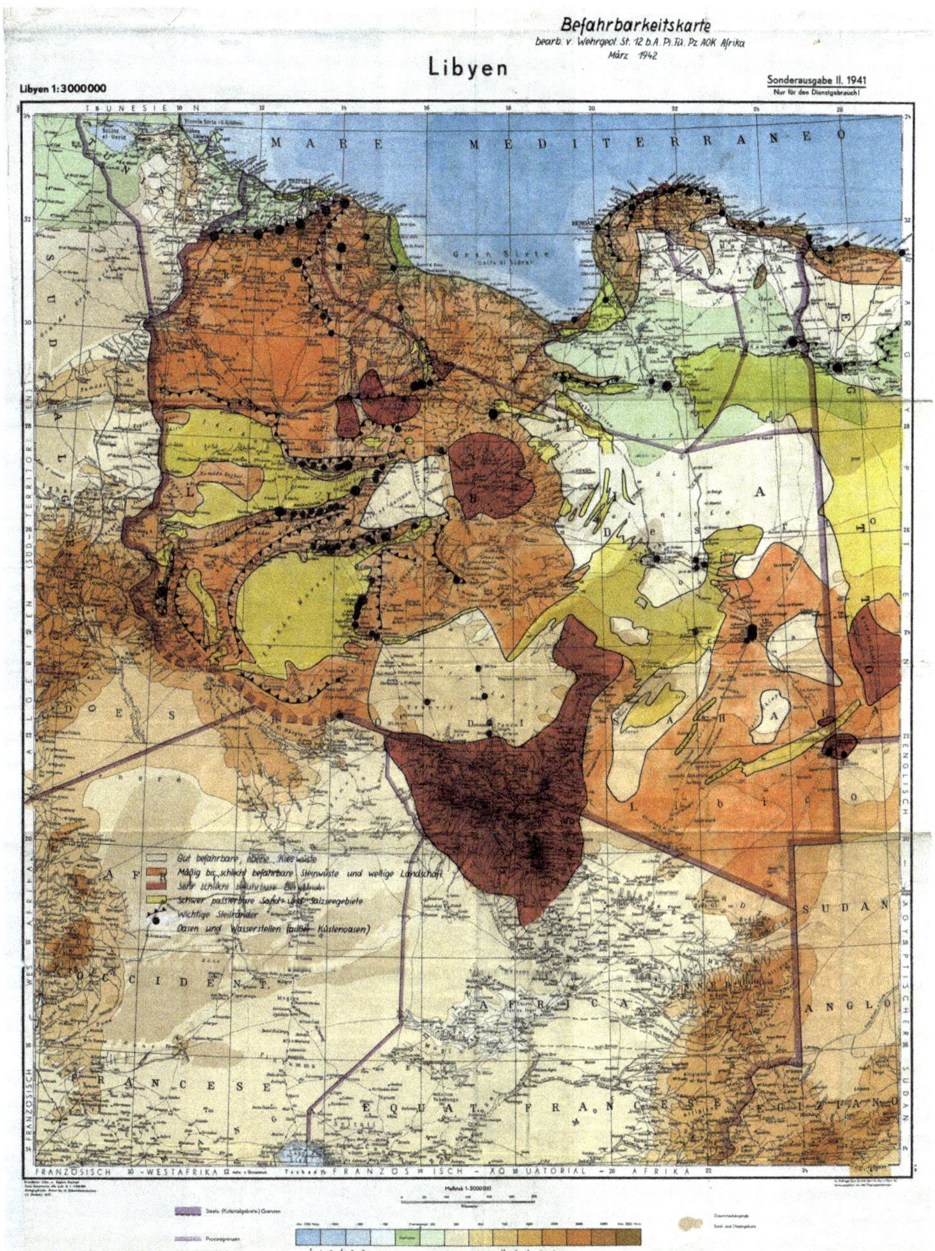

Fig. 6. German trafficability map of Libya, original scale 1 : 3 million. The map was created on the basis of soil and geological maps, aerial photographic interpretations and targeted reconnaissance patrols. Key: blank, well trafficable, planar gravel desert; orange, moderate to poor trafficability, stone desert and wavy landscape; red, very poor trafficability, highlands; yellow, poor trafficability, sand and salt lake areas; line with triangles, major slopes; black spots, oases and water wells (except coastal oases). Key in bottom margin indicates by colour the relief on the topographical base map, from −1000 m (blue) to >2500 m (brown). From Kuckelkorn (1942).

production of analogue trafficability maps created for large parts of Central Europe by various NATO countries until the mid-1990s in preparation for armoured warfare against a potential attack by forces of the Union of Soviet Socialist Republics and its allies. On most of these maps, soil moisture was

Fig. 7. Volkswagen Kübelwagen (German military utility vehicle) stuck in saline sabkha soil after rainfall in Libya in 1941. After Willig (1997); original in Kuckelkorn (1941).

reduced to two or three categories (e.g. dry, moist, wet). Most of the maps were limited to the common main battle tanks in use at the time, such as the Leopard 2 for Bundeswehr maps.

The only way to thoroughly include all the important factors in the trafficability forecast is by means of computer-assisted analysis. The research carried out for the CI/VCI approach for trafficability prediction eventually led to the NATO Reference Mobility Model (NRMM; Wong 2008). The first version of this computer-based model was used in 1979 (Haley *et al.* 1979). It was based on its predecessors,

Fig. 8. Bogged-down tanks during two Bundeswehr exercises in Germany. The picture on the left was taken during 1985, the one on the right during 1987. Photographs from the collection of D. Willig.

the US Army Material Command '71 Mobility Model (US Army Tank-Automative Command 1973) and the Army Material Command '74 Mobility Model (Jurkat *et al.* 1975). A second version of the NRMM was developed in 1992 (Ahlvin & Haley 1992) and has been continuously updated (Jayakumar & Dasch 2016). Early computer models, at least in their first editions, were often complicated and far from intuitive to use (Jones *et al.* 2011), which restricted their use mostly to the developers. However, enhanced versions of the models are still in use and render accurate trafficability predictions. In-depth trafficability assessments, including additional input parameters, were not available in the German Bundeswehr until GIS tools increasingly emerged at the end of the 1990s.

Conclusions

Today's possibilities for using digital data and computer-assisted data processing have opened up a variety of further options for forecasting CCM and provide an important contribution to advising the forces during exercises and operations. GIS-based tools are especially suited to processing the data and predicting CCM. There is currently a wide choice of software for military and civilian purposes to forecast trafficability.

Existing models are currently being revised and modernized. The NRMM, for example, will soon get a new physical approach for trafficability prediction. The project known as Next Generation NRMM (Dasch *et al.* 2016; Jayakumar & Dasch 2016) is a collaboration among 13 NATO member states (the USA is the leading nation; Germany and the UK are among the members) with the aim of creating a new, non-empirical, physics-based trafficability approach. Next Generation NRMM will use finite and discrete element modelling tools for the evaluation of complex terramechanics as well as GIS-based mapping tools (Dasch *et al.* 2016; Jayakumar & Dasch 2016).

The German Bundeswehr has developed its own trafficability prediction tool based on the CI/VCI approach also used in NRMM, with the aim of creating a quick and easy-to-use GIS-based software tool for the CCM mapping of large areas worldwide (Zieger *et al.* 2015). Its core is the use of a hydrological soil model in combination with three-day weather forecasts from the ICOsahedral Nonhydrostatic (ICON) weather model. It is currently under further development and evaluation by the Bundeswehr Geoinformation Service.

The potential of modern technology in the area of trafficability prediction is far from exhausted. The availability of reliable and high-resolution spatial data remains a fundamental problem, especially in the area of soil and weather. Common and free to use soil data usually have a spatial resolution of 500 m, although some military data have a resolution of up to 30 m, whereas common weather forecasts tend to have resolutions from 7 to 20 km. As elsewhere, a model can only be as good as the data on which it is based. The prediction of CCM is still a scientific challenge and requires interdisciplinary approaches, including pedology, meteorology, and vehicle engineering and material sciences.

References

AHLVIN, R.B. & HALEY, P.W. 1992. *NATO Reference Mobility Model, Edition II, NRMM II Users Guide.* Technical Report **GL-92-19**. US Army Corps of Engineers, Washington, DC.

AMERICAN SOCIETY FOR TESTING AND MATERIALS (ASTM) 2011. *Standard Practice for Classification of Soils for Engineering Purposes (Unified Soil Classification System).* ASTM **D2487-11**. ASTM International, West Conshohocken, PA.

AUSTRALIAN WAR MEMORIAL n.d. www.awm.gov.au/collection/E00962/ [last accessed 14 November 2016].

BAER, C.H. 1917. *Der Völkerkrieg: Eine Chronik der Ereignisse seit dem 1. Juli 1914 – Dritter Band, Erstes Kriegshalbjahr, Von Juli 1914 – Februar 1915.* Julius Hoffmann, Stuttgart.

BARKER, J. 2005. *Agincourt: The King, the Campaign, the Battle.* Abacus, London.

DASCH, J., JAYAKUMAR, P. *ET AL.* 2016. *ET-148 Next-Generation NATO Reference Mobility Model (NRMM).* Science and Technology Organization Technical Report **ET-148**. North Atlantic Treaty Organization, Brussels.

EICHRODT, A.W. 2003. *Development of a Spatial Trafficability Evaluation System.* Hochschulverlag/ETH Zürich, Zurich/Singen.

HALEY, P.W., JURKAT, M.P. & BRADY, P.M., JR. 1979. *NATO Reference Mobility Model, Edition I, Users Guide, Volume I.* U.S. Technical Report **12503**. Army Tank-Automotive Research & Development Command, Warren, MI.

HALSALL, T.J. 2002. Battles on chalk: the geology of battle in southern England during the First Civil War, 1643–1644. *In*: DOYLE, P. & BENNETT, R. (eds) *Fields of Battle: Terrain in Military History.* Kluwer Academic, Dordrecht, 33–50.

HÄUSLER, H. 2003. *Wehrgeologie im nordafrikanischen Wüstenkrieg (1941–1943).* MilGeo, **13**. Bundesministerium für Landesverteidigung, Vienna.

HÄUSLER, H. 2006. Historic maps of terrain evaluation. *In*: MANG, R. & HÄUSLER, H. (eds) *International Handbook Military Geography.* Truppendienst, Ministry of Defence, Vienna, 257–271.

HEADQUARTERS DEPARTMENT OF THE ARMY 1999. *Materials Testing.* Field Manual 5-472 NAVFAC MO 330 AFJMAN 32-1221(I). US Army, Washington, DC.

INSTITUTION OF ROYAL ENGINEERS 1922. *The Work of the Royal Engineers in the European War, 1914–1919. Work in the Field Under the Engineer-in-Chief,*

B.E.F. Geological Work on the Western Front. Institution of Royal Engineers, Chatham.

JAYAKUMAR, P. & DASCH, J. 2016. *The Next Generation NATO Reference Mobility Model Development*. Science and Technology Organization AVT-248-Research Task Group Public Release. North Atlantic Treaty Organization, Brussels.

JONES, R., CIOBOTARU, T. & GALWAY, M. 2011. *NATO Reference Mobility Modelling. North Atlantic Treaty Organisation, Neuilly-Sur-Seine Cedex, France*. Research and Technology Organisation Technical Report **TR-AVT-107**. North Atlantic Treaty Organization, Brussels.

JURKAT, P., NUTTALL, C. & HALEY, P. 1975. *The AMC '74 Mobility Model*. Technical Report **11921 (LL-149)**. US Army Tank-Automative Command, Warren, MI.

KUCKELKORN, L. 1941. *Fotosammlung Wehrgeologie Libyen 1941*. Heringen Collection Archive. Bundeswehr Geoinformation Center, Euskirchen.

KUCKELKORN, L. 1942. *Befahrbarkeitskarte Libyen 1:3.000.000. Karte der Wehrgeologenstelle (12) vom März 1942*. Heringen Collection Archive. Bundeswehr Geoinformation Center, Euskirchen.

KUCKELKORN, L. 1962. *Entwicklung und Aufgaben der deutschen Wehrgeologie bis 1944*. Vortrag vor Pionieren am 09.03.1962 [transcript from the private archive of D. Willig].

MORDZIOL, C. 1938. *Einführung in die Wehrgeologie*. Otto Salle, Frankfurt.

STREMME, H. 1944. *Auswertung Bodentypenkarten für wehrgeologische Zwecke*. Bericht der Wehrgeologenstelle (2) vom 27.03.1944. Heringen Collection Archive, Bundeswehr Geoinformation Center, Euskirchen, Germany.

US ARMY TANK-AUTOMATIVE COMMAND 1973. *The AMC '71 Mobility Model*. Vols I & II. Technical Report **11789 (LL-143)**. US Army Tank-Automative Command, Warren, MI.

VON BÜLOW, K., KRANZ, W. & SONNE, E. 1938. *Wehrgeologie*. Quelle & Meyer, Leipzig.

WILLIG, D. 1997. *Wehrgeologie als spezielle Angewandte Geologie – Anforderungen, Entwicklung und Aufgabenspektrum dargestellt an ausgewählten Beispielen*. Dr. rer. Nat. dissertation, Bayrische Julius-Maximilians-Universität.

WONG, J.Y. 2008. *Theory of Ground Vehicles*. Wiley, Hoboken, NJ.

ZIEGER, P., BIERMANN, J. & HEDEL, R. 2015. Bewertung der Befahrbarkeit von Böden mit CCMod2 – ein Service auch für die Landwirtschaft. *In*: RUCKELSHAUSEN, A., SCHWARZ, H.-P. & THEUVSEN, B. (eds) *Informatik in der Land-, Forst- und Ernährungswirtschaft*. GI Edition Proceedings, **238**. Gesellschaft für Informatik, Bonn, 205–208.

Index

Page numbers in *italics* refer to Figures. Page numbers in **bold** refer to Tables